Introduction to
PHYSIOLOGY

VOLUME 5

Introduction to
PHYSIOLOGY

VOLUME 5
CONTROL OF REPRODUCTION

HUGH DAVSON

Physiology Department, King's College
London, England

M. B. SEGAL

Sherrington School of Physiology, St. Thomas's Hospital
London, England

1980

ACADEMIC PRESS: LONDON TORONTO SYDNEY
GRUNE & STRATTON: NEW YORK SAN FRANCISCO

ACADEMIC PRESS INC. (LONDON) LTD.
24/28 Oval Road
London NW1

United States Edition published by
GRUNE & STRATTON INC.
111 Fifth Avenue
New York, New York 10003

Copyright © 1980 by
HUGH DAVSON

All Rights Reserved
No part of this book may be reproduced in any form by photostat, microfilm or any other
means, without permission from the publishers

British Library Cataloguing in Publication Data

Davson, Hugh
    Introduction to physiology.
    Vol. 5: Control of reproduction
    1. Vertebrates—Physiology
    I. Title II. Segal, Malcolm Beverley
    III. Control of reproduction
    596'.01   QP31.2   74–5668

    ISBN (Academic Press): 0–12–791005–0
    ISBN (Grune & Stratton): 0–8089–0900–2

VANDERBILT UNIVERSITY
MEDICAL CENTER LIBRARY

JUN 1 1981

NASHVILLE, TENNESSEE
37232

Printed in Great Britain by
Latimer Trend & Company Ltd Plymouth

CONTENTS

Chapter 1 Introduction

Chapter 2 Male Trophic Hormones and the Ovarian Cycle in the Female

v

Chapter 5 Pregnancy: Maintenance and Prevention

Chapter 6 Pregnancy: Parturition and Labour

ACKNOWLEDGMENTS

The authors would like to acknowledge with gratitude the valuable help provided by Dr. D. A. Begley, of King's College London, who kindly commented on the manuscript and read the proofs. Also one of us (H.D.) is happy to thank the Leverhulme Trust Fund for an Emeritus Fellowship that contributed materially to the cost of preparing the manuscript for this volume.

CHAPTER 1

Introduction

The co-ordination of the several factors concerned in the sexual reproduction of two individuals to produce one or more progeny requires control over a variety of factors; thus, mature ova and sperm must be produced at the same time, and at this time the female must be in a state to receive the sperm and the male to copulate, i.e. the female must be receptive, or on heat (in oestrus) close to the time at which she produces a new ovum. The timing of the events must be such that, at the end of gestation, there will be adequate food available for mother and offspring; and this is doubtless the basis for the breeding season in many lower animals. The fertilized ovum requires nutrition, and this is achieved by implantation of the blastocyst in the wall of the uterus; for successful implantation the uterus must be prepared—the decidual reaction—and subsequent nutrition must be achieved by formation of a placenta, whilst the mammary glands must be prepared for their later function of feeding the infant. During pregnancy the cyclically occurring events taking place in the non-pregnant female, included in the term *oestrous cycle*, or, in primates, the *menstrual cycle*, must be suppressed; and finally the relaxed state of the uterine muscle, necessary for retention of the foetus in the womb, must be succeeded by a highly active state leading to delivery.

STEROID HORMONES

Castration

The control mechanisms are mediated through the steroid hormones secreted mainly by the gonads and, if pregnancy takes place, in many species by the placenta. In the male, the primary function of its secreted androgens is to bring about a state of "maleness" in the developing foetus and young animal, and to sustain this maleness throughout adult life. Thus, castration of the young male, before puberty, prevents the development of its accessory sex organs so that the castrated male retains many female features; for example, the human male voice remains unbroken, and the effects of the operation on male animals were early recognized, the castrated ox being more amenable than the bull, the capon producing more fat than the cock, and so on. These effects can be antagonized by injections of androgens, such as testosterone. Castration of the adult male has well defined effects, indicating the necessity for the continuous secretion of androgens for the maintenance of maleness. Thus, Table I from Ramirez and McCann (1965) shows the changes in weight in the seminal vesicles and ventral prostate gland of castrated rats following administration of testosterone at various doses. Castration itself causes a profound drop in weight of the

TABLE I

Changes in weight of seminal vesicles and ventral prostate gland of rats after castration and replacement therapy with testosterone propionate, Tp (Ramirez and McCann, 1965)

Type of rat	Dose of Tp μg/100g/day	Seminal vesicles (mg)	Ventral prostate (mg)
Intact	0	1025 $\pm$ 36*	341 $\pm$ 44
Castrated	0	71 $\pm$ 6	27 $\pm$ 1
Castrated	5	240 $\pm$ 2	50 $\pm$ 4
Castrated	25	397 $\pm$ 17	97 $\pm$ 13
Castrated	100	1147 $\pm$ 60	403 $\pm$ 17
Castrated	150	1416 $\pm$ 70	419 $\pm$ 26

* Mean $\pm$ S.E.

seminal vesicles, for example, from 1025 mg to 71 mg in mature animals, and from 13·9 to 5·0 mg in immature animals. This fall in weight is prevented by administration of the hormone.

Coitus

A more positive indication of the role of male steroid hormone secretion is revealed by the striking changes in blood-level of testosterone during sexual excitement and coitus. Figure 1.1 illustrates the rise in luteotrophic hormone and the subsequent rise in testosterone in a bull at sight of a cow and the ensuing copulation.*

* The relation between blood testosterone and sexual activity in the male is probably more complex than suggested by this finding in the bull; in guinea-pigs the level of blood testosterone is not related to high or low degree of sexual activity; however, animals with high activity responded with a rise in blood testosterone on exposure to an oestrous female which was large by comparison with that in animals of low activity (Harding and Feder, 1976). As we shall see, androgens can be converted in the female or male, to oestrogens, and it has frequently been argued that an effect of androgen is mediated by prior conversion to oestrogen. For example the testosterone-induced mounting behaviour of castrated male rats can be

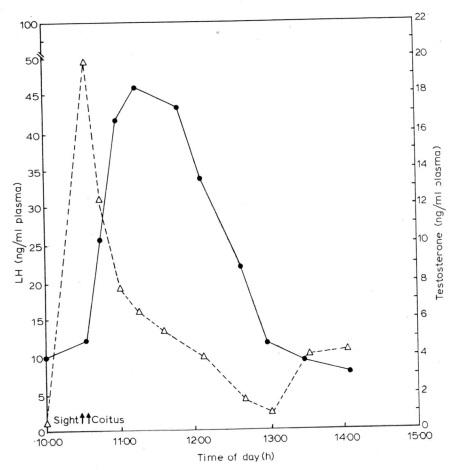

Fig. 1.1. The effect of coitus on the concentrations of luteinizing hormone (LH; △ – – – – – △) and testosterone (●————●) in the peripheral blood of a bull. (Katongole *et al.*, *J. Endocrinol.*)

Non-Reproductive Tissue

In addition to the effects on the primary and accessory sexual organs, the androgens affect many other tissues of the body; thus, the kidney, adrenal gland, liver, pancreas, thyroid, thymus, salivary glands, skin pigmentation, red blood cell formation are all affected. Their influence, moreover, is not confined to the male, except in so far as the female lacks the appropriate target organs.

This is especially obvious in the condition of hirsutism or excessive

reduced if the testosterone is applied to the hypothalamus together with an aromatization-inhibitor, androst-1,4,6-triene-3,17-dione (ADT). It must be emphasized, however, that less than 0·1 per cent of the testosterone would be converted to oestradiol, and it may be that the much larger amounts of dihydrotestosterone formed would co-operate with oestradiol (Christensen and Clemens, 1975).

growth of hair in sites where such growth is a secondary male characteristic; there seems little doubt that the condition is associated with increased androgens, such as testosterone and dihydrotestosterone, in the blood of females so affected (Kuttenn *et al.*, 1977).

Female Hormones

By contrast to that of the male, the secretion of the female sex hormones is cyclical, corresponding with the cyclical events of the oestrous and menstrual cycles and the production of ova. The female steroid hormones belong to two main classes, the *oestrogens*, secreted mainly by the follicle during the "follicular stage" of the menstrual or oestrous cycle, and the *progestins* or *progestogens*, secreted by the corpus luteum during the luteal phase which, if pregnancy follows ovulation, continues its secretory activity throughout pregnancy. This activity is supplemented to a greater or less extent, according to species, by the secretion of placental hormones, so that in many species removal of the ovaries does not cause termination of pregnancy, which continues to term (p. 202).

Chemical Aspects

The sex hormones, are, like the adrenal cortical hormones to which they are closely related, steroids derived biosynthetically from cholesterol, which itself is synthesized from acetate.

Synthetic Pathways

The main pathways of synthesis are indicated in Fig. 1.2; the major androgen is testosterone; its reduced dihydro-compound is actually more effective weight for weight and in most target organs for this androgen there is an enzyme, 5α-testosterone dehydrogenase, that is capable of transforming testosterone rapidly to the reduced form, so that it is highly probable that it is in this form that it exerts its action on the cells of many target organs.

The human ovary secretes mainly 17-β-oestradiol (E_2), oestrone (E_1), progesterone and 17-α-hydroxyprogesterone.* It will be seen that the synthetic pathway involves the formation of both male and female sex hormones so that, in fact, the female gonad does, indeed, produce small quantities of androgens such as testosterone and the male gonad produces oestrogens; furthermore, the ovarian cells that produce the

---

* The main oestrogen of the goat is 17α-oestradiol. 20α-hydroxy-progesterone is formed from progesterone in peripheral metabolism; it is also secreted by the monkey ovary, especially in the later, luteal, phase of its menstrual cycle (Hayward, 1963). 17α-hydroxyprogesterone, although produced in the ovary, is relatively inactive.

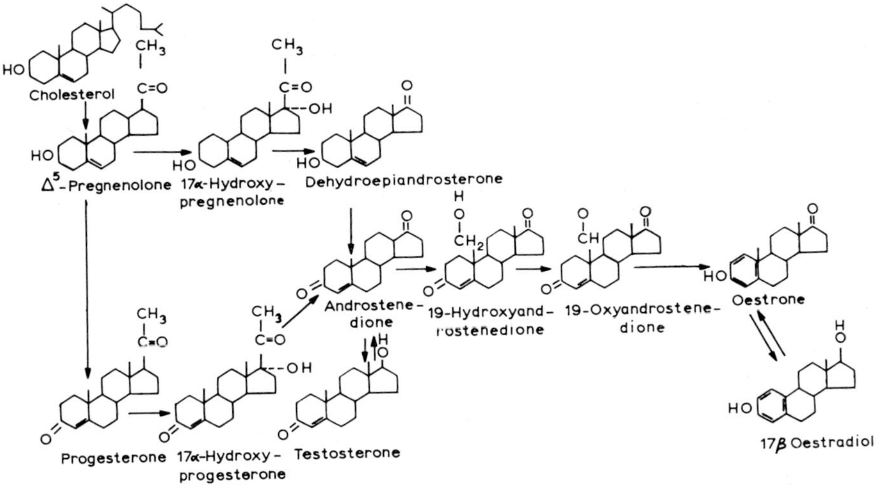

Fig. 1.2. Steroid synthetic pathways in human ovaries and testes. The major secretory product of the testis is testosterone; the ovary secretes 17β-oestradiol, progesterone and 17α-hydroxy-progesterone. (Odell and Moyer, *Physiology of Reproduction.*)

oestrogen type of hormone also produce the progesterone type, so that we must speak rather of a main product or products, of a given tissue; thus the main product of the follicular cells of the ovary is oestradiol whilst that of Leydig cells of the testes is testosterone, presumably because the enzyme compositions favour these stages in synthesis.† The similarity with the adrenal cortical hormones has already been emphasized; this extends to the use of a common synthetic pathway, so that sex hormones are, indeed, produced and secreted by the adrenal gland; this is especially true of progesterone, which is the precursor of both androgens and oestrogens as well as of corticosterone.

Metabolic Changes

When secreted into the blood the hormones may undergo chemical change so that the androgens, oestrogens, etc., isolated from urine or tissue may not be the same as the hormones originally secreted by the glands. Thus, the turnover of progesterone is quite rapid, with a half-life of about 30 minutes; it is converted into the inactive pregnandiol

† The finding of oestrogen in normal male urine could be interpreted as a metabolic conversion from testosterone, rather than release from testes; however, Kelch *et al.* (1972) cannulated the spermatic veins of 8 normal men and obtained a concentration of oestradiol of 1050 ± 57 pg/ml compared with only 20 ± 1·6 pg/ml in blood from a peripheral vein.

which appears in the urine as pregnandiol glucuronide, so that the concentration of this in the urine is a good index to the rate of secretion of progesterone at any time. Oestradiol has a much longer half-life, perhaps as long as 24 hours; in the liver it is converted to oestriol (E_3), having a third OH-group and lacking oestrogenic activity. Of the androgens, as indicated above, testosterone may be converted by the target tissues into 5α-dihydrotestosterone which is much more active than testosterone; the urinary excretory product is androstenediol. A less specific mode of altering the steroids is through conjugation with glucuronic acid or sulphate; this renders them water-soluble and permits their excretion by the kidney.

Protein Carriers

It will be recalled that the adrenal cortical hormones are carried by a protein, transcortin, or corticosteroid-binding $α_2$-globulin (CBG), for which they have a specific affinity much greater than that of, say, serum albumin.

SSBG and TeBP

Of the sex steroids it is known that progesterone is carried by the same protein. So far as the others are concerned, Mercier-Bodard et al. (1970) have reviewed the literature and described in some detail the characteristics of a "sex steroid-binding globulin" (SSBG) which is probably similar to the "testosterone-binding protein" (TeBP) of human pregnancy serum described by Gueriguian and Pearlman (1968) and by Hansson et al. (1974). The carrier-protein is not specific for testosterone, and binds oestradiol, the affinity constants being $1·2$ and $0·5.10^9$ M^{-1} for testosterone and oestradiol respectively (Mercier-Bodard et al., 1970). According to Anderson (1976) if the binding by testosterone is put at 100, 5α-dihydrotestosterone (DHT) binds nearly three times as well and oestradiol about one-third as well. Δ-5 andro-stenediol binds as well as testosterone. Androstenedione and dehydroe-piandrosterone hardly bind at all. The molecular weights of these sex steroid-binding proteins (SBP and TeBP) are higher than that of corticosterone-binding protein, being about 100,000 compared with about 66,000.

Progesterone Carrier

Of some interest is the appearance of a glycoprotein in the blood of the pregnant guinea-pig that binds specifically to progesterone; the association constant of 9.10^8 M^{-1} reveals a high affinity comparing with only $1·6.10^7$ M^{-1} for testosterone. This does not occur in non-pregnant animals and cannot be induced in these by injections of oestrogens or progestins, and so it may well be synthesized in the placenta (Milgrom et al., 1972).

Physiological Role

The role of the binding proteins in plasma may be simply that of carriers, as discussed in Vol. 2; as suggested by Anderson (1976), however, it is possible that variations in the quantity of a binding protein that binds androgens and oestrogens to different extents might represent a mode of controlling the dominance of oestrogenic and androgenic activity at a given target organ, since there is no doubt that the activity of a steroid is governed by the concentration of the free molecule in the neighbourhood of the target cell. An important finding in this respect is that the production of the binding protein is influenced by the concentration of steroid hormones, oestrogen tending to stimulate production whilst testosterone tends to inhibit production. Thus, the binding protein could, perhaps, act as an amplifier for testosterone action, tending to increase the amount of free androgen in the blood; it would operate in the opposite way on oestrogen action (Burke and Anderson, 1972). In this way, the virilizing in the polycystic ovary syndrome could be attributed to the low levels of SSBG in the blood.

Receptors

As we shall see, the target organs for the steroid hormones, e.g. the uterus or prostate, contain specific proteins or "receptors" with which the hormone binds on entering the cell; these "receptors" are different from the plasma carrier proteins described above, characteristically exhibiting a far greater specificity for a given steroid.

Control

In general, as indicated in Vol. 2, the steroid bound to its carrier acts as a reserve, the active form being that dissociated from the carrier at the target-site. Control over steroid activity may well be exerted, not only through alterations in secretion by the appropriate cells, but also by the amounts of carrier protein released into the blood (see, for example, Gueriguian et al., 1974). In addition, control could be modified by alterations in the number and affinity of specific "receptors" with which the hormone must interact before it exerts its action within its target cell.

Binding Sites or Receptors

The results of modern studies of steroid hormone action indicate that the hormone first binds in the target cell to a protein that has a high specific affinity for it; as a result of this initial step, the hormone can now be translocated within the cell to the nucleus where it may exchange places with another—nuclear—receptor. These studies* have been based on isolation of protein material from the target cells with specific binding capacity for the hormone, and also by observation of uptake of the radioactively labelled hormone by the tissue, either by extraction or by radioautography.

Target Tissues

The "target organs" for the sex steroids are the gonads, themselves, and the secondary reproductive organs such as the prostate, uterus, mammary glands, and so on. In addition to these organs, the steroid hormones are able to influence the secretion of their trophic hormones, synthesized by the anterior pituitary, or hypophysis, and they may do this either directly by influencing the secreting cells, or indirectly by modifying the excitability of the hypothalamic neurones whose secretions, *releasing hormones*, are carried by the portal system to the anterior lobe of the pituitary. It is reasonable, therefore, to examine both the pituitary and the hypothalamus for binding sites or receptor material.

Uptake of Labelled Oestradiol

Some results of Eisenfeld and Axelrod (1966), obtained by injections of labelled oestradiol and extracting the labelled material from target organs, are shown in Fig. 1.3; uptake of oestradiol into the female target-organs, such as the uterus, was very high, but it was also high in the pituitary and the hypothalamus, regions of feedback of this hormone. It is interesting in this connection that the uptakes of this "female steroid hormone" into both male and female pituitaries and hypothalami were very similar.

Autoradiography. Stumpf has developed highly sensitive radioautographic techniques for the localization of ^{3}H-labelled steroid in the tissues; the silver grains deposited by this technique occurred over the nuclei of target tissues, such as vagina, oviduct, testis and a mammary tumour, but in the non-target tissues, such as liver and adrenal gland, there was no concentration over the nuclei. In the uterus, for example, epithelial cells, connective-tissue cells and muscle cells all showed

---

* The interaction of labelled steroid hormone with its target tissue, leading to the concept of the steroid receptor, was described first by Jensen and Jacobsen (1962).

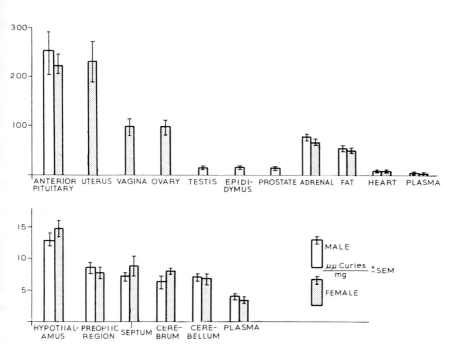

Fig. 1.3. Uptake of isotopically labelled oestradiol in various tissues of male and female rats one hour after administration. (Eisenfeld and Axelrod, *Endocrinology.*)

accumulation in their nuclei, with only very little in the cytoplasm. In the pituitary, there was accumulation in acidophils, basophils and chromophobes. By making silver-grain counts Stumpf (1968) computed that some 70 per cent of the uptake was over the nuclei of uterine cells.

Of special interest is the accumulation by the brain; this was examined in some detail by Stumpf (1970) and Fig. 1.4 illustrates the regions of accumulation identified by autoradiography; the hypothalamic concentrations are striking; parts of the "limbic brain", such as the amygdala and hippocampus, are also sites of accumulation. The accumulation was obvious 2 hr after injection of the ^{3}H-oestradiol, and could be prevented by simultaneous injection of unlabelled material.

Isolation of Binding Sites

By homogenizing a target-tissue, such as the uterus, two main fractions showing specific accumulation of steroid hormone have been found, namely the *nuclear fraction* or, more specifically, the nuclear chromatin or DNA-protein complex extracted from the interphase nucleus, and the *cytosol.*, i.e. the supernatant remaining from the

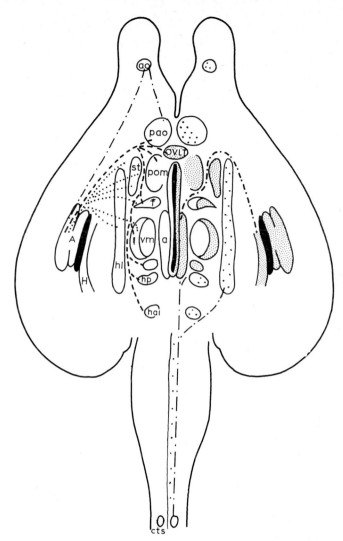

Fig. 1.4. Oestrogen-neurone systems in the rat brain revealed by uptake of labelled oestradiol. Schematic drawing of a hypothetical horizontal plane of the rat brain showing the distribution of oestrogen-neurones and probably related nerve fibre tracts. Oestrogen-neurones, which exist in selective areas of the periventricular brain, are indicated by stipples (right portion of the picture). The intensity of the stipples reflects the frequency of the occurrence of oestradiol-concentrating neurons in the different areas.

Pathways: – – stria terminalis; . . . . . ventral amygdalofugal pathway; –.–. portion of the fornix, probably medial corticohypothalamic tract; —.—. left: tractus olfactorius, right: tractus longitudinalis perependymalis.

Abbreviations: A, amygdala; a, n. arcuatus; ao, n. olfactorius anterior; cts, n. commissuralis tractus solitarii; f, n. paraventricularis; H, hippocampus; hab, n. habenulae lateralis; hl, lateral hypothalamus; hp, n. posterior hypothalami; vm, n. ventromedialis hypothalami, pars lateralis; OVLT, organum vasculosum laminae terminalis (representing also the organon subfornicale and n. triangularis septi); pao, para-olfactory region; pom, n. preopticus medialis; pmv, n. premamillaris ventralis; st, n. interstitialis striae terminalis. The black areas indicate the ventricular system. (Stumpf, *Amer. J. Anat.*)

homogenate after centrifuging particulate matter down at 100,000 g. By causing these two fractions to sediment in a sucrose-gradient they may be isolated and their binding characteristics with the steroid hormone may be determined by techniques described earlier (Vol. 2).

Binding Constant. Thus, we may determine the binding constant, i.e. the reciprocal of the equilibrium constant in the binding reaction:

$$\text{Carrier} \quad + \quad \text{Hormone} \quad \rightleftharpoons \quad \text{Complex}$$
$$[C] \qquad\qquad [H] \qquad\qquad [CH]$$

$$K_{Eq} = \frac{[C] \times [H]}{[CH]} \qquad \text{Affinity} - 1/K_{Eq}$$

If this affinity is very high compared, say, with the binding by plasma proteins, this strongly suggests a specific role for the material. In addition, competition with non-radioactive hormone, e.g. competition between ^{3}H-oestradiol and unlabelled oestradiol, for binding, indicates the reversible nature of the binding process, and by studying the types of substance that are able to compete, in addition to the non-labelled material, the true specificity may be assessed. Thus, Eisenfeld and Axelrod (1966) found that only oestrogens or anti-oestrogens would compete with ^{3}H-oestradiol for binding with their preparation. For example, oestrone and oestriol competed for binding, provided that this was in true target-tissues, so that uptake by non-target tissues, such as the heart, was unaffected. Progesterone, testosterone and hydrocortisone were without effect.*

Using these techniques, specific binding proteins have been isolated from the cytoplasm and nucleus of many tissues, such as uterus, prostate, vagina and chicken oviduct (Table II). The complexing substances from cytosol and from nucleus are proteins, since they are dissociated by proteases, but not by nucleases, and that from the cytosol usually has a sedimentation coefficient in the region of 9 Svedberg unit (S) whereas the nuclear complexing material is smaller, about 4-5 S.

* Androgens as well as oestrogens seem to be involved in controlling uterine growth (Lerner, 1964), oestrogen probably stimulating both growth and secretion whilst androgen stimulates only growth. Since testosterone does not compete with oestradiol for receptor sites it presumably acts by a separate mechanism involving perhaps its own uterine carrier, and in fact Giannopoulos (1971) has isolated a protein from the uterine cytosol with a high degree of specificity for androsterone. *In vivo* studies (Giannopoulos, 1973) show that ^{3}H-testosterone is taken up by the nuclei of the immature uterus and then gradually released over the ensuing 2 hr. Progesterone, cortisone and the anti-androgen, cyproterone, did not influence binding, whilst oestradiol-17β competed a little.

TABLE II

Similarities in the physicochemical properties of cytoplasmic receptor-steroid complexes isolated from cytoplasmic extracts as 105,000 g supernatant. (Mainwaring, 1975)

Steroid	Tissue	s_{20}	f/f_0	pI	MW
1. Oestradiol-17β	Calf uterus	8·0	ND	5·8	ND
		8·6	1·65	6·2	236,000
	Rat uterus	8·0	1·69	5·8	220,000
2. Dexamethasone	HTC cells	6–8	ND	ND	ND
	Liver	7·0	ND	ND (unstable)	200,000
3. Progesterone	Guinea-pig uterus	7·0	1·51	5·8; 4·5 (trace)	210,000
	Chick oviduct	8·0; 5·0	ND	4·0; 4·5	100,000; 357,000
4. 5a-Dihydrotes-tosterone	Male accessory sexual glands	8·0	1·96	5·8	290,000
5. Aldosterone	Rat kidney	8·5	ND	ND	ND

ND indicates not determined. HTC, hepatoma tissue culture cells.

Secondary-Target Tissues. Of special interest was the binding to what we may describe as "secondary target-tissues"; thus the uterus is a main target-tissue of the ovarian hormones; the steroid hormones exert a feedback on the hypothalamus and pituitary which may be called secondary targets. Eisenfeld (1970) extracted oestradiol-receptors from hypothalamus, anterior pituitary and uterus, and their relative concentrations were 1060, 4500 and 16,700 respectively; again Kahwanago et al. (1970) found a high concentration in anterior pituitary and median eminence; the concentration in the hypothalamus was smaller but greater than in cerebral cortex.

Translocation

The results of many studies suggest that the first step in the action of a steroid hormone on its target cell is to bind to a cytoplasmic carrier-protein, which is probably the 9 S material. As a complex with this it is carried to the nucleus where it is transferred across the nuclear membrane to the chromatin, the transfer being by means of the smaller 4–5 S material. Having attached to the nuclear chromatin, a modification in the cell's protein synthetic machinery takes place leading ultimately to the specific target-action of the hormone, such as multiplication of granulosa cells or their transformation into typical progesterone-secreting luteal cells.

Two-Step Mechanism. If the binding of the steroid to the cytoplasmic complexing material is the first step in the process of uptake by the nucleus, we may expect the appearance of labelled steroid in the 9 S material to precede that in the smaller, nuclear, material, and this has been demonstrated several times. Thus, Jensen *et al.* (1968) exposed isolated uteri to ³H-oestradiol at 2°C and found a high percentage of uptake by the 9·5 S cytosol material, provided exposure-time was short; but this decreased as the exposure-time increased whilst uptake into the nuclear complex increased. If, after incubating for a short time in the medium containing ³H-oestradiol, the uterus was placed in a medium without the ³H-oestradiol, there was a steady migration of the labelled material from cytoplasm to nucleus. More than half the radioactive 9·5 S material extractable from the cytoplasm disappeared and could be found in 5 S material from the nucleus; and autoradiographic studies confirmed that, when radioactivity is largely in the 9·5 S material, i.e. after short exposures at 2°C, it is, in fact, in the cytoplasm. If the tissue was exposed for 60 minutes at 37°C the radioactivity was mainly in the nuclei. In Fig. 1.5 we see a similar type of

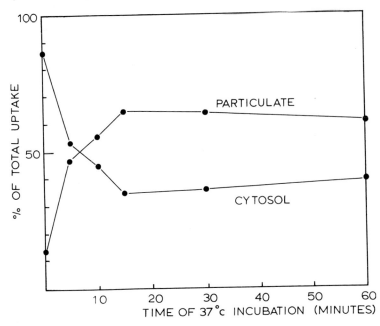

Fig. 1.5. Reciprocal type of distribution of labelled oestradiol between particulate fraction and cytosol. In each group, two uteri were incubated at 0° for 1 min. in medium containing 0·02 μg of ³H-oestradiol-17β. The uteri were then transferred to fresh medium without any oestradiol, and incubated at 37° for various times as indicated. (Shyamala and Gorski, *J. biol. Chem.*)

reciprocal distribution of radioactivity between cytosol and nucleus derived from a comparable study by Shyamala and Gorski (1969).

Importance of Synthesis

The amounts of receptor in a given target cell are by no means constant, and, in fact, as we shall see, the responsiveness of the target-tissue may well be controlled to some extent by variations in the amount of receptor material per cell. An example of the importance of synthesis of new material is given by the effects of oestrogen on the amount of œstrogen-receptor in the cytoplasm of the uterus; Cidlowski and Muldoon (1974) showed that the initial effect was to decrease the concentration of cytoplasmic material whilst that in the nucleus increased; after an hour, however, the cytoplasmic material returned to normal and finally it increased above its original value by 50 per cent. The recovery of cytoplasmic material and subsequent increase could be prevented by the inhibitor of protein synthesis, cycloheximide.*

Nuclear Binding. The Acceptor

The binding of the steroid hormone in the nucleus of the target cell depends on the presence of a specific "acceptor" attached to the nuclear chromatin. The actual acceptor material separated by Puca et al. (1975) from calf uterus appeared to be a basic protein of molecular weight about 70,000 daltons. This protein was different from the nuclear histones, and its activity was not destroyed by treatment with DNAase, in fact new acceptor sites appeared on removal of the associated DNA. Thus, the DNA of nuclear chromatin apparently binds the acceptor protein, and in so doing reduces the number of acceptor sites without, however, affecting their affinity for the receptor-steroid complex. The technique of study developed by Puca et al. involved binding the nuclear acceptor covalently to Sepharose to give an insoluble matrix suitable for affinity chromatography, and in this way the relative affinity of the acceptor material for steroid-receptor complexes was measured.

---

* The relations between cytosol-receptor and nucleus-acceptor complex formation have been examined recently by Mester and Baulieu (1975) who have reviewed earlier work, e.g. that of Sarff and Gorski (1971) in which the effects of cycloheximide were described. They describe an effect of large doses of oestradiol, which cause a rapid and large decrease in cytosol complex with associated increase in nuclear binding; this is followed by a rapid return to the cytosol that is not blocked by cycloheximide, and it is only the later and more complete recovery that is so blocked. With lower, physiological, doses, the reversal of the 50 per cent decrease in cytosol binding is inhibited by cycloheximide.

Specificity

An important point established by this technique was that the acceptor bound to the steroid hormone-receptor complex, but not to either component of this complex.* Thus, Mainwaring *et al.* (1976) extracted the androgen acceptor from rat prostate and measured its binding of $5\alpha(H^3)$ dihydrotestosterone-receptor complex derived from cytoplasmic extracts of prostate, liver, kidney and spleen. Only prostate receptor-steroid complex bound significantly; moreover, the sex steroid binding globulin (SSBG), which binds 5α-dihydrotestosterone with an affinity comparable with that of the receptor protein itself, was quite unable to transfer the steroid to the acceptor. When different steroids were compared, the high degree of specificity of the prostatic acceptor for 5α-dihydrotestosterone was revealed (Table III) both in the binding

TABLE III

Specificity of 5α-dihydrotestosterone in the interaction between the androgen-receptor complex and immobilized acceptor sites. Prostate cytoplasmic extracts were labelled either with 0·5 nM ^{3}H-labelled steroids alone or with 0·5 nM ^{3}H-dihydrotestosterone in the presence of a 20-fold excess of non-radioactive competitor. The extracts were passed through Sepharose columns containing immobilized nuclear extracts to determine interaction between androgen-receptor complex and nuclear acceptor (Mainwaring *et al.*, 1976)

^{3}H-labelled steroid (0·5 nM)	Non-radioactive competitor (10 nM)	Bound ^{3}H-labelled steroid (d.p.m./500 μg of immobilized protein)
5α-Dihydro-testosterone	—	8980 ± 420
Testosterone	—	2160 ± 140
Dexamethasone	—	100 ± 40
Progesterone	—	200 ± 20
Androsterone	—	440 ± 60
5α-Dihydro-testosterone	{ 5α-Dihydrotestosterone	1380 ± 60
	5β-Dihydrotestosterone	8800 ± 200
	Dexamethasone	9340 ± 260
	Anti-androgen*	5040 ± 260

* 6α-Bromo-17β-hydroxy-17α-methyl-4-oxa-5α-androstan-3-one.

* This was originally established by Steggles *et al.* (1971), who also emphasized the tissue-specificity for interaction between nuclear chromatin and steroid-receptor complexes.

of the labelled steroid and the ineffectiveness of unlabelled steroids, such as 5β-dihydrotestosterone, to complete for binding. It is of some interest that the anti-androgen reduced the binding a little but by no means completely. When nuclear extracts from different tissues were examined for their ability to bind cytoplasmic androgen-receptor complex, the results in Table IV were obtained, indicating some specificity, in that the prostate extract was most efficient.

TABLE IV

Showing specificity in the transfer of 5α-dihydrotestosterone from its cytoplasmic receptor complex to immobilized acceptor sites (Mainwaring *et al.*, 1976)

Source of cytoplasmic extract	Bound 5α-[³H]-dihydrotestosterone (d.p.m./500 μg of immobilized protein)
Prostate	8280 ± 440
Prostate [0–33 per cent-satd.-$(NH_4)_2SO_4$ ppt]	7440 ± 220
Prostate [33–66 per cent-satd.-$(NH_4)_2SO_4$ ppt]	640 ± 80
Liver	440 ± 40
Kidney	300 ± 40
Spleen	240 ± 20
Serum	440 ± 40
None; free 5α-[³H]dihydrotestosterone	220 ± 20
Prostate + 1 mM-*N*-ethylmaleimide	840 ± 20

Temperature Dependence

The transfer from the receptor complex to the acceptor molecule in the nucleus not only depends on the existence of the receptor-steroid complex but also on its "activation"; thus Jensen and DeSombre (1973) showed that the oestrogen-receptor complex maintained at 0°C was in the 3·8 S form, and in this condition had a low affinity for the nuclear acceptor; on warming to 37°C it was transformed to the 5·3 S form and in this condition had a high affinity for the acceptor.

Role of DNA

As discussed by Mainwaring (1975) there are alternate views of the role of DNA in the acceptor function; thus DNA can undoubtedly react with acceptor complexes, but the binding lacks the high degree of specificity found with the nuclear protein acceptor material. The nucleus-associated proteins could control the acceptor function of chromatin in either a passive or active fashion; thus on the basis of the "passive" theory, DNA is the acceptor whilst the proteins restrict the sites available for binding; according to the active theory, acceptor activity belongs to the nuclear proteins with little if any involvement of DNA. As Mainwaring points out, if the basic acceptor proteins have an intrinsic affinity for DNA, then the two approaches could be reconciled.

Specificity

The specificity shown by a steroid hormone presumably depends on the affinity it has for a given receptor in the cytoplasm of the target cell; thus the development of receptors in the target cell permits this to take up the hormone from its plasma binding-globulin. A second stage at which specificity is revealed is in the transfer of the receptor-steroid complex to the nuclear acceptor. The general problem of specificity, viewed from these aspects, has been discussed by Mainwaring (1975); the matter is of considerable practical significance in view of the requirements for anti-steroid hormones which presumably compete for receptor or acceptor sites. Figure 5.11 (p. 279) illustrates the close similarities between active and inactive steroids, such as oestradiol-17β and oestradiol-17a, and antagonists, such as 16-epioestriol and nafoxidine.

Metabolism of Steroid

A feature that may govern, or at any rate influence, specificity is the ability of the tissue to convert the steroid to an active form; thus in many target-tissues it seems that the active androgen is the 5-a-reduced steroid, the reduction of testosterone being brought about by a 5a-reductase; in many accessory sexual glands this enzyme is highly active and can convert all the testosterone to 5a-dihydrotestosterone (androstanolone) that is required. Thus the failure to reduce testosterone may limit its activity; this seems to be true of immature prostate and testis; in mature tissues, however, there is little doubt that testosterone itself can be taken up by the receptor and transferred to the nuclear acceptor sites (Jung and Baulieu, 1972; Rennie and Bruchowsky, 1972).

Hypothetical Scheme

Figure 1.6 illustrates a scheme for the mechanism of action of the steroid hormones; the model places emphasis on the importance of the nuclear binding of the steroids, resulting in the enhancement of processes involving DNA as a template. Alternative (a) suggests that nuclear processes may be stimulated by hormones without the involvement of a cytoplasmic receptor system, and alternative (b) imputes direct hormonal stimulation of metabolic processes without involvement of the receptor system at all.

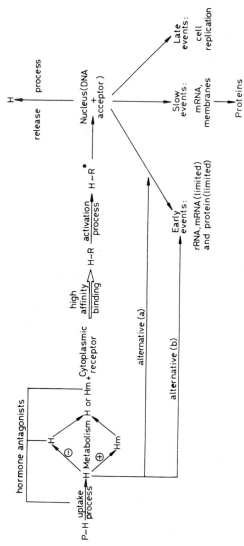

Fig. 1.6. Mechanism of action of steroid hormones, for explanation see text. Abbreviations: P, steroid-binding protein in plasma; H, steroid hormone; Hm, hormone metabolite; ⊖ and ⊕ indicate absence or requirement of intracellular metabolism of the hormone; R, receptor protein; R*, activated configuration of receptor. (Mainwaring, *Vitamins & Hormones.*)

The Ovary

It will be recalled that the ovary consists in an agglomeration of follicles in various stages of development and decay (atresia); the material between these follicles is described as "interstitial" and, although the importance of an interstitial cell-type has been questioned, there is little doubt that this exists and contributes significantly to the steroid secretions of the ovary as a whole. The follicle consists, essentially, of a germ cell, or ovum, immediately enclosed by a cellular layer which proliferates to become the *granulosa cells* of the secondary and more mature follicles. Development of the follicle consists in the formation of an enclosing *theca*, so that the ovum plus granulosa cells finally float in a fluid-filled cavity (Fig. 2.5, p. 58). Ovulation consists in rupture of the follicle, which then becomes a *corpus luteum* through multiplication of the granulosa cells which are now described as *luteal* cells.

The Ovarian Secreting Cells

The nature of the cells that secrete steroid hormones in the ovary is by no means clearly understood; the matter has been lucidly discussed by Mossman *et al.* (1964) and more recently by Dahl (1970) in relation to his electron microscopical studies of the hen ovary. If we accept the definitions of Mossman *et al.*, there are three distinct types of gland cells in adult mammalian ovaries, namely (1) interstitial gland cells formed from the theca interna of atretic follicles; (2) thecal gland cells formed from the theca interna of large secondary follicles and degenerating vesicular (Graafian) follicles of all sizes except those that are essentially "ripe" before atresia sets in. In these ripe, or nearly ripe, follicles the majority of the theca interna cells have already differentiated into functional *thecal gland cells*. These cells thus fail to produce interstitial gland tissue which, as indicated above, is characteristic of tissue derived from atresic unripe follicles. Finally, (3) we have the luteal cells formed from the granulosa cells of ovulated follicles, i.e. from corpora lutea.

Interstitial Cells

Mossman *et al.* emphasize the misleading description of the secretory tissue as "interstitial", indicating as this word does, merely something in the spaces between other and more important elements. In their study they concentrated on the interstitial gland tissue; they showed by planimetric measurements of sections that in non-pregnancy or early pregnancy in the human this occupied only about 1 per cent of the total ovarian tissue, but in the last trimester it had risen to 4–6 per cent. Luteal gland tissue at the height of its development in the first trimester occupied some 31 per cent of the total. Histochemical evidence indicates that the interstitial gland cells secrete steroid substances; and it has been suggested that the main secretions are oestrogens.

The theca of the Graafian follicle is formed from the surrounding stromal cells of the ovary; the transformation of the stromal cells into theca interna cells is preceded by what Balboni calls a bastic phase in which the cells develop the internal organelles such as endoplasmic reticulum characteristic of secreting cells.

Theca Interna

To return to the thecal gland cells, the theca interna is usually present around secondary follicles and growing vesicular follicles, and is composed of rapidly multiplying cells with a rich capillary plexus. During the final pre-ovulatory growth period, the cells differentiate into steroid secretors, and provide the oestrogens that are secreted during pro-oestrus and oestrus. The theca, then, is now a functional gland. When the follicle ovulates, these cells may persist around the developing corpus, but they degenerate rapidly and disappear. If the follicle undergoes atresia without ovulation, then the thecal gland cells degenerate completely. If a follicle undergoes atresia after its theca interna has developed, but before the final ripening and formation of a functional thecal gland, then the still embryonic theca interna differentiates into an interstitial gland.

Sites of Synthesis

As to the sites of oestrogen versus progesterone synthesis, there seems little doubt that the granulosa cells of the luteinized follicle are responsible for progesterone; thus Short (1961) tapped fluid from follicles of the mare, quantities as high as 100 ml being obtained from some. Those containing only granulosa cells, i.e. those that had luteinized without ovulating—a common feature in the mare—only contained progesterone, so he concluded that thecal cells must be the source of oestrogens, probably by converting progesterone to oestrogen. However, Ryan et al. (1968) scraped the granulosa cells free from the theca and examined the synthetic capacity of both theca and granulosa cells from acetate. Both cell-types could synthesize oestrogen although the granulosa cells produced more progesterone than oestrogen. Thus the granulosa behaves like the corporus luteum, and the theca like the whole follicle. Again, Savard and his colleagues* have measured the synthesis of steroids by different parts of the ovary; isolated corpora lutea produced mainly progesterone from ^{14}C-labelled acetate, a process that was stimulated by gonadotrophic hormones—namely HCG and LH. When they separated what they called the "stromal tissue" from the ovary, by first removing the corporus luteum and then the outermost layer of the ovary, they found that the main synthetic products were *androgens*—testosterone, dehydro-epiandrosterone and $\triangle^4$-androstenedione. In fact, as the authors remarked, the synthetic activity of the "stromal tissue" of the ovary resembled that of the intact testis.

Follicles. To assess the hormones produced by the follicles, Ryan and Smith (1961) aspirated the fluids from multiple follicular cysts of the human ovary and minced the cyst linings stripped from their

* See Hammerstein et al. (1964), Rice et al. (1964) and Rice and Savard (1966). The histology of the ovary with respect to its secretions has been discussed in detail by Scully and Cohen (1964), and Balboni (1976).

capsules. From acetate both oestrone and oestradiol were synthesized, whilst progesterone also acted as a substrate for these hormones, presumably by way of androgens (Fig. 1.2).

Three Ovarian Tissues. Thus, according to Savard, the three main components of ovarian tissue produce three separate types of steroid hormone; the follicles produce oestrone (E_1) and oestradiol (E_2), the corpora lutea produce progesterone, and the "interstitial tissue", comprised of "follicles in various stages of development and atresia with their thecal linings, the stromal tissue from which the thecal cells are said to be derived, and scattered interstitial cells, said to be derived from stromal and thecal elements", produces androgens.

Male Steroid Secretion

The synthesis of the principal male androgen, testosterone, by the isolated testis has been demonstrated by providing the tissue with ^{14}C-labelled acetate and isolating the radioactive steroid (Savard *et al.*, 1956). By perfusing the testis *in vitro* Savard *et al.* (1961) demonstrated not only the synthesis of testosterone but also its release into the venous effluent. The cells responsible for androgen synthesis are undoubtedly the interstitial cells of Leydig, although the extent to which the cells of the seminiferous tubules e.g. Sertoli cells are also capable of androgen synthesis has not been completely resolved (Steinberger and Steinberger, 1972). As with the ovary, the steroids secreted by the testis are not exclusive to the sex, so that appreciable amounts of 17-hydroxy-progesterone are present, this being, in fact, a precursor to oestrogen secretion. Oestradiol itself is also produced by the testis and may well serve a physiological function; thus, according to Stumpf (1969) ^{3}H-labelled oestradiol accumulates in the nuclei of interstitial cells of the immature rat testis, whilst specific receptor material has been isolated from rat's testis interstitial tissue (Brinkman *et al.*, 1972); this accounts for uptake of labelled oestradiol by both cytoplasm and nucleus, but labelling of the nucleus requires the presence of the cytoplasmic material (Mulder *et al.*, 1973).

"*Aromatization*". Testosterone and androstenedione can be metabolized to oestrogens; in fact, as Fig. 1.2 indicates, these male steroids are on the direct synthetic pathway to oestrogens. It has been argued that in some situations the effects of androgens are brought about by prior conversion to oestrogen. Thus, in the male, the feedback lowering of the levels of gonadotrophic hormones in the blood, revealed, say, by injections of steroids, is actually more effective if oestrogens are given than if androgen is injected (Peterson *et al.*, 1968). Unless it should be concluded that all actions of androgen are mediated through oestrogen, we must note that many of the actions of testosterone are more effectively induced by the ring-A-reduced metabolite, 5-dihydrotestosterone. This compound cannot be "aromatized", so that its androgenic action is due to itself.

Two-Cell Hypothesis

So far as the production of oestrogens is concerned, we must appreciate that they represent the last stage in the steroid synthetic pathway (Fig. 1.2); thus the C_{21} compounds—pregnenolone and progesterone—are formed directly from cholesterol, which itself is formed from acetate. The subsequent steps in oestrogen synthesis involve, first, the removal of the side-chain to form C_{19} compounds, androstenedione, testosterone, etc. These C_{19} androgens are subsequently converted to oestrogens by aromatization of one of the rings. There is some evidence that the syntheses of both testosterone and oestradiol require the co-operation of two types of cell, the one producing the C_{21} precursors and the other carrying out the later stage. Falck (1959) found that transplants of pure cell systems of the ovary, be they interstitial cells, theca interna cells, corpus luteum cells or granulosa cells, would not produce oestrogens, as measured by their effects on transplanted vaginal tissue. Thus he dissected from the rat's ovary the various components and implanted them in the anterior chamber of the rat's eye. Only when combined transplants were made, namely of theca interna cells and granulosa cells, or interstitial cells and granulosa cells was oestrogen formed. Subsequent work suggests that the granulosa cells of the ovary are only able to synthesize oestrogen if they are supplied by androgen precursors from other cells, such as the theca interna cells (see, e.g. Short, 1964). A similar relation is encountered in the testis. The testis produces oestrogens, which were considered to be produced by the interstitial cells of Leydig; however, Dorrington and Armstrong (1975) cultured Sertoli cells and found that they would synthesize oestrogen provided testosterone was in the culture medium. Sertoli cells were unable to synthesize androgens from pregnenolone, so that in vivo they must rely on an extraneous source of androgens to synthesize oestrogens. Like the granulosa cell, therefore, being similarly derived from the same cell line in embryogenesis, the Sertoli cell relies on an extraneous source of oestrogen-precursor for its oestrogen production, in this case the interstitial cells of Leydig. As we shall see, this two-cell hypothesis provides a rational basis for the different actions of FSH and LH in stimulating steroid hormone production in the gonads (p. 128).

CONTROL OF THE SECRETION OF THE STEROID HORMONES

Gonadotrophic Hormones

The gonadotrophic hormones of the anterior pituitary, or adenohypophysis, exert a controlling influence on the secretion of the sex hormones. They do this through their trophic influence on the tissues synthesizing and secreting the steroid hormones, governing their development and maintenance, and through the promotion of synthesis and release of steroid hormones. For example, follicle stimulating hormone, FSH, promotes the development of the Graafian follicle and thus of its powers to synthesize oestrogen, and, in the developed state, it promotes synthesis and secretion into the blood-stream.

ICSH, FSH and LH

Androgens are secreted in response to the liberation of interstitial cell stimulating hormone, ICSH, whilst the follicular (oestrogens) and luteal hormones (progestogens) are secreted in response to follicle stimulating hormone, FSH, and luteinizing hormone, LH. ICSH of the male and LH of the female are identical and are now referred to as LH. FSH is secreted by male and female pituitaries. The secretion of these gonadotrophic hormones is governed by the hypothalamus. Thus, the characteristically cyclical activity of the gonads in the female, as contrasted with that of the male, is the result of an inherent rhythmicity in the activity of the female hypothalamus which, through its releasing factors, excites the cyclical liberation of the female gonadotrophic hormones, FSH and LH, which in their turn cause the secretion of oestrogen and progesterone. In the male, on the other hand, the secretion of androgens, and thus of LH and FSH, is more uniform.

THE HYPOTHALAMUS-PITUITARY SYSTEM

Pituitary Gland

The basic features of the hypothalamic-pituitary system have been described in Vol. 2, and may be briefly recapitulated here. The anterior lobe of the pituitary develops as an upgrowth of epithelial tissues from the primitive pharynx and meets a downward growth from the base of the brain; this latter, neural, component becomes the infundibular process or posterior lobe (neurophyophysis) and the neural part of the pituitary stalk (Fig. 1.7). Thus the glandular tissue is made up of the *pars distalis* (PD) and *pars tuberalis* (PT), and is referred to as the anterior pituitary or adenohypophysis; it contains the cells secreting the hormones of the anterior pituitary-gonadotrophins, adrenocorticotrophic hormone, and so on.

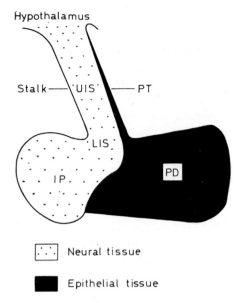

Fig. 1.7. The human pituitary gland in sagittal section. "Neural tissue" indicates the neurohypophysis; "epithelial tissue" the adenohypophysis. IP, infundibular process (posterior lobe); LIS, lower infundibular stem; PD, pars distalis (anterior lobe); PT, pars tuberalis; UIS, upper infundibular stem. (Daniel, *J. clin. Path.*)

Posterior Pituitary

The secretion of the hormones of the posterior pituitary, or *neurophypophysis*, is brought about by neurosecretory cells in the hypothalamus, the most prominent groups of these cells being the supra-optic and paraventricular nuclei. The cells secrete the hormones vasopressin and oxytocin, and a binding protein, neurophysin; these are carried down the axons of the secreting neurones in the hypothalamo-neurohypohyophyseal tract to be liberated in relation to capillaries of the infundibular process.

Adenohypophyseal Releasing Hormones

The secretions of the adenohypophysis, or anterior pituitary, are controlled by *releasing factors* or *releasing hormones* (RF or RH) which are secreted by neurones in the nuclei in the basal part of the hypothalamus, especially in the tuber cinereum (around the infundibular recess). These neurones secrete the releasing (and inhibiting) hormones from their terminals in relation to coiled capillaries forming the primary capillary bed of the long and short portal vessels. The releasing or inhibitory hormones reaching these capillaries are carried in this portal system to the *pars distalis*, or *anterior pituitary*, where they cause secretory cells to synthesize and release their specific hormones (Figs. 1.8 and 1.9). The

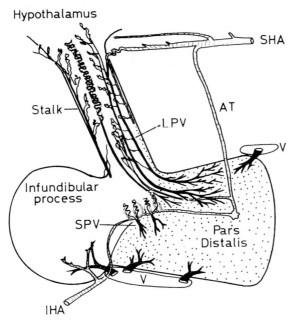

Fig. 1.8. Sagittal section of human pituitary gland showing main features of the blood supply. Note that sinusoids of the pars distalis are supplied by two groups of portal vessels: long portal vessels, LPV, draining characteristic capillary loops in the upper infundibular stem (neural tissue of the stalk) and short portal vessels, SPV, draining similar capillary loops in lower infundibular stem. AT, artery of the trabecula; IHA, inferior hypophyseal artery; SIIA, superior hypophyseal artery; V, venous sinus. (Daniel, *J. clin. Path.*)

portal vessels running in the pituitary stalk are classified as long and short, and are derived from different arteries, the long portal vessels springing from the arterial ring supplied by the superior hypophyseal arteries, and the short portal vessels are derived from the inferior hypophyseal arteries; the regions supplied by the two systems are indicated in Fig. 1.8.*

* A retrograde flow of blood from anterior pituitary back to the hypothalamus has been demonstrated (Török, 1964) and the same is true for the posterior pituitary to the hypothalamus (Page *et al.*, 1976); this latter route provides a connection between posterior and anterior hypophyses and accounts for the finding by Zimmerman (1973) of vasopressin, a posterior hypophyseal hormone, in the portal blood. Oliver *et al.* (1977) have discussed this communication between the anterior and posterior pituitaries in the light of their own study on gonadotrophic hormones in the portal vessels.

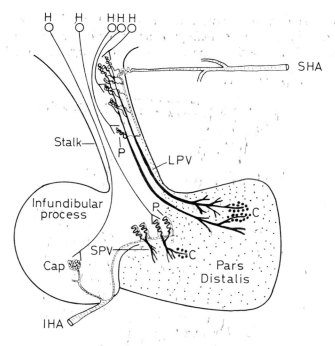

Fig. 1.9. Diagram of human pituitary gland in sagittal plane showing neurovascular pathways by which nerve cells in certain hypothalamic nuclei (H) control output of anterior and posterior pituitary hormones. Axon on left ending on the capillary bed (Cap) in the infundibular process represents the tract from the large nerve cells of the supraoptic and paraventricular nuclei, which are concerned with posterior pituitary function. The other axons have their origin in nerve cells in the so-called hypophysiotrophic area of the hypothalamus and end on the primary capillary bed (P) feeding the portal vessels which supply the pars distalis. Here their neurohormones are transmitted into the blood stream and are then carried through the long and the short portal vessels (LPV, SPV) to the epithelial cells (C) in a given area of the pars distalis to control the output of hormones from these cells. (Daniel, *J. clin. Path.*)

Hypophysiotrophic Area

The part of the hypothalamus concerned with the anterior pituitary secretions has been called by Halász *et al.* (1962) the hypophysiotrophic area (Fig. 1.10) and it is in this area that grafts of the anterior pituitary are able to support their full function, presumably because of the tonic liberation of releasing factors in their neighbourhood. These grafts are successful not only in maintaining the integrity of the secretory cells but also in maintaining adequate function of their target cells, e.g. the adrenal, thyroid, and gonads.* That the pituitary can maintain its function was shown by transplanting tissue from one female under the median eminence of another hypophysectomized female and observing that ovulation, pregnancy and milk secretion all occurred in the recipient. As we shall see, transplanting adult male pituitary tissue into female hypothalamus permitted the hypophysectomized female to begin oestrus and become pregnant, indicating the similarity of the gonadotrophic hormones in the two sexes and, more important, the role of the female *hypothalamus* in determining cyclic behaviour.

Autonomy

This region of the hypothalamus possesses some degree of autonomy, since complete isolation by surgical means from the rest of the central nervous system, leaving its pituitary connections intact, is not followed by gonadal degeneration—the tonic secretion required for maintenance being sustained. However, the separation from more rostral regions of the brain, in particular the preoptic anterior hypothalamic region (POAH), does release the hypophysiotrophic area from control mechanisms that govern the cyclical activity in the female.

The Gonadotrophic Hormones

The gonadotrophic hormones, such as FSH, LH and human chorionic gonadatrophin (hCG) are glycoproteins, which may contain up to 30 per cent of carbohydrate attached covalently to the amino-acid chains; the sugar residues are sialic acid, N-acetylglucosamine and fucose; and these may be connected through asparagine and serine linkages. The molecular weights of human FSH and LH are about 30,000, whilst that of human chorionic gonadatrophin, because, mainly, of its much greater sugar content, is about 40,000.

Subunits

The interesting feature of the gonadatrophic hormones, which they share with thyroid stimulating hormone (TSH), is that they are composed of two subunits, a and β, linked together by non-covalent bonds, so that they may be reversibly dissociated with 8-molal urea. In a given species, the a-subunit has the same, or very nearly the same, amino-acid sequence in all the hormones; thus the a-subunit of human FSH, hFSHa, is a chain of 89 amino-acid residues, and is identical with that

---

* It might be argued that, because of the proximity of the grafts to the third ventricle, they received their neurosecretions from the cerebrospinal fluid; Halász *et al.* point out, however, that when grafts were made outside the hypophysiotrophic area and actually bulged into the IIIrd ventricle, the integrity of their basophilic secretory cells was not maintained.

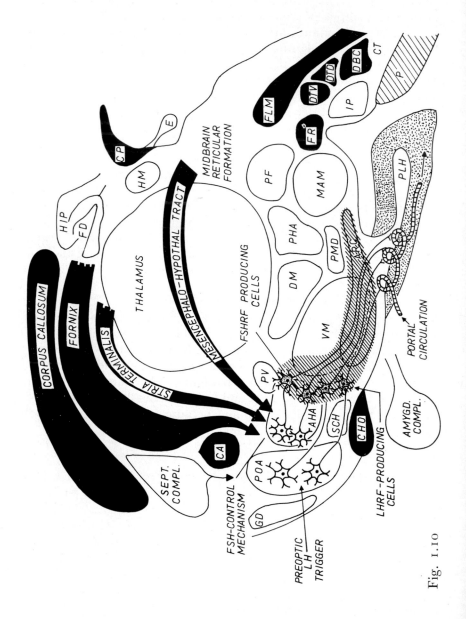

Fig. 1.10

of hLHα (Shome and Parlow, 1974a), and is nearly identical with that of hCGα except for three extra residues on the amino-terminus of hCGα; hTSHα is probably identical with hLHα (Sairam and Li, 1973). When α-subunits from the same hormone, but from different species, were compared, considerable differences were found, the amino-terminal portion varying with the species, whereas the COOH-terminal portion of all glycoprotein α-subunits regardless of species, was invariable (Fig. 1.11). Thus, we may say that the α-subunit is species-specific but lacks hormone-specificity, being very nearly the same in all the glycoproteins. When α and β-subunits of different hormones are recombined it is found that, whereas the individual subunits, either α or β, have no, or very little, activity, the recombined subunits are active and the *activity is that corresponding to the β-subunit.** For example, Pierce *et al.* (1971) combined bovine β-TSH with human α-chorionic gonadotrophin (α-hCG) and obtained a product with *thyrotrophic* activity; when α-hCG was combined with the bovine β-LH subunit, a hormone with *gonadotrophic* activity was obtained.† Thus, whereas the α-subunit is species-

* Shownkeen *et al.* (1976) have broken FSH into its subunits and recombined them; they have confirmed the low activity of the subunits so that the α-subunit had only 1 per cent of the activity of the intact hormones whilst the β-subunit had 2–10 per cent. Both subunits contained sialic acid.

† The differences in the α-subunits between species reflect amino-acid substitutions in parts of the sequences that are not involved in interactions between the subunits, or between hormones and their target-organ receptors, since the α-subunits of different species may substitute for each other in combination with β-subunits, e.g. when human α-hCG combines with bovine β-TSH, and the combination has virtually unimpaired hormonal activity (Pierce *et al.*, 1971).

Fig. 1.10. Illustrating the hypothalamic and limbic control mechanisms in secretion of gonadotrophic hormones. The hypophysiotrophic area is cross-hatched, and the neurones in this area that release releasing and inhibitory factors terminate in the capillary loops of the portal system. These neurones are considered to be responsible for a tonic secretion of releasing hormones, FSH-RH and LH-RH, and are indicated by dotted bodies. Perikarya of the neurones belonging to the hypothalamic "cyclic" mechanism are represented by clear bodies located in the pre-optic and anterior hypothalamic area (PO/AH). *Abbreviations used:* AHA, anterior hypothalamic area; Amygd. compl., amygdaloid complex; ARC, arcuate nucleus; CA, anterior commissure; CHO, optic chiasma; CP, posterior commissure; CT, nucleus centralis tegmenti; DBC, decussatio brachiorum conjunctivorum; DM, dorsomedial nucleus; DTD, decussatio tegmenti dorsalis; DTV, decussatio tegmenti ventralis; E, pineal gland (epiphysis); FD, fascia dentata; FLM, fasciculus longitudinalis medialis; FR, fasciculus retroflexus; GD, gyrus diagonalis; HIP, hippocampus; HM, medial habenular nucleus; IP, interpeduncular nucleus; MAM, mammillary nucleus; P, pons; PF, nucleus parafascicularis thalami; PHA, posterior hypothalamic area; PLH, posterior lobe of the hypophysis; PMD, dorsal premammillary nucleus; POA, preoptic area; PV, paraventricular nucleus; SCH, suprachiasmatic nucleus; Sept. compl., septal complex; VM, ventromedial nucleus. (Flerko, *The Hypothalamus*).

```
                        10
H·CG-α
H·LH-α          H –ALA–PRO–ASX–VAL–GLX–ASX–CYS–PRO–GLX–CYS–THR–LEU–GLX–GLX–ASX-
                                H-VAL
P·LH-α                  H-THR-MET                    GLY                    LYS     LYS
B·TSH-α         H-PHE-PRO-ASP-GLY-GLU-PHE-THR-MET    GLY                    LYS     LYS

                        20                           30
H·CG-α
H·LH-α          PRO-PHE-PHE-SER-GLX-PRO-GLY-ALA-PRO-ILE-LEU-GLN-CYS-MET-GLY-CYS-CYS-PHE-SER-ARG-
P·LH-α          LYS TYR                                             TYR
B·TSH-α         LYS TYR                              LYS ASX        TYR

                        40                          50
H·CG-α
H·LH-α          ALA-TYR-PRO-THR-PRO-LEU-ARG-SER-LYS-LYS-THR-MET-LEU-VAL-GLN-LYS-ASN(CHO)-VAL-THR-
P·LH-α                          ALA                                 PRO                  ILE
B·TSH-α                         ALA                                 PRO                  ILE

                        60                                         70
H·CG-α
H·LH-α          SER-GLX-SER-THR-CYS-CYS-VAL-ALA-LYS-SER-TYR-ASN-ARG-VAL-THR-VAL-MET-GLY-GLY-PHE-
P·LH-α          ALA(THR,CYS,CYS)                ALA PHE THR LYS ALA                    ASX ALA
B·TSH-α         ALA                             ALA PHE THR LYS ALA                    ASN VAL

                        80                          90
H·CG-α
H·LH-α          LYS-VAL-GLX-ASN(CHO)-HIS-THR-ALA-CYS-HIS-CYS-SER-THR-CYS-TYR-TYR-HIS-LYS-SER-OH
P·LH-α                                           SER CYS                              -OH
        SER-    GLX (CYS,HIS,CYS, HIS, THR,CYS)
P·LH-α                  ARG                                                           -OH
B·TSH-α         GLU             ARG                                                   -OH
```

Fig. 1.11. Comparison of the amino-acid sequences of human and porcine LH-α and bovine TSH-α with hCG-α. Only residues not identical with those in hCG-α are shown. Bovine and ovine LH-α have identical amino-acid sequences with that of TSH-α. The portions of the pLH-α sequence enclosed in parentheses were determined by composition only. (Bellisario et al., J. biol. Chem.)

specific the β-subunit is *hormone-specific*. When β-subunits of the same species but different hormones are compared there is a considerable homology in the amino-acid sequences but by no means the virtual identity found with the α-subunits; thus Shome and Parlow (1974b) compared hFSHβ with hTSHβ and hLHβ; the FSH subunit was more similar to that of TSH, having 49 identical positions, than to that of hLH with only 39 identical positions. In general, according to Saxena and Rathnam (1976), the homology amongst β-subunits from different hormones is mainly confined to the NH$_2$-terminal 32 residues, so that the *hormone-specificity* is revealed in the remaining portion of the amino-acid chain.

β-Subunit Homologies. If the β-subunit determines the hormonal activity, then we may expect to find considerable homologies between β-subunits of the *same hormone* derived from *different species*, and in fact, the interspecies variations in β-subunits are less than those of α-subunits. Figure 1.12 illustrates the sequences of β-subunits of the same hormone, LH, derived from human, porcine and bovine/ovine‡ pituitaries; included in the comparison is the β-subunit from human chorionic gonadotrophin (hCG) which hormonally has strong analogies with LH. The analogy between hLHβ and hCGβ is strong, but hCGβ has an additional 28 residues at the COOH-terminus. There are considerable analogies amongst the others, and it is interesting that, when a change occurs from the human subunit, this is usually the same in pig, ox and sheep.

Human Chorionic Gonadotrophin (hCG).

This hormone, separated from pregnant women's urine, has largely luteinizing hormone LH activity when measured in the female, and interstitial cell stimulating activity, ICSA, when studied in the male; the preparation also contains some follicle-stimulating activity, FSA, which could be due to contamination with pituitary FSH. However, Louvet *et al.* (1976) have shown that, when hCG is dissociated into its two subunits, its follicle-stimulating activity—FSA—disappears as with its other activity as described earlier. On recombination, however, 80 per cent of its LH-activity and 76 per cent of its FS-activity were regained, and the slopes of the dose-response curves for the recombined preparation, as measured by typical assays for LH and FSH, were identical with those of the original preparation. Thus the LH and FSH-activities of hCG are an intrinsic property of the hormone and not due to separate hormones.

‡ Bovine and ovine LH have identical amino-acid sequences.

10

```
  H CG-β  H-SER-LYS-GLN-PRO-LEU-ARG-PRO-ARG-CYS-ARG-PRO-ILE-ASN(CHO)-ALA-THR-LEU- X -ALA-VAL-GLU-
  H LH-β R-      ARG                              TRP  GLX                   X              ALA
  P LH-β R-      ARG-GLY                          LEU  GLX                  ARG             ALA
O/B LH-β R-      ARG-GLY                          LEU  GLN                   X              ALA
```

20   30

```
  H CG-β  LYS-GLU-GLY-CYS-PRO-VAL-CYS-ILE-THR-VAL-ASN(CHO)-THR-THR-ILE-CYS-ALA-GLY-TYR-CYS-PRO-
  H LH-β
  P LH-β      ASX     ALA                        PHE      THR                           SER
O/B LH-β              ALA                        PHE      THR                           SER
```

40   50   60

```
  H CG-β  THR-MET-THR-ARG-VAL-LEU-GLN-GLY-VAL-LEU-PRO-ALA-LEU-PRO-GLX-LEU-VAL-CYS-ASN-TYR-ARG-
  H LH-β      ARG(MET)LEU                      ALA        PRO VAL                  THR
  P LH-β  SER ARG                      PRO ALA ALA        PRO VAL                  THR
O/B LH-β  SER LYS                      PRO VAL ILE        PRO MET              ARG     THR     HIS
```

70   80

```
  H CG-β  ASP-VAL-ARG-PHE-GLU-SER-ILE-ARG-LEU-PRO-GLY-CYS-PRO-ARG-GLY-VAL-ASN-PRO-VAL-VAL-SER-
  H LH-β
  P LH-β      GLU LEU ILE                  ALA                 PRO                  THR
O/B LH-β      GLU LEU                      VAL                 PRO              ASP     MET
```

90   100   110   120

```
  H CG-β  TYR-ALA-VAL-ALA-LEU,SER-CYS-GLN-CYS.ALA-LEU-CYS-ARG(ARG)-SER-THR-THR-ASP-CYS-GLY-GLY-PRO
  H LH-β      PHE PRO              ARG  GLY PRO                          SER                PRO GLY
  P LH-β      PHE PRO              HIS  GLY PRO         LEU  SER SER                        PRO GLY
O/B LH-β      PHE PRO              HIS  GLY PRO         LEU  SER                            PRO GLY
```

120

```
  H CG-β  LYS-ASP-HIS-PRO-LEU-THR-CYS-ASP-ASP-PRO-ARG-PHE-GLN-ASP-SER(CHO)-SER-SER-LYS-ALA-PRO
  H LH-β           (GLX,ASX,SER,LYS)GLY-OH                                                  -PRO
  P LH-β  ARG ALA GLX      ALA      PRO LEU PRO GLY          LEU    LEU-OH                    PRO
O/B LH-β  ARG THR GLU      ALA      PRO LEU PRO          ILE        LEU-OH                    SER
                                                                                             LEU
                                                                                             PRO
                                                                                             SER(CHO)
```

140

```
GLN PRO LEU ILE ... PRO THR ASX PRO PRO GLY PRO LEU ARG SER(CHO) PRO
```

Carbohydrate Residues

These may constitute a considerable proportion of the total molecular weight of the subunit; for example the a-subunit of hCG has a molecular weight of 14,900 made up of protein (10,200) and carbohydrate (4700) the latter consisting in two bulky units attached to asparagine residues. The β-subunit has a molecular weight of 23,000 made up of 16,000 protein and 7000 carbohydrate, the latter consisting of two oligo-saccharide units attached to asparagine residues and three attached to serine residues (Carlsen *et al.*, 1973). Removal of the carbohydrate, e.g. by treatment with neuraminidase to "desialate", caused loss of activity when tested *in vivo*; however, immunoreactivity was actually enhanced, and the loss of biological activity was shown to be due to the reduced stability of the hormone, the loss of sialic acid permitting its rapid incorporation into liver cells (Morell *et al.*, 1971). Thus an *in vitro* assay on granulosa cells showed that desialated hCG was fully effective in luteinizing these. By contrast, the subunits, when tested individually, had only one hundredth of normal hormonal activity (Channing and Kammerman, 1973).*

Free Subunits. Maternal blood from pregnant women, or extracts of placenta, contain considerable quantities of free a-subunits of hCG; in the later stages of pregnancy the proportion of the a-subunit increases so that it may represent ten times the concentration of hCG in placenta. According to Vaitukaitis (1974) there was no evidence of the β-subunit, and this suggests that control over synthesis of the gonadotrophin is through synthesis of the β-subunit. In maternal *plasma*, the hCG far exceeds the a-subunit in concentration throughout pregnancy.

Pre-Hormones

A number of protein hormones are initially synthesized as larger molecules from which the active hormone is later split off; thus insulin-

---

* By employing appropriate methods of separation it can be shown that there is some "microheterogeneity" in gonadaotrophic hormones due to slight variations in amino-acid sequence; according to Graesslin *et al.* (1973), six forms of hCG can be distinguished, differing in the β- rather than the a-subunit. Moyle *et al.* (1975) showed that hCG could stimulate steroidogenesis and cAMP accumulation in rat testicular tissue; removal of carbohydrate reduced the ability to stimulate cAMP accumulation but not to stimulate steroidogenesis.

Fig. 1.12. Comparison between the amino-acid sequences of the β-subunits from human chorionic gonadotrophin (hCG-β), human luteinizing hormone (hLH-β), porcine luteinizing hormone (pLH-β) and ovine and bovine luteinizing hormones (o/bLH-β) whose sequences are identical. Residues are identical with those shown for the hCG-β sequence unless otherwise indicated; X indicates a deletion. (Carlsen *et al.*, *J. biol. Chem.*)

mRNA induces the synthesis of a protein 21 amino acids larger than the proinsulin. The same is true for human placental lactogen and pro-lactin, the material synthesized by mRNA in cell-free systems being some 2000 to 4000 daltons heavier than 'the secretory forms (Boime *et al.*, 1975; Evans and Rosenfeld, 1976). According to Evans *et al.* (1977) it is unlikely that the very large preprolactins reported in the literature are direct products of synthesis, but result from protein-protein inter-action.

Sites of Origin of Gonadotrophic Hormones

Histochemical studies of the pituitary indicate, in general, that its hormones are synthesized and secreted by separate types of cell—"one cell one hormone". However, the evidence regarding LH and FSH is conflicting, and it would appear from recent work involving the immunocytochemical localization of the hormones* that the two can be present in the same pituitary cell. Barnes† (1962) classified the gonadotrophic cells as "FSH-cells" with a smooth oval contour and "LH-cells" with angular profiles, a classification that was maintained by Kurusomi and Oota (1968), who found that in persistent oestrus, which is associated with a normal FSH but impaired LH secretion, the putative FSH-cells were normal; in persistent dioestrus, on the other hand, the putative FSH-cells were few in number and atrophic whilst the LH-cells were normal. However, when specific immunocyto-chemical methods were employed, Nakane (1970), using an anti-FSH, found that both cell-types, i.e. the oval and angulated, reacted with anti-FSH; however, the oval cells formerly considered to be FSH-cells also reacted to anti-LH, so that this cell-type contained both hormones. The angulated cells, on the other hand, reacted only with anti-FSH. The cells containing both hormones were concentrated in the periphery of the gland, where they might well be more accessible to the releasing hormone. More recently, Moriarty (1975) has confirmed that pituitary cells, characterized morphologically as FSH-cells, react with an anti-LH serum. He found a sex difference in the reactions of pituitary "LH-cells" and "FSH-cells", as defined morphologically on the Kurosumi-Oota basis; thus in the female pituitary both LH-cells and

---

* Nakane applied a rabbit antiserum to the pituitary hormone in question to the pituitary tissue; this would precipitate on the hormone-containing cell and its localization was carried out by applying a peroxidase-labelled sheep anti-rabbit gamma-globulin, the peroxidase subsequently providing the electron-dense material that permitted its visualization in the electron microscope.

† Five types of cells, called α, β, γ, δ and ϵ were described; the two gonadotroph-containing cells were β and γ; and from variations in the degree of granulation of these cells during the oestrous cycle, Barnes concluded that the β-cell contained FSH and the γ-cell LH. The ϵ-cell contained prolactin and became prominent during pregnancy and lactation.

FSH-cells stained with anti-LH_β whilst, in the male, staining by anti-LH_β occurred primarily with FSH-cells. Thus, it appeared that, in the female, the LH-cells, defined on the basis of morphology, were, indeed, LH-secreting cells, whereas the FSH-cells of either sex, defined morphologically, could be secretors of either LH or FSH. Moriarty suggested that female pituitaries had LH-secreting cells of two distinct morphological types, whilst in the male there was only one.

The Releasing Hormones

It seems that a single gonadotrophin releasing hormone—GnLH—is responsible for the release of both LH and FSH; it is usually referred to as LHRH. It is a decapeptide amide and has been synthesized.* Figure 1.13 illustrates its amino-acid sequence along with those of the simple thyrotrophin releasing hormone—TRH—and the more complex

TRH - PYRO - GLU - HIS - PRO - NH_2

GnRH - PYRO - GLU - HIS - TRP - SER - TYR - GLY - LEU - ARG - PRO - GLY - NH_2

GHIH - H - ALA - GLY - CYS - LYS - ASN - PHE - PHE - TRP - LYS THR - PHE - THR - SER - CYS - OH

Fig. 1.13. Structural formulae of thyrotrophin releasing hormone, TRH, gonadotrophin releasing hormone, GnRH or LHRH, and somatostatin or growth hormone inhibiting hormone, GHIH. (Wilber *et al.*, *Rec. Progr. Horm. Res.*)

growth hormone inhibiting hormone—GHIH—or somatostatin. According to Schally *et al.* (1972), the release of LH and FSH from the isolated pituitary by three samples of the releasing hormone, synthesized by different techniques, was quantitatively the same and equal to that produced by a highly purified porcine preparation. Thus, if the releasing hormone is administered to the anoestrous ewe, the same rise in both LH and FSH occurs as that which takes place during the "preovulatory surges" (p. 62) of these trophic hormones during the

* It is worth recording that a synthetic analogue of LHRH has been prepared with some nineteen times the LH-releasing activity for LH and 17 times that for FSH when compared with the natural decapeptide. It is a nonapeptide with replacement of glycine in position 6 by D-serine protected by a tertiary butyl group and replacement of the CO_2H-terminal glycinamide by ethylamide: [D-Ser(TBU)⁶, des Gly-$NH_2$¹³] LHRH ethylamide. It is immunologically distinct from LHRH, so that it can cause release of gonadotrophins in the immunized animal (Fraser and Sandow, 1977).

oestrous cycle. When the releasing factor is injected into the constant pro-oestrous rabbit, i.e. in the rabbit that is awaiting copulation in order to ovulate, then ovulation occurs (White *et al.*, 1973). When infused into a portal vessel of the rat's hypophyseal stalk, it caused release of both LH and FSH (Ondo *et al.*, 1973). The fact that a single releasing hormone is involved in the release of the two trophic hormones is, at first sight, surprising; however, as we have seen above, the two hormones can apparently be synthesized within the same cell of the anterior pituitary.*

Influence of Steroids

There is no doubt that sensitivity of the pituitary to the releasing hormone is strongly influenced by the level of steroid hormones in the blood, and it is likely that, in this way, the differential synthesis and secretion of the two gonadotrophins, LH and FSH, may be achieved. Thus, according to Martini (1974), pre-treatment of a castrated male rat with oestrogen favours release of LH by the releasing hormone, LHRH, whereas pre-treatment with testosterone favours release of FSH.

TABLE V

Concentrations of gonadotrophin releasing hormone, LHRH, in the rat's central nervous system. (Wilber *et al.*, 1976)

Tissue	LHRH (ng/g wet wt.)	
	Male	Female
Hypothalamus	205	112
Pituitary	167	178
Midbrain	84	11
Cerebellum	32	9
Brain-stem	7	14
Cerebral cortex	6	3

Distribution

As Table V shows, the gonadotrophin releasing hormone is not confined to the hypothalamus and pituitary, the midbrain being a relatively rich source.

Distribution in the Hypothalamus. Modern studies, employing radioimmuno-assay of thin sections of hypothalamic tissue, have delineated the distribution of the

* Schally *et al.* (1972), at the end of a valuable review of the releasing hormone, suggest that its name should be changed to "regulating hormone" as more indicative of its function, not only in releasing the gonadotrophins, but also in influencing their synthesis within the pituitary cells.

releasing hormone in the hypothalamus of several species; in the rat, for example, according to Palkovits *et al.* (1974) and Wheaton *et al.* (1975), the highest concentrations are in the arcuate nucleus and median eminence, regions probably responsible for tonic liberation of the releasing hormone; in addition, the medial basal tissue overlying the optic chiasma contains the hormone, and this may mediate the effects seen with electrical stimulation (p. 112). Another peak of activity occurred in the organum vasculosum of the lamina terminalis (OVLT), a region where it is suggested that the neurones accumulate the hormone and release it into the cerebrospinal fluid whence it is delivered to the portal system (Ben-Jonathan *et al.*, 1974, Baker *et al.*, 1975). In other species, such as the monkey (Silverman *et al.*, 1977) and guinea-pig (Silverman, 1976), the distribution is wider.

    Location in Neurones. Using a rabbit antiserum to LHRH, Setalo *et al.* (1975) were able to identify a system of delicate nerve fibres, presumably containing the releasing hormone, in the median eminence of the rat corresponding exactly with the classically described tubero-infundibular tract; this was traced to the retrochiasmatic area including the arcuate nucleus. These authors failed to identify the releasing hormone in the cell bodies of hypothalamic or other neurones, and they question the validity of other studies, notably those of Barry *et al.* (1973), in which the releasing hormone was apparently identified in cell bodies. They also failed to identify the hormone in tanicytes and in the organum vasculosum of the lamina terminalis (OVLT), as claimed by Zimmerman *et al.* (1974).* So far as subcellular distribution is concerned, Taber and Karavolas (1975) found that activity in homogenates of medial basal hypothalami of rats was contained in the mitochondrial fraction; from this, a fraction consisting of electron-dense vesicles was separated containing the releasing activity.

Prolactin

    This pituitary hormone has some gonadotrophic action in the rat where it is luteotrophic (p. 75), but as its name implies, the hormone is primarily concerned with lactation.

Inhibitory Factor

    Hypothalamic control over its release is through an inhibitory factor —PIF—that is tonically secreted and serves to maintain low plasma levels; increases in secretion are achieved by suppression of hypo-

---

* The distribution of the gonadotrophin releasing hormone (LHRH) in the central nervous system has been reviewed by Elde *et al.* (1978) and Brownstein (1978). There seems no doubt from the study of Kizer *et al.* (1976) on the rat that the organum vasculosum of the lamina terminalis (OVLT) contains a high proportion of the total in the brain; these authors found significant amounts in all the circumventricular organs, namely median eminence, OVLT, subfornical organ, area postrema and subcommissural organ. These are all regions that are in close relation to the cerebrospinal fluid.

    In the human brain Okon and Koch (1977) found the largest amount of LH in the pituitary stalk (52 ng) and the next most concentrated region was the OVLT (9·4 ng), and they concluded that the OVLT acted as a secretory organ; they felt that the secreted HRLH would be unlikely to be carried to the pituitary in the cerebrospinal fluid, and suggested a direct vascular route, analogous with the portal circulation. They remarked that the OVLT would be likely to be involved in any experimental lesions designed to destroy the PO/AH, so that the effects of lesions here on reproductive control might well be due to interruption of secretion of LHRH by the OVLT.

thalamic release of the PIF. A transplanted pituitary will secrete prolactin vigorously, being freed from hypothalamic control, whereas the opposite occurs with the gonadotrophs LH and FSH.

Releasing Factor

There is some evidence for the existence of a releasing hormone or factor—PRF or PRH—but care must be taken to distinguish any stimulation of prolactin release from that brought about by thyroid releasing hormone—TRH—and the posterior pituitary hormone vasopressin. Thus, extracts of hypothalamus can induce milk formation in oestrogen-primed rats, but the possibility that this is due to a specific releasing factor has been seriously questioned since it is known that synthetic thyrotrophin-releasing hormone—TRH—will release prolactin from rat pituitary tumour cells in culture and is a potent releasing hormone in man and other species. Moreover, vasopressin, also present in hypothalamic extracts, is a prolactin-releasing agent. The evidence bearing on this point has been summarized by Boyd et al. (1976), who have extracted hypothalamic fragments of the pig whilst inactivating the TRH and vasopressin by incubation with rat serum. The preparations still contained prolactin-releasing activity, as measured by its release in oestrogen-progesterone pretreated male rats; and studies of the responses to different doses ruled out the possibility that the release could be due to residual TRH. Thus, maximal effects with the preparation could be obtained with a dose containing as little as 2·6 ng of PRH, whereas more than 400 ng of TRH were required for this maximal activity. The matter will be discussed later in connexion with the role of biogenic amines in control of reproductive behaviour.

Cerebrospinal Fluid

It has been suggested that releasing factors, finding their way into the cerebrospinal fluid, might influence release of pituitary hormones by being taken up into the portal vessels of the median eminence. Injection of LHRH into the IIIrd ventricle of the rat does actually cause release of LH into the blood (Ondo et al., 1973), but the actual manner in which the releasing hormone is carried to the pituitary cells is not completely clear. Thus, the release of LH, observed by Ben-Jonathan et al. (1974) after ventricular injection of LHRH, was very prolonged, so that even after two hours the concentration of LH in the blood was 40–50 times that of controls. Hence, it is likely that the releasing hormone was being slowly carried out of the cerebrospinal fluid into the general circulation and reached the pituitary cells and portal system by a roundabout route. When the portal blood was tapped after intraventricular injection of LHRH, the rise in concentration was rapid, but only 5 per cent of the total amount injected was recovered by this route. When labelled releasing hormone (TRH) was injected intraventricularly, Oliver et al. (1975) found activity in the portal blood, and considered that the releasing hormone had diffused across the ventricular wall and through the median eminence to reach the portal system. Thus, it seems that releasing hormones injected into the cerebrospinal fluid can, indeed, find their way to the portal system, but the process, viewed as a means of influencing pituitary secretion, must be very inefficient compared with the direct release into the portal system. In fact, analysis of the cerebrospinal fluid, taken from the third ventricle under a variety of experimental conditions designed to promote its liberation from hypothalamic neurones, failed to demonstrate the presence of gonadotrophic releasing hormone (LHRH) in the fluid (Cramer and Barraclough, 1975). According to Wilber et al. (1976), however, there are measurable concentrations of TRH in human lumbar fluid (ca 40 pg/ml).

Wider Role. The quite extensive localization of the releasing hormones throughout the central nervous system, together with a variety of physiological and behavioural experiments, have led to the view that the releasing hormones have a wider function than that originally envisaged (Wilber et al., 1976). The neurones containing them, and releasing them at their terminals, may be classed as *peptidergic neurones* capable of activating neurones and effectors in other regions than the pituitary-hypothalamic system. Some aspects of this wider function will be discussed in Chapter 2.

REFERENCES

Anderson, D. C. (1976). The role of sex hormone binding globulin in health and disease. In *The Endocrine Function of the Human Ovary*. Ed. Jones, V. H. T., Serio, M. and Giusti, G. Academic Press, London and New York.

Baker, B. L., Dermody, W. C. and Reel, J. R. (1975). Distribution of gonadotrophin-releasing hormone in the rat brain as observed with immunocytochemistry. *Endocrinology*, **97**, 125–135.

Balboni, G. C. (1976). Histology of the ovary. In *The Endocrine Function of the Human Ovary*. Eds. V. H. T. Jones, M. Serio and G. Giusti, pp. 141–158. Academic Press, London and N.Y.

Barnes, B. G. (1962). Electron microscope studies on the secretory cytology of the mouse anterior pituitary. *Endocrinology*, **71**, 618–628.

Barry, J., Dubois, M. P. and Poulain, P. (1973). LRF producing cells of the mammalian hypothalamus. *Z. Zellforsch.*, **146**, 351–366.

Bellisario, R., Carlsen, R. B. and Bahl, O. P. (1973). Human chorionic gonado-tropin. Linear amino acid sequence of the α-subunit. *J. biol. Chem.*, **248**, 6796–6807.

Ben-Jonathan, N., Mical, R. S. and Porter, J. C. (1974). Transport of LRF from CSF to hypophyseal portal and systemic blood and the release of LH. *Endocrinology*, **95**, 18–25.

Boime, I., Boguslawski, S. and Caine, J. (1975). The translation of a human pla-cental lactogen mRNA in heterologous cell-free systems. *Biochem Biophys. Res. Comm.*, **62**, 103–109.

Boyd, A. E., Spencer, E., Jackson, I. M. D. and Reichlin, S. (1976). Prolactin re-leasing factor (PRF) in porcine hypothalamic extract distinct from TRH. *Endo-crinology*, **99**, 861–871.

Brinkmann, A. O. *et al.* (1972). An oestradiol receptor in rat testis interstitial tissue. *FEBS Letters*, **26**, 301–305.

Brownstein, M. J. (1978). Are hypothalamic hormones central neurotransmitters? In *Centrally Acting Peptides*. Ed. J. Hughes. Macmillan, Lond., pp. 37–47.

Burke, C. W., Anderson, D. C. (1972). Interrelationships of unbound testosterone and oestradiol in human serum at 37°C, and a biological role for sex hormone binding globulin. *J. Endocrinol.*, **53**, xxvi–xxvii.

Carlsen, R. B., Bahl, O. P. and Swaminathan, N. (1973). Human chorionic gonado-tropin. Linear amino acid sequence of the β-subunit. *J. biol. Chem.*, **248**, 6810–6825.

Channing. C. P. and Kammerman, S. (1973). Effects of hCG, asialo-hCG and the subunits of hCG upon luteinization of monkey granulosa cell cultures. *Endo-crinology*, **93**, 1035–1043.

Christensen, L. W. and Clemens, L. G. (1975). Blockade of testosterone-induced mounting behaviour in the male rat with intracranial application of the aro-matization-inhibitor, androst-1,4,6-triene-3,17-dione. *Endocrinology*, **97**, 1545–1551.

Cidlowski, J. A. and Muldoon, T. G. (1974). Estrogenic regulation of cytoplasmic receptor populations in estrogen-responsive tissues of the rat. *Endocrinology*, **95**, 1621–1629.

Cramer, O. M. and Barraclough, C. A. (1975). Failure to detect luteinizing hormone-releasing hormone in third ventricle cerebral spinal fluid under a variety of experi-mental conditions. *Endocrinology*, **96**, 913–924.

Dahl, E. (1970). The ultrastructure of the thecal gland of the domestic fowl. *Z. Zellforsch.*, **109**, 195–211.

Daniel, P. M. (1976). Anatomy of the hypothalamus and pituitary gland. *J. clin. Path.*, Suppl. **7**, 1–7.

Dorrington, J. H. and Armstrong, D. J. (1975). Follicle-stimulating hormone stimulates estradiol-17β synthesis in cultured Sertoli cells. *Proc. Nat. Acad. Sci. Wash.*, **72**, 2677–2681.

Eisenfeld, A. J. (1970). ^{3}H-Estradiol: *in vitro* binding to macromolecules from the rat hypothalamus, anterior pituitary and uterus. *Endocrinology*, **86**, 1313–1318.

Eisenfeld, A. J. and Axelrod, J. (1966). Effect of steroid hormones, ovariectomy, estrogen pretreatment, sex and immaturity on the distribution of ^{3}H-estradiol. *Endocrinology*, **79**, 38–42.

Elde, R. *et al.* (1978). Immunohistochemical localization of peptides in the nervous system. In *Centrally Acting Peptides*. Ed. J. Hughes. Macmillan, Lond., pp. 17–35.

Evans, G. A., Hucko, J. and Rosenfeld, M. G. (1977). Preprolactin represents the initial product of prolactin mRNA production. *Endocrinology*, **101**, 1807–1814.

Evans, G. A. and Rosenfeld, M. G. (1976). Cell-free synthesis of a prolactin precursor directed by mRNA from cultured rat pituitary cells. *J. biol. Chem.*, **251**, 2842–2847.

Falck, B. (1959). Site of production of oestrogen in rat ovary as studied in micro-transplants. *Acta physiol. scand.*, **47**, Suppl. 163.

Flerko, B. (1970). Control of follicle-stimulating hormone and luteinizing hormone secretion. In *The Hypothalamus*. Eds. L. Martini, M. Motta and F. Fraschini. Academic Press. N.Y. pp. 351–363.

Fraser, H. M. and Sandow, J. (1977). Gonodotropin release by a highly active analogue of luteinizing hormone releasing hormone in rats immunized against luteinizing hormone releasing hormone. *J. Endocrinol.*, **74**, 291–296.

Giannopoulos, G. (1971). Binding of testosterone to cytoplasmic components of the immature rat uterus. *Biochem. Biophys Res. Comm.*, **44**, 943–951.

Giannopoulos, G. (1973). Binding of testosterone to uterus components of the immature rat. *J. biol. Chem.*, **248**, 1004–1010.

Graesslin, D., Wiese, H. C. and Braendle, W. (1973). The microheterogeneity of human chorionic gonadotropin (HCG) reflected in the β-subunits. *FEBS Letters*, **31**, 214–216.

Gueriguian, J. L. and Pearlman, W. H. (1968). Some properties of a testosterone-binding component of human pregnancy serum. *J. biol. Chem.*, **243**, 5226–5233.

Gueriguian, J. L., Sawyer, M. E. and Pearlman, W. H. (1974). A comparative study of progesterone- and cortisol-binding activity in the uterus and serum of pregnant and non-pregnant women. *J. Endocrinol.*, **61**, 331–345.

Halasz, B., Pupp, L. and Uhlarik, S. (1962). Hypophysiotropic area in the hypothalamus. *J. Endocrinol.*, **25**, 147–154.

Hammerstein, J. H., Rice, B. F. and Savard, K. (1964). Steroid formation in the human ovary, I. *J. clin. Endocrinol.*, **24**, 597–605.

Hansson, V. *et al.* (1974). Preliminary characterization of a binding protein for androgen in rabbit serum. Comparison with the testosterone-binding globulin (TeBG) in human serum. *Endocrinology*, **95**, 690–700.

Harding, C. F. and Feder, H. H. (1976). Relation between individual differences in sexual behaviour and plasma testosterone levels in the guinea-pig. *Endocrinology*, **98**, 1198–1205.

Jensen, E. V. *et al.* (1968). A two-step mechanism for the interaction of estradiol with rat uterus. *Proc. Nat. Acad. Sci. Wash.*, **59**, 632–638.

Jensen, E. V. and De Sombre, E. R. (1973). Estrogen-receptor interaction. *Science*, **182**, 126–134.

Jensen, E. V. and Jacobson, H. I. (1962). Basic guides to the mechanism of estrogen action. *Rec. Prog. Horm. Res.*, **18**, 387–408.

Jung, I. and E.-E. Baulieu (1972). Testosterone cytosol "receptor" in the rabbit levator ani muscle. *Nature*, **237**, 24–26.

Kahwanago, I., Heinrichs, W. L. and Hermann, W. L. (1970). Estrodiol "receptors" in hypothalamus and anterior pituitary gland. *Endocrinology*, **86**, 1319–1326.

Katongole, C. B., Naftolin, F. and Short, R. V. (1971). Relations between blood levels of luteinizing hormone and testosterone in bulls, and the effects of sexual stimulation. *J. Endocr.*, **50**, 457–466.

Kelch, R. P. *et al.* (1972). Estradiol and testosterone secretion by human, simian, and canine testes. *J. clin. Invest.*, **51**, 824–830.

Kizer, J. S., Palkovits, M. and Brownstein M. J. (1976). Releasing factors in the circumventricular organs of the rat brain. *Endocrinology*, **98**, 311–317.

Kurusomi, K. and Oota, Y. (1968). Electron microscopy of two types of gonadotrophs in the anterior pituitary glands of persistent estrous and diestrous rats. *Z. Zellforsch. mikrosk. Anat.*, **85**, 34–46.

Kuttenn, F., Mowszowicz, I., Schaison, G. and Mauvais-Jarvis, P. (1977). Androgen production and skin metabolism in hirsutism. *J. Endocrinol.*, **75**, 83-91.

Lerner, L. J. (1964). Hormone antagonists and inhibitors of specific activities of estrogen and androgen. *Rec. Progr. Horm. Res.*, **20**, 435-476.

Louvet, J.-P. *et al.* (1976). Follicle stimulating activity of human chorionic gonadodotropin: effect of dissociation and recombination of subunits. *Endocrinology*, **99**, 1126-1128.

Mainwaring, W. I. P. (1975). Steroid hormone receptors: A survey. *Vitam & Horm.* **33**, 223-245.

Mainwaring, W. I. P., Symes, E. K. and Higgins, S. J. (1976). Nuclear components responsible for the retention of steroid-receptor complexes, especially from the standpoint of the specificity of hormonal responses. *Biochem. J.*, **156**, 129-141.

Martini, L. (1974). Gonadotropin releasing factors: recent physiological findings. In *Recent Progress in Reproductive Physiology*. Eds. P. G. Crosignani and V. H. T. James, pp. 295-321. Academic Press, London and N.Y.

Mercier-Bodard, C., Alfsen, A. and Baulieu, E. E. (1970). Sex steroid binding plasma protein (SBP). *Acta endocr.*, Suppl. **147**, 204-221.

Mester, J. and Baulieu, E. E. (1975). Dynamics of oestrogen-receptor distribution between the cytosol and nuclear fractions of immature rat uterus after oestradiol administration. *Biochem. J.*, **146**, 617-623.

Milgrom, E., Atger, M., Perrot, M. and Baulieu, E.-E. (1972). Progesterone in uterus and plasma. VI. Uterine progesterone receptors during the estrous cycle and implantation in the guinea-pig. *Endocrinology*, **90**, 1071-1078.

Morell, A. G. *et al.* (1971). The role of sialic acid in determining the survival of glycoproteins in the circulation. *J. biol. Chem.*, **246**, 1461-1467.

Moriarty, G. C. (1975). Electron microscopic-immunocytochemical studies of rat pituitary gonadotrophs: a sex difference in morphology and cytochemistry of LH cells. *Endocrinology*, **97**, 1215-1225.

Mossman, W. W., Koering, M. J. and Ferry, D. (1964). Cyclic changes of interstitial gland tissue of the human ovary. *Amer. J. Anat.*, **115**, 235-256.

Moyle, W. R., Bahl, O. P. and März, L. (1975). Role of the carbohydrate of human chorionic gonadotropin in the mechanism of hormone action. *J. biol. Chem.*, **250**, 9163-9169.

Mulder, E., Brinkmann, A. O., Lamers-Stahlhofen, G. J. M. and van der Molen, H. J. (1973). Binding of oestradiol by the nuclear fraction of rat testis interstitial tissue. *FEBS Letters*, **31**, 131-136.

Nakane, P. K. (1970). Classification of anterior pituitary cell types with immunoenzyme histochemistry. *J. Histochem. Cytochem.*, **18**, 9-20.

Okon, E. and Koch, Y. (1977). Localization of gonadotropin-releasing hormone in the circumventricular organs of human brain. *Nature*, **268**, 445-447.

Oliver, C., Ben-Jonathan, N., Mical, R. S. and Porter, J. C. (1975). Transport of thyrotropin-releasing hormone from cerebrospinal fluid to hypophyseal portal blood and the release of thyrotropin. *Endocrinology*, **97**, 1138-1143.

Oliver, C., Mical, R. S. and Porter, J. C. (1977). Hypothalamic-pituitary vasculature: evidence for retrograde blood flow in the pituitary stalk. *Endocronology*, **101**, 598-604.

Ondo, J. G., Eskay, R. L., Mical, R. S. and Porter, J. C. (1973). Effect of synthetic LRF infused into a hypophyseal portal vessel on gonatropin release. *Endocrinology*, **93**, 205-209.

Page, R. B., Munger, B. L. and Bergland, R. M. (1976). Scanning microscopy of pituitary vascular casts. *Amer. J. Anat.*, **146**, 273-285.

Palkovits, M. *et al.* (1974). Luteinizing hormone-releasing hormone (LH-RH) content of the hypothalamic nuclei in rat. *Endocrinology*, **95**, 554-558.

Peterson, N. T., Midgley, A. R. and Jaffe R. B. (1968). Regulation of human gonadotropins III. *J. clin. Endocr. Metab.*, **28**, 1473–1478.

Philibert, D. and Raynaud, J.-P. (1973). Progesterone binding in the immature mouse and rat uterus. *Steroids*, **22**, 89.

Pierce, J. G., Bahl, O. P., Cornell, J. S. and Swaminathan, N. (1971). Biologically active hormones prepared by recombination of the α chain of human chorionic gonadotropin and the hormone-specific chain of bovine thyrotropin or of bovine luteinizing hormone. *J. biol. Chem.*, **246**, 2321–2324.

Puca, G. A., Nola, E., Hibner, U., Cicala, G. and Sica, V. (1975). Interaction of the estradiol receptor from calf uterus with nuclear acceptor sites. *J. biol. Chem.*, **250**, 6452–6459.

Ramirez, V. D. and McCann, S. M. (1965). Inhibitory effect of testosterone on luteinizing hormone secretion in immature and adult rats. *Endocrinology*, **76**, 214–417.

Rennie, P. and Bruchovsky, N. (1973). Studies on the relationship between androgen receptors and the transport of androgens in rat prostate. *J. biol. Chem.*, **248**, 3288–3297.

Rice, B. F., Hammerstein, J. and Savard, K. (1964). Steroid formation in the human ovary II. *J. clin. Endocrinol.*, **24**, 606–615.

Rice, B. F. and Savard, K. (1966). Steroid hormone formation in the human ovary. IV. *J. clin. Endocrinol.*, **26**, 593–609.

Ryan, K. J. and Smith, O. W. (1961). Biogenesis of estrogens by the human ovary. I. Conversion of acetate-1-C^{14} to estrone and estradiol II. Conversion of progesterone-4-C^{14} to estrone and estradiol. III. Conversion of cholesterol-4-C^{14} to estrone. IV. Formation of neutral steroid intermediates. *J. biol. Chem.* **236**, 705–709; 710–714; 2204–2206; 2207–2212.

Ryan, K. J., Petro, Z. and Kaiser, J. (1968). Steroid formation by isolated and recombined ovarian granulosa and thecal cells. *J. clin. Endocr.*, **28**, 355–358.

Sairam, M. R. and Li, C. H. (1973). Human pituitary thyrotrophin: isolation and chemical characterization of its subunits. *Biochem. Biophys. Res. Comm.*, **51**, 336–342.

Sarff, M. and Gorski, J. (1971). Control of estrogen-binding protein concentration under basal conditions and after estrogen administration. *Biochemistry*, **10**, 2557–2563.

Savard, K., Dorfman, R. I., Baggett, B. and Engel, L. L. (1956). Biosynthesis of androgens from progesterone by human testicular tissue *in vitro*. *J. clin. Endocrinol.*, **16**, 1629–1630.

Savard, K., Mason, N. R., Ingram, J. T. and Gassner, F. X. (1961). The androgens of bovine spermatic venous blood. *Endocrinology*, **69**, 324–330.

Saxena, B. B. and Rathnam, P. (1976). The amino acid sequence of the β-subunit of follicle-stimulating hormone from human pituitary glands. *J. biol. Chem.*, **251**, 993–1005.

Schally, A. V., Kastin, A. J. and Arimura, A. (1972). FSH-releasing hormone and LH-releasing hormone. *Vitam. & Horm.*, **30**, 83–164.

Scully, R. E. and Cohen, R. G. (1964). Oxidative-enzyme activity in normal and pathologic human ovaries. *Obstet. Gynecol.*, **24**, 667–680.

Setalo, G., Vigh, S., Schally, A. V., Arimura, A. and Flerkó, B. (1975). LH-RH-containing neural elements in the rat hypothalamus. *Endocrinology*, **96**, 135–142.

Shome, B. and Parlow, A. F. (1974a). Human follicle stimulating hormone (hFSH): first proposal of the amino acid sequence of the α-subunit (hFSHα) and first demonstration of its identity with the α-subunit of human luteinizing hormone (hLHα). *J. clin. Endocrinol.*, **39**, 199–202.

Shome, B. and Parlow, A. F. (1974b). Human follicle stimulating hormone: first proposal of the amino acid sequence of the hormone specific, β subunit (hFSHβ). *J. clin. Endocr.*, **39**, 203–205.

Short, R. V. (1961). Steroid concentrations in the follicular fluid of mares at various stages of the reproductive cycle. *J. Endocrinol.*, **22**, 153–163.

Short, R. V. (1964). Ovarian steroid synthesis and secretion *in vivo*. *Rec. Prog. Horm. Res.*, **20**, 303–340.

Shownkeen, R. C., Hartree, A. S., Stewart, F., Mashila, K. and Stevens, V. C. (1976). Purification and properties of the subunits of human pituitary follicle-stimulating hormone. *J. Endocrinol.*, **69**, 263–273.

Shyamala, G. and Gorski, J. (1969). Estrogen receptors in the rat uterus. *J. biol. Chem.*, **244**, 1097–1103.

Silverman, A. J. (1976). Distribution of luteinizing hormone-releasing hormone (LHRH) in the guinea-pig brain. *Endocrinology*, **99**, 30–41.

Silverman, A. J., Antunes, J. L., Ferin, M. and Zimmerman, E. A. (1977). The distribution of luteinizing hormone-releasing hormone (LHRH) in the hypothalamus of the rhesus monkey. *Endocrinology*, **101**, 134–142.

Steggles, A. W., Spelsberg, T. C., Glasser, S. R. and O'Malley, B. W. (1971). Soluble complexes between steroid hormones and target-tissue receptors bind specifically to target-tissue chromatin. *Proc. Nat. Acad. Sci. Wash.*, **68**, 1479–1482.

Steinberger, E. and Steinberger, A. (1972). Testis: basic and clinical aspects. In *Reproductive Biology*. Eds. H. Balin and S. Glaser. *Excerp. Med. Amsterdam*, pp. 144–267.

Stumpf, W. E. (1968). Subcellular distribution of ^{3}H-estradiol in rat uterus by quantitative autoradiography—a comparison between ^{3}H-estradiol and ^{3}H-norethynodrel. *Endocrinology*, **83**, 777–782.

Stumpf, W. E. (1969). Nuclear concentration of ^{3}H-estradiol in target tissues. Dry-mount auto radiography of vagina oviduct, ovary, testis, mammary tumor, liver and adrenal. *Endocrinology*, **85**, 31–37.

Stumpf, W. E. (1970). Estrogen-neurons and estrogen-neuron systems in the periventricular brain. *Amer. J. Anat.*, **129**, 207–217.

Taber, C. A. and Karavolas, H. J. (1975). Subcellular localization of LH releasing activity in the rat hypothalamus. *Endocrinology*, **96**, 446–452.

Török, B. (1964). Structure of the vascular connections of the hypothalamo-hypophysial region. *Acta anat.*, **59**, 84–99.

Vaitukaitis, J. L. (1974). Changing placental concentrations of human chorionic gonadotropin and its subunits during gestation. *J. clin. Endocr. Metab.*, **38**, 755–760.

Wheaton, J. E., Krulich, L. and McCann, S. M. (1975). Localization of luteinizing hormone-releasing hormone in the preoptic area and hypothalamus of the rat using radioimmunoassay. *Endocrinology*, **97**, 30–38.

White, W. F., Hedlund, M. T., Rippel, R. H., Arnold, W. and Flouret, G. R. (1973). Chemical and biological properties of gonadotrophin-releasing hormone synthesized by the solid-phase method. *Endocrinology*, **93**, 96–106.

Wilber, J. F. *et al.* (1976). Gonadotropin-releasing hormone and thyrotropin-releasing hormone: distribution and effects in the central nervous system. *Rec. Progr. Horm. Res.* **32**, 117–153.

Zimmerman, E. A. *et al.* (1973). Vasopressin and neurophysin: high concentrations in monkey hypophyseal portal blood. *Science*, **182**, 925–927.

Zimmerman, E. A., Hsu, K. C., Ferin, M. and Kozlowski, G. P. (1974). Localization of gonadotropin-releasing hormone (GnRH) in the hypothalamus of the mouse by immunoperoxidase technique. *Endocrinology*, **95**, 1–8.

CHAPTER 2

Male Trophic Hormones and the Ovarian Cycle in the Female

MALE REPRODUCTIVE ACTIVITY

The physiology of reproduction is largely a matter of female physiology, and the various features of this latter aspect will concern us almost exclusively in the following pages. The male, of course, is necessary in the initial stage, namely in courting behaviour, aggressiveness to other males, and in coitus; and the maintenance of these functions relies on hormonal control.

Gonadotrophins

The gonadotrophic hormones of the male are, as indicated, ICSH (LH) and FSH, and their functions are considered to be those of maintaining the normal activities of the reproductive apparatus, both primary—the seminiferous tubules and interstitial cells—and secondary —prostatic secretions, and so on. The generally held view concerning the functions of the two male gonadotrophins is that ICSH (LH) controls the steroid secretions of the interstitial cells of Leydig, whilst FSH controls the functioning of the reproductive cells of the seminiferous tubules. Thus, ^{125}I-labelled LH, but not ^{125}I-labelled FSH, binds specifically to interstitial cells (De Kretser et al., 1971) and ^{3}H-labelled FSH is bound to the plasma membranes of Sertoli cells in the peripheral region of the seminiferous tubules (Orth and Christensen, 1977). The cells also produce cAMP in response to FSH, which also stimulates the production of androgen-binding protein (ABP, p. 50) by the seminiferous epithelium. We must appreciate, however, that in the long term the steroid secretions of the Leydig cells—mainly testosterone—themselves exert a trophic action on the primary and accessory reproductive tissues, in fact a high level of testosterone is necessary for spermatogenesis (Desjardins et al., 1973) so that the gonadotrophic hormones may exert both a primary action on their target tissues and a secondary action by virtue of their influence on steroid secretion.

Influence of Androgens

A complicating factor in the analysis of gonadotrophic hormone action is the effect of the testis secretions—androgens—on the reproductive apparatus. Thus, there is no doubt that a high level of testosterone within the testicular lymph, and thus surrounding the seminiferous tubules responsible for spermatogenesis, is necessary for this latter process, so that the regression of the tubules following hypophysectomy, and its ensuing abolition of spermatogenesis, could be due both to a primary effect of the loss of the trophic hormone—FSH—and the secondary effect of the loss of LH and consequent failure to stimulate secretion of testosterone by the interstitial cells of Leydig. A further complicating factor, when analysing, say, the effects of hypophysectomy on male gonadal function, is the feedback of the androgen, e.g. testosterone, on the gonadotrophic secreting system—pituitary-hypothalamus. Thus, as we shall see, injections of testosterone will inhibit the secretion of LH and thus cause a negative feedback on testosterone secretion by the testis.

Biphasic Effect. For example, Desjardins et al. (1973) pointed out that the action of testosterone on the rabbit's seminiferous tubules was biphasic; at low levels of injection of exogenous testosterone (< 5 mg/kg/day) spermatogenesis was arrested; and this was due to suppression of gonadotrophin secretion through negative feedback. At high levels (> 10 mg) spermatogenesis was permitted, although gonadotrophin secretion was still suppressed. In their own work, where plasma levels of testosterone were well controlled, the weight of the seminal vesicles correlated well with the plasma testosterone level, in spite of the low level of LH.

Effect of LH

That LH can influence secretion of testosterone by the testis can be demonstrated by adding the hormone to slices of testis incubated with ^{14}C-acetate, when the rate of synthesis of labelled testosterone is increased (Hall and Eik-Nes, 1962). Again, Dufau et al. (1972), incubated isolated decapsulated testis in Krebs-Ringer and measured the production of testosterone by radioimmunoassay; FSH had no effect at all on production whilst LH or human chorionic gonadotrophin (hCG), stimulated production (Fig. 2.1), nor yet did FSH potentiate the effects of these latter hormones.

cAMP Formation. Further proof of the dominant, if not exclusive,

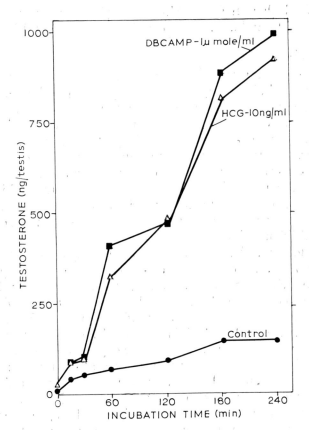

Fig. 2.1. Production of testosterone by incubated rat testis. In the controls (●——●), testosterone production rose slowly to reach a plateau at 80 min. Addition of hCG (△—— △) or dibutyryl cAMP (■——■) caused striking increases in testosterone production. (Dufau *et al.*, *Endocrinology*.)

role of LH in controlling secretion of androgens is provided by examining the influence of the gonadotrophic hormones on cyclic AMP formation by different components of the reproductive system, since both LH and FSH act through stimulation of adenyl cyclase in their target cells (p. 500). Dorrington and Fritz (1974) separated interstitial cells from seminal vesicles and showed that LH stimulated cAMP production, whereas FSH had no effect, nor yet did FSH synergize with LH, confirming the study of Cooke *et al.* (1972).

Embryonic Development. Finally, the appearance of specific LH-type receptors in the rabbit's testis correlates closely with Leydig-cell differentiation, and synthesis of testosterone. Thus, between Days 17 and 19 of life there is a sharp increase in testosterone content of the embryonic testis and this is paralleled by a sharp rise in LH-binding.

Effect of FSH

This trophic hormone is considered to control the functions of the seminiferous tubules, thereby exerting a primary influence on spermatogenesis. Maturation of the cells takes place in the epididymis (in the upper cauda in the rat) and the maintenance of the lining of this accessory organ is thus an important aspect in male fertility. Hypophysectomy causes regression of the seminal vesicles with abolition of spermatogenesis, and this can be prevented by administration of FSH, thereby indicating the importance of this trophic hormone in spermatogenesis. Dorrington and Fritz showed that isolated seminiferous tubules responded to FSH by increased cAMP production in the presence of theophylline.

When they studied the tubules of the hypophysectomized rat, 30–35 days after the operation when the epithelium had fully regressed and only Sertoli cells were present, cAMP production was stimulated, suggesting that the action of FSH might be on the secretions of these cells. The ratio of the amounts of cAMP produced with FSH over that produced without: cAMP with FSH/cAMP without FSH, increased with time after hypophysectomy, i.e. with the degree of loss of the spermatogenic seminiferous tubule cells and the increase in proportion of Sertoli cells, further suggesting the exclusive action on Sertoli cells. LH had no effect on cAMP production by the Sertoli cells,* and this agrees with the finding of Means and Vaitukaitis (1972) that ^{3}H-labelled FSH bound only to the seminiferous epithelium and not to the interstitial cells of Leydig.

Role of Androgen

Early studies of hypophysectomized rats, e.g. those of Woods and Simpson (1961), indicated the importance of both LH and FSH in maintaining spermatogenesis in hypophysectomized animals; thus when LH was administered immediately after hypophysectomy, before regression of the seminiferous tubules had begun, testicular weight and the histology of the seminiferous tubules and Leydig cells remained normal. If regression was allowed to take place by delaying treatment, this was not completely reversed by LH, whereas a combination of this with FSH was effective. Subsequent studies have emphasized the

---

* Kuehl et al. (1970), using a preparation of seminiferous tubules separated from the testis by use of collagenase, found a stimulatory action of LH on the seminiferous tubules; Dorrington and Fritz suggest that the preparation contained some interstitial cells.

importance of androgen in the maintenance of spermatogenesis, and it is likely that the effects of LH described above were due to stimulated secretion of the steroid rather than to a trophic action on the seminiferous tubules. In general, it would seem that androgen administration immediately after hypophysectomy permits the maintenance of spermatogenesis; if regression is allowed to proceed before treatment, then FSH is necessary in addition to androgen (see, for example, Steinberger, 1971).

Antibody Treatment

The effects of treatment of the male rat with an anti-LH serum might well have been due to the inhibition of androgen secretion; it caused atrophy of the testes and accessory organs; the seminiferous tubules atrophied and spermatogenesis ceased, whilst the number of cytoplasmic granules in the interstitial cells decreased (Wakabayashi and Tamoaki, 1966). These effects could be due to a primary failure of testosterone secretion, but of course the experiment does not rule out the possibility that LH has a trophic action on the spermatogenic process directly.

Talaat and Laurence (1969), using an ovine anti-FSH preparation, found no effects on the rat's gonadal function. However, Madhwa Raj and Dym (1976) felt that an ovine preparation might not substitute for an antiserum specifically prepared against rat FSH; they made such a preparation and found that the size of the testis, when this was administered to immature rats, was reduced by 50 per cent compared with control animals; there was no change in the accessory organs, nor yet in the level of testosterone in the blood. The Leydig cells, responsible for testosterone secretion, appeared normal but the number of spermatocytes and spermatids in the seminiferous tubules was reduced. These experiments are consistent with the notion that LH exerts its main, or exclusive, action on the interstitial cells of Leydig, so that its effects on other cells are secondary to the secretion of androgen. The effects of FSH are trophic to the seminiferous tubules, the main influence being apparently on the Sertoli cells which, by their influence on the composition of the intratubular fluid, control male fertility through the spermatogenesis process directly, and the maturation process in the epididymis.

Androgen-Binding Protein

An important role of FSH in control of testicular function, and thus of fertility in the male, is revealed in the study of the androgen-binding protein (ABP) described by Hansson, since it seems that this is secreted by the Sertoli cell, which is apparently the main target for FSH. Thus the study of secretion of ABP may well be equivalent to the study of seminiferous tubule function; hence the secretion of ABP could provide an important experimental model for the study of FSH activity.

A binding protein that is distinct from the cytoplasmic receptor material was discovered by Hansson in 100,000 **g** supernatants from the rat epididymis (Hansson and Tveter, 1971); in the testis the same protein could be found, provided the large amounts of endogenous testosterone, which would mask binding by the labelled androgen, were first removed. This material could be shown to pass from testis to the epididymis, having been secreted into the testicular fluid (French and Ritzen, 1973). As with epididymal and prostate receptor material, the affinity of ABP for dihydrotestosterone was greater than that for testosterone. Its molecular weight was smaller than that of the receptor material, being some 87,000 compared with greater than 200,000 for epididymal receptor. The site of production seemed to be the seminiferous tubules rather than the interstitial compartment, the Sertoli cells being responsible. After hypophysectomy ABP disappears from both testis and epididymis, to reappear after treatment with gonadotrophic hormones; the gonadotrophin responsible for synthesis by the Sertoli cells seems to be exclusively FSH* (Hansson et al., 1975); and thus it could be that the pituitary controlled the transport of androgen from the testis by producing a transport protein, ABP, that increased its efficiency of transport and carried it to the seminiferous tubules. An important point to appreciate is that the cells of the testis, e.g. the seminiferous tubules and interstitial cells of Leydig, are exposed to very much higher concentrations of androgen than those in the blood plasma by virtue of the transport of these androgens in this bound form, other tissues relying on the blood-borne androgen.

Role of Androgen-Binding Protein. A possible scheme for the role of ABP in the testis, and its relation to LH and FSH secretion, is indicated in Fig. 2.2. According to this (Hansson et al., 1974), both the gonadotrophic hormones exert their effects on spermatogenesis directly, LH through its effect on androgen secretion by the Leydig cells and FSH by its effect on the secretion of ABP by the Sertoli cells; the increased secretion increases the efficacy with which androgen will be

---

* Tindall and Means (1976) have shown that testosterone stimulates production of ABP by rat's testis, apparently not acting through cAMP (Means et al., 1976). They found that LH was more effective than FSH and since there was no synergy between the two gonadotrophs they concluded that the action of FSH was due to LH impurities. However, Fritz et al. (1976) have confirmed the synthesis of ABP by Sertoli cells, a process that was accelerated by FSH or dibutyryl cAMP.

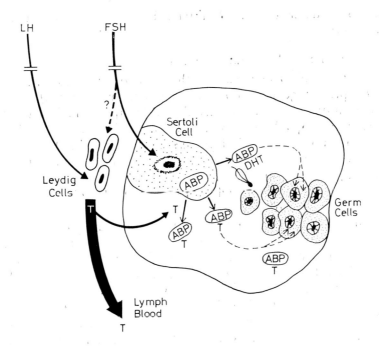

Fig. 2.2. Schematic drawing of androgen action in the testis. Testosterone (T) is secreted from the Leydig cells in response to LH stimulation. Most of the testosterone is transported from the testis bound to different binding proteins in blood and lymph. ABP is produced by the Sertoli cells in response to FSH stimulation and secreted into the seminiferous tubules. ABP within the seminiferous tubules generates a diffusion potential causing a net inflow of testosterone and accumulation of androgens in close proximity to the anrodgen-dependent cells within the germinal epithelium. (Hansson *et al.*, *J. Reprod. Fert.*)

carried to the cells of the seminiferous epithelium. Hansson *et al.* (1974) showed that these cells have androgen-specific receptors in their cytoplasm, and acceptor-material in their nuclei; the cyoplasmic receptor is different from ABP so that the tubular cells are definitely target cells for androgen. Thus, according to Hansson *et al.*, LH stimulates spermatogenesis through its effect on secretion of testosterone; FSH works by stimulating synthesis of ABP by the Sertoli cells, thereby ensuring the transport of the androgen to the androgen-dependent cells of the seminiferous tubule and epididymis. In the tubular cells the testosterone is probably converted to 5α-dihydrotestosterone, which in this tissue is probably the effective androgen; this attaches to the receptor and is carried to the nucleus where it initiates the processes leading to sperm maturation.*

Interdependence of Testicular Tissues

Lee *et al.* (1976) have concluded that it is no longer possible to consider the spermatogenic and intertubular compartments of the testis as independent units; thus both interstitial cells and seminiferous tubules are apparently capable of steroid metabolism (Rivarola *et al.*, 1973),† and by immunoreactive techniques Bubenik *et al.* (1975) have demonstrated the presence of androgen in the cells of the seminiferous tubules and the adjacent tubular wall, but the origin of this is unknown and it could be derived from the Leydig cells and carried via the blood or the seminal vesicular fluid.

Inhibin. As we shall see, the secretion of FSH by the pituitary can be controlled by a protein secretion of the Sertoli cells of the testis—*inhibin*—whilst testosterone itself, through its negative feedback on the pituitary-hypothalamus (p. 129) would exert control on both LH and FSH secretion. It could be, as argued by Steinburger and Chowdhury (1974), that the non-steroidal factor—inhibin—controlled the *synthesis* of FSH whilst its *release* might be controlled by testosterone. These authors found that testosterone, given to orchiectomized rats, caused a fall in *plasma* FSH but a rise in *pituitary* FSH, suggesting a negative feedback on release.

Prolactin

A number of studies have suggested that prolactin (p. 360) controls prostate function; for example an anti-prolactin decreased the prostatic weight in rabbits (Asano *et al.*, 1971). Again, atrophy is more marked after hypophysectomy than after castration; whilst treatment of hypophysectomized rats with prolactin enhanced the response of the prostate to testosterone. Kledzik *et al.* (1976) isolated a particulate membrane fraction from the rat's ventral prostate which had specific high-affinity sites for prolactin, whilst other hormones such as TSH, LH, FSH and GH, all failed to compete significantly with prolactin. Castration reduced the binding of this preparation to about one sixth, whilst

---

* Hansson *et al.* (1976) have emphasized the strong similarities between Sertoli secretion of androgen-binding protein (ABP) and maintenance of spermatogenesis. Both processes can be maintained by androgen in the hypophysectomized animal provided treatment is begun at once, i.e., before regression of the tubular epithelium; with both processes, delay in androgen treatment means that FSH is required in addition to androgen.

† This is true to the extent that seminiferous tubules can convert testosterone to its dihydroderivative, and can convert progesterone and pregnenolone to testosterone, but there is no evidence that they can synthesize testosterone from cholesterol or acetate (Hall *et al.*, 1969).

treatment of the castrated animal with testosterone returned this to normal. Using an immunohistochemical technique, Witorsch and Smith (1977) demonstrated an androgen-dependent intracellular binding of prolactin in the rat's ventral prostate.

Testosterone. So far as the rat is concerned, it seems that the synergy between LH and prolactin is manifest in the enhanced secretion of testosterone; thus, in hypophysectomized rats the blood testosterone levels were low; administration of prolactin, alone, raised the level a little; administration of LH raised it higher but only when LH and prolactin were administered together did the blood testosterone levels rise to normal or above (Hafiez et al., 1972).

Mating Behaviour in the Male

In an earlier chapter we have mentioned the role of male androgens in promoting sexual behaviour, and shown the rise in blood testosterone of the bull in anticipation of mating. A similar mating-induced rise in blood testosterone has been observed in rats and has been investigated in some detail by Kamel et al. (1977); as Fig. 2.3 shows, there is a steady rise in testosterone level after mating whilst prolactin and LH show steep rises and subsequent slow falls. The level of FSH is unaffected. The rise did not depend on successful copulation, so that attempts to copulate with unreceptive females produced similar changes in the male blood. The authors argue that failure of males to mate may be due to a failure to raise their hormone levels on contact with the female rather than on abnormally low resting levels, and they point to Harding and Feber's observation on guinea-pigs that males with a low level of sexual activity had normal levels of blood testosterone, but that this failed to rise on contact with the female. According to Pfaff (1973) LH-releasing hormone (LHRH) activates mating behaviour in castrated, steroid-primed rats, so that the initial result, in the male, of exposure to the female, might be an increased secretion of the releasing hormone; the consequent increase in level of LH, synergizing with the raised prolactin level, would promote secretion of testosterone (Bartke et al., 1976) which, in the long run, would maintain the reproductive capacity of the male. If, as is considered, hypothalamic neurones containing LHRH communicate with other parts of the brain as well as with the anterior pituitary (p. 151), these LHRH-neurones might influence behaviour directly. In this way, the single hormone would affect hormonal secretion and behaviour.

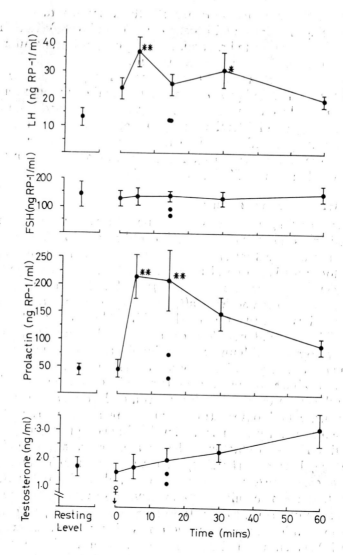

Fig. 2.3. Mean serum levels of LH, FSH, prolactin and testosterone in sexually experienced male rats at various times during mating. Vertical lines indicate standard error and asterisks indicate statistical significance of the difference from resting levels. The isolated circles represent the hormone levels of two rats that did not mate.
(Kamel *et al.*, *Endocrinology*.)

Spermiation

The release of the spermatids or spermatozoa into the lumen of the seminal vesicles is described as spermiation; the process, as described by Vitale-Calpi and Burgos (1970, a), consists of a swelling and rupture of the Sertoli cell. More specifically, the endoplasmic reticulum is widened and the recess in the apical cytoplasm of the Sertoli cell, containing the spermatid, is erased. In some species such as the guinea-pig, coitus is associated with spermiation, and this may be associated with large increases in circulating LH in the male; at any rate this occurs in the rat (Taleisneik *et al.* 1966) and hamster (Donoso and Santolaya, 1969). Ultrastructurally, the increased spermiation after copulation is revealed as a generally more intense form of the cytological changes observed in the resting condition, so that whereas release of spermatids occurred in the resting state from Stage VIII, after copulation much younger spermatids were released belonging to stages as early as Stage VI. Thus, the basal level of LH secretion probably enables the release of spermatids at Stage VIII to occur spontaneously, but release from earlier stages requires the LH-surge associated with copulation (Vitale-Calpe and Burgos, 1970, b).

Oestrous or Ovarian Cycle

The basic feature of the cycle is the development within the ovary of the Graafian follicle, so that it is more appropriate to refer to the cycle as the *ovarian* rather than the oestrous cycle, the latter term referring to the period of *oestrus*, or heat, that accompanies the production of the mature ovum in non-primate species. After the release of the ovum, development of the corpus luteum proceeds rapidly, the body behaving as an organ of hormonal secretion, the secreted progestins having, among other functions, the preparation of the uterus for implantation of the developing ovum. If the ovum has been fertilized, then implantation occurs; if not, the corpus luteum ultimately involutes and the system becomes ready for a new cycle. In general, we may distinguish two phases in the cycle, that involving the development of the mature ovum within the follicle leading to its release, and that following release of the ovum and leading to its implantation; and the two phases are said to be dominated by secretion of *oestrogens* and *progestins* respectively. We have seen that the features of this cycle vary widely in different mammalian species, but the variations are essentially in detail, the two main features being always present.

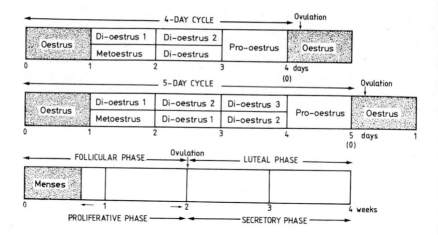

Fig. 2.4. Illustrating the rat's four- and five-day oestrous cycles and the primate menstrual cycle.

Rat's Cycle

Thus at the one extreme we may consider the rat's cycle (Fig. 2.4), which lasts only for four, or five, days according to the strain. The day of *oestrus*, or heat, is ascertained by the presence of cornified cells on the vaginal epithelial surface; the period following is called *metoestrus*, and represents the preparatory phase for the new cycle of ovulation, with increasing size of follicles. The following period is called *di-oestrus*, with the uterus showing maximal size; this is followed by *pro-oestrus*, during which the follicles swell preparatory to liberation of the ovum and the formation of the corpora lutea. Mating behaviour, typically recognized by lordosis, begins in the evening of pro-oestrus, with ovulation occurring at about 1 a.m. of the day of oestrus. If pregnancy does not ensue, the cycle begins again.

Menstrual Cycle

At the other extreme we have the 28-day menstrual cycle in the human or lower primates (Fig. 2.4); its details have been described before and here we need only emphasize its similar division into a phase leading up to ovulation (follicular) and a postovulatory (luteal) phase, which is one of preparation of the uterus for reception of the ovum should it be fertilized. Thus, we must appreciate that it is unlikely that the mere act of meeting a spermatozoon in the oviduct could be signalled to the central nervous system, so that the uterus "acts blindly" for a period immediately following ovulation "on the assumption" that fertilization has occurred, and it is then ready to receive a fertilized ovum by the time it has migrated to it. This postovulatory phase is essentially one of *pseudopregnancy* and lasts, in the primate, for several days until the *absence* of a signal leads to involution of the uterine changes and of the corpus luteum (luteolysis).

Pseudopregnancy

A similar state of pseudopregnancy occurs in other oestrous cycles, and it may be strongly enhanced either naturally, as in the dog, or experimentally.

Dog. Thus, in the dog, there is only one oestrous cycle during the breeding season, characterized by lengthy pro-oestrus and oestrus lasting each from 3–12 days. Spontaneous ovulation occurs during the first three days of oestrus, and in the absence of fertilization, a pseudopregnancy lasting from 30 to 90 days may develop followed by anoestrus of 3 to 4 months' duration. During this period of pseudopregnancy, the uterine and mammary gland developments are quite similar to those associated with pregnancy, and near the normal time of parturition many dogs exhibit nest building, experience lactation, and may continue to lactate for up to 60 days. Thus, in this case the "absence of signal" takes a very long time to reverse the pseudopregnancy condition due, as we shall see, to the sustained secretion of progestins from the corpora lutea.

Experimental Pseudopregnancy. Experimentally, a sustained pseudopregnancy may be obtained in the rat either by electrical stimulation of the cervix or by allowing it to copulate with a sterile (e.g. vasectomized) rat.

Induced Ovulators. In those animals that only ovulate after copulation, e.g. the rabbit and cat, then failure to copulate leaves the cycle uncompleted, and a state of pseudopregnancy may last for some time. In these species, copulation is followed by profound changes in the electroencephalogram, and it is doubtless a change in the state of the forebrain that acts as the stimulus to the release of gonadrophins leading to ovulation. Since copulation has already occurred, the brain needs no further signal to establish the conditions favourable for sustaining pregnancy so that the prolonged condition of pseudopregnancy that follows a sterile mating is understandable.*

* The reproductive process in the cat has been summarized recently by Verhaage *et al.* (1976); as indicated, it is an induced ovulator; unlike the rabbit it is a seasonal reproducer, exhibiting oestrus, or heat, between January and March and May and June. If the female fails to copulate when on heat it goes into a state of *polyoestrus*, exhibiting oestrus every 2–3 weeks. Copulation with an infertile male leads to a lengthy pseudopregnancy which may last or 30 to 73 days.

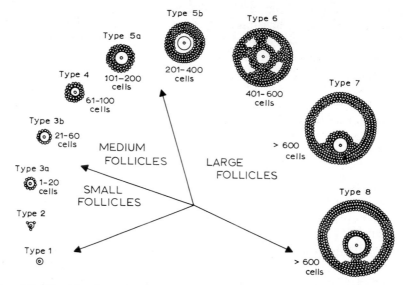

Fig. 2.5. Classification of the different stages in development of follicles in the mouse, according to the number of granulosa cells in the largest cross-section. Type 8 is seen only within 13 hr of ovulation. (Schwartz, *Biol. Reprod.* from Pederson.)

Ovulation

The basic features of the process of ovulation have been described in Volume 2 and here we may recall that the öocytes, before becoming mature ova, are enclosed by a cellular layer, or layers to become follicles which may pass through successive stages of development to become the mature follicles capable of releasing the ovum in response to the appropriate stimulus. As Fig. 2.5 shows, the changes consist essentially in the proliferation of granulosa cells; secretion of follicular liquor, and a definite organization of the theca cells into an investing layer around the follicle. An enlargement of the antrum continues, producing a large fluid-filled structure containing the ovum. Types 6 and 7 are mature follicles, and Type 8 is only seen within some 13 hours of ovulation. As discussed earlier, the theca and granulosa cells of mature follicles synthesize and secrete ovarian steroid hormones, notably the oestrogens. In addition the interstitial cells of the ovary are a source of steroid hormones.

Short-Gestation Periods. The ability to induce pseudopregnancy in animals is peculiar to short-gestation period species, such as the rabbit, rat, mouse. As Davies and Ryan (1972) have pointed out, the pseudopregnancy indicates that control over the duration of pregnancy in these species is exclusively, or nearly exclusively, a function of the mother, so that in the rabbit or ferret, the act of copulation sets in train a series of events controlled by the female hypothalamus. Thus, in the ferret the period of pseudopregnancy is exactly as long as that of normal pregnancy, indicating that the presence of the embryo and foetus fail to influence the maternal adjustments involved in pregnancy. In long-gestational animals, like the primates, a purely maternal control over a process lasting 9 months, uninfluenced by events taking place in the uterus, would involve considerable difficulties, and feedback from the developing embryo becomes of importance and, as we shall see, eventually becomes the dominant factor in control.

Atresia

The newborn animal has its full complement of follicles, so that during life the number decreases, either by atresia or by the process of follicular rupture during ovulation; atresia occurs after a certain degree of development has occurred; thus, types 1 to 3a consist in a non-proliferating pool, and in order that atresia may take place one of these in the non-proliferating pool must pass into the proliferating pool—Type 3b and larger—from which atresia takes place.

Anterior Pituitary

The maintenance of many of these changes during post-natal life depends on the anterior pituitary, so that removal of this early in life means that ovarian follicles never achieve an increase in size beyond a few cell-layers, being limited to the small and medium sizes illustrated in Fig. 2.5. Since the follicles that secrete oestrogens are only the mature ones, the follicular secretion of this class of hormones is abolished by hypophysectomy, which is also accompanied by degeneration of the ovarian interstitial cells and abolition of their secretion of ovarian hormones. The pituitary is also necessary for the final stage, namely ovulation, so that hypophysectomy in the mature female abolishes this.

FSH and LH. The hormones secreted by the anterior pituitary responsible for these two phases of ovarian activity, namely follicular maturation and ovulation, are described as *follicle stimulating hormone FSH*—and *luteinizing hormone—LH*—the latter name describing the changes in the follicle that precede and follow its rupture to form the corpus luteum.

Corpora Lutea. The morphological changes that normally precede rupture of the follicle are not always followed by ovulation, in which case well developed corpora lutea enclosing the ovum are found, as for example in pseudopregnancy (p. 201). The morphological changes in luteinization consist in hyperplasia, cytoplasmic granulation and accumulation of lipid droplets within the granulosa cells. Associated with these changes the synthetic machinery of the cells is activated, through the medium of cAMP, and this leads to the synthesis of progesterone. In order that the follicle may rupture, moreover, the cells must synthesize an enzyme or enzymes that weaken the theca and adjacent tunica albuginea on the ovarian surface allowing rupture to occur (Espey and Rondell, 1968).

Binding Sites

The binding sites for LH on the follicular granulosa cells are different from those for FSH (Kammerman et al., 1972; Nimrod et al., 1976). During maturation of the Graafian follicle the greater sensitivity to LH that develops is accompanied by a striking increase in the binding capacity of the granulosa cells; thus Channing and Kammerman (1973) found the following relative figures for granulosa cells derived from small, medium and large follicles: 7554, 12,234 and 624,300. According to Rao et al. (1977), the trophic hormones, FSH, LH, and prolactin influence the number of receptors specific for their hormones; thus, FSH acts on granulosa cells of pre-antral follicles in the presence of oestrogen to promote follicular development, and it increases the content of receptors for FSH and LH. LH acts on the same cells of the mature antral follicle to induce luteinization, and it decreases the content of FSH and LH receptors, but increases that of prolactin receptors. (Fig. 3.7, p. 192.)

Secretion of Hormones During the Oestrous Cycle

Modern techniques of radioimmunoassay, which have been extended to the determination of minute concentrations of both the trophic and the steroid hormones, have permitted exact descriptions of the course of secretion of these hormones during the oestrous, or ovarian, cycle, and, if this occurs, during pregnancy. By sampling the ovarian vein in experimental animals (and in humans, e.g. Baird and Fraser, 1974), moreover, the course of secretion may be more accurately ascertained than by studies on peripheral blood. Figure 2.6 shows the measured changes in the blood plasma of the 4-day cycle rat throughout a single cycle. The most striking feature is the peak rate of secretion, as manifest in the blood concentration, of all the hormones at a period close to the day of ovulation; and a great deal of research has been carried out in order to determine, within narrow limits, the order of events, in terms of gonadotrophic and steroid hormone secretion, that lead to and follow ovulation.

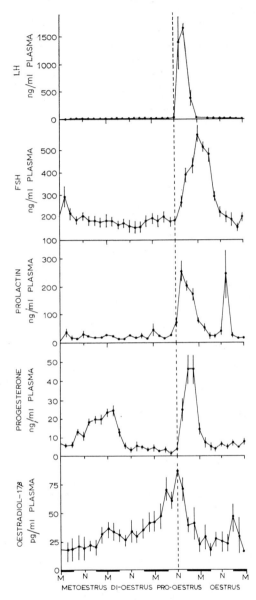

Fig. 2.6. Mean plasma concentrations of hormones of the rat during the 4-day oestrous cycle. Measurements were at 3-hr intervals beginning at 03.00 hr of metoestrus. (Butcher *et al.*, *Endocrinology*.)

The Pre-Ovulatory Surge

The gonadotrophic hormone of ovulation is LH or luteotrophin, inducing the rupture of the follicle and subsequent luteinization of the corpus luteum. Ovulation occurs after the sharp rise in LH secretion; thus, the careful studies of Goldman *et al.* (1969) established an average time of 3 a.m. on the morning of oestrus in rats (Fig. 2.7). As Figure 2.8 shows, the rise in LH attained its peak in the same group of animals in the evening of pro-oestrus. In general, a similar surge of secretion of FSH takes place. In the rat, as Fig. 2.6 shows, prolactin, a hormone predominantly concerned with lactation, also exhibits a peak, suggesting a role in ovulation for the hormone in this species (p. 75).

Progesterone

Progesterone is secreted by the granulosa cells of the corpus luteum under the primary influence of the luteinizing hormone, LH. In the course of an ovarian cycle, then, the plasma level of the hormone may be expected to rise in the luteal, i.e. post-ovulatory, stage. The changes in the rat described by Fig. 2.6 show a peak closely related to the peak in LH, but this may well be due to secretion by the corpora lutea of the previous cycle, so that the slower and later rise, in di-oestrus, represents the rise following ovulation due to the newly formed corpora lutea. Figure 2.9 illustrates, on the same graph, the changes in steroid and LH secretion of the ewe, with a cycle of 16 to 17 days, Day 0 being the day of beginning oestrus. The concentration of LH remains low throughout the cycle except for the peak at oestrus; progesterone secretion rises slowly after oestrus, being presumably secreted by the newly forming

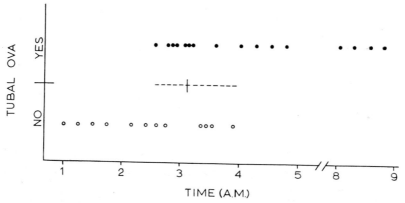

Fig. 2.7. The time of ovulation in rats on day of pro-oestrus. (Goldman *et al.*, *Endocrinology*.)

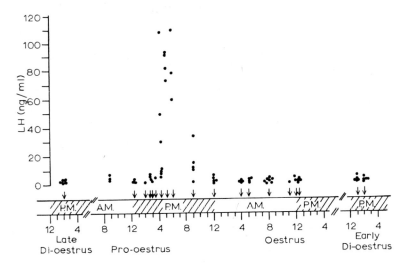

Fig. 2.8. Concentrations of LH in the plasma of cycling rats during the oestrous cycle. Each dot represents the value for plasma LH in one animal. The arrows indicate the approximate times at which the animals were bled by cardiac puncture, and the phases of the cycle are indicated below the time-scale. Note that the LH-surge occurs in the evening of pro-oestrus. (Goldman *et al.*, *Endocrinology*.)

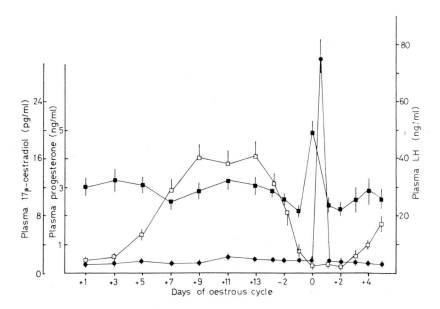

Fig. 2.9. Mean concentration in jugular vein plasma of oestradiol-17β (■), progesterone (□) and LH (●) in six ewes during the oestrous cycle. Day of oestrus is Day-0. Vertical lines are standard errors. (Pant *et al.*, *J. Endocrinol.*)

corpus luteum, and it rises to a plateau between Days 9 to 13; the fall following this represents the luteolytic phase, i.e. the phase of regression of the corpus luteum that occurs unless fertilization of the ovum and implantation of the blastocyst take place. The secretion of the hormone stimulates changes in the uterus preparatory to implantation, and, in the rat at any rate, it induces receptivity; thus Feder *et al.* (1968) were able to associate the onset of lordosis with the onset of progesterone secretion. As we shall see, the secretion of the gonadotrophic hormones, LH and FSH, is influenced by feedback from their target organs, and it is likely that the secretion of progesterone tends to exert a negative

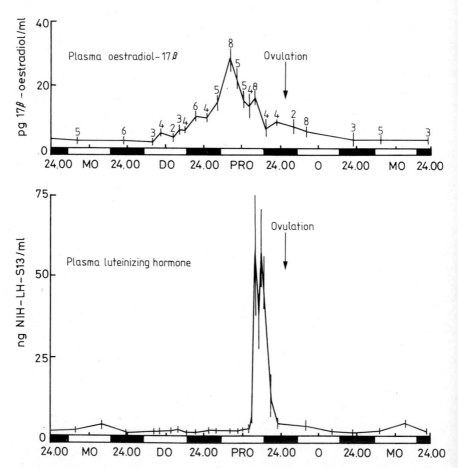

Fig. 2.10. Changes in the rat's plasma concentrations of oestradiol and LH during the 4-day cycle. MO, metoestrus; DO, di-oestrus; PRO, pro-oestrus; O, oestrus. (Brown-Grant *et al.*, *J. Endocrinol.*)

feedback on the pituitary, thereby reducing the secretion of FSH and LH.

Oestrogen

The changes in blood oestrogen during the rat's 4-day cycle are illustrated more precisely in relation to ovulation in Fig. 2.10; the rise in oestrogen in pro-oestrus is clear and it obviously precedes both ovulation and the pre-ovulatory surge of LH, so that, as we shall see, the rise in oestrogen has been considered to be the trigger that sets off the LH-rise which then leads to the progesterone rise. In the sheep, as Fig. 2.9 shows, there is a similar pro-oestrous rise in plasma oestrogen.

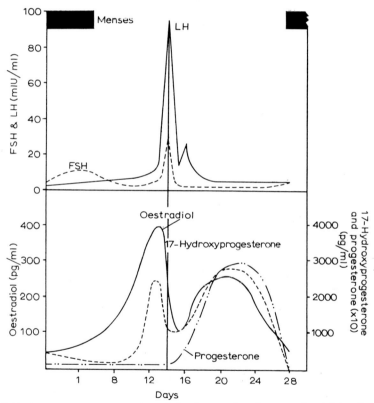

Fig. 2.11. Schematic representation of the fluctuations in serum luteinizing hormone (LH), follicle stimulating hormone (FSH), progesterone, 17-hydroxyprogesterone, and oestradiol during the normal menstrual cycle in women. Note that both oestradiol and 17-hydroxyprogesterone rise before the LH-FSH ovulation surge. In order to show progesterone and 17-hydroxyprogesterone on the same scale, progesterone concentrations were divided by 10; that is, they are actually ten times greater than shown. (Odell and Moyer, *Physiology of Reproduction.*)

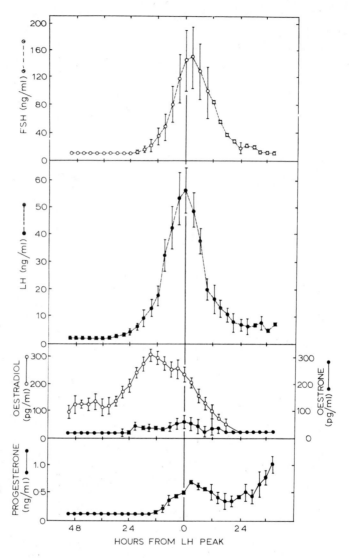

Fig. 2.12. Mean concentrations of LH, FSH, oestradiol and progesterone in plasma of monkeys around the period of the LH-surge. The results were normalized to the time at which the peak occurred (○). Wieck *et al.*, *Endocrinology*.)

Primates

The changes in the blood during the menstrual cycle of the human female are shown schematically in Fig. 2.11; these are similar to those in the non-primate oestrous cycle, the mid-cycle peak of luteotrophs being associated with ovulation. We may note that there is a pre-ovulatory peak of the relatively inactive progestogen—17-hydroxy-progesterone; this is a precursor of progesterone, and it seems that the gonadotrophic hormone, LH, governs the interconversion of the two. The relation between the hormones in the monkey has been studied with considerable precision over the hours immediately before and after the mid-cycle pre-ovulatory surge (Fig. 2.12). The pre-ovulatory peaks of gonadotrophs are preceded, as with the non-primates, by a rise in oestrogen; the fall in steroids is presumably due to the abrupt fall in FSH and LH secretion. Experimental removal of the granulosa cells of the monkey's large pre-ovulatory follicle did not affect the rate of secretion of oestrogen into ovarian blood, so that the thecal cells are presumably the source (Channing and Coudert, 1976). A careful study of oestrogen secretion into ovarian blood of the human female by Baird and Fraser (1974) revealed a similar surge of oestrogen 1–2 days prior to ovulation (Fig. 2.13); this study brings out a biphasic secretion of

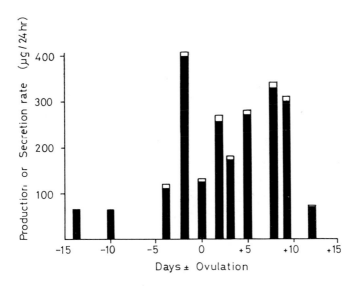

Fig. 2.13. Blood production of, and ovarian rate of secretion of oestradiol by women during the menstrual cycle. The production of oestradiol for each subject is indicated by the total height of each column; white portions indicate the oestradiol derived from oestrone. (Baird and Fraser, *J. clin. Endocrinol.*)

oestrogen, the large pre-ovulatory rise is followed by a fall at ovulation, when it is considered that ovarian secretion may be very small indeed (Baird and Guevara, 1969). This is followed by a rise to peak in the mid-luteal phase of the cycle, indicating secretion during both follicular and luteal phases of the ovary.

Episodic Bursts. As Fig. 2.14 shows, the levels of gonadotrophic hormone in the plasma do not fluctuate smoothly but episodically, with outbursts especially at the midcycle peak and in the postovulatory phase. The changes in the plasma levels of FSH are very much smaller, but there seems little doubt that the release of both hormones occurs simultaneously; and this is not surprising since, as we have seen, the two gonadotrophic hormones can apparently be found in the same type of cell and are released in response to the same releasing hormone.

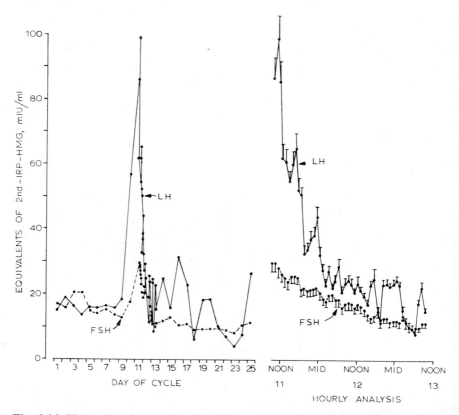

Fig. 2.14. Illustrating the episodic bursts of secretion of gonadotrophic hormones in women. The hourly analysis is that of the concentrations during the descending phase of the LH-peak observed at mid-cycle. (Midgley and Jaffe, *J. clin. Endocrinol.*)

Critical Period

The period during which the rise in LH-secretion, with its attendant rise in plasma progesterone concentration, takes place is short and is called the *critical period*. If a barbiturate is injected at 13.30 hours on the day of pro-oestrus, ovulation is blocked in 18 out of 18 rats, whereas if it is given only 5 hours later, at 18.30 hours, only 3 out of 20 are blocked (Feder *et al.*, 1971). The blockage resulting from injections at 13.30 hours on pro-oestrus day lasted for 24 hours, and was paralleled by a corresponding delay in the rise in progesterone concentration. The onset of sexual receptivity occurred at the time of the rise in progesterone concentration.*

Heat. It is important that receptivity in the female be timed with ovulation; if the rise in progesterone secretion is the cause of heat, then we may expect that ovariectomy carried out just before this rise will inhibit heat. In fact, Barfield and Lisk (1974) found just this; thus the rise in progesterone in the rat occurs on the evening of the day of pro-oestrus, the rise in oestrogen occurring earlier, in the morning. Ovariectomy at 15.30 hours on the day of pro-oestrus prevented receptivity, but if this was delayed until 17.30 hours the expected period of receptivity for this cycle took place.

LH and Progesterone. In general, then, the release of progesterone before ovulation may be regarded as the hormonal agent for inducing female receptivity; the rise is so closely associated in time with the pre-ovulatory surge of gonadotrophic hormones that it has been difficult to decide which occurs first, so that the sharp rise in blood-progesterone has been invoked as the ovulating trigger, its secretion at a time of high blood-oestrogen evoking a large secretion of LH. The study of Uchida *et al.* (1969) in the rat showed clearly, however, that the rise in LH preceded the rise in progesterone of the blood in the ovarian vein (Fig. 2.15). It will be seen from the Figure that there are, in fact, two peaks of progesterone concentration, the large peak on the evening of pro-oestrus and a smaller one in early di-oestrus, the latter being attributed to the secretion of the corpora lutea formed at the previous ovulation, as discussed earlier. Thus, far from acting as a stimulator of

---

* Sexual receptivity can be induced in ovariectomized rats by high doses of oestrogen (Davidson *et al.*, 1968), but progesterone acts synergistically with oestrogen under these conditions, reducing the concentration of oestrogen required. In the intact 4-day cycle rat, too, sexual receptivity may be induced by oestradiol injections on any day except the second following spontaneous oestrus (Södersten and Hansen, 1977). We may note that, according to Powers (1970), ovariectomy abolishes mating if it is carried out before the steep rise in progesterone concentration in pro-oestrus.

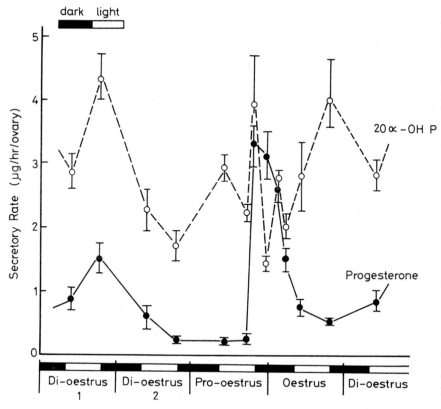

Fig. 2.15. Ovarian secretion of progesterone and 20α-hydroxypregn-4-en-3-one (20α-OH-P) during the oestrous cycle in rats. (Uchida *et al.*, *Endocrinol. Jap.*)

LH-release, the progesterone acts as an inhibitor in a negative feed-back.*

Progravid or Pseudopregnant Rat

When the rat copulates with a sterile male a *progravid* or *pseudopregnant* condition ensues; this seems to be analogous with the postovulatory, or luteal, phase in, say, the monkey or human, a phase in which the secretion of progesterone dominates. In the rat the progravid period lasts for some 11 days; as Fig. 2.16 shows, the condition is, in-

---

* In the hamster, too, Bosley and Leavitt (1972) showed that lordosis required the presence of the ovaries on the day of pro-oestrus, and was associated with a rise in blood-progesterone. This could be blocked by phenobarbital during the "critical period", a block that could be removed by a dose of 1 mg of progesterone. They concluded that the gonadotrophic hormones stimulated progesterone secretion necessary for lordosis and loss of uterine fluid.

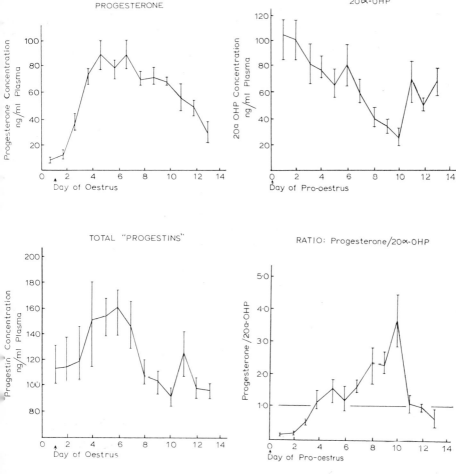

Fig. 2.16. Daily changes in peripheral plasma progestin concentrations during the progravid phase of the rat reproductive cycle. It is suggested that the abrupt fall in the ratio of progesterone/20a-OHP indicates luteolysis. (Bartosik and Szarowski, *Endocrinology*.)

deed, associated with a rise and fall of plasma progestin concentration lasting throughout the pseudopregnant period (Bartosik and Szarowski, 1973). The concentration of the metabolic derivative of progesterone —20a-OPH—steadily declined, suggesting an inverse relation between the two, and this lends support to the hypothesis of Wiest *et al.* (1968), according to which the level of progestational potency of the ovarian secretions is controlled through conversion of progesterone to the 20a-OPH through an enzyme, active in the ovary, 20a-OHP-SDH. Thus, during the rat's oestrous cycle there is a 4- to 5-day pattern of enzyme activity with a minimum during di-oestrus and a maximum following the pre-ovulatory surge of LH and continuing into oestrus. Histochemical studies indicated that the enzyme was absent in new corpora lutea of pregnancy, whereas it was highly active in older involuting corpora lutea (Balogh, 1964).

Induced Ovulation

In the rabbit and cat, among other species, ovulation is not spontaneous but is deferred until the act of copulation; in this case copulation is followed by a depletion in the pituitary content of LH, presumably as a result of increased secretion, since several workers, e.g. Kanematsu *et al.* (1974), have described a surge in plasma LH associated with induced ovulation in the rabbit. Figure 2.17 illustrates the rise in plasma LH after copulation and compares this with the effects of electrical stimulation in the region of the amygdala and with the effects of the releasing hormone, LHRH. We may note that, although the LH is

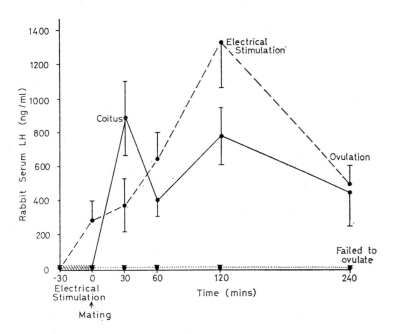

Fig. 2.17. Comparative effects of coitus and electrical stimulation of the amygdala on serum LH concentrations in oestrogen-primed female rabbits which ovulated following stimulation. Note the double peak following coitus. Also shown are the effects of electrical stimulation of other areas of the brain in 11 rabbits that failed to ovulate. (Kanematsu *et al.*, *Endocrinology.*)

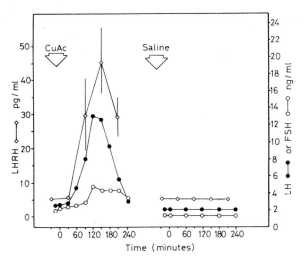

Fig. 2.18. Showing the effects of copper acetate on the pituitary stalk plasma concentrations of LH or FSH (right-hand ordinates) and LHRH of female rabbits. (Tsou *et al.*, *Endocrinology.*)

undoubtedly secreted in response to hypothalamic release of LHRH, the level of FSH remains about the same (Dufy-Barbe *et al.*, 1973). Thus, the trigger to ovulation in this case is a hypothalamic secretion of LHRH, which could presumably be measured in the pituitary stalk plasma. However, sampling of this would be difficult without anaesthesia and Tsou *et al.* (1977) made use of the fact that an intravenous injection of copper acetate evokes the typical post-coital pattern of gonadotrophic hormone release, accompanied by ovulation, acting through the hypothalamus. As Fig. 2.18 shows, intravenous copper acetate causes a rise in LHRH in the stalk plasma associated with a corresponding rise in LH in peripheral plasma. The secretion of LH thus acts as the "final common path" for ovulation in these species as in the spontaneously ovulating animals.*

Role of 20α-dihydroxyprogesterone

According to Hilliard *et al.* (1967) 20-α-hydroxyprogesterone (20-α-P) serves as a positive feedback agent amplifying the amount of LH released after mating in the rabbit; however Goodman and Neill (1976) were unable to relate changes in plasma 20-α-P to the post-coital surge. That some ovarian hormone is necessary to amplify

---

* Interestingly, coitus is associated with release of LH from the copulating bull's pituitary gland (Katongole *et al.*, 1971). In the rabbit, too, there is an increased plasma level of LH at copulation, but by no means as large as that in the female (Younglai *et al.*, 1976). The rat is a cyclical animal and does not require copulation to induce ovulation; nevertheless copulation is accompanied by a rise in plasma LH in both male and female with a fall in pituitary content (Taleisnik *et al.*, 1966).

the effects of the LH-surge was shown by the fact that the ovariectomized rabbit, pretreated with oestrogen, exhibited the post-coital LH-surge, but only one out of ten animals ovulated; 20-a-P administered gave no significant increase in the LH-surge. If the ovaries were removed 15 minutes after copulation a normal LH-surge was obtained; consequently, ovarian hormones are required before mating, but once the surge has been initiated it can proceed autonomously without supplementation of ovarian hormones such as 20-a-P.

Inhibition of Secretion of Oestrogen

In general, the concentration of oestrogen in the plasma falls after oestrus; in the ewe this is especially striking, and the experiments of Moor (1974) suggest that the LH-surge at this period of the cycle is responsible for the lowered oestrogen-secretion. Thus, Fig. 2.19 shows

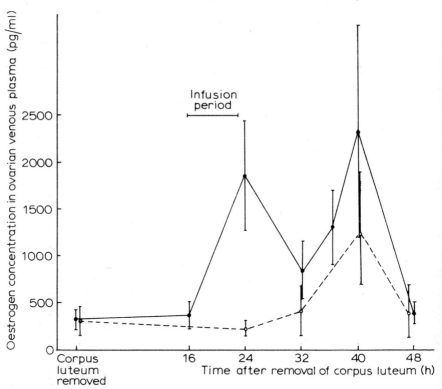

Fig. 2.19. Effects of infusion of LH into the jugular vein of sheep from which the corpus luteum had been removed on the concentration of oestrogen in the vein draining the ovary with the largest follicle. In the control animals, infused with saline, there are two peaks of oestrogen concentration following removal of the corpus luteum (●——●); the infusion of LH (○ ---- ○) suppresses the first rise in oestrogen concentration, but not the second which is due to development of a new corpus luteum. (Moor, *J. Endocrinol.*)

the effects of an infusion of LH on the level of oestrogen in the blood-plasma of a ewe from which the corpus luteum had been removed; it will be seen that the LH infusion suppresses the rise in oestrogen that takes place between the 16th and 32nd hours after removal of the corpus luteum. The later rise in oestrogen secretion is due to the development of a new corpus luteum; and this time the secretion is not affected appreciably by the LH-infusion, presumably because the newly formed corpus luteum had not been influenced by the LH-infusion.

Prolactin

This pituitary hormone, together with growth hormone (GH), is the controlling factor in lactation; however, in the rat, it also plays an important role in maintenance of the corpus luteum and its secretion of progesterone; in other words the luteinizing hormone, LH, is not the only gonadotrophic hormone exercising this control.

Luteotrophic Action. This function was revealed by studies on auto-transplantation of the anterior pituitary into a region remote from the hypothalamus, e.g. the anterior chamber of the eye, when it is uninfluenced by the gonadotrophic releasing hormones. The cells responsible for secretion of these hormones, LH and FSH, are basophilic and they degenerate, since the releasing hormones liberated from the median eminence apparently exert a trophic action on these cells as well as causing them to release their hormones. The acidophilic cells, orangeophilic in the azan stain and presumed to be the source of prolactin, now become the predominant type of granulated cell in the transplant. If the transplant is made into a female with newly formed corpora lutea, these remain intact and are similar in size and histological appearance to those seen in pseudopregnancy. Thus the transplanted pituitary, with atrophic gonadotrophic-secreting cells but active prolactin-secreting cells, is able to maintain the corpora lutea and presumably progesterone secretion. McLean and Nikitovich-Winer (1973) injected anti-LH and anti-prolactin immune bodies into pseudopregnant rats (made pseudopregnant by electrical stimulation of the cervix which inhibited ovulation). Both of these factors abolished the pseudopregnancy and decidual reaction if they were given immediately after stimulation, thus indicating the importance of both hormones for establishing luteal function. However, the efficacy of the antiprolactin depended on the time at which it was administered, so that if the anti-prolactin was administered 8–16 hr after the initiation of pseudopregnancy it no longer abolished this; anti-LH was effective at all times, so that the authors concluded that prolactin was important for the initiation of luteal function whereas LH was required for this and also for maintenance. This seems reasonable, as the primary role of LH is to stimulate luteal steroidogenesis, and it will therefore be necessary for the maintenance of luteal function; prolactin's primary role seems to be the preservation of progesterone by inhibiting its catabolism.*

Synthesis of LH-Receptors. This may be to reduce the importance of the hormone, however, and it may well be that a significant role is to maintain LH-activity by inducing the synthesis of LH-receptors. According to Smith *et al.* (1975) and Holt *et al.* (1976), prolactin provides the principal luteotrophic stimulus for prolonging progesterone secretion from Day 2 through Day 7 of pregnancy or pseudo-pregnancy, whereas LH is essential from Days 8–12. Although prolactin is ineffective as an acute stimulator of steroid secretion, it can augment the power of LH to promote progesterone secretion (Armstrong *et al.*, 1969). It probably does this by increasing the number of LH-receptors in the corpora lutea; thus Holt *et al.* (1976) induced pseudopregnancy in rats by giving repeated doses of FSH from Day 25 of birth to Day 27, and following this with a single dose of LH. Administration of prolactin increased the concentration of LH-receptors in the corpora lutea.†

In general, analysis of the times during pregnancy at which hypophysectomy or an anti-LH serum will induce abortion leads to the conclusion that the luteotrophic effect of LH requires the co-operation of either prolactin or of placental gonadotrophin (Ford and Yoshinaga, 1975).‡

* It must be emphasized that there are major species differences with respect to the gonadotrophic action of prolactin; it seems unlikely that it has this function in other species than the rat, mouse and ewe. We may note that lysergic acid derivatives will inhibit prolactin release, but under these conditions rats ovulate normally and have normal oestrous cycles (Döhler and Wuttke, 1974). We may note with respect to LH function that the LH concentration in blood of women is very low in the luteal phase of the menstrual cycle, so that the corpus luteum may not be dependent on this hormone.

† The pseudopregnancy resulting from infertile coitus in the rat is associated with two daily surges of secretion of prolactin (Freeman *et al.*, 1974); blocking these surges with ergot alkaloids reduced the binding of LH by corpora lutea. Again implantation of prolactin in the median eminence, which blocks secretion of the hormone, terminates pseudopregnancy (Dang and Voogt, 1977) frequently inducing ovulation.

‡ The role of prolactin in the reproduction of the human female has been discussed by Rolland *et al.* (1976); the plasma levels do not vary in relation to the menstrual cycle, but there seems to be a biphasic action on ovarian steroidogenesis, promoting this in granulosa cells (McNatty *et al.*, 1974), but, when the plasma levels of the hormone are high, during post-partum lactation, for example, actually *inhibiting* steroidogenesis. Although the plasma levels of prolactin do not exhibit a cyclical pattern, the specific binding capacity of ovarian tissue varies; thus corpora lutea of pregnancy have very high binding capacity; granulosa cells from follicles have a binding capacity according to their degree of maturity. We may note that in the human male, hypogonadism is frequently associated with hyperprolactinaemia, and this tallies with a reduced sensitivity of the pituitary to releasing hormone—LHRH—under these conditions in the rat (Winters and Loriaux, 1978).

The Pre-Ovulatory Surge and its Trigger

The sudden rise in secretion of the pituitary gonadotrophic hormones, especially of LH, which occurs in the rat, during a critical period of about two hours in pro-oestrus, and which is a general feature of the ovulating cycle in all species, has attracted an enormous amount of research, and we must be careful, when considering some of this, not to become so involved in the details that we fail to distinguish the wood for the trees. It is essentially the cyclic onset of this surge that determines the cyclical character of the whole reproductive or ovulatory process; thus any event that inhibits this pro-oestrous rise in LH secretion brings the cycle to a stop; this may occur naturally in the case of sterile copulation in the rat, or in the absence of copulation in those species that depend on this for their ovulatory trigger, such as the rabbit.

Persistent Oestrus. Experimentally a variety of devices, such as anaesthesia with barbiturates a short time before the expected rise in LH secretion, or injections of steroids that, through a negative feedback, inhibit the release of gonadotropic hormones, may be employed. In the rat, for example, such an inhibition brings about a prolonged state of oestrus—*persistent oestrus*—with persistent vaginal cornification and a continued succession of large ovarian follicles, the ovarian secretion of oestrogen being maintained beyond its normal duration presumably because of the absence of the rise in progesterone that, in normal circumstances, would tend to suppress FSH secretion.*

Hypothalamus. This cyclicity, as we shall see, is essentially a characteristic of the hypothalamus and not an intrinsic feature of the endocrine glands, either gonads or pituitary. Because the steroid ovarian hormones are able to influence the secretion of their trophic hormones—feedback—both at the pituitary and hypothalamic levels, the pattern of their concentrations probably governs the threshold of excitability of neurones within the hypothalamus; and it would seem that, when the proper hormonal milieu is attained, the so-called "intrinsic trigger" is able to activate the release of FSH and LH from the pituitary, giving rise to the pre-ovulatory surge that leads to ovulation (Barraclough *et al.*, 1971).

Intrinsic Rhythmicity. To recognize that the timing of the cycle, in fact its completion, can depend on the concentrations of hormones

---

* In the rat a reflexly induced ovulation, similar to that occurring normally in the rabbit, may be brought about by first inducing a state of persistent oestrus through continuous exposure of the animal to light; the animal in this state has an interrupted cycle and remains continuously receptive to the male. Within ten minutes of mating the plasma LH-level rises and this is followed by ovulation. There was little change in FSH, but this was high in the plasma because of the persistent oestrus (Davidson *et al.*, 1973).

secreted during the cycle, is one thing, but to proceed from this to suggest that the ovaries themselves can act as the "Zeitgeber" for the LH-surge (Everett and Sawyer, 1950) may be to go too far. Instead, we must emphasize the intrinsic rhythmicity of the central nervous system whose periodicity is constantly being modified by feedback through the blood; as a result of this feedback the most important reproductive event in the cycle, namely ovulation, takes place when the chances of fertilization and subsequent implantation of the ovum are at their highest. The trigger is, almost certainly, the balance of steroid hormone concentration in the blood at a certain period, so that if this balance is disturbed the cycle is brought to an end, but eventually the inherent rhythmicity of the hypothalamus is able to assert itself, and a new cycle will begin.

The Influence of the Steroid Hormones

Before entering into the detailed relations between the hypothalamus, on the one hand, and the pituitary and gonads on the other, it will be profitable to consider a few examples of the way in which the concentrations of the circulating steroid hormones may influence cyclical activity.

Ovariectomy

The preovulatory surge of release of gonadatrophic hormone may be prevented by ovariectomy. Schwartz (1964) removed the ovaries from either 4-day or 5-day cycling rats at different periods during the oestrous cycle, in accordance with the schedules indicated in Fig. 2.20; she established that the ovary must be *in situ* between 4 p.m. of diooestrus (the day before pro-oestrus when the surge occurs) and pro-oestrus at 10 a.m. if the cornification of the vagina, increase in uterine weight and discharge of pituitary hormones are to occur. She pointed out that the extra day of the 5-day cycle compared with the 4-day cycle was due to a delay in the steroid discharge from the ovary, necessary to trigger the release of the ovulating surge of LH. This follows from the fact that the necessary period for *in situ* presence of the ovary was the same, in relation to oestrus, in both cycles.

Oestrogen and LH-Surge

During this period of the cycle the dominant ovarian hormone to be secreted is oestrogen, so that the studies of Schwartz, more recently repeated by Mann and Barraclough (1973), emphasize that on the afternoon of dioestrus Day 2 and the morning of pro-oestrus, it is the secretion of oestrogen that is largely responsible for the LH-surge, at

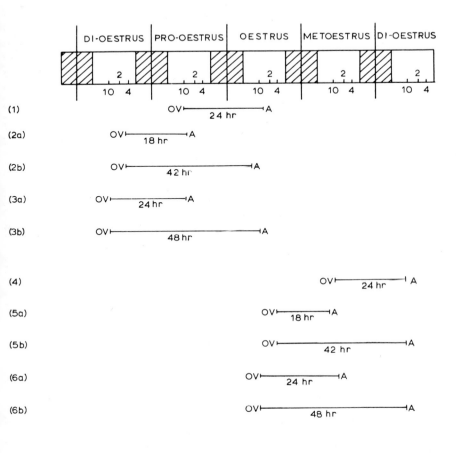

OV   Time of Ovariectomy
A   Time of Autopsy

Fig. 2.20. Schedule for removal of ovaries of cycling rats and subsequent autopsy to determine whether ovulation occurred. Time of cyclic release of LH is between 2 pm and 4pm on the afternoon of pro-oestrus. Hatched area equals lights off. OV, ovariectomy; A, autopsy. Groups 1–3b are the basic experimental animals with groups 4–6b providing "time after ovariectomy" controls. (Schwartz, *Amer. J. Physiol.*)

any rate so far as the ovarian source of the hormones is concerned. Interestingly, the increased secretion of oestrogen at this time is brought about by secretion of LH by the pituitary so that, in a sense, LH provokes its own "surge". Thus, Freeman *et al.* (1972) measured the surge of prolactin, which in the rat is synchronous with that of LH; administration of an antiserum to LH at 10.00 of di-oestrus 2 but not at 10.00 of pro-oestrus, blocked the prolactin surge by blocking the rise in oestrogen that would otherwise have occurred. It blocked the uterine ballooning as would be expected of an inhibition of oestrogen secretion. Both these effects of the anti-LH were counteracted by the powerful oestrogen, diethylstilboestrol. In a similar way De la Cruz *et al.* (1976) blocked ovulation in the hamster by injection of anti-LHRH at any time during the cycle; the effect lasted 12–13 days, commensurate with the persistence of the anti-LH action. Serum oestradiol levels were reduced, although not completely suppressed.

Progesterone. Lest we should conclude, however, that oestrogen is the exclusive controller of the trigger, we must note that the studies on plasma concentrations of oestrogen and progesterone indicate a rise in progesterone before the large pre-ovulatory peak, a rise that may be due to secretion by existing corpora lutea in the ovary formed in previous cycles (Kirton *et al.*, 1970), or may be due to adrenal secretions (p. 88). Certainly, progesterone administration on the morning of pro-oestrus, i.e. when the LH-surge is expected, will trigger off this surge in ovariectomized rats treated with oestrogen, the LH-surge coinciding with the time of injection of progesterone (Mann and Barraclough, 1973).

Antibody Treatment

Antibodies to oestradiol and progesterone may be prepared by injecting these into ewes;* administration of anti-oestradiol to 4-day cycle rats on the day of dioestrus 2 inhibited ovulation and the expected ballooning of the uterus and cornification of the vaginal mucosa. There was apparently no damage to the ovaries of these rats since administration of human chorionic gonadotrophin restored ovulation. Anti-progesterone did not block ovulation but the peripheral effects of progesterone were blocked, so that ballooning of the uterus in oestrus persisted. When anti-progesterone was injected on day 1 of di-oestrus, this period was prolonged by two days in 50 per cent of animals (Ferin *et al.*, 1969). In the monkey injections of anti-oestradiol caused the animals to become anovulatory; in this case the level of progesterone in

---

* The steroid hormones form complexes with serum albumin, and it is essentially the antibody to the conjugated steroid with bovine serum albumin that was employed.

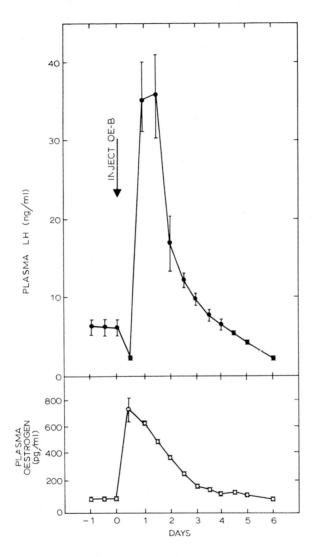

Fig. 2.21. Effects of a single dose of oestradiol benzoate on the plasma oestrogen and plasma LH of monkeys primed with oestrogen by way of silastic capsules containing 17-β oestradiol. (Karsch *et al.*, *Endocrinology*.)

the blood remained low and no corpora lutea were formed in the ovaries (Ferin *et al.*, 1974). The anti-oestrogen did not affect the target-organs, so that when human chorionic gonadotrophin (hCG) was administered the animals ovulated normally (Sundaram *et al.*, 1973).

Oestrogen-Progesterone Synergy in the Primate

In the monkey, the effects of ovariectomy, and their neutralization by steroid hormone injections, have been examined by Karsch *et al.* (1973). They developed a technique of implantation of steroid sub-cutaneously in silastic capsules, so that the slow release permitted the maintenance of a steady plasma level for periods of months. When 17-β-oestradiol was implanted in this way in ovariectomized animals, the level of LH in the plasma was maintained at a value characteristic of the early follicular phase of the menstrual cycle, indicating a negative feedback. As Fig. 2.21 shows, when a single subcutaneous dose of oestrogen was superimposed on this background level there was a burst of LH-release reminiscent of the preovulatory surge. Thus oestrogen clearly exerts both inhibitory and excitatory actions on the gonado-trophin releasing system, a *negative* feedback maintaining LH-secretion at an appropriate level and then a *positive* feedback promoting LH-secretion; and it is essentially the combination of these two actions imposed on a natural rhythmicity of the hypothalamic-pituitary sec-reting system that brings about the pre-ovulatory LH-surge. When implants containing both oestradiol and progesterone were included in the study, a very prominent synergism between the two ovarian hor-mones was found. This is illustrated by Fig. 2.22. In the top block are the plasma oestradiol values, which were held between 50 and 80 pg/ml, a level that occurred before ovariectomy. These levels were about the same whether oestradiol or oestradiol plus progesterone were im-planted. In the middle block the progesterone levels are shown; after overiectomy the concentration falls to zero if the implant contains only oestradiol, but with the combined implant a steady value, correspond-ing with the pre-ovariectomy one, was sustained. In the bottom block the concentrations of LH are shown. In untreated animals the con-centration rises, indicating the absence of negative feedback. When only an oestradiol implant was made, the achieved concentration of oestradiol was insufficient to hold back the level of LH at its pre-ovariectomy level, but if both oestradiol and progesterone were implanted the level of LH remained low.

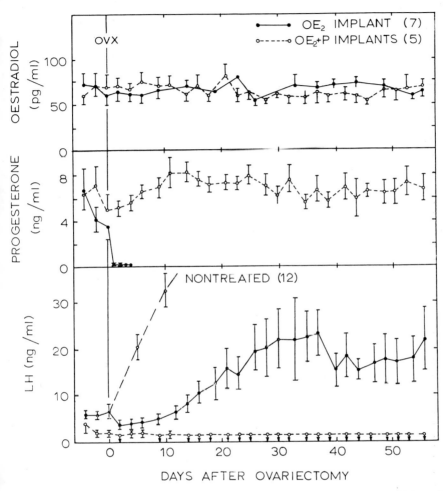

Fig. 2.22. Illustrating the effects of combined administration of oestradiol and progesterone on the levels of oestradiol, progesterone and LH in the blood of ovariectomized monkeys. OE 2, oestradiol; P, progesterone. Ovariectomy causes a rise in plasma LH due to removal of negative feedback; oestradiol partly prevents this whilst combined oestradiol plus progesterone completely suppresses the rise. (Karsch *et al.*, *Endocrinology*.)

Progesterone and Cycle Length

The early studies of Everett (1948) showed that the normal 5-day cycle of the rat could be converted into a 4-day cycle by injection of progesterone on Di-oestrus Day 3, in the sense that ovulation was brought forward by 24 hours; in this case progesterone was exerting an excitatory action. An inhibitory action may be shown by injecting progesterone into the normal 4-day cycling rat on Di-oestrus Day 1 in which case oestrus is retarded by a day; if injections are given every Di-oestrus Day 1 the animal has an artificial 5-day cycle indefinitely. With this artificial 5-day cycle, induced by an injection of progesterone on Di-oestrus Day 1, if progesterone is given on Di-oestrus Day 3, ovulation is accelerated by 24 hours.

Importance of Oestrogen. Thus, progesterone can inhibit or excite ovulation according to the phase of the cycle, and this is related to the level of oestrogen in the blood; when this is high, as in Di-oestrus Day 3 in the 5-day cycle rat, it facilitates ovulation, whereas, in the early stage, when oestrogen secretion is probably slight, it inhibits ovulation. Barfield and Lisk (1970) have examined the effects of progesterone on the advancement of oestrus in the 4-day-cycle rat. They studied spayed animals and showed that heat may be induced by progesterone provided that the animal has been previously primed with oestrogen, and this is the natural course of development of heat, the progesterone secretion being imposed on a background of oestrogen. Applying this information to the intact rat they showed that, provided progesterone was given to the animal when its oestrogen level was high, an advance in oestrous behaviour could be induced, and the latest time at which it was effective was 03.00 hours of pro-oestrus.* In a similar study, in which plasma LH was measured, Brown-Grant and Naftolin (1972) observed an initial decrease in plasma level of the gonadotroph when progesterone was injected at 9.00 a.m. on pro-oestrus, but within 5 hr the injection had triggered an abrupt and large rise in plasma LH-level (> 20 ng/ml), adequate to induce ovulation; this could be prevented by barbiturate anaesthesia, indicating the involvement of neural mechanisms (p. 113).†

---

* Injections of oestrogen alone will also advance the day of ovulation in 5-day-cycle rats given in early di-oestrus (D-1) (Everett, 1948; Brown-Grant, 1969).

† Joslyn et al. (1976) have discussed the possibilities of advancing ovulation by progesterone injections alone; their own studies on the guinea-pig emphasize the importance of oestrogen-priming; thus if they gave progesterone alone at the time when the guinea-pig shows its cyclical surge of this steroid, i.e. on Day 15 of the 16-day cycle, ovulation could often be advanced; given earlier, before oestrogen-priming is complete, it delayed ovulation. The delays could be quite striking, ovulation occurring on Day 29 when progesterone was given on Day 16, for example.

Pituitary Involvement. That the pituitary is directly involved in these feedback actions of progesterone, as well as the hypothalamus, was shown by Martin *et al.* (1974) when they demonstrated increases or decreases in release of LH in response to the releasing hormone, LHRH, when progesterone advanced or retarded the day of ovulation. Similarly, advancement of the day of oestrus through oestradiol injections involved increased pituitary sensitivity to the releasing hormone. Earlier Arimura and Schally (1970) had shown that exogenous progesterone would decrease the release of LH in response to LHRH from 1·5–3·6 ng/ml to 0·5–1·6 ng/ml.

Secretion of Steroids by Normal Women

DeJong *et al.* (1974) have analysed ovarian blood of normal cycling females; this gives a valuable clue to rate of synthesis of the given hormones; moreover, by comparing blood from the ovulating ovary with that from the non-ovulating one, the contribution of follicular maturation and corpus luteum formation can be assessed. The hormones identified in ovarian blood were oestradiol, oestrone, androstenedione, testosterone, dehydroepiandrosterone and progesterone; Fig. 2.23 shows the comparison with ovarian and peripheral bloods on Day 11, i.e. during the follicular phase; here the right ovary was the active one with a large (3 cm) follicle. It will be seen that progesterone is being secreted by the ovary at this time; Fig. 2.24 shows the picture on Day 25; this is dominated by the large progesterone secretion. In general, the secretions of ovarian oestradiol, oestrone, androstenedione and progesterone varied together with the cycle; the low concentrations of testosterone, and dehydroepiandrosterone found in the follicular fluid, and the small differences between ovaries, indicate that the ovaries are not an important source of these androgens. The relatively high concentration of testosterone in mid-cycle peripheral blood could be due to conversion from androstenedione. In general, the authors concluded that the ovaries contributed greater than 90 per cent of the oestradiol in peripheral blood; their secretions of oestrone, androstenedione and progesterone contributed significantly to the peripheral levels.

Menopausal Women

In menopausal women Odell and Swerdloff (1968) were able to mimic the pre-ovulatory surge of FSH and LH by injections of progesterone followed by injections of oestradiol; the initial progesterone injections reduced the levels of FSH and LH from their normally high values in post-menopausal women (p. 471) and the subsequent oestrogen injection, together with the progesterone, apparently acts as a hypothalamic stimulus to cause a rise in LH and FSH secretion by a positive feedback.

As to whether the normal ovulatory trigger is, indeed, operated by the level of progesterone in the blood, rather than the balance between this and oestrogen, it is difficult to say positively. Odell and Swerdloff have argued that if the trigger is operated through an inherent cyclicity

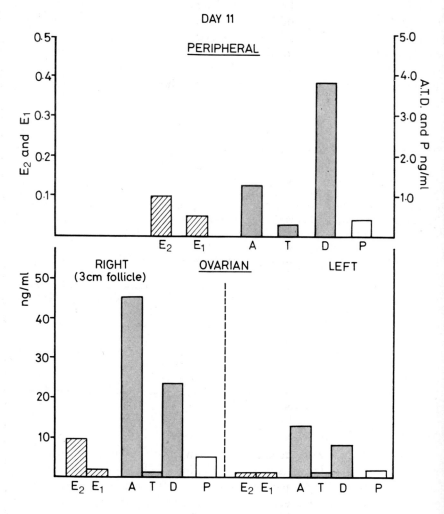

Fig. 2.23. Concentrations of oestradiol-17β (E₂), oestrone (E₁), androstenedione (A), testosterone (T), dehydroepiandrosterone (D) and progesterone (P) in peripheral and ovarian venous plasma from both ovaries on Day 11 of the cycle in a normal menstruating woman. (De Jongh *et al.*, *Acta endocrinol.*)

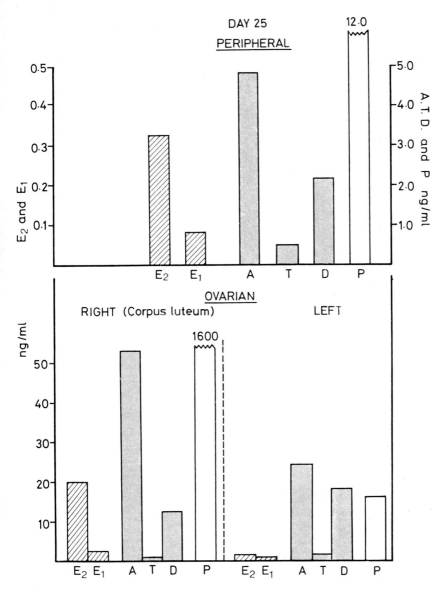

Fig. 2.24. Concentrations of oestradiol-17β (E₂), Oestrone (E₁), androstenedione (A), testosterone (T), dehydroepiandrosterone (D) and progesterone (P) in peripheral and ovarian venous blood from both ovaries on Day 25 of the cycle in a normal menstruating woman. (De Jongh *et al.*, *Acta endocrinol.*)

in the hypothalamus it is likely that, during the course of many cycles, one might be found in which the pre-ovulatory surge occurred inappropriately in respect to the state of the developing Graafian follicle and so be ineffectual; in fact their studies on women failed to reveal any such error, so that they suggested that some message from the ovary was involved whereby the hypothalamus was informed that the follicle was ready to ovulate. Since the large pre-ovulatory rise in progesterone definitely follows the LH-FSH surge, whereas the rise in oestrogen precedes it, we must rule this hormone out as the messenger, but of course other progestins, e.g. 17-α-hydroxy progesterone or perhaps the progesterone secreted by the adrenal gland, might play this role.

Adrenal Sources of Progesterone

It is well established that the adrenal gland synthesizes progesterone, which is the precursor of the androgens, oestrogens and of corticosteroids; the adrenal release of progesterone is controlled by ACTH (Resko 1969), and it has been suggested that the slow rise in progesterone level in the plasma, preceding the LH-surge, might be due to adrenal secretion, and it could be that adrenal sources of progesterone were important for determining some of the manifestatons of reproductive behaviour. Thus, in the ovariectomized oestrogen-primed rat, administration of ACTH can stimulate production of sufficient progesterone to induce heat (Feder and Ruf, 1969). Dexamethasone,* a synthetic steroid, suppresses progesterone secretion by the adrenal but not by the ovary, and this has been found to suppress the early onset of heat resulting from acute ovariectomy at pro-oestrus (Barfield and Lisk, 1974).

Thus it has been suggested that the slow rise in progesterone level in the plasma preceding the LH-surge, might be due to adrenal secretion. For example, Mann and Barraclough (1973) cannulated the ovarian vein of the 4-day-cycle rat and determined the plasma levels, which were indicative of secretion rate by the ovary, during the cycle. They found that the secretion rate was fairly constant throughout the cycle, except for the large pre-ovulatory rise. The levels in the general circulation, however, exhibited a diurnal fluctuation, due to adrenal secretion, the lowest level being at 10.00 to 14.00 hours, and the peak occurring at 01.00–05.00 hours. Only during the large pro-oestrous rise was this rhythm disturbed. They suggested that the morning peak

* Dexamethasone's actions cannot be entirely explained by its inhibition of adrenal steroid secretion; it apparently synergizes with oestrogen in inducing heat, for example, in ovariectomized animals (Lisk and Reuter, 1976).

might well act as the trigger for the pre-ovulatory rise in LH-secretion. In this case we should expect adrenalectomy to block ovulation. Feder *et al.* (1971) had shown that the slow rise in progesterone in the plasma was associated with a corresponding rise in corticosterone, and when the animal was dosed with dexamethasone, in order to block adrenal activity, ovulation was indeed blocked, a blockage that was reversed by administration of ACTH or progesterone. When animals were adrenalectomized, some 50 per cent of these failed to ovulate. As to whether the adrenal secretion of progesterone acts as the final trigger to the LH-surge must remain an open question; certainly the injection of ACTH into oestrogen-primed ovariectomized rats can induce mating behaviour (Resko, 1969; Feder and Ruf, 1969) whilst adrenalectomy delays lordosis (Nequin and Schwartz, 1971). On the other hand, Davidson *et al.* (1968) found that the oestrogen-induced lordosis in ovariectomized rats took place in spite of adrenalectomy.

Luteolysis

If fertilization of the ovum does not take place the corpus luteum regresses—luteolysis—and the changes in the uterus and oviduct that were, in effect, preparatory to receiving the fertilized ovum, regress.

Transplanted Corpora Lutea

There is considerable evidence that the life of the corpus luteum is not pre-determined, but depends on external factors, and this is true both of the normally cycling animal, the pseudopregnant animal or the pregnant animal. Thus Scott and Rennie (1970) transplanted 2-day-old corpora lutea to the tissue under the kidney capsule, where they remained normal. If they were transferred to recipients at oestrus, their life-span was normal, about 17 days, similar to the duration of pseudopregnancy of the rabbit. If they were transferred into 12-day pseudopregnant rabbits, their life-span was reduced to about 11 days, and both the ovarian and transplanted corpora lutea regressed at the same time. Hysterectomy at the time of transplantation prolonged the life-span of each set of corpora lutea to 23–27 days.*

* That the luteolytic effect can be derived from the conceptus was shown by Rowson and Moor (1967) who infused extracts of homogenates of 14–15-day sheep embryos into the uterus of the cycling sheep and prolonged the oestrous cycle; the effect could not be obtained from the 25-day embryo, nor yet from that of the pig.

Earlier Moor and Rowson (1964) found that they could transfer a 12-day sheep embryo to the uterus of a non-pregnant ewe on the 12th day of its cycle; a 13-day embryo transferred on the 13th day of the cycle usually failed to implant and the animal returned to oestrus. Thus, at about 12 days a luteolysin appears that prevents the secretion of progesterone necessary for successful implantation. They found that hysterectomy, performed as late as the 15th day of the cycle, could arrest the involutionary changes in luteal cells, provided that the corpus luteum was functional at the time of hysterectomy.

Uterine Distension

Mere distension of the uterus also exerts, in the guinea-pig, a curtail-ment of luteal activity (Donovan and Traczyk, 1962), so that if two glass beads are introduced into the two horns of the uterus the oestrous cycle is extended for some three days; increasing the number of beads increased the duration of cycle (Bland and Donovan, 1966). Each horn exerted a local action on its ovary so that a reflex action through the pituitary is unlikely.

Oestrogen

The luteolytic agent in the primate may well be the oestrogen secreted by the corpus luteum; thus Butler *et al.* (1975) analysed the corpus luteum and ovarian blood plasma during the luteal phase of the monkey's menstrual cycle; during the late luteal phase—8–13 days after the ovulatory LH-surge—the concentration of oestrogen in the ovary increased fourfold and this was paralleled by a corresponding in-crease in concentration in the ovarian vein homolateral with the ovulating Graafian follicle. The increased oestrogen in this vein was due to oestrone, whilst the oestradiol concentrations in the two veins were the same, so that the oestrogen produced by the corpus luteum in the monkey is predominantly oestrone, by contrast with the human where it is oestradiol (Mikhail, 1970).

Premature Menstruation. Further evidence suggesting that oestrogen is the luteolytic agent in the primate corpus luteum is the observation of Hoffmann (1960) that injection of oestrogen into the ovary of a woman in the luteal phase of the cycle induced premature menstruation with premature regression of the corpus luteum, and the same was observed by Karsch and Sutton (1976) in the monkey, the concentration of progesterone in the monkey's blood falling to that characteristic of the early follicular phase of the cycle, bringing an early onset of menstruation.

Isolated Luteal Cells. Stouffer *et al.* (1977) showed that high con-centrations (1–10 μg/ml) of oestradiol reduced progesterone production of isolated luteal cells *in vitro*; lower concentrations (less than 0·1 μg/ml) had no direct effect but they did antagonize the stimulatory action of human chorionic gonadatrophin (hCG). It is possible, then, that in the normal menstrual cycle of the primate, oestrogen contributes to, or determines, regression of the corpus luteum; after implantation in the pregnant animal it could be that secretion of chorionic gonadotrophin (p. 202) "rescued" the corpus luteum from impending oestrogen-induced functional regression (Stouffer *et al.*, 1977). Certainly the rise in serum progesterone during the luteal phase of the fertile menstrual cycle of the monkey occurred 24–28 hr after monkey chorionic gonadotrophin (mCG) was detected in the blood and urine of the pregnant animal (Hodgen et al., 1974).

Prostaglandin F_2a as Luteolysin

It was discovered by Loeb (1923) that removal of the uterus—hysterectomy—caused prolongation of the luteal phase of the ovarian cycle in guinea-pigs; and subsequent work has shown that this effect is due to the removal of a uterine "luteolysin" which, in several species, has been identified as prostaglandin F_2a. McCracken et al. (1971) autotransplanted a sheep's ovary in the neck so that its blood supply was through the carotid and jugular. By infusing PGF_2a into the carotid they caused an immediate decline of progesterone secretion into the jugular blood. PGE_1 was without effect. The species in which PGF_2a has been implicated in luteolysis are the rat, hamster, rabbit, guinea-pig, pig and sheep (McNatty et al., 1975).

Fluctuations in Blood Concentration. If prostaglandin is the natural luteolytic agent, we may expect to find fluctuations in its concentration in the blood during the normal ovarian cycle. Because of its rapid destruction in the body it should be sought in the utero-ovarian blood rather than in the peripheral circulation. Figure 2.25 shows the

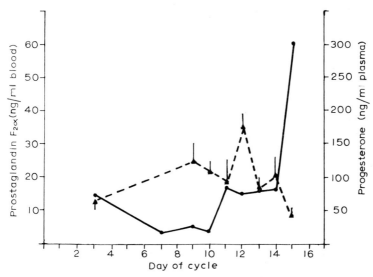

Fig. 2.25. The relationship between the concentrations of PGF_2a (—) and progesterone (– – – –) in utero-ovarian venous blood of the guinea-pig. (Blatchley et al., J. Physiol.)

concentrations of $PGF_2\alpha$ and progesterone in utero-ovarian blood during the course of the guinea-pig's cycle; there is a striking increase in $PGF_2\alpha$ concentration at the time that the concentration of progesterone is on the wane.

Prostaglandin Injections. Injections of $PGF_2\alpha$ into cycling guinea-pigs at the time when a rise in plasma progesterone was expected—Days 6–8—caused a progressive fall in plasma progesterone concentration; the effects were usually transitory, however, so that cycle-length need not be affected by this treatment (Blatchley and Donovan, 1976).

Effect of Anti-$PGF_2\alpha$. When the ewe was actively immunized against $PGF_2\alpha$ the cyclic oestrous behaviour was blocked due to persistence of the corpus luteum and the raised plasma level of progesterone. Removing the corpus luteum led to a return to cyclic behaviour (Scaramuzzi and Baird, 1976).

Primates

In the primate the regression of the corpus luteum of the menstrual cycle will occur in the absence of a uterus (Fraser et al., 1973) so that an ovarian source of prostaglandin has been suggested if this does indeed, contribute to, or determine, the process of luteolysis in the primate. Luteal cells can certainly synthesize $PGF_2\alpha$ (Challis et al., 1976) and $PGF_2\alpha$ can inhibit the production of progesterone by human luteal cells in culture (McNatty et al., 1975) whilst direct injection into the human corpus luteum causes a pronounced fall in blood-progesterone associated with uterine bleeding (Korda et al., 1975). However, according to a recent study on the monkey by Manaugh and Novy (1976), the prostaglandin synthesis inhibitor, indomethacin, administered during the luteal phase of menstruation of a non-fertile period, failed to affect progesterone secretion and the duration of the luteal phase, and they concluded that prostaglandins were not the physiological luteolysins in primates.*

---

* The involvement of prostaglandin in luteolysis has been discussed in detail by Challis et al. (1976) and Swanston et al. (1977); the matter is highly complex and we must distinguish the inhibition of progesterone secretion, occurring rapidly, from the morphological involution which occurs much later (Umo, 1975). Also it has been found that the inhibitory effect of $PGF_{2\alpha}$ on progesterone synthesis is smallest at the time when synthetic capacity is maximal, so that the early corpus luteum may be protected in some way from the inhibitory effect of the prostaglandin (McNatty et al., 1975). Henderson and McNatty (1977) have suggested a see-saw interaction between LH, which stimulates progesterone secretion, and $PGF_{2\alpha}$, which inhibits, the interaction consisting primarily in a competition for binding sites in the granulosa cell membrane. Thus, the fall in LH secretion in the luteal phase of the cycle permits the luteolytic action of $PGF_{2\alpha}$.

Local Action

The uterine influence on the life of corpus luteum of the normally cycling ewe was demonstrated by the extension of its life after hysterectomy, as Loeb has shown in the guinea-pig (Wiltbank and Casida, 1956). This extension was very striking and could last as long as the normal gestation period of the ewe (Moor and Rowson, 1964). That the influence was local, and apparently not mediated through the general circulation, was suggested by several findings; e.g. in a case of congenital absence of a uterine horn the ewe exhibited luteal retention of the adjacent ovary and failed to go into oestrus for 58 days (McCracken and Caldwell, 1969). Again transplantation of the ovary to the

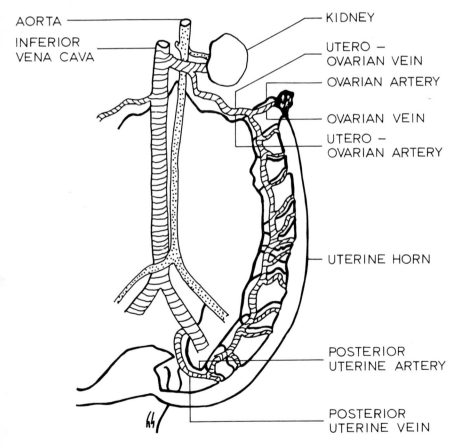

Fig. 2.26. The vascular system of the female reproductive organs of the rat. (Pharriss, *Persp. Biol. Med.*)

neck leaving the uterus *in situ* caused persistence of the corpus luteum (Goding *et al.*, 1967). The mechanism of transport from uterus to ovary is, however, a puzzle. According to McCracken *et al.* (1971), the luteolytic factor, PGF_2a, would be carried along the utero-ovarian vein (Fig. 2.26) and the PGF_2a might diffuse out of this vein into the ovarian artery by a countercurrent mechanism, although diffusion through an arterial wall is a concept not easy to accept. However, according to Hansel (1971), there is a remarkable network of small venules surrounding the artery. Experimentally the artery and vein were separated from each other, in which case the corpora lutea persisted; again, ^{3}H-labelled PGF_2a was infused into a uterine vein and the radioactivity in ovarian arterial plasma was compared with that in plasma collected from the adjacent artery, and it was found to be some 30 times higher, indicating that some transport between vein and artery was possible.

Rupture of Follicle

A possible non-luteolytic role for PGF_2a in the primate is suggested by the work of Swanston *et al.* (1977). They measured the steroid hormones and PGF_2a in human corpora lutea at different stages of the menstrual cycle; the highest concentration of PGF_2a occurred in the post-ovulatory corpus luteum, and in corpora albicantia, with a relatively constant level throughout the luteal phase, so that it could be that the prostaglandin is concerned in ovulation, as suggested by other studies, being possibly involved in the mechanical rupture of the follicle (Lindner *et al.*, 1974).

Mechanism of Luteolytic Action

Hydroxysteroid Dehydrogenase

The failure of some investigators* to find an *in vitro* inhibition of steroidogenesis by the corpus luteum on incubation with PGF_2a suggests that the action may be different from simple inhibition of synthesis. Fuchs and Mok (1974) pointed out that the negative correlation between secretion of progesterone and 20-a-OHP, its derivative formed by the catalytic action of the enzyme hydroxysteroid dehydrogenase—

---

* Pharriss and Wyngarden (1969) found prostaglandins to be potent stimulators of steroidogenesis by corpora lutea *in vitro* in obvious contrast with their luteolytic action *in vivo*. However, O'Grady *et al.* (1972) have described a culturing procedure for corpora lutea, taken from the pregnant rabbit, that maintains synthesis of progestin. This was markedly inhibited by $PGF_{2\alpha}$ in a concentration of 10 μg/ml. The *in vivo* study of Auletta *et al.* (1973) showed a dual action of $PGF_{2\alpha}$ on progesterone secretion into the ovarian vein of the monkey; low doses (10 ng/min) actually stimulated output, whilst larger doses (50 μg/min) caused 90 per cent inhibition.

20α-OH-SDH—described by Wiest *et al.* (1968), might well give the clue to the luteolytic action of prostaglandin. Thus $PGF_2\alpha$ increases the activity of the enzyme 20α-OH-SDH, and this may be assessed histochemically; they showed, in pregnant rats, that infusions of $PGF_2\alpha$ only terminated pregnancy when a strong inhibition of enzymatic activity was observed; luteinizing hormone, LH, prevents the interruption of pregnancy and also the appearance of SDH-activity. The stimulation of enzymatic activity was highly effective during two limited periods, namely Days 9–12 and Days 18–20, and it was only during these two periods that pregnancy was interrupted by moderate doses of $PGF_2\alpha$.*

Ovarian Ischaemia

Pharriss (1970) suggested that the mechanism of action of prostaglandin was through a powerful vasoconstrictive action on the ovarian vein, reducing blood-flow and thus inhibiting progesterone synthesis. He emphasized the local action of prostaglandin, in the sense that unilateral hysterectomy in bicornual uteri resulted in regression of corpora lutea on the intact side and maintenance on the hysterectomized side, indicating that the luteolytic agent was not carried in the general circulation. The circulation in the uterine region of the rat is illustrated in Fig. 2.26, and it will be seen that in this species the anterior uterus and ovary share a common vein; Pharriss suggested that local release of prostaglandin might cause a sufficiently powerful venoconstriction as to result in an ischaemic degeneration in the corpora lutea. This vascular explanation of prostaglandin action is, however, not supported by recent studies; thus Bruce and Hillier (1974) measured ovarian blood-flow, by the microsphere technique, in pregnant rabbits during infusions of $PGF_2\alpha$ sufficient to cause a fall in blood-progesterone until delivery. Regression of the corpora lutea was not accompanied by a reduced ovarian blood-flow. Again, the studies of McNatty *et al.* (1975), referred to earlier, have shown that $PGF_2\alpha$ can, indeed, inhibit progesterone secretion by isolated granulosa cells from human corpora lutea. Finally, $PGF_2\alpha$ will exert a luteolytic action on a corpus luteum, transplanted beneath the kidney capsule and with its vascular supply unaffected by $PGF_2\alpha$ (Keyes and Bullock, 1974). Nevertheless, the local transport of prostaglandin in the ovary from the uterus by way of the uterine vein seems to be well established (see, for example, Blatchley *et al.*, 1972).

* Behrman *et al.* (1971) measured the ovarian secretion *in vivo* of 8-day pregnant rats; $PGF_2\alpha$ caused a decrease of 50–80 per cent in progesterone secretion whilst the secretion of 20-α-OHP rose some 2·8 to 3·7-fold. When rats were treated with an anti-LH serum, the effects were very similar to those of $PGF_2\alpha$, suggesting that the prostaglandin antagonizes the gonadotrophin.

Influence on LH-Binding

A possible mechanism for the luteolytic action of prostaglandin might be through inhibition of LH-uptake by the corpora lutea; Behrman and Hichens, working on pseudopregnant rats, found a rapid block of uptake of ^{125}I-labelled human chorionic gonadotrophin by the luteal tissue, accompanied by a drop in progesterone content. In a later study (Grinwich *et al.*, 1976) have considered the balance between LH and prolactin activities on the one hand, and prostaglandin activity on the other, the gonadotrophins and prostaglandin working through gonadotrophin receptor sites of the corpus luteum. Table I shows that $PGF_2\alpha$ causes, in pseudopregnant rats, a decrease in LH-receptor capacity, accompanied by a fall in serum-progesterone, indicating

TABLE I

Effect of $PGF_2\alpha$ on corpus luteum binding capacity (nghCG/mg corpus luteum) and serum progesterone and 20α-ol in rats (Mean $\pm$ S.E.). (Grinwich *et al.*, 1976)

Time Hours	LH receptor binding capacity ng/mg	Progesterone ng/ml	20α-ol ng/ml	Prog/20α-ol
0	3.2 ± 0.3	40.5 ± 12.4	14.5 ± 2.2	2.4 ± 0.5
2	2.7 ± 0.6	$12.1 \pm 0.6\dagger$	11.2 ± 5.6	1.9 ± 0.5
8	$2.2 \pm 0.5*$	$14.9 \pm 1.4\dagger$	$37.8 \pm 4.7\ddagger$	$0.4 \pm 0.1\ddagger$
24	$0.9 \pm 0.2\ddagger$	$10.5 \pm 0.5\dagger$	$52.6 \pm 8.5\ddagger$	$0.2 \pm 0.1\ddagger$

* $P < 0.10$; $\dagger$ $P < 0.05$; $\ddagger$ $P < 0.01$.
Each event is the average of 5 rats.

luteolysis; the increase in 20α-ol is also symptomatic of luteolysis. The elevation of 20α-ol indicates an inhibition of prolactin action, and earlier work has shown that prolactin can prevent the luteolysis due to $PGF_2\alpha$; Grinwich *et al.* showed that administration of prolactin with $PGF_2\alpha$ prevented the loss of LH-receptors and fall in serum progesterone produced by $PGF_2\alpha$ alone. In general then, these experiments indicate a close association between loss of LH-receptors and the fall in serum progesterone and elevation of serum 20α-ol associated with functional luteolysis, effects that appear to be related to the loss of activity of prolactin.*

Luteal and Endometrial PGF Secretion

In the pig, $PGF_2\alpha$ is probably a natural luteolysin, the concentration

---

* Ergocryptine blocks pituitary prolactin secretion; it decreased progesterone secretion and LH-receptors, the effects on LH-receptors coming on much later than the fall in progesterone secretion. Thus ergocryptine's action is more complex than that of $PGF_2\alpha$.

in the utero-ovarian venous plasma being high at the time when luteolysis occurs during the oestrous cycle (Gleeson *et al.*, 1974). Removal of the uterus prolongs the oestrous cycle in those species in which PGF_2a has been demonstrated to be luteolytic, the prostaglandin synthesized by the endometrium of the uterus being apparently carried by a counter-current mechanism to the ovary (McCracken *et al.*, 1971). Patek and Watson (1976) measured the *in vitro* synthesis of PGF_2a by endometrium and luteal tissue at different phases of the oestrous cycle and found that both tissues synthesized the prostaglandin, with greatest activity during the mid-to-late luteal phase, a period corresponding with the maximal concentration in the utero-ovarian vein measured *in vivo* (Gleeson *et al.*, 1974). It was not possible to state whether the two secretions were coincident or which event occurred first; however, the fact that luteal synthesis is maximal in the mid-to-late phase of the luteal cycle, i.e. when luteolysis is occurring, indicates a control over luteal function through luteal prostaglandin synthesis.

PGF_2a and PGE_2

These prostaglandins have opposite influences on the corpus luteum; thus PGE_2 in primates and non-primates mimics the action of the trophic hormone, stimulating the production of progesterone (see, for example, Marsh and LeMaire, 1974). McNatty *et al.* (1975) have compared the effects of the two prostaglandins on human granulosa cells in tissue-culture; PGF_2a inhibited progesterone production by the cells, an effect that could be reduced by prior exposure to LH and FSH some six days before. If the prostaglandin (PGF_2a) and the trophic hormones occupy the same binding sites, acting on the adenyl cyclase system to stimulate cAMP production in the cells, then the antagonism is understandable. The point of attack of PGE_2 is probably different; it increases progesterone production by the cells, but this is unaffected by PGF_2a, and there is no synergism with LH and FSH. The fact that PGF_2a can cause regression of synthetic activity in the corpus luteal cells in tissue-culture emphasizes that it can act independently of any vascular changes (p. 95).*

---

* Henderson *et al.* (1977) infused the ovary of the ewe with $PGF_{2\alpha}$, $PGF_{2\alpha} + PGE_2$, and PGE_2 alone and measured the progesterone concentrations in the ovarian vein. $PGF_{2\alpha}$ alone decreased progesterone secretion but simultaneous infusion of PGE_2 prevented this until after the infusion. PGE_2 did not increase progesterone secretion but this was probably because secretion was maximal at the time (Day 11). The ability of PGE_2 to antagonize $PGF_{2\alpha}$ is probably due to its stimulation of cAMP formation through a membrane receptor different from the LH-receptor. Thus $PGF_{2\alpha}$ competes with LH for its receptor and so inhibits progesterone secretion; PGE_2 increases progesterone secretion by an independent action on the adenylate cyclase system.

Anti-Oestrogenic Action of Progesterone

Progesterone is secreted by the corpus luteum, which forms after ovulation; and its secretion is closely related to the morphological regression of the uterus and oviduct characteristic of the terminal phase of the menstrual cycle. Brenner *et al.* (1974) have examined the changes taking place in the oviduct during the cycle in monkeys. At mid-cycle the epithelium is fully ciliated and secretory, whereas in the late luteal phase the epithelium has atrophied, secretion is minimal and most of the cilia have disappeared from the epithelial cells. In the early follicular phase, coinciding with menstruation, the epithelium becomes metabolically active, ciliogenesis begins, and by the eighth day the oviduct has become fully ciliated and secretory. During these changes, although oestrogen has been shown to be necessary for oviduct function, the concentration of this hormone remains constant except for the day preceding the LH-surge. On the other hand, the concentration of progesterone in the blood is negatively correlated with the morphological changes, in the sense that regression is high when the concentration of progesterone is high. Progesterone seems therefore to be acting as an anti-oestrogen. Brenner *et al.* (1974) spayed monkeys, causing regression of the oviducts; these were restored to normal in the spayed animal by injections of oestrogen and, by establishing a cycle of oestrogen injections followed by progesterone injections, an artificial menstrual cycle was created. Figure 2.27 illustrates the changes in steroid concentrations in the blood and also the degree of ciliation of the oviduct epithelium and concentrations of oestrogen receptors in the oviduct. It would appear that morphological regression of the oviduct is brought about by reduction in the ability of the oviduct to bind oestrogen, and not through changes in oestradiol levels in the blood, so progesterone is acting as an anti-oestrogen by reducing the concentration of oestradiol receptors in the tissue.

Maturation of the Ovum

It will be recalled that the mammalian öocyte begins its first maturation division in prenatal or early post-natal life; thus just before or after birth, according to the species, the germ cell has reached the stage of diplotene, and at this stage meiosis is arrested and the öocyte remains in this "dictyate state" throughout infancy and for a variable period beyond the onset of puberty. In the adult, during each oestrous cycle, a number of öocytes complete their first reduction division, resulting in the abstriction of the first polar body shortly before ovulation. This resumption of meiosis and its progress to the metaphase of the second meiotic division is called "maturation" (Lindner *et al.*, 1974). Completion of the second meiotic division with extrusion of the second polar body occurs only after sperm penetration.

Gonadotrophic Hormones

This process of maturation depends on the secretion of gonadotrophic hormones, and the question arises as to what is the factor that arrests the meiotic process. Is there some inhibitory substance secreted by the Graafian follicle, so that maturation results from a release from inhibi-

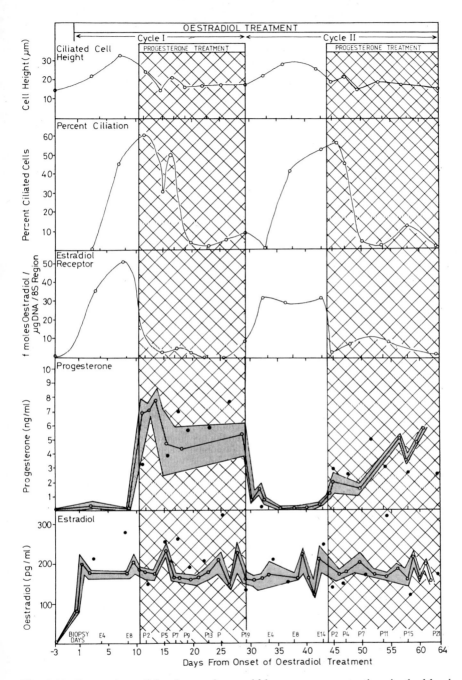

Fig. 2.27. Comparison of the changes in steroid hormone concentrations in the blood of spayed monkeys with the histological features of the oviduct epithelium and the concentration of oestradiol receptors. (Brenner *et al.*, *Endocrinology*.)

tion? In support of this view is the finding of Pincus and Enzmann (1935) that a denuded öocyte will mature *in vitro*, whereas within the follicle it must be stimulated by gonadotrophic hormone. Thus, when the follicle is removed and maintained *in vitro*, the probability of its maturation depends on the stage in the ovarian cycle at which it was removed, the greatest degree of "commitment" to maturation being when removed at about 19 hr of pro-oestrus, when the level of LH in the blood is at its highest. Follicles taken at an early stage could be made to mature by addition of LH or human chorionic gonadotrophin (hCG) to the medium whereas prolactin was ineffective. FSH was also effective, even when treated with an anti-LH serum* to remove the chance of LH-contamination. Under normal conditions it seems that there is insufficient FSH circulating to bring about maturation, so that co-operation with LH is necessary. Both gonadotrophs induce formation of cAMP in the follicle but theophylline, which, by inhibiting the destruction of cAMP by phosphodiesterase, should favour LH action, did not do so, perhaps because it exerted a toxic action of its own. It is very likely that the action of the gonadatrophic hormones, LH and FSH, is a direct "trophic" action, rather than the result of promoting progesterone or other steroid secretion; thus an inhibitor of steroid synthesis failed to affect the efficacy of LH, whilst addition of progesterone to the medium did not cause maturation.

Prostaglandins

These substances, particularly those of the E-series, stimulate cAMP formation by the ovary as well as ovarian progesterone synthesis and lutcinization of cultured granulosa cells. Lindner *et al.* (1974) found that PGE_2 stimulated cAMP formation in their *in vitro* cultured follicles and completed the first meiotic division of follicle-enclosed öocytes as effectively as LH; $PGF_2\alpha$ was less effective. Thus, substances that increase cAMP accumulation in the follicle, namely LH, FSH, prostaglandin and cholera toxin (Zor *et al.*, 1972) all promote maturation of the öocyte. However, different routes for achieving maturation by gonadotroph and prostaglandin seem to be involved; thus treatment with an inhibitor of PG synthesis, indomethacin, did not inhibit LH action, either on maturation or cAMP accumulation; when given systemically to pro-oestrous rats, it had no effect on maturation although it blocked follicular rupture. Thus LH acts independently of prostaglandin mediation.

To revert to the existence of an inhibitor of maturation of the

---

* Actually an antiserum to the β-subunit of LH, which was used in the hope that it would be more specific than an antiserum to the whole hormone.

öocyte, Tsafriri and his colleagues have shown that porcine follicular fluid will inhibit the maturation of rat öocytes *in vitro*; this inhibition can be overcome by addition of LH to the medium. As the follicle becomes older, the öocyte becomes less sensitive to the inhibitor (Tsafriri *et al.*, 1977).

Rupture of the Follicle

A role for prostaglandins in this aspect of reproduction seems certain; thus, administered in the critical period, indomethacin blocked ovulation in spite of the presence of a normal pro-oestrous LH-surge, and all of the ova entrapped in the unruptured follicles had undergone maturation. Administration of PGE_2 to indomethacin-treated pro-oestrous rats induced ovulation. Thus, according to Lindner *et al.*, prostaglandin may have an essential role in ovulation in that it is necessary for rupture; certainly LH administration increases the PGE_2-content of the ovary by 8- to 10-fold.

Inhibitor of Luteinization

Nabaldanov suggested that the follicular fluid contained an inhibitor of luteinization, and Ledwitz-Rigby (1977) employed the large pre-ovulatory follicles which Channing (1970) had shown would luteinize spontaneously in culture. The granulosa cells from these follicles will normally develop lipid droplets, secrete progesterone and accumulate cAMP in response to LH; these processes were inhibited by follicular fluid when it was removed from small (1–2 mm) antral follicles; the fluid from large follicles (6–12 mm) was, however, without effect, so that it may be that, as the follicle grows to become a large pre-ovulatory follicle, it loses an inhibitory factor that holds the luteinization process under control. Failure to lose the inhibitor might thus lead to atresia.

HYPOTHALAMICO-PITUITARY-GONADAL RELATIONSHIPS

It is time, now, to examine the behaviour of the hypothalamus in relation to the ovarian cycle in some detail. Having done this we may return to the problem of the initiation of ovulation.

Some examples of the interrelationships between the three main stations controlling the level of circulating steroid gonadal hormones in the blood have already been given, notably in the apparent negative feedback of the circulating ovarian hormones that tends to reduce the secretion of the pituitary trophic hormones, LH and FSH, and a positive feedback of oestrogen promoting further secretion which probably constitutes the ultimate step in precipitating the pre-ovulatory surge.

There are two obvious sites at which the level of steroid hormone could influence gonadotrophic hormone secretion, namely at the hypothalamic level, inhibiting activity of the neurosecretory cells of the hypophysiotrophic region, or at the pituitary level, modifying sensitivity of the secretory cells to releasing hormones. So far as the hypothalamus is concerned, moreover, we must consider the relationship between the anterior preoptic region (PO/AH) and the hypophysiotrophic area, and the possibility of both humoral and neural feedbacks in this circuit. As we should expect, a large variety of experimental techniques have been employed in piecing together the various components that lead to the co-ordinated activities involved in reproduction.

The Hypothalamus

The dominant role of the hypothalamus in governing reproductive behaviour was emphasized by the work of Sawyer and of G. W. Harris, the technique developed by the latter author for grafting pituitary tissue into the hypothalamus permitting the unequivocal demonstration of the fundamental control exerted by this neural structure.

Transplanted Ovaries

Nowhere is the importance of the hypothalamus better demonstrated than when analysing the cause for the characteristic difference between the male and female secretion of gonadotrophic hormones; in the female this is cyclical and in the male it is continuous. The first clue to this difference was given by the grafting experiments of Pfeiffer (1936); he transplanted ovaries into males castrated at different times in their neonatal development, and found that the ability of the ovary to develop and form corpora lutea depended on the time during which it was exposed to the circulating androgens in the male recipient. Thus, if the recipient rat was castrated at birth, the ovaries developed and produced corpora lutea. If castration was delayed, the typical development failed. If a female rat was ovariectomized at birth and later, in adult life, an ovary was transplanted (into the eye) this was capable of producing corpora lutea. This work suggested that exposure to the male steroid secretions somehow prevented the transplanted ovary from fulfilling all its functions.

Transplanted Testes

Again, when testes were implanted into newborn females their ovaries, when adult, contained only follicles and no corpora lutea; there was persistent vaginal cornification—called *persistent oestrus*—indicating the continuous secretion of FSH but the absence of the cyclical surge of LH that induces ovulation. Pfeiffer concluded that male and female hypophyses were undifferentiated at birth, both having a latent capacity to secrete gonadotrophic hormones in a cyclical fashion, but if, during a critical period, the hypophysis was exposed to testicular hormones it lost this capacity for cyclical behaviour. Later, when the role of the hypothalamus in controlling secretion of gonadotrophic hormones became clear, the Pfeiffer hypothesis was modified, since it became clear that the androgens acted on the hypothalamus during the pre-natal period, and in certain animals with short gestation periods such as the rat, in the early post-natal period, converting the hypothalamus into a male structure.

Hypothalmic Control

Thus Harris* transplanted a male pituitary beneath the hypothalamus of a hypophysectomized female rat, and the animal went through the normal oestrous cycle, indicating that the female hypothalamus was exerting the cyclical control of secretion of gonadotrophic hormones. The neonatal critical period during which the male steroid acts is quite short in the rat; for example, Harris transplanted ovaries into the aqueous humour of the eye of male rats, where they are capable of maintaining normal activity; if the recipient males were castrated after more than 24 hours after birth only a few developed corpora lutea, and if 3–4 days were allowed to pass before castration, no ovary showed corpora lutea development.†

Bodily Activity

The difference between male and female extends to their periods of activity throughout the diurnal cycle; Fig. 2.28 shows the activity plotted from day to day. I is that of a normal male rat and II that of a normal female rat; it will be seen that in the female there are peaks coinciding with oestrus. III represents a male rat castrated when 8–10 hr old and transplanted with ovarian and vaginal tissue when 63 days of age. It will be seen that in this last animal characteristically female episodes of activity occur with a periodicity apparently corresponding with that of the female. We may presume that the hypothalamus in this castrated male rat is liberating gonadotrophic releasing hormone cyclically and, by providing the animal with the necessary target organ,

* For a review of this classical work, see Harris (1964).

† The critical period lasts longer in the albino rats studied by Yasaki (1960). It must be appreciated that an effect of testosterone can be demonstrated in the neonatal rat because of the immature state in which it is born; to demonstrate an influence on the hypothalamus of, say, the guinea-pig, the androsterone would have to be injected into the pregnant female.

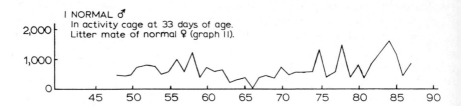

I NORMAL ♂
In activity cage at 33 days of age.
Litter mate of normal ♀ (graph II).

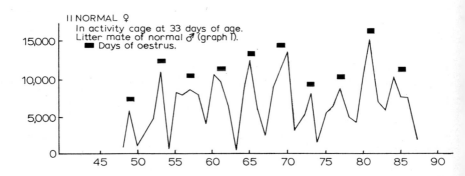

II NORMAL ♀
In activity cage at 33 days of age.
Litter mate of normal ♂ (graph I).
■ Days of oestrus.

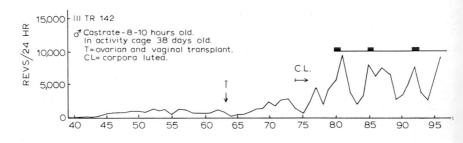

III TR 142
♂ Castrate-8-10 hours old.
In activity cage 38 days old.
T=ovarian and vaginal transplant.
CL= corpora lutea.
C.L.

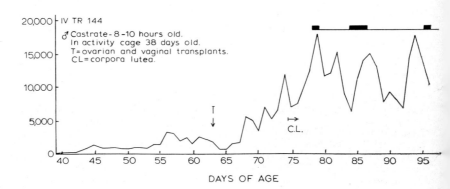

IV TR 144
♂ Castrate-8-10 hours old.
In activity cage 38 days old.
T=ovarian and vaginal transplants.
CL=corpora lutea.
C.L.

REVS/24 HR

DAYS OF AGE

namely ovarian tissue, this is reflected in secretion of ovarian hormones that modify the animal's activity.

Male Androgen Secretion

It seems, then, that, during development *in utero* and in the immediate post-partum period, the male gonads secrete androgens, and that this secretion governs not only maturation of the hypothalamus but also the determination of the sexual characters—i.e. the gonads, and accessory organs.* The histological evidence indicates that testicular interstitial cells are acting during the neonatal period of organization and regress shortly before puberty. As Table II shows (Resko *et al.*,

TABLE II

Androgen concentrations in plasma (mg/100 ml) and testis (μ/g/tissue) of developing rat. (After Resko *et al.*, 1968)

Age (days)	Testosterone		Androstenedione	
	Plasma	Testis	Plasma	Testis
1	0·027	0·329	ND	0·113
5	0·021	0·103	ND	0·044
10	0·009	0·075	ND	0·008
15	0·010	trace	ND	ND
30	0·015	0·003	0·013	0·007
40	0·064	0·003	0·018	0·005
60	0·110	0·029	0·095	0·003
90	0·204	0·082	0·012	0·008

* Elger (1966) treated pregnant rabbits with the anti-androgen cyproterone, and examined the effects on the progeny. The effects on the female were not striking, as we should expect. In the male, the differentiation of the gonads, the descent of the testes and the regression of the Mullerian ducts in the tubal and uterine sectors were undisturbed, indicating that *in utero* these features of male development were genetically determined and did not require secretion of androgens. The Wolffian ducts and ductuli efferentes had regressed completely, as in the female; the prostate complex was completely suppressed and in many cases the external genitalia were completely feminized. Later Goldman and Baker (1971) showed that metabolites of testosterone, such as dihydrotestosterone, would cause masculinization, as measured by the ano-genital distance, effects that were antagonized by cyproterone.

Fig. 2.28. Illustrating the activity cycles, and, in 3 of the 4 graphs, the days of vaginal oestrus (black rectangles) of I, a normal male rat, II, a normal female rat, III, a male rat castrated when 8–10 hr old and transplanted with ovarian and vaginal tissue, when 63 days of age, IV, another male rat, also castrated when 8–10 hr old and transplanted with ovarian and vaginal tissue when 63 days of age. Note that the normal male rat shows a low and irregular level of activity, as compared with the normal female rat, which shows peaks in the activity cycles which correspond closely to the days of oestrus; the male rats castrated soon after birth and implanted with ovarian tissue when 63 days old show an increase in activity following implantation, with an apparent cyclicity in activity associated with vaginal oestrus. (Harris, *Endocrinology.*)

1968) the concentration of testosterone in the plasma and testes of male rats is initially high and falls rapidly after birth; the concentration of another androgen, androstenedione, does not vary so much and seems not to be important in masculation. The low concentration of androgen in plasma and testes of rats following the relatively high concentrations at birth are indicative of the prepuberal stage when sexual activity remains latent; as we shall see, the onset of puberty is associated with a rise in secretion of androgens in the male, and of oestrogens in the female.

Androgenization of the Female

The hormone responsible for determining the maleness of the hypothalamus is testosterone, so that injections of this into castrated neonatal males will give them "male" hypothalami. When testosterone is injected into the newborn female a similar maleness develops characterized anatomically in an enlargement of the clitoris, and physiologically in the abolition of cyclical activity; the animal is permanently sterile, failing to ovulate and form corpora lutea due to the failure of the pituitary to release LH and FSH cyclically (Gorski and Wagner, 1965); the condition is one of persistent oestrus with maintained vaginal cornification. By injecting androgen into neonatal rats on successive days after birth Barraclough (1961) found that, if the injection was given between Days 2–5, all rats were permanently sterile; if injected at day 10 only four out of ten were sterile, and he concluded that ten days represented the limiting period for testosterone to produce sterility.* In Fig. 2.29 the incidence of sterility induced by an injection of testosterone has been plotted against age at which the injection was given. The effects depended on the dose, so that 1 mg gave 100 per cent sterility up to nine days after birth. A dose of 10 μg gave a generally lower incidence and the value depended on whether the rats were examined 45 days (continuous line) or 100 days after the injection.†

Inhibition of Testosterone Action. Pituitary transplantation ex-

---

* A condition of persistent oestrus can be induced in weanling rats by oestrogens (Gorski, 1963), and also by human chorionic gonadotrophin (hCG); it might be thought that the effect of hCG would be due to a stimulation of androgen secretion by the ovary, which then exerted its masculinizing action. However, Wollman and Hamilton (1967) showed that the anti-testosterone, cyproterone, whilst it abolished the masculinizing effects of testosterone, did not abolish the persistent oestrus due to hCG. Thus the persistent oestrus caused by hCG was not due to production of androgens by the ovary, but presumably to the hypersecretion of oestrogens.

† In the primate prenatal treatment with androgen induces a pseudohermaphroditic female with definite male behaviour (Goy and Resko, 1972). That the male foetus produces androgens is shown by the higher concentration of testosterone in its umbilical blood than that of a female foetus.

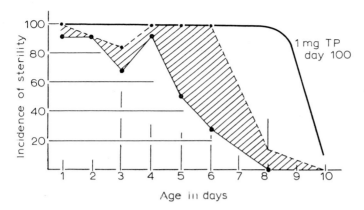

Fig. 2.29. Influence of age at which testosterone is injected into female rats on incidence of sterility. The continuous heavy line indicates effect of 1 mg of testosterone propionate (TP). The other curves indicate effect of 10 μg, the observed effects depending on day of examination, the dashed line representing examination 100 days after injection and the continuous line 45 days after injection. (Arai, *Neuroendocrine Control*, after Gorski.)

periments have made it certain that the effects of male steroid hormones arc exerted on the hypothalamus rather than on the pituitary. Anaesthesia with barbiturates, which blocks ovulation in the rat if carried out on the day of pro-oestrus (p. 113), also blocks the androgenizing effect of testosterone, both of these effects being presumably on the brain rather than on the pituitary. Arai and Gorsky (1968) found that there was a very definite critical period during which anaesthesia would antagonize the effects of testosterone. Thus, anaesthesia given some six hours after testosterone treatment gave some protection, but if this was postponed for twelve hours it was ineffective (Fig. 2.30). They considered that an initial period of three hours was necessary for the uptake of the androgen, and then a longer period, perhaps lasting for nine hours, was required for intraneuronal differentiation; and during all this time anaesthesia was able to block the effect of the testosterone. The inhibitory effect of barbiturates may be due to its generally depressant action on the central nervous system; so far as androgenization is concerned, metrazol, which excites the central nervous system, counteracts the effects of barbiturates (Arai and Gorski, 1968).

Rabbit. The effect of neonatal androgenization is essentially an upset of the hypothalamic timing mechanism of the oestrous cycle, rather than an impairment of the ovulation process *per se*, since ovulation can be caused in these "sterile" animals by electrical stimulation of the hypothalamus. The rabbit has no inborn timing mechanism,

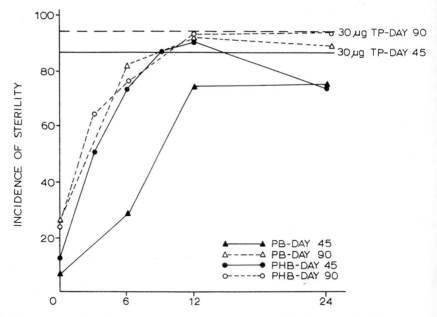

Fig. 2.30. Incidence of sterility at 45 and 90 days of age following the injection of 30 μg of testosterone propionate (TP) alone, simultaneously with, or at various times after pentobarbital (PB) or phenobarbital (PhB) in the 5-day-old female rat. (Arai and Gorski, *Endocrinology.*)

ovulation being induced by copulation. It is of interest, therefore, that neonatal treatment with testosterone causes only a slight retardation of the onset of mating behaviour. Campbell (1965) suggests that the difference is related to the development of the hypophyseal portal system; this occurs about 5-days post natally in the rat at a time when the animal is maximally sensitive to injected steroids. In rabbits it occurs pre-natally; surgical castration of the male at 19 days of gestation caused feminization of the genital tract.

"Aromatization" of Androgen. It has been found that oestrogens are actually more potent in effecting sexual differentiation of the rodent brain than androgens (Gorski and Barraclough, 1963) and since Naftolin *et al.* (1971) and Reddy *et al.* (1974) have shown that rat hypothalamus, incubated with androstenedione, produces oestrone, it has been suggested that androgenization of the female through injections of androgens might be effected by prior conversion to oestrogen. The evidence in favour of "aromatization" within the central nervous system, as well as in peripheral tissues, is strong and has been summarized by Naftolin *et al.* (1975); thus pretreatment of newborn females with an anti-oestrogen (MER-25) blocked the sterilizing action of testosterone, whilst 19-hydroxytestosterone, an intermediate in the aromatization of testosterone to oestradiol (Fig. 1.2), is as potent a sterilizing agent as testosterone in the neonatal female rat (McDonald and Doughty, 1974).* It can be asked, however, why androgenization does not occur naturally if the foetus is exposed to oestrogens of maternal origin; Reddy *et al.* suggest that the specific binding of oestrogen to a carrier protein, present in embryonic blood, inhibits its action on the hypothalamus; androgens are not bound to this protein (Nunez *et al.*, 1971; Soloff *et al.*, 1972) so they would have access to the hypothalamus where they would be converted to oestrogen free from binding protein.†

Feminization of the Male

Feminization of the male cannot be achieved by mere transplantation of an ovary or injections of ovarian hormones; first the male must be castrated within a few days of birth (Yasaki, 1960) but the ovary may be transplanted later. However, treatment of intact males with oestrogens is not without effect; the testes develop poorly, the accessory reproductive organs, such as the prostate and seminal vesicles, are deficient, and the animals lack the normal sex drive (Mayer and Thevenot-Duloc, 1962; Harris and Levine, 1965). In the rat, this oestrogen-treatment must be carried out before 14 days if these effects are to be manifest. Thus, as Harris and Levine concluded, rats of both sexes are born with a sexually undifferentiated system which is of the female pattern; during the first few days of neonatal life this pattern becomes fixed in type in the female. In the male, during the first few days of life, the normal mechanisms underlying the future pattern of sexual behaviour and gonadotrophin secretion are organized by the internal secretions of the immature testes into those of the normal male.

* Oestrogens are capable of reproducing other "androgen actions" in brain differentiation, including timing of puberty (Eckstein *et al.*, 1973), gonadotrophin control (McCann and Ramirez, 1964) and initiation of sexual behaviour in male castrates (Feder *et al.*, 1974). In this last situation, oestrogen does not restore the complete copulatory act to the rat, so that intromission and ejaculation are rare whereas testosterone administration permits these. If didydrotestosterone, which by itself is inadequate to initiate sexual behaviour in castrates, and is not aromatized, is administered with oestradiol then the complete reproductive act is achieved. Oestrogen is unable to maintain the spines of the glans penis, so that an androgen is necessary for this. Thus, if the aromatization hypothesis is correct, the oestrogen formed from testosterone must act in conjunction with the non-aromatizable androgen, DHT, to maintain full sexual performance.

† The binding protein, with very high affinity for oestrogen and negligible affinity for testosterone, has a high concentration in embryonic serum; at 21 days old, the binding is reduced to about one-tenth, and in the pregnant female it is negligible. Corticosterone also binds but not so strongly.

Primates. The primate seems to be less influenced by pre- and early post-natal administration of androgens; thus cyclicity in the female monkey was unaffected by large doses of testosterone at the day of birth (Treloar *et al.*, 1972). Again, the LH-surge induced by oestrogen can be obtained in male monkeys gonadectomized in adulthood, although this is not possible in similarly treated gonadectomized rats. Thus, exposure of the hypothalamus to androgens in the foetal and post-natal period does not prevent the differentiation of the control system in primates. This does not mean that pre-natal treatment with androgens is without influence on the primate; the female monkeys studied by Goy and Resko (1972) showed pronouncedly masculine behaviour; the mothers were given repeated injections of testosterone beginning on Day 24 of pregnancy.

Sex Differentiation in Utero

The androgenizing effects of implants of the male gonads or of injections of androgens n newborn females, observed in rats and mice, and the development of female hypothalamic characteristics in males by castration at birth, were possible to demonstrate because in these species sex differentiation is not irreversibly completed before birth. In species born at a later stage of development the experimental interferences must be carried out *in utero*. However, even in such species as the rat and mouse, sexual differentiation has already proceeded a long way by birth, even if the typically masculine hypothalamus has not developed; thus, the gonads have developed sufficiently for it to be possible to recognize testes and ovaries, and the question arises as to whether this process of sexual differentiation has been under the influence of the gonadal secretions from a very early stage. The pioneering studies of Jost (see, for example, Jost *et al.*, 1973 for a review) have shown that this is, indeed, true. The *genetic* sex is, of course, determined by the presence of the Y-chromosome in the male, and this leads to the first stages of differentiation of the gonadal primordium so that in the male embryo at a stage when there is no histologically identifiable development of the presumptive ovary in the female there are clear signs of testicular differentiation, so that all germ cells are enclosed in clear compact and cord-like masses of cells which are the early Sertoli cells. By contrast, at the same period (about 13 days in the rat) the presumptive ovary of the female foetus is completely undifferentiated.

Intra-uterine Castration. Studies on intra-uterine castration of males, or the injections of androgens into females, have shown without doubt that female organogenesis, i.e. the development of vagina, uterus, and so on, results from the absence of the foetal testis, the presence or absence of developing ovaries being unimportant; thus the testis is a very important organ of internal secretion. Femaleness, then, depends on an intrinsic organization of the primordia and is obtained *in vitro* in the absence of hormones (Jost and Bergerard, 1949; Picon, 1969). Thus every structure would be feminine if not prevented from being so by testicular hormones.

Male and Female Homologies. According to Jost the same control might well be exerted on the development of the ovary, i.e. the genetic absence of the Y-chromosome might prevent the development of the male gonad and thus allow the development of the ovary. Figure 2.31 shows the homology between the male and female reproductive systems from the undifferentiated condition when there are

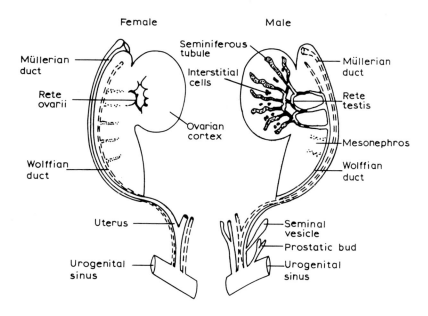

Fig. 2.31. Composite scheme showing some homologies in the development of male and female organs from the undifferentiated condition characterized by the presence on the mesonephros of a double set of ducts (Wolffian and Müllerian ducts) and of a gonadal primordium. In the female or in the castrated male rabbit foetuses, only the Müllerian ducts persist, and a female system develops. In males the testis is responsible for the disappearance of the Müllerian ducts and for the development of the male characters. The differentiation of the urogenital sinus and external genitalia is not shown. (Jost, *Rec. Progr. Hormone Res.*)

present a double set of ducts—Wolffian and Müllerian—and a gonadal primordium. In the female, or the castrated male foetus, only the Müllerian ducts persist so that a female system ensues. In males the testis is necessary for the disappearance of the Müllerian ducts and for the development of male characteristics.

Exogenous Androgens. Studies employing testosterone or methyltestosterone, given to castrated male rabbit foetuses showed that these androgens could replace the testis in masculinizing most of the foetal structures but were not able to induce retrogression of the Müllerian, or female, ducts; hence an additional factor than androgens is probably necessary, called by Jost a Müllerian inhibitor.

Testicular Feminization. This could explain the pathological human condition of testicular feminization; here the Müllerian ducts have disappeared, but the genital tract has not been masculinized, a condition that could be attributed to the failure to secrete androgens (or to an androgen-insensitivity of the tissues), whilst the Müllerian inhibitor was secreted. A similar condition is obtained when foetal rabbits are treated with the anti-androgen, cyproterone acetate; in the male foetuses all male characteristics were prevented from developing, but the Müllerian ducts were absent (Elger,

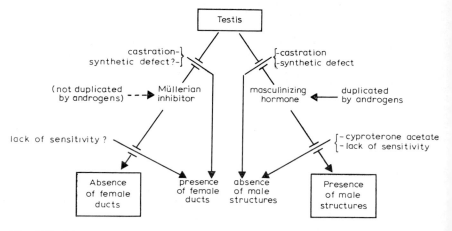

Fig. 2.32. Scheme summarizing the testicular control of the differentiation of the sex-ducts and sex characters. Some conditions capable of interfering with the normal testicular activity are indicated (crossing arrows). The data concerning cyproterone acetate refer to experiments made on rabbit foetuses; in rats the effect is not the same. (Jost, *Rec. Progr. Hormone Res.*)

1966).* The general scheme for intra-uterine development with some of the ways in which it can go wrong is illustrated in Fig. 2.32.

Electrical Stimulation of the Hypothalamus

The importance of the hypothalamus for ovulation was demonstrated by Harris (1948) who was able to study the unanaesthetized animal by implanting the secondary coil of a stimulator in the hypothalamus and, after the animal had recovered, stimulation was brought about by passing a current through a large primary coil surrounding the animal's head. The region that induced ovulation was the tuber cinereum (Fig. 2.33), whereas stimulations of the pars distalis, pars intermedia or infundibular stem of the pituitary were without effect. The close relation to the median eminence of the most sensitive regions suggests that Harris was stimulating the small-celled neurones whose secretions are emptied into the portal system. Later Harris *et al.* (1971) were able to identify the origin of the neurones terminating in the primary

* The consequences of injecting testosterone into pregnant females on their *female* progeny are striking; in rats the clitoris becomes essentially a male phallus, and when, in adulthood, these hermaphrodite-type females are injected with testosterone they will exhibit a pattern of masculine coital behaviour indistinguishable from that of a genetic male. In monkeys, too, the female progeny of animals treated with androgens exhibit essentially male behaviour. The freemartin is a genetic female cow that has developed *in utero in parabiosis* with at least one male twin. The interchange of blood means that the female gonads develope under the influence of male steroids. They are, in fact, females that have been partly masculinized.

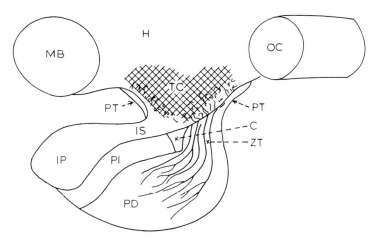

Fig. 2.33. Diagram of a sagittal section through the hypothalamus and pituitary gland of a rabbit. C, connective tissue trabeculae; H, hypothalamus; IP, infundibular process; IS, infundibular stem; MB, mamillary body; OC, optic chiasma; PD, pars distalis; PI, pars intermedia; PT, pars tuberalis; TC, tuber cinereum; ZT, zona tuberalis. The area cross-hatched represents the region in which electrical excitation was followed by ovulation. The diagram also illustrates the hypophysial portal vessels, starting as a multitude of sinusoidal loops in the tuber cinereum, uniting to form the wide trunks of the portal vessels in the zona tuberalis, which pass caudo-ventrally and break up into the sinusoids of the pars distalis. (Harris, *J. Physiol.*)

capillary plexus of the portal system; they placed recording electrodes in different hypothalamic regions and stimulating electrodes at the junction of the median eminence and the pituitary stalk. Thus by stimulating antidromically the source of the neurones at the junction could be found. In fact, most of the antidromically activated units were in the arcuate nucleus, less than 0·8 mm from the midline. Some were found in the dorsal premamillary, ventromedial and anterior peri-ventricular nuclei.

Barbiturate Block

In experimental studies designed to localize precisely the sites necessary for evoking the ovulatory response, use is made of the finding that treatment of the rat with a barbiturate such as pentobarbitone (Nembutal) blocks the oestrous cycle by preventing ovulation. As Fig. 2.34 shows, administration of Nembutal, before the critical period, blocks the pre-ovulatory surge of LH-secretion. In this condition electrical stimulation can cause ovulation and the continuation of the oestrous cycle (Critchlow, 1958). The effect of Nembutal is to block neural mechanisms involving polysynaptic pathways, so we may assume

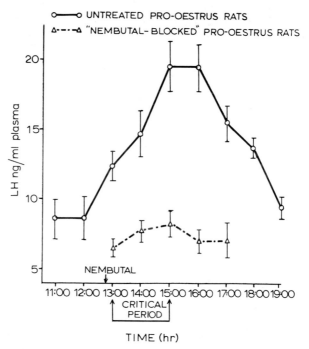

Fig. 2.34. Illustrating the block of ovulation by Nembutal as revealed in the inhibition of the pre-ovulatory surge of LH in the blood. (Cramer and Barraclough, *Endocrinology.*)

that the electrical stimulation avoided these complex pathways, exciting the late stage.

Ovulating Centres. The regions of the hypothalamus that may induce ovulation in the barbiturate-blocked rat, when stimulated, include both the hypophysiotrophic area, typified by the arcuate nucleus, and the preoptic anterior hypothalamic (PO/AH) area (Teresawa and Sawyer, 1969a). When the PO/AH area of the hypothalamus was stimulated, and records were taken from the arcuate-median eminence area, there was an increase in activity here; when the stimulus was carried out during the "critical period" of pro-oestrus, all the animals that showed this increase in electrical activity ovulated, whereas those that failed to respond also failed to ovulate. After ovulation the system became refractory to PO/AH stimulation (Teresawa and Sawyer, 1969b).*

---

* Everett and Radford (1961) showed that, when they stimulated with steel electrodes, the actual stimulus seemed to be the deposition of $FeCl_3$ at the electrode site, since the same current passed through platinum electrodes failed to excite. A microinjection of $FeCl_3$ has

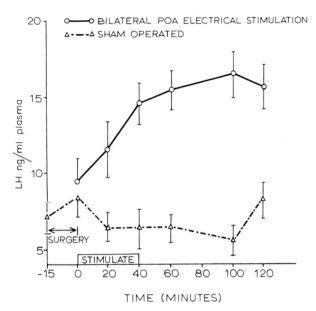

Fig. 2.35. Effects of stimulation of the PO/AH area in rats on plasma LH. Also shown are effects of surgery and sham electrode placement into the PO/AH of Nembutal-blocked pro-oestrous rats. (Cramer and Barraclough, *Endocrinology*.)

Requirement of Prolonged Stimulation

The result of electrical stimulation is first to induce the discharge of LH; this is shown by Fig. 2.35 where the preoptic anterior hypothalamic region (PO/AH) was stimulated continuously over a period of 40 minutes. A remarkable feature of the response to electrical stimulation, whether carried out by electrochemical irritative lesions or by true electrical pulses, as in these experiments, is the long delay, of the order of 40 minutes, between the beginning of stimulation and the peak response, a delay that may represent the time required for mobilization of releasing factor or of biogenic amines, such as dopamine (p. 478). At any rate, it appears that, to obtain an "ovulatory quota"

the same effect, namely causing ovulation in the pentobarbitone-blocked rat. These authors therefore speak of "electrochemical stimulation" when the result is a deposit of Fe at the electrode-site, and its usefulness consists in the fact that the irritative lesion, which constitutes the stimulus, lasts for some time. It had generally been considered that the ferrous ions caused a state of excitation in the hypothalamic neurones; however, there is no doubt from Dyer and Burnet's (1976) study that deposition of ferrous ions actually depresses the spontaneous activity of central neurones, either in the cerebral cortex or rostral hypothalamus. In a more recent study Dyer (1978) has shown that electrochemical deposition of Fe^{2+} into the paraventricular nucleus or pituitary-stalk fails to cause release of oxytocin, i.e. it does not stimulate peptidergic neurones of the hypothalamus.

of gonadotrophin, probably some 40 minutes of stimulation are required. Another factor is the area over which the stimulus is spread; if it is not wide enough there is insufficient release to induce ovulation.*

Release of Releasing Hormone

The concentration of the gonadotrophic releasing decapeptide, LHRH, increases in the blood in association with the ovulatory LH-surge, indicating the primary hypothalamic activity. Electrical stimulation of the hypothalamus, e.g. the medial preoptic area, causes an increase in LH-releasing hormone concentration in the pituitary stalk blood, i.e. the blood passing in the portal system to the pituitary (Fink and Jamieson, 1976). Blockade of this pre-ovulatory surge may be achieved by injection of anti-LHRH serum in rats (Arimura et al., 1974) when administered on the morning or afternoon of pro-oestrus, and, as Table III shows, this is accompanied by inhibition of ovulation.

TABLE III

Effects of anti-LHRH on ovulation and serum gonadotrophin levels (Arimura et al., 1974)

Injection	No. injected	No. ovulated	Mean No. Ova.	Serum LH (ng/ml)	Serum FSH (ng/ml)
Normal rabbit serum	4	4	$12 \pm 1{\cdot}3$	$58 \pm 13{\cdot}2$	720 ± 108
Anti-LHRH	4	0	0	$0{\cdot}8$	145 ± 21

The FSH-surge is also blocked, emphasizing the probable identity of the releasing hormones or at any rate, if there are different hormones, their similar antigenic determinant, which is the C-terminal tetrapeptide of the LHRH decapeptide, Leu-Arg-Pro-Gly. NH$_2$.

Morphine Block

Barraclough and Sawyer (1955) found that morphine, administered as a single dose just prior to the critical period on the afternoon of pro-oestrus, blocked ovulation that evening; the block could be overcome by stimulation of the median eminence, and it was concluded that morphia was blocking the neurogenic release of LH and FSH from the

---

* In ovariectomized rats there is an episodic release of LH, brought about by episodic release of LHRH; Gallo and Osland (1976) were able to inhibit this release by stimulation of the arcuate nucleus. When the rats were primed with oestrogen, however, the same electrical stimulation caused release of LH. This is one further example of the effect of steroid hormones on sensitivity of the hypothalamo-hypophyseal axis to steroid hormones.

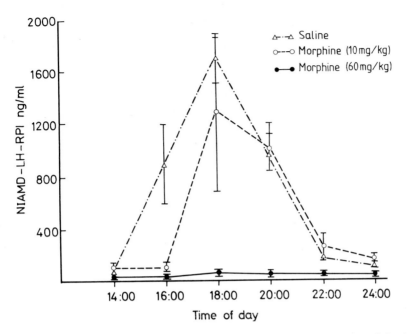

Fig. 2.36. Time-course of plasma LH levels following saline or morphine injection at 1400 h on the afternoon of pro-oestrus. (Pang *et al.*, *Endocrinology.*)

anterior pituitary. Figure 3.36 illustrates the results of a recent study by Pang *et al.* (1977) where the steep pre-ovulatory rise in plasma LH is completely blocked by a dose of 60 mg/kg of morphine, whereas the smaller dose of 10 mg/kg has only a small effect. In general, when ovulation was blocked by a given dose, LH secretion was completely suppressed; secretion of FSH was only partly suppressed. The specific anti-opiate naloxone antagonized the effects of morphine, suggesting that the morphine was acting at a neural site, and since opiate receptors are present in the hypothalamus at the same regions as those governing gonadotrophin release, it is very likely that morphine's action is on the hypothalamus. This is further supported by the observation that LHRH caused release of LH and ovulation in morphine-blocked rats.

FSH and Prolactin

The studies illustrated by Fig. 2.6 (p. 61), indicate that FSH and prolactin participate in the pre-ovulatory surge of pituitary hormones. However, when the hypothalamus was stimulated in different regions,

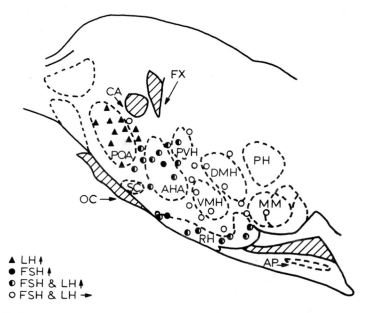

Fig. 2.37. Location of electrodes for stimulating rat's brain projected on a para-sagittal section. Symbols indicate the type of hormonal response to electrical stimulation. OC = optic chiasm; CA = anterior commissure; SC = suprachiasmatic nucleus; FX = fornix; PVH = paraventricular nucleus; AHA = anterior hypothalamic area; DMH = dorsomedial nucleus; VMH = ventromedial nucleus; RH = arcuate nucleus; PH = posterior hypothalamic nucleus; MM = medial mamillary nucleus; AP = anterior pituitary; POA = preoptic area. (Kalra *et al.*, *Endocrinology*.)

it was impossible to evoke prolactin secretion (Kalra *et al.*, 1971).*
Stimulation of the median eminence-arcuate region, and of the preoptic anterior hypothalamus (PO/AH) caused ovulation and a 20–50-fold rise in LH secretion; it was interesting that FSH did not rise with the more rostral PO/AH stimulation whereas stimulation of an anterior hypothalamic area (Fig. 2.37) could cause release of FSH only. Since stimuli that only caused a rise in LH were effective in causing ovulation, it seems that release of the other gonadotrophins is not necessary for this.

Androgenized Rat

When the same regions were stimulated in the androgen-sterilized rat, ovulation could not be induced (Barraclough and Gorski, 1961); how-

* Quadri *et al.* (1977) obtained an increased secretion of prolactin in the ovariectomized monkey on electrical stimulation of the medial basal hypothalamus (MBH), i.e. the ventromedial complex; this was enhanced by treatment with oestradiol.

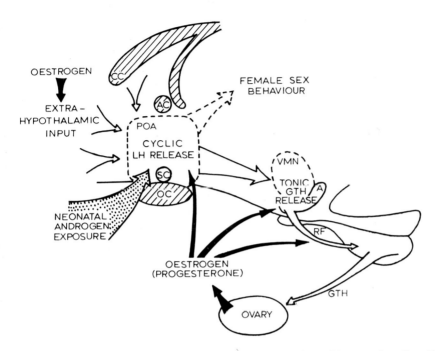

Fig. 2.38. Illustrating dual control mechanisms for gonadotrophin secretion. Localization is on a parasagittal section of the brain. Dark arrows indicate oestrogen or oestrogen to progesterone ratios. Stippled arrow indicates the permanent action of neonatal androgen exposure. Abbreviations: A, arcuate nucleus; AC, anterior commissure; CC, corpus callosum; GTH, gonadrotrophin; OC, optic chiasm; POA, preoptic area; RF, releasing factors; SC, suprachiasmatic nucleus; and VMN, ventromedial nucleus. (Gorski, *Amer. J. Anat.*)

ever, if the animal was treated with progesterone—the animal was said to be progesterone-primed—then a significant difference between the two hypothalamic sites emerged; stimulation of the ventromedial-arcuate complex, i.e. in the hypophysiotrophic area, caused ovulation, but stimulation of the preoptic area did not.*

Tonic and Cyclic Secretions

On the basis of these experiments, Gorski (1970) proposed a dual control over pituitary gonatrophin secretion. A basic tonic control, common to male and female, would be exerted through the ventro-

---

* This is true when the rat has been sterilized by a high dosage of androgen; when sterility is induced with a low dosage, 10 μg compared with 1·25 mg, then progesterone priming allows stimulation of the PO/AH area to induce ovulation as well as stimulation of the arcuate region (Gorski and Barraclough, 1963).

medial arcuate complex; this would be adequate to sustain normal gonadotrophic function but would be unable to exert the cyclic stimulus to gonatrophic hormone release required for ovulation, i.e. the pre-ovulatory surge of LH would be missing. In the androgenized female this region is permanently blocked; in the progesterone-primed animal it is assumed that the ventromedial-arcuate region was able to exert its influence, an influence that was suppressed by the persistent oestrogen secretion that occurs in the Nembutal-blocked animal, a suppression that is relieved by the antagonistic action of progesterone. The basic scheme is illustrated schematically by Fig. 2.38; here feedback loops have been incorporated suggesting that the level of oestrogen and progestin in the blood may control the activities of the hypothalamic centres and the pituitary gland itself. These feedback mechanisms will be discussed later. On the basis of this dual control system we should expect that destruction of the PO/AH system would leave basic secretion of the gonatrophic hormones unaffected, but abolish the cyclic rise in LH secretion.*

Hypothalamic Lesions

The classical study of Hillarp (1949) demonstrated that lesions in the anterior preoptic area caused sterility, and this was associated with abundant follicles in the ovaries, but the absence of corpora lutea, i.e. the sterility was probably the result of a failure of the cyclic mechanism to operate. Small localized lesions caudal to the paraventricular nucleus caused the same effects, and these were probably due to interruption of the pathway towards the median eminence. Lesions in the median eminence itself that destroyed the connections between the hypothalamus and anterior hypophysis caused genital atrophy in guinea-pigs, indicating the cessation of tonic release of gonadotrophic hormone (Dey, 1941, 1943).

De-Afferentation

Complete de-afferentation of the hypophysiotrophic area or MBH, as a result of which it cannot be affected by input from the PO/AH region (Fig. 2.39), leaves the target organs for the gonadal hormones, e.g. the gonads and accessory organs such as the uterus and seminal

---

* More recently Kubo et al. (1975) have pointed out that the sensitivity of the ovary to LH is greatly reduced by androgenization, whereas the sensitivity of the pituitary to LHRH is unaffected. Thus, measurement of the effects of androgenization by the ovulatory response to electrochemical stimulation is an unsure mode of demonstrating a reduced sensitivity of the preoptic hypothalamic area (POA). In fact, when plasma levels of LH in response to POA stimulation were measured, androgenization had no effect, so that we must look elsewhere for the cause of the reduced ovulatory response.

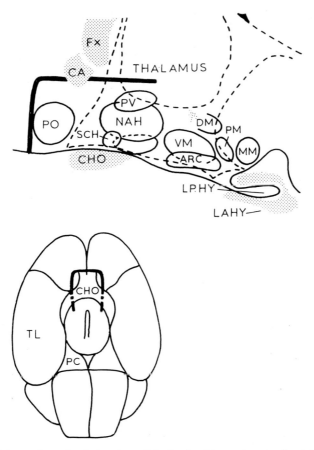

Fig. 2.39. Schematic sagittal drawing of the brain, illustrating deafferentation of the preoptic area and the same cut as seen from the base of the brain. Heavy lines indicate the cut. Fx, fornix; LAHY, anterior lobe of hypophysis; LPHY, posterior lobe of hypophysis; MM, medial mammillary nucleus; NAH, anterior hypothalamic area; PC, cerebral peduncle; PM, premammillary nucleus; PO, preoptic area; SO, supraoptic nucleus; TL, temporal lobe. (Flerko, *The Hypothalamus.*)

vesicles, unaffected, indicating that tonic secretion of the gonadotrophic hormones continues (Halasz and Gorski, 1967) whilst the cyclic activity leading to the LH-surge and ovulation is abolished giving a state of persistent oestrus. When the "hypothalamic island" produced by this deafferentation was extended rostrally to include the PO/AH region, cyclic ovulation was restored (Köves and Halasz, 1970). These authors point out that cyclic activity of the PO/AH region does not depend on an input from other parts of the brain.

Primates. In the primate, however, surgical disconnection of the medial basal hypothalamic region (MBH) from the preoptic anterior region (PO/AH) did not prevent the occurrence of spontaneous and oestrogen-induced ovulating surges of gonadatrophins (Krey *et al.*, 1975). Norman *et al.* (1976) failed to confirm this, bilateral lesions of the PO/AH region abolishing spontaneous and oestrogen-induced LH-surges as in rats. Hess *et al.* (1977), however, have confirmed the earlier work, making a hypothalamic peninsula, posteriorly continuous with the brain-stem, by aspirating the cerebral hemispheres and all brain structures anterior and dorsal to the optic chiasma. Subcutaneous injections of oestradiol benzoate induced surges of LH and FSH.

Krey *et al.* pointed out that the cyclicity in the rat was linked to the light-dark periodicity, whereas in the primate this is not true, so that it could be that the importance of the preoptic hypothalamic area in the rat is due to this linkage.*

Discrete Lesions. Plant *et al.* (1978) placed lesions in the arcuate nucleus of monkeys by passing radio-frequency current through stereotactically placed electrodes in the rhesus monkey. When the MBH lesions were confined to the arcuate nucleus, there was cessation of tonic secretion of LH and FSH and the positive feedback of oestradiol on LH release was also blocked. So far as prolactin secretion was concerned it appeared that when the lesions were strictly confined to the arcuate nucleus, secretion of this hormone was unaffected, whilst extension of the lesion to regions anterior or dorsal to the arcuate nucleus produced well defined rises in secretion, presumably through blocking of PIF release.

Reflex Ovulation. The copulatory induced ovulation in the rabbit can be inhibited by hypothalamic lesions comparable with those employed by Halasz to inhibit cyclic ovulation, namely a virtually complete de-afferentation of the medial basal hypothalamus (MBH); if only semicircular cuts were made, inhibition did not occur, suggesting that a mechanism of widely dispersed centres gives rise to stimuli converging on the MBH from all or nearly all directions (Voloschin and Gallardo, 1976).†

---

* The reader is referred to an interesting review of experiments on the control of gonadotrophin secretion in the monkey by Knobil (1974); he mentions that male monkeys respond to positive feedback of oestrogen with an LH-surge similar to that in females, in marked contrast to the rat. He concludes: ". . . . primates, in contrast to a number of other mammals, have evolved simpler solutions to the physiological problems posed by control of gonadal function, as members of this genus become decreasingly dependent on the dictates of their environment for the timing of reproductive processes".

† These authors describe a variety of insults that failed to inhibit copulation-induced ovulation, including anaesthesia of the vulva and vagina, removal of sacral cords, and so on.

Medial Preoptic Area

When Clemens *et al.* (1976) made hypothalamic lesions in rats that were restricted to the medial preoptic area (MPOA) they obtained a state in which the animals showed repeated periods of pseudopregnancy; between pseudopregnancies, on the day of oestrus, ova were detected in the oviducts, so that ovulation, *per se*, had not been prevented by this anterior hypothalamic type of lesion, and we may conclude that this region is concerned with the *timing* of the ovulation cycle. Clemens *et al.* have suggested that the MPOA might normally prevent daily rises in prolactin secretion in the rat, through an inhibitory influence, and that this might permit the surges to be restricted to the 4- or 5-day cycle. Lesions extending into the anterior hypothalamus resulted in persistent oestrus, as found by others.

Unit Activity

By what may now be described as classical techniques, recordings may be made of single neurones—units—in the hypothalamus by stereotactically inserted electrodes. As Cross (1973) has emphasized, it is important to identify the relations of a hypothalamic unit with another region with which a connection is supposed; thus, according to the current hypotheses, the preoptic units concerned with gonatrophin secretion might make synaptic connections with the hypophysiotrophic area, which may be regarded as part of the final common pathway for control over the pituitary.

Antidromic Identification

By inserting a stimulating electrode in this hypophysiotrophic area, as in Fig. 2.40 antidromic impulses should be recorded in those neurones of the preoptic area that make connection in the stimulated area or whose axons pass through. In this way, Dyer and Cross (1972) established the existence of neurones in the PO/AH region that were directly related to the arcuate-ventromedial area. The neurones that connected directly were described by Dyer (1973) as Type A cells, to distinguish them from Type B, which were activated orthodromically by the ventromedial stimulus (Fig. 2.40); Type C units failed to respond. That the neurones stimulated in the ventromedial nucleus were concerned with reproductive activity was demonstrated by electrical stimulation, when increases in mammary pressure were obtained equivalent to the intravenous administration of oxytocin. The increase in firing-rate previously observed in neurones of the preoptic/anterior hypothalamic (PO/AH) region was confined to Types B and C.

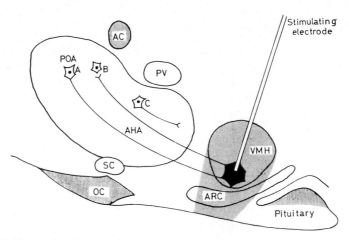

Fig. 2.40. Illustrating three main types of neurone from which recordings were made in the preoptic and anterior hypothalamic areas. The light-hatched area represents the probable maximal extent of effective stimulation for all rats. The dark-hatched area at the tip of the electrode shows the effective spread of 1 mA biphasic pulses. (AC, anterior commissure; AHA, anterior hypothalamic area; ARC, arcuate nucleus; OC, optic chiasm; POA, preoptic area; SC, suprachiasmatic nucleus; VMH, ventromedial nucleus.) (Dyer, *J. Physiol.*)

Follicle-Stimulating and Luteinizing Hormones

These two gonadotrophic hormones are secreted, apparently, by the same basophilic cells of the anterior pituitary in response to a single releasing hormone and, in general, they are released simultaneously so that in the oestrous cycle, for example, changes in concentration of LH are usually paralleled by changes in concentration of the other, FSH. Classically, the follicle stimulating hormone, FSH, is said to maintain follicular growth and induce secretion of oestrogens from the follicle cells, and in accordance with this view, atrophy of the follicles and failure of secretion of oestrogen are consequences of hypophysectomy or of localized lesions in the median eminence. The luteinizing hormone, LH, is considered to be primarily involved in the induction of ovulation and the associated development of the corpus luteum, characterized by changes in the morphology of the granulosa cells indicative of secretion of progesterone, which becomes the dominant steroid hormone during the progravid phase of the ovarian cycle, promoting changes in the uterus that prepare it for implantation.

Isolated Granulosa Cells

However, this rigid separation of function is probably not valid, both hormones showing changes in secretion in any phase of oestrus, so that it is better to speak of LH- or FSH-domination rather than of LH- or FSH-action. This is exemplified by the study of granulosa cells of the monkey's ovary in tissue-culture; these transform spontaneously in the culture to luteal cells provided they are harvested from large pre-ovulatory follicles; the transformation is recognized cytologically by accumulation of eosinophilic granules and lipid droplets, an increase in size, and the secretion of large amounts of progesterone. This spontaneous luteinization, independent of addition of any trophic hormone to the culture-medium, suggests that the granulosa cells had already been programmed for luteinization before being removed from the ovary, probably as a result of the rise in circulating LH taking place before removal, since it would only occur if the cells had been harvested at the 12th to 14th days of the ovarian cycle. It was only the large follicles that developed in this way, cells from small follicles not luteinizing. Moreover, these large follicle cells would respond to either LH or FSH alone, the hormones prolonging the life of the culture. Secretion of progesterone by the cells from small follicles was always very low by comparison with that from the large follicles; secretion was promoted by LH and FSH, but the best effect was obtained by a combination of LH and FSH.

Secretion of Oestrogen

The association of FSH exclusively with the secretion of oestrogen by the ovary is incorrect, the classical study of Greep *et al.* (1942), confirmed by Lostroh and Johnson (1966),* having shown that in the hypophysectomized rat FSH alone would not support oestrogen secretion, but a combination of this with LH would. As we shall see, this dual control is related to the different points of control over steroid biosynthesis exerted by the gonadotrophs.

Separate Identities

So interwoven are the actions of the two gonadotrophins that the separate chemical identities of the two have been questioned; and it has been suggested that the preparations from the anterior pituitary, called FSH and LH, are, in fact, preparation of a single hormone. Alternatively, it has been suggested that the secretions of the two hormones are so tightly coupled that they behave as one. This latter hypothesis, if either must be adopted, is preferable to the former since there is little doubt that the materials separated from the pituitary, and assayed by methods that are specific for FSH-action and LH-action,† are different substances as revealed by their physical properties and their amino-acid compositions. Furthermore, changes in the ratio of

---

* Greep *et al.* (1942) found that in hypophysectomized immature female rats thylakentrin (FSH) brought about growth of Graafian follicles without accompanying luteinization and effected no growth of the uterus or cornification of the vaginal epithelium. All these manifestations of oestrogen action appeared when metakentrin (LH-ICSH) was administered. Lostroh and Johnson (1966) found that FSH alone could not induce secretion of oestrogen; it increased ovarian weight, growth of follicles and induced antrum and cavity formation in developing follicles.

† The early techniques of assay quite reasonably employed the measurement of the two putative functions, namely follicle stimulation for FSH and, for LH (or ICSH in the male) interstitial cell repair. These assays, because of their cumbersomeness and insensitivity, were replaced by the HCG-augmentation test for FSH, and the ovarian ascorbic acid depletion (OAAD) test for LH. In the former (Steelman and Pohley, 1953) the increase in ovarian weight induced by the test material, together with a fixed dose of human chorionic gonadotrophin (hCG), is measured in immature assay rats. In the OAAD test of Parlow (Parlow; 1961; Parlow and Reichert, 1964), immature rats are sequentially treated with pregnant mare serum gonadotrophin (PMSG) and hCG, which causes heavy luteinization of the ovaries; treatment with LH results, within 4 hr, in a depletion of the ovarian ascorbic acid, the extent being proportional to the LH-content of the sample. Because of the insensitivity of these assays, modern work, especially on plasma levels, depends on the radioimmunoassay techniques.

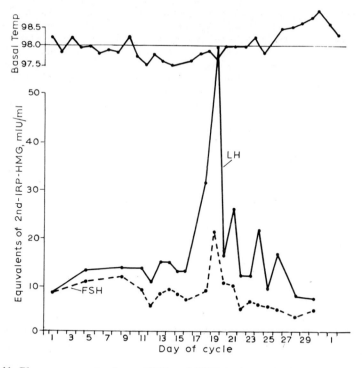

Fig. 2.41. Plasma concentrations of LH and FSH during a single menstrual cycle in a woman. (Midgley and Jaffe, *J. clin. Endocrinol.*)

FSH/LH within the pituitary gland, or in the plasma, take place normally during the oestrous cycle; thus Fig. 2.41 shows the changes in plasma concentrations of FSH and LH during a single menstrual cycle in a woman; it is quite clear that, although the pre-ovulatory peak is shared by both, the concentration of LH can vary independently of that of FSH. Again, Goy's (1973) study of the effects of testosterone implants into monkeys showed that these, by a negative feedback, could reduce the concentration of LH to zero, but left the concentration of FSH unchanged. Yet again, Harrington and Elton (1969) showed that FSH alone will cause ovulation in chlorpromazine-treated rats, but this, of course, could have been due to the presence of LH; however, treatment with chymotrypsin leaves the activity of an FSH preparation unchanged whilst destroying LH activity, and Harrington et al. (1970) showed that, when they treated their FSH preparation with chymotrypsin, the blockage of ovulation, caused by previous injection of chlorpromazine prior to the "critical period" in rats, was overcome.

More recently, Tsafriri et al. (1976) have removed all LH from their standard FSH preparations by treatment with an anti-LHβ, i.e. an antiserum to the β-subunit of the luteinizing hormone. With this, they were able to overcome Nembutal-block in pro-oestrous rats and also the block caused by treatment with an anti-releasing hormone preparation.

Anti-FSH. A possible argument against the separate identity of FSH is the finding of Schwartz *et al.* (1973) that an anti-FSH serum injected into female rats failed to influence the events in the oestrous cycle, blocking neither ovulation nor oestrogen secretion. They pointed out, however, that the general idea that the secretion of gonadotrophic hormones should be considered exclusively in relation to the cycle in which they were secreted might well be wrong, so that the maturation of the follicles that ovulated in a given cycle might well have been "programmed" by the secretion of FSH in the previous cycle. In support of this there was some evidence that, after multiple injections of the antiserum, some follicles did not ovulate normally.

LH-Surge and Succeeding Cycle. This point has been taken up by Rani and Moudgal (1977) who found that an anti-FSH, given on the day of the pro-oestrous LH-surge, did, in fact, affect the process of follicle maturation in the *succeeding* cycle, although it did not affect the LH-surge. This could have been due to the persistence of the anti-serum through the next cycle, but when the persisting antiserum was removed immediately after it had exerted its effect, by giving exogenous FSH, maturation was still inhibited, as revealed by failure of ovulation at the next cycle. Thus, the surge of gonadotrophins, LH and FSH, is apparently necessary not only to induce ovulation but also to permit maturation of the follicle required for the next cycle. Interestingly, LH was necessary for maturation as well as FSH.

Combined Actions. In general, Schwartz (1974) has concluded that both FSH and LH are, indeed, separate functional and chemical identities;* both FSH and LH are needed continuously at low levels to enhance follicular development, FSH being responsible for the earlier stages of maturation with LH being necessary for the later stages when oestrogen secretion occurs. So far as ovulation is concerned, either LH or FSH can induce ovulation when presented as a "surge", although only LH causes an acute rise in progesterone secretion. As to why two hormones are involved, having such strong overlap, Schwartz argues that the FSH is of value in exerting a "holding" operation, providing a continuous supply of follicles capable of responding to an episodic release of the ovulating hormone. Thus, the system is always "go" except during seasonal anoestrus, and needs only the induction of a surge of LH to trigger ovulation (Schwartz, 1974).

Aromatization

We have alluded to the possibility that androgens, such as testosterone and androstenedione, are converted by the ovary to oestrogen and thus are able to fulfil an oestrogenic role in the female; and this may well be the cue to the combined requirement for FSH and LH in, say, the secretion of oestrogen in the female. Thus, neither purified FSH nor purified LH, administered alone to hypophysectomized rats, is capable of eliciting significant secretion of oestrogen, as manifest in stimulation of uterine growth; in combination the two gonadotrophs stimulate uterine growth.

---

* Especially important in this respect is to show that the two materials are immunologically different, since modern studies depend heavily on radioimmunoassay; according to Chen and Ely (1971), FSH, as determined by the HCG augmentation test, is immunologically separate from LH.
When separate anti-FSH and anti-LH antibodies are prepared and injected into female rats and rabbits, the effects are similar, the animals becoming anoestrous and exhibiting non-mating behaviour or sterile matings (Talaat and Laurence, 1969). The effects on the ovary are, however, different; anti-LH giving rise to a dearth of luteal tissue with evident follicular development, whilst anti-FSH prevented maturation of growing follicles, with persistent corpora lutea.

TABLE IV

Effects of FSH, LH, and androgens, alone and in various combinations, on uterine weights in hypophysectomized rats (Armstrong and Papkoff, 1976)

Exp. no.	Gonadotrophin (μg/day, 3 days)	Steroid (5 mg/day, 3 days)			
		None	Testosterone	Androstenedione	DHT
1	None	17·2 ± 0·4	80·2 ± 2·7		
	FSH-HP (2·5)	17·2 ± 0·6	101·7 ± 5·3		
2	None	17·2 ± 0·5	82·3 ± 2·1	59·0 ± 1·5	
	FSH-HP (2·5)	26·4 ± 3·9	111·1 ± 4·1	111·7 ± 3·7	
3	None	32·0 ± 2·4	91·4 ± 2·5		90·2 ± 1·9
	FSH-HP (2·5)	38·2 ± 8·3	118·0 ± 3·6		91·0 ± 4·5
4	None	25·7 ± 1·2		86·5 ± 8·1	
	FSH-HP (2·5)	25·8 ± 1·2		130·3 ± 4·8	
	LH-HP (1·0)	32·8 ± 2·6		85·2 ± 10·8	
	FSH-HP + LH-HP	48·4 ± 1·2			
5	None	27·7 ± 2·3	115·1 ± 6·1	91·5 ± 7·1	
	FSH-HP (2·5)	26·0 ± 1·7	126·3 ± 7·5	140·8 ± 7·0	
	NIH-FSH (50)		106·1 ± 0·9		
	NIH-FSH (100)		149·5 ± 0·5		

LH is able to promote steroid biosynthesis in many gonadal tissues by stimulating the conversion of cholesterol to pregnenolone; FSH has also been shown recently to be concerned in steroid synthesis, its point of attack, however, being the stimulation of the aromatization reaction in Sertoli cells (Dorrington and Armstrong, 1975), and in granulosa cells (Dorrington et al., 1975). Thus Dorrington et al. (1975) found that FSH would only induce synthesis of oestrogen in cultured granulosa cells if androgen was present in the medium, an effect not produced by LH. Since LH stimulates secretion of androgens by immature rat ovaries (Louvet et al., 1975),* the suggestion naturally arose that the synergism between the two gonadotrophs was made possible by these separate modes of attack; thus LH would stimulate synthesis of pregnenolone by the theca cells of the ovary (or the Leydig cells of the testis) whilst FSH would stimulate conversion of this precursor to oestradiol in the granulosa cells of the ovary, which have FSH-receptors (or in the Sertoli cells of the testis). Support for this "two-cell" theory of gonadotroph action *in vivo* is provided by the study of Armstrong and Papkoff (1976), who estimated oestrogen secretion in hypophysectomized rats by changes in uterine weight. As Table IV shows, FSH, given alone, had either no, or only a small, effect on uterine weight; if the effect of LH is to promote conversion of androgen to oestrogen, then giving testosterone or androstenedione in addition to FSH should promote growth; and this indeed happens. (We may note that the

* Louvet et al. (1975) found that LH (hCG) was behaving as an anti-oestrogen with respect to follicular maturation, behaviour which they suspected was due to induction of androgen synthesis; the effect of the hCG was blocked by the anti-androgen, cyproterone acetate.

androgens themselves promote uterine growth.) When a non-aromatizable androgen dihydrotestosterone (DHT), was given, there was no augmentation of the increased uterine growth when FSH was given.†

Androgen Secretion

El Safoury and Bartke (1974) found that FSH potentiated the action of small doses of LH in promoting synthesis of testosterone by hypophysectomized adult male rats, and they have suggested that in this way FSH could be involved in androgen secretion by the rat testis.

Feedback Phenomena

Examples of the influence of the gonadal steroid hormones on other parts of the body than their "true" target organs have already been seen, for example in the modification of the neonatal female hypo-thalamus that imprints a masculine, or non-cycling, character on it, through injections of testosterone or implantation of male gonads at a critical period; or, in males, the requirement of circulating androgens for the correct organization of their male behaviour (Harris and Levine, 1965). We must now examine the phenomena in more detail.

Trophic Hormones

This feedback is especially important in a system where the actual secretion of the hormone is under the control of a "trophic" hormone, which not only maintains the normal physiological state of the hormone secretor, e.g. Leydig cells of the testis, but also directly influences the rate of synthesis and secretion of its hormones. Let us imagine, first, a negative feedback directly on to the secreting organ, say the testis; when the testis secretes its testosterone, the concentration in the blood rises and this, by negative feedback on the testis, exerts a restraining influence on further secretion. Injections of exogenous hormone would raise the blood-level and reduce the rate of secretion even further. When the secreting organ is controlled by another gland, e.g. the pituitary, then an increase in secretion of the trophic hormone causes increased secretion of testosterone; if the consequent increase in blood concentration fed back on the pituitary in a negative fashion, the same

---

† Armstrong and Papkoff also studied the effects of the hormone combinations on ovarian weight, an index to trophic action; FSH, alone, caused an increase; combined with testosterone there was a greater increase in weight, due perhaps to the synthesis of oestradiol, which itself promotes granulosa cell proliferation. They found that the non-aromatizable DHT actually antagonized the effects of FSH on ovarian weight. Since the ovaries of immature rats contain abundant 5α-reductase activity, it is possible that DHT, synthesized in the ovary, causes inhibition of follicular development, acting as an antagonist to FSH. The authors suggest that the balance of the two opposing forces of oestrogen and DHT action on the FSH-response might determine whether granulosa cell proliferation continued, ultimately leading to ovulation, or stopped, leading to atresia.

effect would be achieved, namely reduced secretion of the trophic hormone, followed by reduced secretion of testosterone. Injection of exogenous testosterone has an inhibitory feedback action on the trophic gland, reducing the secretion of the trophic hormone, which reduces the secretion of the testosterone. Thus, the negative feedback, and the inhibitory influence of exogenous hormone, operate in the same way, reducing the secretory activity of the primary gland.

Inhibition of Trophic Gland. However, the exogenous steroid, testosterone, is doing more than this if applied repeatedly, since it inhibits the trophic gland's secretions, the effects of which are not only to promote secretion of the steroid hormone but to maintain the viability and synthetic capacity of the testis. Thus, the feedback may cause atrophy of the testis and prevent it from carrying out its normal secretory activity. Because of this action on the trophic aspects of the pituitary gland, the feedback processes have great practical importance since they permit the investigator, or the clinician, to modify the secretion of the pituitary hormones by altering the concentrations of

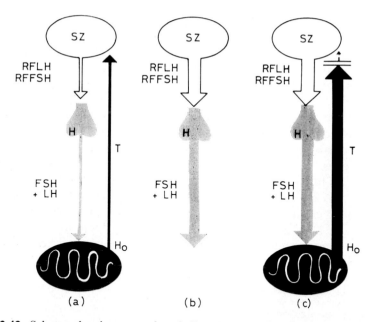

Fig. 2.42. Scheme showing normal and disturbed feedback mechanism in males. (a) Normal feedback mechanism; (b) feedback mechanism after castration; (c) feedback mechanism after administration of a "pure" antiandrogen (cyproterone). SZ, sexual centre; H, hypophysis; Ho, testis; RFLH, LH-releasing factor; RFFSH, FSH-releasing factor; LH, luteinizing hormone; FSH, follicle-stimulating hormone; T, testosterone. (Neumann and Schenck, *J. Reprod. Fert.*)

oestrogens, progestogens, or androgens in the body, either with naturally occurring steroid hormones or with synthetic substitutes. There is
little doubt, for example, that many anti-fertility agents exert their
effects primarily through this feedback on to the pituitary.

Gonadectomy

An obvious sign of feedback is in the response to castration, either
orchidectomy in the male or ovariectomy in the female. In both cases
there is a rise in the level of gonadotrophic hormones, an effect that can
be compensated by injections of the steroid hormone. The mechanism
of this rise is illustrated schematically in Fig. 2.42. Gay and Midgley
(1969) castrated adult male rats and, as Fig. 2.43 shows, the concentration of LH in the plasma rose rapidly by some 20–50 fold; a similar
rise occurs after ovariectomy in females (Fig. 2.44). In males, the response may be reversed by treatment with testosterone (Gay and
Denvers, 1971), but the whole effects of castration in the male, including
changes in weight of the accessory sex organs, seem not to be completely

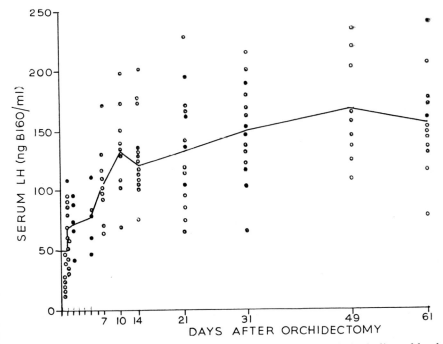

Fig. 2.43. Effects of orchidectomy on serum LH in rats. Open circles indicate blood
drawn by cardiac puncture and closed circles by exsanguination. Intact male controls (not shown) usually had less than 5 ng of B160/ml of serum. (Gay and Midgley,
Endocrinology.)

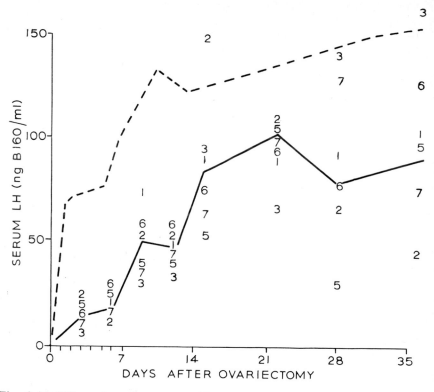

Fig. 2.44. Effects of ovariectomy on blood-LH in female rats. Each number represents the concentration in blood from a single rat and the solid line connects the mean values for these six rats. The broken line indicates, for comparison, values in orchidectomized males at the same time-intervals. (Gay and Midgley, *Endocrinology.*)

restored, so that Ramirez and McCann (1965) have postulated an additional factor, possibly an oestrogen, normally secreted by the testis. Certainly, in the male dog, oestradiol is far more effective than testosterone and other androgens in suppressing secretion of LH, induced by injections of the releasing hormone. Associated with the rise in plasma LH, there is a rise in the concentration in the pituitary, indicating an enhanced synthesis of the hormone in the absence of steroid feedback. According to Halasz and Gorski (1967), this happens in the neurally isolated hypothalamus, i.e. in an animal in which the rostral connections of the hypophysiotrophic area have been severed (Fig. 2.39) so that the feedback is apparently on to this area rather than the PO/AH area, or else lower down on the pituitary itself. It seems likely that both regions are affected by steroids, however, since Ramirez *et al.* (1964)

found that implants of oestradiol crystals into either the median eminence or the anterior pituitary would prevent the castrational rise in
plasma LH; as these authors emphasize, however, great care in inter-
pretation of this sort of experiment must be exercised, since an implant
into the median eminence might well spill over into the portal circulation and affect the pituitary directly rather than through the neurones
of the median eminence.

Compensatory Hypertrophy. When one ovary is removed, there
is hypertrophy of the other, and an obvious cause could be the increased level of FSH following a lowered level of oestrogen in the
blood; in other words, the fall in concentration of circulating oestrogen
would release the anterior pituitary from a negative feedback. The site
of action of this feedback seems to be the anterior preoptic area since
D'Angelo and Kravatz (1960) and Flerkó and Bardos (1961) found
that, in persistent oestrus rats, with lesions in this area, there was no
hypertrophy of the remaining ovary, whereas in control animals the
weight had doubled.

Ovulation. The role of negative feedback in the response to unilateral ovariectomy is interesting; thus normally in the golden hamster,
for example, the ovary, in the early stages of ovulation, develops twice
as many follicles as are going to mature, so that, during a cycle, half of
these regress (atresia). This regression can be prevented by giving
FSH (actually PMS), so that as many as 20–60 ova may be released.
With the unilateral ovariectomy, carried out before Day 3 of the cycle,
the remaining ovary compensates, so that the follicles that would have
regressed no longer do so, and the ovulation is normal in number.
Presumably the release of the pituitary from a negative feedback, following removal of one ovary, causes an increased secretion of FSH that
prevents atresia (Grady and Greenwald, 1968).

Importance of Reproductive State. An increase in blood-LH
does not always follow ovariectomy in the rat; and whether or not it
does depends on the extent to which the pituitary has been able to dispose of the oestrogen taken up during the oestrous cycle. Thus if the
operation is carried out at pro-oestrus, there is no immediate rise in
blood-LH, which is deferred until four days later, whereas if it is done
at di-oestrus there is; in the latter case the blood level of oestrogen has
been low for sufficient time to enable the pituitary to dispose of the
oestrogen taken up earlier (Tapper et al., 1972). In the male, the rise in
blood-LH occurs immediately at castration; the neonatally androgenized rat is continuously under the influence of oestrogen, so that the
response to ovariectomy is very sluggish.

Releasing Hormone. The negative feedback that leads to a reduced secretion of gonadotrophic hormones, revealed as an increase in circulating levels after ovariectomy in the female rat, is associated with a diminished LHRH content of the hypothalamus indicating, presumably, an increased secretion (Kalra, 1976). Exogenous oestradiol prevented these changes. The most pronounced effects were on the median eminence and arcuate nucleus, and were accompanied by increases in both pituitary and plasma LH (Kobayashi *et al.*, 1978). Interestingly, the amounts of releasing hormone in the organum vasculosum of the lamina terminalis (OVLT) were not affected by ovariectomy or exogenous oestradiol benzoate.

Changes in Excitability during the Oestrous Cycle

Clearly, if the steroid hormones exert a feedback on the hypothalamic neurones, we may expect cyclical changes in excitability in the female. Kalra and McCann (1973) made use of the fact that electrical stimulation of the medial preoptic area, or the median eminence-arcuate region of the hypothalamus, on the afternoon of pro-oestrus, caused ovulation in Nembutal-blocked rats, due to release of gonadotrophic hormones FSH and LH (e.g. Teresawa and Sawyer, 1969). They found that the greatest release of LH, in response to stimulating either hypothalamic site, was obtained on the morning of pro-oestrus; on other days the release was much smaller and failed to evoke premature ovulation. The stimulation at 23.00 hours on the day of pro-oestrus, i.e. later than the expected LH-surge, caused the release of large amounts, which brought the LH-level in the plasma nearly to that observed in early pro-oestrus, only if the afternoon surge had been blocked by Nembutal. Thus, in the normal unblocked animal, the pituitary stores are depleted after this surge, so that electrical stimulation, a little later, cannot evoke a large release; if the surge is blocked, the stores remain and are available for release in response to electrical stimulation.

PO/AH Units. When individual units in the PO/AH area were examined for variations in discharge-rate during the oestrous cycle, Dyer *et al.* (1972) found an unquestionable increase in the frequency of firing during pro-oestrus, and noted that this was much greater in the ventral PO/AH than in the dorsal region (Fig. 2.45).

Effect of Oestrogen

Experimentally, the concentration of oestrogen in the blood is most easily altered by intravenous administration of oestradiol; as Fig. 2.46 shows, such an injection can profoundly modify the discharge in single units of the hypothalamus; in Fig. 2.46a the unit is in the arcuate

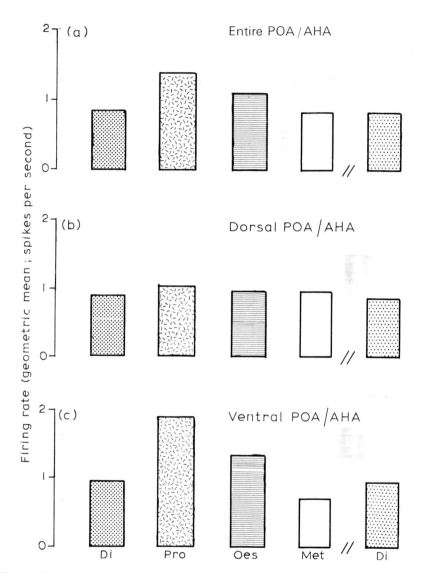

Fig. 2.45. Firing rates of single neurones of the PO/AH area during the oestrous cycle. (a) Entire PO/AH area; note increase in mean firing rate from di-oestrus to pro-oestrus. (b) Units from dorsal half of anterior hypothalamus; note uniform mean rates. (c) Units from ventral half of anterior hypothalamus; note increased mean firing rate at pro-oestrus and that there is a significant difference for the two zones on this day. POA, pre-optic area; AHA, anterior hypothalamic area; Di, di-oestrus; Pro, pro-oestrus; Oes, oestrus; Met, metoestrus. (Dyer *et al.*, *J. Endocrinol.*)

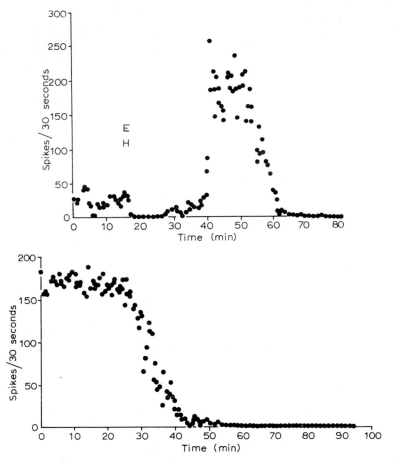

Fig. 2.46. (a) Showing increase in firing rate of unit in arcuate nucleus following intravenous administration of 50 μg of oestradiol-17β to a rat. The mark E indicates oestradiol injection. (b) Decrease in firing rate of unit in pre-optic nucleus of rat following injection of 50 μg of oestradiol-17β. (Yagi and Sawaki, *Neuroendocrine Control.*)

nucleus, i.e. in the lower tonic centre, and the result is increased firing. The unit in the preoptic nucleus, illustrated by Fig. 2.46b, showed a striking decrease in firing rate, a decrease that was maintained for as long as recording was made (up to 150 minutes).

Positive Feedback

The trigger for the pre-ovulatory surge in gonadotrophin release seems to be a correct balance of ovarian steroid hormones in the blood; thus, oestrogen-priming, followed by a dose of progesterone will cause a peak secretion of gonadotrophic hormones in the ovariectomized animal (p. 82). Alternatively, injection of an anti-oestrogen on Day 2 of the 4-day cycle rat will prevent ovulation (Ferin *et al.*, 1969), an inhibition that is not due to loss of power to ovulate, since administration of hCG could bring it about. Hence, we may speak of a positive feedback (or a positive facilitation, Davidson, 1969) of gonadotrophic hormone release. Taleisnik *et al.* (1970), by means of a retrochiasmatic cut in the hypothalamus, cut off the arcuate-ventromedial area from the PO/AH region, and found that negative feedback, recognized by the rise in plasma LH following ovariectomy, was still present. The positive feedback exerted by progesterone on LH release was abolished, however, so that they concluded that the PO/AH region was that concerned with the cyclic release of gonadotrophins, and this, of course, is consistent with the observation that destruction of this area abolishes cyclical activity in the female (D'Angelo and Kravatz, 1960).*

Anti-Ovulatory Action by Progesterone. In general, the action of progesterone is *anti*-ovulatory, and this is a primary function in the maintenance of pregnancy; this anti-ovulatory action may be observed in rabbits by the abolition of ovulation following copulation (Makepiece *et al.*, 1937); and the absence of ovulation associated with pseudopregnancy is doubtless due to the high level of circulating progesterone. A stimulatory effect has already been described, namely the advancement of the day of ovulation in cycling rats (p. 84), an effect that depends on the concentration of oestrogens in the blood at the time. The dual effects are illustrated by Fig. 2.47, where the concentration of LH in the blood of ovariectomized rats has been plotted against time. At the arrow marked E, an injection of oestrogen was given, which reduced the level of LH through the well established negative feedback of oestrogen. This depression lasted till the afternoon of the fourth day after the injection, when the LH-level returned to its high value characteristic of the ovariectomized animal. If progesterone was injected on the third day after the oestrogen injection (arrow marked P), there was a rise in LH in the plasma (positive feedback), but the level returned to a low value, which was maintained, whereas in the absence of the progesterone injection it would have risen. Thus, the period of positive feedback is followed by one of negative feedback, the two actions presumably depending on the level of oestrogen at the time, the positive effect occurring when the oestrogen concentration was high and the negative effect after the concentration had fallen.

---

* The pathway from the pre-optic anterior hypothalamic region (PO/AH) to the hypophysiotrophic area has been examined by Tejasen and Everett (1967) by making sections that led ultimately to complete de-afferentation.

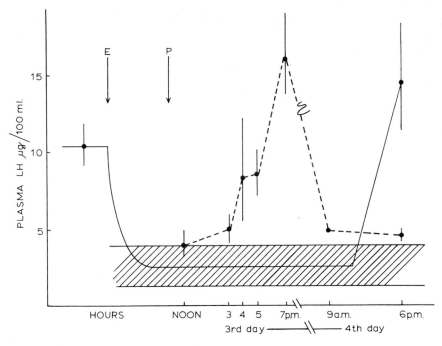

Fig. 2.47. Effect of progesterone on plasma LH activity in ovariectomized rats treated with oestrogen. Solid line, plasma-LH in ovariectomized rats; dashed line, plasma-LH in ovariectomized rats treated with oestrogen and subsequently with progesterone. The striped area denotes the extreme values found after the injection of oestrogen. (Caligaris *et al.*, *Acta endocrinol.*)

Site of Action of the Feedback Mechanisms

There is no doubt from the transplantation studies of Harris that the hypothalamus is a target for the action of steroid hormones, and the question that has been posed frequently is whether all feedback phenomena are mediated by way of the hypothalamus, and therefore only secondarily through the anterior pituitary, or whether both hypothalamus and pituitary are susceptible to the concentrations of steroid in the blood. Certainly a feedback on to the anterior pituitary is observed when the actions of the thyroid hormones are studied, and the same applies to the adrenal steroids, so that a direct action of gonadal steroids on the pituitary, affecting its susceptibility to releasing hormones, would not be something new.

Pituitary Sensitivity. The concentrations of the gonadal steroid hormones vary during the oestrous cycle, the concentration of oestradiol, for example, reaching a peak during pro-oestrus; if the sensitivity of the pituitary to releasing hormones, e.g. LHRH, is dependent on the concentration of oestrogen in the blood, we may expect the response to intravenous infusions of the hormone to vary during the cycle. Cooper *et al.* (1973) injected an extract of median eminence containing the releasing hormone for LH into the jugular veins of rats at different times during the oestrous cycle, measuring the concentration of LH in the plasma. As Fig. 2.48 shows, the maximum effect of the hormone occurs in pro-oestrus at a time when the plasma level of oestrogen is probably at its peak, so that this experiment indicates a potentiation of LH-release by oestrogens when these are secreted naturally. Figure 2.49 illustrates a similar cycle of sensitivity of the pituitary as measured by the changes of FSH concentration in response to intravenous injections of the releasing hormone.* That oestrogen is the necessary hormone governing this sensitivity was suggested by the finding that an

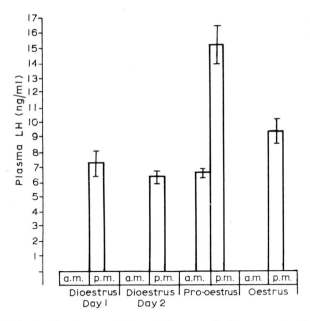

Fig. 2.48. Variation in the sensitivity of the rat pituitary to releasing hormone during the oestrous cycle. The plasma-LH in response to median eminence extract is highest at pro-oestrus. (Cooper *et al.*, *J. Endocrinol.*)

* Zeballos and McCann (1975) examined pituitary sensitivity to releasing hormone in respect of both LH and FSH release; whereas the maximal responsiveness, measured as LH discharge, occurred at 5 p.m. of pro-oestrus at the time of the naturally occurring gonadotrophin surge, the responsiveness with respect to FSH secretion was smaller and fell to its di-oestrous low value long before the ovulatory rise in FSH secretion fell.

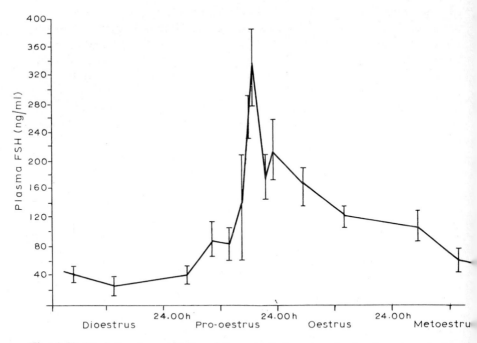

Fig. 2.49. Variation in sensitivity of the rat pituitary to releasing hormone during the oestrous cycle measured by changes in plasma-FSH. Animals were anaesthetized with Nembutal 30–60 min before injection of LHRH. (Aiyer *et al.*, *J. Endocrinol.*)

anti-oestrogen (I.C.I. 46474) reduced the response, whilst the increased responsiveness followed the same time-curves (after a delay) as the oestrogen titres (Tapper *et al.*, 1972). Moreover, Arimura and Schally (1971), had shown that exogenous oestradiol, injected into the 5-day cycle rat on di-oestrus Day 1, followed by an injection of porcine LHRH into the carotid artery, caused a large increase in plasma LH, whereas only a slight rise was brought about by the same dose of releasing hormone in the non-oestrogen primed rat.†

Cultured Pituitary Cells. Pretreatment of the rat's cultured anterior pituitary cells with 17-β-oestradiol at a concentration of 10^{-9} M decreased the concentration of LH-releasing hormone required to produce half-maximal stimulation of LH release from 3·0 to 1·6.10^{-10} M, whilst the basal release, not requiring LH-releasing hormone, increased

† Legan and Karsch (1975) have confirmed the increased responsiveness of the pituitary to releasing hormone during pro-oestrus; however, implants of oestrogen in ovariectomized rats that caused rises in blood-LH concentrations failed to increase responsiveness to LH-RH, and they suggest that other ovarian steroids may be important, e.g. progesterone, which facilitates secretion of LH in rats (Uchida *et al.*, 1972; Brown-Grant and Naftolin, 1972).

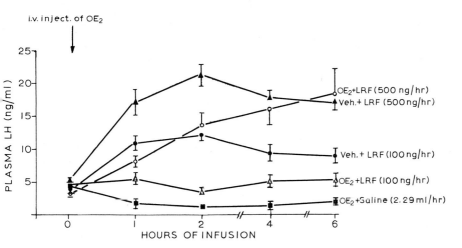

Fig. 2.50. Effects of oestradiol benzoate (OE2) on response to LHRH infusion in the rat. Veh is the saline vehicle for OE2. (Libertun *et al.*, *Endocrinology*.)

by more than twice. By contrast, androgens such as testosterone decreased the responsiveness of the cells, an effect that was only partially reversed by oestradiol treatment. (Drouin *et al.*, 1976).

Biphasic Response. The situation is necessarily complex, however, because oestradiol, for example, seems to have a biphasic effect on the release of LH by the pituitary in response to the releasing hormone, LHRH. According to the study of Libertun *et al.* (1974), simultaneous infusion of LHRH and oestradiol caused a *diminution* in the release of LH compared with infusion of LHRH alone* (Fig. 2.50). However, if the oestrogen had been given some 6 hr before the infusion of releasing hormone began, there was a pronounced facilitation of LH release (Fig. 2.51). Thus the time of exposure to a high level of oestrogen in the blood is obviously of importance in determining the sensitivity of the pituitary to its releasing hormone. This biphasicity in responsiveness to blood-oestradiol has been amply confirmed; thus Vilchez-Martinez *et al.* (1974) found that oestradiol benzoate depressed the pituitary response to LHRH on Di-oestrus Day 1 if given 2–9 hr before the injection, but increased it if given 14–24 hr after. At 48 hr, the responses were the same, with or without oestradiol. Again, Appelbaum and Taleisnik (1976) measured the release of LH from isolated half-pituitaries in response to LHRH; the glands were removed from ovari-

---

* Negro-Vilar *et al.* (1973) seem to have been the first to demonstrate that oestradiol could inhibit the response of the pituitary to releasing hormone. They studied ovariectomized animals pre-treated with oestrogen to keep the basic level of LH down to normal.

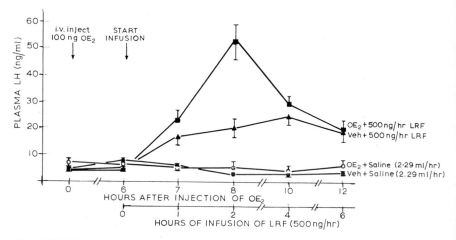

Fig. 2.51. The effect of an injection of oestradiol (OE2) six hours before beginning infusion of LH-releasing hormone on plasma-LH. The injection of oestradiol permits a large response to releasing hormone. (Libertun *et al.*, *Endocrinology.*)

ectomized rats. Addition of the releasing hormone increased the spontaneous release of LH into the medium. If oestradiol benzoate was given 24 hr before removal of the pituitary, the response to the releasing hormone was increased; if the oestradiol was given just two hours before removal, then it decreased the response to the releasing hormone. Thus a short exposure to oestrogen seems to inhibit, and a longer exposure to increase, responsiveness to the releasing hormone.

Human Studies. Yen *et al.* (1972) found that there was an increase in the responsiveness of the human female pituitary to LHRH, measured as an increase in LH-secretion, during the late follicular phase of the menstrual cycle, the maximum response being at mid-cycle. Secretion of FSH was not affected in the same way. Later Jaffe and Keye (1974) showed that oestrogen injections into cycling women would increase the pituitary responsiveness to releasing hormone provided these were given for several days, beginning on the first day of the cycle, when plasma levels of oestrogen are low. The total amount given, to produce the increased sensitivity, was equivalent to the amount secreted during the late follicular phase of the cycle. In their earlier study (Keye and Jaffe, 1974) they had found that a single injection of oestradiol to women caused a diminution in the amount of LH released by administered LHRH.

Effects of Ovariectomy. Aiyer and Fink (1974) found that the slower rise in sensitivity could be prevented by ovariectomy in the morning of di-oestrus and was restored by exogenous oestrogen. Thus, this initial rise in sensitivity was due to the rise in blood-oestrogen that precedes the pre-ovulatory LH-surge. However, administration of oestradiol failed to restore the abrupt rise in sensitivity, which had likewise been abolished by ovariectomy; this could be restored by previous treatment with progesterone, so that it may be that the second phase of increased sensitivity depends on progesterone, acting on an oestrogen-primed pituitary. Thus, although the main increase in plasma progesterone occurs in response to the LH-surge, there is a significant elevation of plasma progesterone before this (p. 69).

Pro-Oestrous Surge of LHRH. If we agree that the LH-surge at pro-oestrus is the consequence of a positive feedback of oestrogen, secreted by the ovary, then, if oestrogen acts on the hypothalamus, as well as to increase the sensitivity of the pituitary to releasing hormone, we may expect it to increase the secretion of releasing hormone by the hypothalamus before the LH-surge. Thus the portal blood should show an LHRH-surge in pro-oestrus. Sarkar *et al.* (1976) actually found this.

Oestrogen Implants. In efforts to discover the sites of action of oestrogen in its feedback phenomena, numerous experiments have been carried out involving implantation of the steroid into the pituitary or hypothalamus. Many of these studies have been discussed by Goodman (1978) in relation to his own studies on the rat. Thus, the obvious objection to hypothalamic implants is the danger that the implanted oestrogen will be carried to the pituitary along the portal circulation. Again, use of oestradiol benzoate raises the question as to whether the neural tissue can split this to oestradiol, the active molecule. Goodman's studies left no doubt that implants in the PO/AH region induced the ovulatory LH-surge in ovariectomized rats, and the effectiveness of different implantation-sites bore no relation to the concentration found in the pituitary, thus showing that carriage from hypothalamus to pituitary was not a decisive factor.

Electrical Stimulation of Hypothalamus. Kalra and Kalra (1974) found that the rise in blood-LH, in response to electrochemical stimulation of the hypothalamus, depended on the phase of the oestrous cycle, the biggest effects being in pro-oestrus. When they studied the responses of the pituitary to injections of releasing hormone, they found that, once again, the greatest release was in pro-oestrus, so that the high sensitivity to electrochemical stimulation could have been due to either a high sensitivity of the hypothalamic neurones or to both this and a high sensitivity of the pituitary neurosecretory cells to releasing hormone. That the hypothalamic neurones are themselves sensitive to steroid levels is well established by single-unit studies (p. 134); when oestradiol was injected into the third ventricle of ovariectomized rats, i.e. in close approximation to the hypothalamus, there was a reduction in the high level of LH that results from ovariectomy (p. 131). The sensitivity of the pituitary to releasing hormone, however, was unaffected, so that here we have strong evidence of a direct inhibitory action of oestradiol on hypothalamic neurones (Orias et al., 1974).*

Secretion of Releasing Hormone

By directly measuring the changes in amount of LH-releasing hormone appearing in the pituitary stalk blood, Fink and Jamieson (1976) revealed changes in sensitivity of the hypothalamic neurones to electrical stimulation. The magnitude of the increment due to electrical stimulation varied with the period of the rat's oestrous cycle, a peak response being obtained between 18.00 hr of di-oestrus and 13.00 hr of pro-oestrus, which corresponds approximately with the surge in blood-oestradiol, which reaches its peak between 08.30 and 16.00 hr of pro-oestrus. As Fig. 2.52 shows, however, the increase in LH-secretion did not run parallel with the increment of secretion of the releasing hormone, so that an additional factor of increasing sensitivity of the hypothalamus to oestrogen must be invoked. By injecting oestradiol-17β into ovariectomized animals, Sherwood et al. (1976) found responses to electrical stimulation of the anterior hypothalamus that were actually greater than in normal unoperated animals; testosterone propionate was also effective, presumably due to "aromatization" since dihydrotestosterone was ineffective. The anti-oestrogen, ICI 46474, did not influence the response to electrical stimulation, suggesting that the site of action of this compound is on the pituitary rather than the hypothalamus.

Priming Action of Releasing Hormone

As Gordon and Reichlin (1974) point out, the increased pituitary sensitivity to releasing hormone, or the increased responsiveness to hypothalamic neurones, described by these studies is unlikely to be the sole cause of the pre-ovulatory surge of LH, and they suggest that the releasing hormone "primes" the pituitary to respond to itself, in the same way that ACTH sensitizes the adrenal cortex to itself.

Evidence for this priming action is strong; thus Aiyer et al. (1973; 1974) found that the response to a second dose of releasing hormone was greater than that to a first when the interval between them was 60 or

---

* Fink and Aiyer (1974) found that the increases in plasma LH and FSH following electrical (as opposed to electrochemical) stimulation of the hypothalamus (PO/AH area) varied with the phase of the cycle, following the pattern of sensitivity to releasing hormone described earlier.

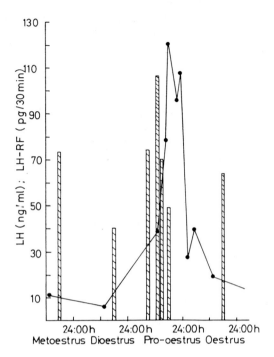

Fig. 2.52. Effects of electrical stimulation of the pre-optic area of the hypothalamus on concentration of LH-releasing hormone in pituitary stalk blood as a function of the oestrous cycle. Hatched bars indicate increments in LHRH in blood whilst the points joined by the line indicate the mean maximal increments of LH in the blood. (Fink and Jamieson, *J. Endocrinol.*)

120 min. Again, Blake (1976) found a much larger response to a give amount of LHRH when this was infused over a period of four hours compared with the response to a single injection of the same amount.

Finally, Fink *et al.* (1976) showed that endogenous releasing hormone, released in response to pre-optic stimulation, likewise exhibited a priming action; thus the LH released in response to a second stimulus separated by an hour from a first was far larger than the response to the first stimulus. Interestingly, the priming could not be shown in dioestrous rats, suggesting the importance of oestrogen in inducing the effect.

Dependence on Protein and RNA Synthesis. When the inhibitor of protein synthesis, cycloheximide, was administered there was no effect on the first response of the rat to releasing hormone, but the priming action, manifest in the second response, was reduced. The priming action could be demonstrated in the isolated pituitary; it was inhibited by cycloxeximide and also by the RNA synthesis inhibitor, actinomycin D (Pickering and Fink, 1976).

Progesterone

Progesterone is capable of blocking the pre-ovulatory LH-surge and ovulation, an observation that formed the basis for development of oral anti-fertility agents. This hormone acts at both hypothalamic and pituitary sites; thus Hilliard *et al.* (1971) blocked the rabbit's ovulatory response to porcine LHRH, or to a median eminence extract, when these were infused directly into the pituitary gland. If the dose of releasing hormone was raised sufficiently, ovulation occurred. Endogenous progestins, released during pregnancy and pseudo-pregnancy, were also completely effective in blocking the ovulatory response to LH–RH.* Using an *in vitro* preparation, Schally *et al.* (1973) showed that release of LH by the isolated pituitary was inhibited not only by progesterone but by two additional progestogens, 5a-dihydro-progesterone and 17a-progesterone. In monkeys, however, the evidence indicates that progesterone affects only the hypothalamic releasing mechanism, since Spies *et al.* (1972) found that the response to injected median eminence extracts (LHRH) i.e. the rise in blood-LH, was unaffected by the stage of the menstrual cycle, so that the large rise in progesterone in the blood during the luteal phase of the cycle, or following an intravenous injection of progesterone, failed to influence the release of LH. Spies and Niswender (1972) conclude from their studies on monkeys that progesterone blocks the pre-ovulatory LH-surge through an anti-oestrogen action, in the sense that it blocks the acute pre-ovulatory rise in secretion of this hormone; thus the blocking effect of progesterone can be overcome by further injection of oestradiol.

Action on Positive Feedback. Dierschke *et al.* (1973) showed that progesterone failed to affect the action of oestradiol, so far as its negative feedback on the hypothalamus-pituitary system was concerned. Thus the decrease in blood-LH of ovariectomized animals, following injection of the oestrogen, was unaffected by progesterone. Hence progesterone seems to exert an anti-oestrogen action only in so far as it blocks the *positive* feedback involved in the pre-ovulatory rise in LH-secretion.† Thus the pathways for negative and positive feed-

* We may note that Debeljuk *et al.* (1972) found that a combination of progesterone and oestradiol was necessary to block the release of LH by its releasing hormone in the rat. Modern antifertility procedures employ both oestrogens and progestins. It must be emphasized that progesterone can facilitate the release of LH (p. 80) and so induce ovulation in the rat if given on the morning of pro-oestrus.

† Dierschke *et al.* (1973) conclude that, in monkeys, progesterone does not exert any facilitatory action on oestrogen during the period of the pre-ovulatory surge, pointing to the fact that implants of oestrogen, alone, can allow ovulation to occur.

back are different. In this connection we may note that Spies *et al.* (1977) found that the rise in LH-secretion of the monkey, following PO/AH stimulation, could be facilitated by oestradiol, whereas the effects of stimulating the mediobasal region were inhibited by the oestrogen.

Suppression of the Hypothalamic Signal. Freeman *et al.* (1976) and Banks and Freeman (1978) have suggested that the surge of progesterone that follows the ovulating LH-surge exerts a negative feedback that is sufficient to suppress the signals for release of LH for several days; their quantitative studies indicated that the rat needed the entire progesterone surge occurring in pro-oestrus to limit the LH-surge to this day.

Testosterone

Swerdloff *et al.* (1972) showed that, in rats, the *in vivo* release of LH and FSH in response to releasing hormone was strongly inhibited by testosterone and dihydrotestosterone, and Schally *et al.* (1973) found that release from isolated pituitaries was likewise inhibited. In the male dog, however, Jones and Boyns (1974) found that injections of testosterone and other androgens failed to inhibit the release of LH in response to synthetic releasing hormone, although oestradiol was highly effective. They computed that, although the testis secretes androgens in quantities some thousand times greater than oestrogens (Kelch *et al.*, 1972), the far greater sensitivity of the pituitary to oestrogens might well mean that control over LH secretion in the male was exerted through testicular oestrogen rather than androgen. Thus, testosterone undergoes "aromatization", i.e. conversion to oestrogen, by homogenates of median eminence, hypothalamus and pituitary (Naftolin *et al.*, 1975); and Sherins and Loriaux (1973), in their study on infusion of testosterone into human subjects, found that the suppression of FSH and LH was accompanied by raised oestradiol levels. The fact that the non-aromatizable dihydrotestosterone can affect LH levels, however, indicates that androgens can have a direct action on gonatrophin secretion (Stewart-Bentley *et al.*, 1974).*

Inhibin. An obvious complicating factor is the presence in the testis of a specific factor, probably secreted by the seminal vesicles, that inhibits secretion of FSH; thus destruction of the seminal vesicles causes a rise in plasma FSH (Bain and Keene, 1975) presumably by releasing the anterior pituitary from the inhibiting action of the substance which was called by McCullagh *inhibin*. That it is secreted by the Sertoli cells was shown by Rich and Kretser (1977) who found that the raised

---

* Reddy *et al.* (1973) found that, whereas the anterior hypothalamus converted androstenedione to oestriol *in vitro*, the pituitary failed to do so.

FSH following damage to the testes of rats ran parallel with a diminished secretion of androgen-binding protein (ABP), which is known to be synthesized by the Sertoli cells (see, for example, Hagenas, 1975). More directly, Steinberger and Steinberger (1976) proved this origin by culturing Sertoli cells and preparing from the medium a factor that reduced the release of FSH by cultured pituitary cells *in vitro*; since the effect was a direct one on the pituitary it seems that the principle—inhibin—operates a direct feedback on this gland rather than through the hypothalamus. As Setchell and Sirinathsinghji (1972) pointed out, the rate of spermatogenesis is remarkably constant for a given animal; and it is likely that inhibin, secreted into the rat's testis fluid and absorbed from this fluid in the head of the epididymis, is the feedback agent responsible for this control. They found that testis fluid from the ram or boar reduced plasma FSH in mice. Again Franchimont *et al.* (1975) found that pretreatment of rats with an extract of bull's seminal plasma reduced the FSH response to LH–RH without affecting that of LH.† Finally Keogh *et al.* (1976) found that extracts of bovine testes reduced the FSH levels of castrated sheep; interestingly, when they employed the original extraction procedure of McCullagh and Schneider (1940), their preparations were inactive.

Female Inhibin. Less direct evidence for a separate inhibitory control over FSH production in the female is provided by Sherman and Korenman (1975) who found high levels of blood-FSH in menopausal women when the LH level was in the normal range for the cycle stage. They suggested that this was due to the diminished number of follicles capable of secreting an inhibitory factor; this must be something other than oestradiol since the latter's inhibitory feedback actions on FSH and LH are usually similar. They consider that a female inhibin would be of value in controlling the number of follicles maturing in a given cycle, and would be of special importance in primates when the number of conceptuses supported during a gestation is limited.

Hypothalamic Implants

The normal maintenance of the gonads depends on a tonic secretion of gonadotrophic hormones, so that destruction of the posterior median eminence region leads to gonadal atrophy, due to inhibited secretion of FSH and LH (Davidson and Ganong, 1960). This tonic secretion is, of course, subject to negative feedback by the steroid

---

† The authors used this observation as evidence against the existence of a single releasing hormone for LH and FSH; however the dissociation between FSH and LH release probably depends on a variable response of the respective pituitary secreting cells to the releasing hormone, in that the synthesis of one hormone can be modified independently of the other.

hormones. Davidson and Sawyer (1961) were able to produce gonadal atrophy of the testes and prostates in dogs by implanting testosterone into the ventral hypothalamus, i.e. close to the median eminence; in these animals ejaculation failed or, if it occurred, the sperm-count was reduced. Pituitary implants of testosterone were usually without effect, the histology of the prostate and testes being normal. Davidson (1969) considered that androgens acted only at the hypothalamic level, by contrast with the oestrogens that acted at both this and the pituitary level.

Pituitary Implants. However, Kingsley and Bogdanove (1973) implanted pellets of three androgens, namely testosterone, dihydro-testosterone, and 7a-methyl-19-nortestosterone, into the pituitary of castrated male rats and observed that the characteristic development of "castration cells" in the tissue that follows castration, and represents a release of the negative androgen feedback (p. 131), was inhibited; this was accompanied by an increase in the FSH/LH ratio in the pituitary gland either by lowering LH concentration or increasing the FSH concentration, or both. That the effects were not due to release of the steroid hormones into the general circulation, where they would have a direct action on the accessory sex organs, was shown by the finding that the latter were unaffected by the implants. It could, of course, be argued that the androgens were being converted to oestrogens, or progestins, locally, but it is known that these hormones do not increase pituitary stores of FSH, so that the action is a direct one even though aromatization of two of the hormones, namely testosterone and 7aMe, to form oestrogens is possible. Earlier, Mittler (1972) had found that treatment of isolated organ-culture anterior pituitary with testosterone raised the FSH/LH ratio in the medium.

Short-Loop Feedback

What has been described as a "short-loop" or "internal" feedback* seems to operate in the secretion of gonadotrophic hormones, in the sense that the hormones, LH and FSH tend to inhibit their own secretion by exerting an influence on the secretion of releasing hormones into the median eminence-stalk region of the hypothalamus. This was surmised by the discovery that injections of gonadotrophic hormones (pregnant mares serum, PMS with both FSH and LH activity, or human chorionic gonadotrophin, hCG) into intact rats caused a diminution in pituitary weight and cell population together with a

---

* Corbin (1966) has defined this feedback as an effect of gonadotrophic hormones on the secretion of their releasing hormones by the hypothalamus—it is a hypophysio-hypothalamic interaction. The various alternative names are given; the phenomenon is by no means peculiar to the gonadotrophic hormones of the pituitary.

lowered concentration of FSH; that this was not due to a primary increase in secretion of ovarian steroids, which then resulted in the "long-loop" negative feedback on the pituitary-hypothalamus, was shown by removing the ovaries, when a similar change occurred in response to the injection of the gonadotrophins. The main evidence indicating a definite inhibition of hypothalamic action was provided by Corbin's studies involving implantation of the gonadotrophins, LH or FSH, into the median eminence; he observed a depression in the LH-content of the gland some 8–10 days after implantation of LH. In the intact animal this was a 51% decrease, suggesting a decreased synthesis by the pituitary, but the animals still ovulated normally (Corbin and Cohen, 1966). When the same experiment was carried out on ovariectomized animals, thus eliminating effects of steroid secretion, the reduction in pituitary LH-content was 31%, the smaller effect being probably due to the elimination of some negative feedback by the ovarian steroids (Corbin, 1966). More definite proof of the inhibitory action is provided by analysis of the median eminence-stalk region for releasing hormone; thus Corbin and Story (1967) implanted FSH into the median eminence and found a reduction of FSH in the pituitary, and of releasing hormone in the median eminence-stalk tissue. In the normal rat, Corbin et al. (1970) were unable to measure any FSH-releasing hormone in the peripheral blood plasma; however, when the pituitary had been removed, thereby removing a negative feedback of circulating FSH on the hypothalamus, the FSH-releasing hormone could be easily measured in the blood, the rise in the plasma being associated with a fall in the median eminence-stalk tissue. The same result was obtained in animals that had been ovariectomized subsequently to hypophysectomy, thereby ruling out the long-loop feedback. When the median eminence area was destroyed, the rise in plasma concentration of releasing hormone due to hypophysectomy was abolished.

Hypothalamic Neurones. Micro-iontophoretic application of LH and FSH to the medial-basal hypothalamus of the ovariectomized rat caused either excitation or inhibition of the hypothalamic neurones (Sanghero et al., 1978); the effects were not confined to those neurones that could be excited antidromically from the median eminence but to others that could not, so that Sanghero et al. suggested that feedback of gonadotrophic hormones on to the hypothalamus did not simply represent a control of the neurosecretory neurones of the hypophyseal stalk, but a more general influence on the neurones of the hypothalamus including those that spread hypothalamic influences to other parts of the central nervous system, such as the amygdala.

Ultra-Short Loop. Hyppo *et al.* (1971) prepared median eminence extracts containing FSH–RH activity, as demonstrated by the reduction in pituitary FSH-content after intra-arterial injections. In castrated hypophysectomized rats the hypothalamic stores of FSH–RH were increased, suggesting a normal inhibitory feedback of circulating hormone on the hypothalamus; when these animals were treated with the hypothalamic extracts containing FSH–RH activity the stores returned to normal. If the effect was, indeed, due to the releasing hormone in the extracts, this suggests a negative feedback of the releasing hormone on the hypothalamus, an "ultra-short feedback loop".

Releasing Hormones as Neurotransmitters

The role of the releasing hormone as a neurotransmitter is supported by several findings; thus, LHRH induced mating behaviour in hypophysectomized ovariectomized female rats (Moss and McCann, 1973; Pfaff, 1973). The thyrotrophic releasing hormone, TRH, is located in several non-hypothalamic regions of the brain (Jackson and Reichlin, 1974) and may act as a central transmitter. Renaud *et al.* (1975) have applied releasing hormones, including LHRH, to the hypothalamus and to various other regions of the central nervous system and found that the depressant effect of the hormones is by no means confined to the hypothalamus, but extends to the cerebral and cerebellar cortices and cuneate nucleus of the medulla. Figure 2.53 is an example of the depressant effects of LHRH on two neurones of the

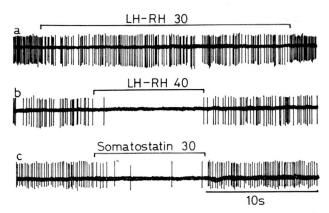

Fig. 2.53. Effects of micro-injection of LHRH and somatostatin on cerebral neurones. (a) and (b) hypothalamic neurones in ventromedial hypothalamus; (c) neurone in parietal cortex. Numbers above each bar indicate micro-iontophoretic current (nA) used to apply each hormone. (Renaud *et al.*, *Nature*.)

ventromedial hypothalamus and of somatostatin, the inhibiting hormone controlling growth hormone release, on a cerebral cortical neurone.

This role of the releasing hormone as an exciter or inhibitor of central neurones suggests the possibility that the secretion by pituitary cells may be influenced indirectly by the releasing hormones, which first activate, say, a hypothalamic neurone. There seems little doubt that a great deal of the LHRH of the brain is contained in neurones with perikarya outside the hypophysiotrophic area (median eminence, etc.) since Brownstein et al. (1976) formed an "island" containing the median eminence, ventromedial nucleus and adjacent structures, and ten days later they removed this and analysed it for LHRH. Over 70–90 per cent. had been lost, indicating that the releasing hormone belonged to neurones with perikarya outside the island. Thus neurones of the median eminence might excite the pituitary cells through a neurogenic mechanism involving another transmitter, e.g. dopamine or acetylcholine, as well as by secreting releasing hormone into the portal system. Kizer et al. (1976) found that the concentrations of dopamine and of choline acetyl transferase in the bovine median eminence correlated well with those of releasing hormones TRH and LHRH. A possible scheme is illustrated by Fig. 2.54 from Wilber et al. (1976).

Recurrent Facilitation and Inhibition

Stimulation of the median eminence antidromically activates tubero-infundibular neurones, presumably those in the medial basal hypothalamus including the arcuate nucleus that are responsible for the release of releasing hormone (Sawaki and Yagi, 1976). Stimulation also causes a recurrent inhibition or facilitation of the same neurones, and this is presumably due to the existence of axon-collaterals from these tubero-infundibular neurones, which run back to synapse with other tubero-infundibular neurones, either facilitating or inhibiting them. The inhibition would, of course, require an inhibitory inter-neurone comparable with the Renshaw cell of the motor neurone's axon-collateral. The inhibition was abolished by picrotoxin, and in this case it was easy to reveal a facilitation of the neurones caused by median eminence stimulation. This facilitation was blocked by a-methyl-*p*-tyrosine suggesting a catecholaminergic excitatory pathway, whilst the effects of picrotoxin suggest a GABA-mediated inhibitory pathway.

We may assume, then, that the neurosecretory hypothalamic neurones terminating in the median eminence liberate releasing

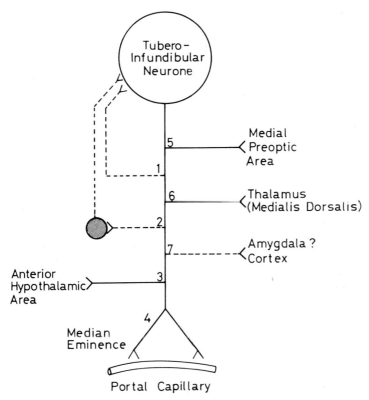

Fig. 2.54. Illustrating possible methods whereby a tubero-infundibular neurone could not only activate the pituitary through liberation of releasing hormone into the portal circulation, but also other regions by virtue of the transmitter action of its releasing hormone. The possibilities of feedback on to this neurone are also indicated. (Wilber *et al.*, *Rec. Progr. Hormone Res.*)

hormone at their terminals; those terminals coming into close relations with the portal capillaries effectively release their hormone into the blood-stream; however collaterals presumably release the same hormone in relation to hypothalamic hormones where it now acts as a synaptic transmitter, activating catecholaminergic or GABA-ergic neurones, mediating a feedback to the hypothalamic neurones causing either facilitation (positive feedback) or inhibition (negative feedback).

Single Unit Discharge

By micro-injection of releasing hormones, Dyer and Dyball (1974) showed that the rate of spontaneous discharge of single units in the PO/AH region of the hypothalamus was reduced, and they suggested that the hypothalamic neurones that release LHRH into the portal circulation also release the hormones, through axon branches, directly on to PO/AH neurones, the LHRH liberated in this manner acting this time as a neurotransmitter in the ultra-short-loop feedback. They pointed out that some of the nerve fibres from the neurones concerned with releasing hormones do not go to the median eminence.

Axon-Collaterals. Thus Renaud (1976) identified neurones in the mediobasal hypothalamus that could be antidromically activated from the median eminence. He found some mediobasal neurones that could be activated not only from the median eminence but also from other hypothalamic and non-hypothalamic areas, namely the anterior hypothalamic area, the medial pre-optic area, and the thalamic nucleus medialis dorsalis. Presumably collaterals from the mediobasal hypothalamic neurones activate these other regions in addition to liberating their releasing hormone in the median eminence; thus, in accordance with Dale's rule, these collaterals will be liberating releasing hormone in relation to other neurones so that the releasing hormone is behaving as a transmitter at a synapse.

Prostaglandin-Induced Gonadotrophin Release

Prostaglandins, especially* PGE_2, stimulate release of luteinizing hormone, LH, whilst inhibitors of PG synthesis are said to block release, or at any rate to inhibit ovulation (Grinwich et al., 1972). When they infused PGE_2 into the lateral ventricles of rats, Eskay et al. (1975) obtained release of LH, FSH and prolactin into the blood, the effects on LH release being the greatest. When the hypophyseal portal blood was collected, it was shown that the prostaglandin caused the release of large amonts of LH-releasing hormone. An anti-LHRH serum blocked the effect of PGE_2 (Chobsieng et al., 1975). When the prostaglandin was infused into the portal vessel there was no increase in LH-release, indicating the absence of direct action on the pituitary.

Sexual Receptivity

This liberation of releasing hormone in response to prostaglandin may be the mechanism whereby the prostaglandin increases receptivity of female rats in the presence of the male since LHRH facilitates this in ovariectomized oestrogen-primed rats (Moss and McCann, 1973). Dudley and Moss (1976) found that, next to progesterone,

---

* PGE_2 is easily the most effective prostaglandin in promoting gonadotrophic hormone release; as indicated by Ojeda et al. (1976), the absence of the double bond at 5,6 to give PGE_1 reduces activity; the absence of the C-11 hydroxyl in the ring structure, as in PGA_2 and PGB_2, causes further loss of activity, and the introduction of a hydroxyl at C-9 as in the F-series causes complete loss of activity.

PGE_2 was most effective, the order being: Progesterone > PGE_2 > PGE_1 > LHRH. By applying PGE_2 to the PO/AH region, which is involved in the display of masculine sex behaviour in the rat, Clemens and Gladue (1977) increased the copulatory activity of castrated male rats previously primed with testosterone propionate.

Site of Action

It is generally believed that prostaglandins act through the nervous system, inducing certain effector nerves to discharge or be inhibited. However, Harms et al. (1976) were unable to prevent the PGE_2-induced hormone release with neuronal blocking drugs, such as pronethalol (β-blocker) phentolamine (α-blocker) pimozide (dopamine-blocker) methysergide (5-HT-blocker) or atropine (cholinergic blocker) so that it must be assumed that PGE_2 acts directly on the LHRH-neurones; at any rate it has been shown by micro-iontophoretic injections, that the prostaglandin can increase the rate of firing of pre-optic and arcuate hypothalamic neurones in the guinea-pig (Poulain and Caretta, 1974). By implanting PGE_2 into definite regions of the hypothalamus, Ojeda et al. (1977) established that important regions were the arcuate-median eminence area and also the preoptic-anterior hypothalamic area (PO/AH), i.e. regions where LH-releasing hormone is concentrated. Implants in these regions resulted in the liberation of both LH and FSH.

Indomethacin and Aspirin

Both these drugs inhibit synthesis of prostaglandins, and they are able to block ovulation; the blockage due to aspirin can be reversed by treatment of the animal with LH or its releasing hormone, indicating that the aspirin is acting at a hypothalamic level. Blockage due to indomethacin cannot be overcome in this way, and so is presumably due to action at the ovarian level, probably interfering with the rupture of the follicle and release of the ovum (Berman et al., 1972). Thus O'Grady et al. (1972) found that the ovaries in the blocked animals had large haemorrhagic follicles in which the ova were retained.* Nevertheless, indomethacin does inhibit the release of LH, in so far as the rise in blood-level following ovariectomy in females, and castration in males, was depressed (Ojeda et al., 1975) suggesting an action at the pituitary or hypothalamic level. Since, however, indomethacin did not affect the release of LH in ovariectomieed rats in response to LHRH, the main effect must have been on the hypothalamic releasing mechanism. The effectiveness of indomethacin, when injected into the third ventricle or implanted in the medial basal hypothalamus (MBH), further indicates a hypothalamic site for its action, presumably preventing synthesis by the neural tissue. In this case the effect of indomethacin occurred after a much shorter delay, indicating that access to the neural tissue from the general

---

* The prolongation of pseudopregnancy in the rabbit observed by O'Grady et al. was presumably due to an anti-luteolytic action of indomethacin, probably by inhibiting synthesis of $PGF_{2\alpha}$, which is luteolytic (p. 91). Resorption of foetuses also occurred.

circulation might be restricted, in fact this restraint may account for failure of others to observe an inhibition of LH release by indomethacin (e.g. Sato *et al.*, 1974). That some of the effect of indomethacin is to block action on the pituitary, rather than the hypothalalamus, is shown by the partial block of release of LH by LHRH in ovariectomized, oestrogen-progesterone primed rats.

Effects of PGE₂ on Steroidogenesis. It is considered that the steroidogenic action of LH on the corpus luteum is mediated through cyclic AMP. PGE_2 stimulates steroidogenesis in corpora lutea *in vitro*, and it also promotes the incorporation of ^{3}H-adenosine into cyclic AMP, due to stimulation of adenyl cyclase activity (Marsh and LeMaire, 1974). Interestingly, the corpora lutea of the menstrual cycle were much more responsive to the prostaglandin than those of pregnancy, and this is also true of the response to LH and chorionic gonadotrophin (LeMaire *et al.*, 1968).

THE OVULATORY CYCLE

Having discussed in detail the various aspects of hormonal control over the fundamental processes concerned in the ovulatory cycle, we may return to a general consideration of the likely mechanism of the prime event in the cycle, namely ovulation, and the subsequent formation of the corpus luteum and finally the initiation of luteolysis.

The Pre-Ovulatory Surge of Gonadotrophins

As summarized by Kalra (1975), a continuous exposure of the central nervous system to circulating oestrogen until the early morning of pro-oestrus is necessary for induction of the LH-surge; the period of "priming" up to di-oestrus 2 without a further rise in oestrogen level will not *cause* ovulation, in fact, according to his study, based on the measurement of LH levels in the 4-day cycling rat, oestrogen secretion is essential until 03.00 hours of pro-oestrus to elicit the pre-ovulatory surge of LH.

Two Stages of Positive Feedback

In general, the stimulatory (positive) feedback of oestrogen on those parts of the central nervous system known to participate in pituitary activation is exerted in two stages, corresponding to the secretion during di-oestrus 2 and pro-oestrus. Thus the first phase is one of priming by a slow rise in oestrogen as observed in di-oestrus 2. During late di-oestrus 2 the second phase begins with a large rise in circulating oestrogen, and sometime between this and pro-oestrus the threshold concentration for trigger-action is reached. In the ovariectomized animal, in fact, a pro-oestrous surge can be brought about by a two-step schedule of injection of oestradiol; thus Caligaris *et al.* (1971) gave a single dose of oestradiol benzoate to ovariectomized rats, and this was sufficient to reduce the high level of LH, characteristic of the ovariectomized animal, to normal for a period of 8 days. A second dose at noon 3 or 4 days later gave a rise for a few hours and, what was interesting, a peak of LH in the blood occurred the next day. If this supplementary dose was given on alternate days after the first priming dose, peaks of LH activity in the blood occurred every day up to 8 days. This positive feedback, moreover, was only observed when the dose was given in the afternoon, and not in the morning, suggesting that there was a circadian, i.e. 24-hourly, rhythm of hypothalamic sensitivity, so that the second dose of oestradiol was effective only at a certain period of the successive days.

Circadian Rhythm. This circadian change in excitability was examined further by Legan *et al.* (1975) and Legan and Karsch (1975); they implanted Silastic capsules containing oestradiol into rats ovariectomized 15 days before and left these in place for $29\frac{1}{2}$ hours, after which they were removed. The levels of plasma oestradiol resulting from the implant are shown as the stippled area in Fig. 2.55; it will be seen from the Figure that the plasma LH level rose for four

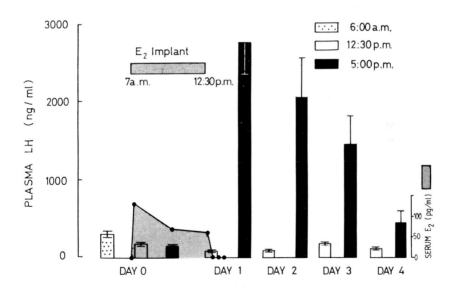

Fig. 2.55. Surges of LH on four successive days after implantation of Silastic capsules containing oestradiol into rats on the 15th day following ovariectomy (Day 0). The implants were left in place for 29·5 hr. The stippled area enclosed by the solid line represents oestradiol concentrations from a separate group of similarly treated rats. (Legan *et al.*, *Endocrinology*.)

successive days, the time of this rise being about 5 p.m. In this case, then, a single dose of oestrogen brings out the rhythmic secretion of LH, a process that can be blocked by treatment with Nembutal; moreover, in this situation, a maintained high level of oestradiol in the blood was not necessary to bring out the rhythm, since the levels had fallen to baseline soon after removal of the implant. This study strongly supports the view of Everett and Sawyer (1950) of a rhythmic hypothalamic neural signal occurring at the same time each day in the intact rat.

Short- and Long-Term Ovariectomy

We have seen that ovariectomy on the day of pro-oestrus fails to block the LH-surge, and it is interesting that on the following days the LH-surge observed in the evening of long-term ovariectomized rats treated with oestradiol, fails to manifest itself. Thus recent and long-term ovariectomy seem to be characterized by different release mechanisms for LH; in the long-term animal, repetitive LH secretion occurs, whereas in the recently ovariectomized animal this has been prevented. The difference is probably related to the levels of oestradiol in the blood in the various conditions; apparently, recently ovariectomized animals require high levels of oestradiol in their blood to maintain the repeated LH-surge, whereas in long-term ovariectomized animals this high concentration is unnecessary. Thus implanting oestradiol immediately after ovariectomy, so as to maintain high plasma oestradiol, permitted repetitive LH-surges for as many as ten days after the implantation. The result emphasizes the importance of adequate sensitization of the hypothalamus-pituitary through oestrogen for LH-signals to become operative, but of course raises the question as to why, in long-term ovariectomized animals, a lower regimen of oestrogen seems to be adequate.

Mimicking the LH-Surge

Blake (1976, a, b) has plotted very carefully the changes in plasma LH and FSH during the rat's pro-oestrous surge of gonadotrophins; he has then successfully mimicked these surges by appropriate infusions of releasing hormone in the phenobarbital-blocked rat.

The pattern of release of the gonadotrophs in response to the releasing hormone was independent of ovariectomy immediately prior to pro-oestrus, so that feedback of steroid hormones on the release from the pituitary is unimportant. This applies to the normal LH-surge occurring spontaneously, so that the evidence indicates that secretion of releasing hormone is brought to an end very soon after the LH-surge, perhaps by exhaustion of hypothalamic stores. Thus, during the ovulatory gonadotrophic surge, the pattern of events is not affected appreciably by feedback of steroid hormones on pituitary or hypothalamus.

Integration of the Events during the Ovulatory Cycle

The ewe has probably been studied more exhaustively even than the rat, and it is worth recapitulating the probable events, and their integration, as deduced by McCracken *et al.* (1971) from their studies involving transplantation of ovary and uterus to the neck where their blood supplies are more accessible (Goding *et al.*, 1967). The scheme is illustrated by Fig. 2.56 and may be explained as follows; 1. Mid-Luteal Phase. During this the corpus luteum secretes progesterone at a constant rate under the tonic influence of pituitary LH. 2. By Day 15, i.e. two days before oestrus, regression of the corpus luteum occurs, as manifest in a fall in blood progesterone. This is due to release of uterine PGF_2a. 3. The subsequent fall in progesterone permits follicular development, either owing to discharge of pituitary FSH or by sensitizing the ovary to existing levels of FSH. 4. (a). Under the influence of LH, and possibly other pituitary hormones, the developing follicle secretes

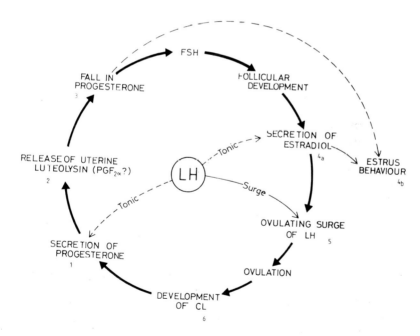

Fig. 2.56. Scheme of the ovulatory cycle in the ewe. (McCracken *et al., Rec. Progr. Hormone Res.*)

oestrogen. 4(b). The combination of falling levels of progesterone and rising levels of oestrogen causes behavioural oestrus. 5. The rise of oestrogen causes the massive LH-surge after a latent period of 12–24 hours; this leads to ovulation with decrease in steroid secretion. 6. The high LH at ovulation induces luteinization, which then leads to progesterone secretion.

Clearly, the details of this cyclical process will differ according to the species, but the fundamental relation between the luteotroph, in this case LH, and the steroid hormones is probably similar, oestrogen provoking the LH-surge, and its ability to do this being dependent on the level of progesterone in the blood. LH has a twofold function, namely an *ovulatory* and a *steroidogenic* one.

REFERENCES
(See pp. 543–545 for additional refereces)

Aiyer, M. S., Chiappa, S. A. and Fink, G. (1974). A priming effect of luteinizing hormone releasing factor in the female rat. *J. Endocrinol*, **62,** 573–588.

Aiyer, M. S., Chiappa, S. A., Fink, G. and Greig, F. (1973). A primary effect of luteinizing hormone releasing factor on the anterior pituitary gland in the female rat. *J. Physiol.*, **234,** 81–82 P.

Aiyer, M. S. and Fink, G. (1974). The role of the sex steroid hormone in modulating the responsiveness of the anterior pituitary gland to luteinizing hormone releasing factor in the female rat. *J. Endocrinol*, **62,** 553–572.

Aiyer, M. S., Fink, and G. Greig, F. (1974). Changes in the sensitivity of the pituitary gland to luteinizing hormone releasing factor during the œstrous cycle of the rat. *J. Endocrinol*, **60,** 47–64.

Appelbaum, M. E. and Taleisnik, S. (1976). Interaction between oestrogen and gonadotrophin-releasing hormone on the release and synthesis of luteinizing hormone and follicle-stimulating hormone from incubated pituitaries. *J. Endocrinol*, **68,** 127–136.

Arai, Y. (1973). Sexual differentiation and development of the hypothalamus and steroid-induced sterility. In *Neuroendocrine Control*. Eds. K. Yagi and S. Yoshida. John Wiley. N.Y., Toronto, London, pp. 27–55.

Arai, Y. and Gorski, R. A. (1968). Critical exposure time for androgenization of the developing hypothalamus in the female rat. *Endocrinology*, **82,** 1010–1014.

Arai, Y. and Gorski, R. A. (1968). Protection against the neural organizing effect of exogenous androgen in the neonatal female rat. *Endocrinology*, **82,** 1005–1009.

Arimura, A., Debeljuk, L. and Schally, A. V. (1974). Blockade of the preovulatory surge of LH and FSH and of ovulation by anti-LH-RH serum in rats. *Endocrinology*, **95,** 323–325.

Arimura, A. and Schally, A. V. (1970). Progesterone suppression of LH-releasing hormone-induced stimulation of LH release in rats. *Endocrinology*, **87,** 653–657.

Arimura, A. and Schally, A. V. (1971). Augmentation of pituitary responsiveness to LH-releasing hormone (LH-RH) by estrogen. *Proc. Soc. exp. Biol. Med. N.Y.*, **136,** 290–293.

Armstrong, D. T., Miller, L. S. and Knudsen, K. A. (1969). Regulation of lipid metabolism and progesterone production in rat corpora lutea and ovarian interstitial elements by prolactin and luteinizing hormone. *Endocrinology*, **85,** 393–401.

Armstrong, D. T. and Papkoff, H. (1976). Stimulation of aromatization of exogenous and endogenous androgens in ovaries of hypophysectomized rats *in vivo* by follicle stimulating hormone. *Endocrinology*, **99**, 1144–1151.

Asano, M., Kanzaki, S., Sekiguchi, E. and Tasaka, T. (1971). Inhibition of prostatic growth in rabbits with antiovine prolactin serum. *J. Urology*, **106**, 248–252.

Bain, J. and Keene, J. (1975). Further evidence for inhibin: *J. Endocrinol.*, **66**, 279–280.

Baird, D. T. and Fraser, I. S. (1974). Blood production and ovarian secretion rates of estradiol-17β and estrone in women throughout the menstrual cycle. *J. clin. Endocr. Metab.*, **38**, 1009–1017.

Baird, D. T. and Guevara, A. (1969). Concentration of unconjugated estrone and estradiol in peripheral plasma in nonpregnant women and in men. *J. clin. Endocrinol.*, **29**, 149–156.

Balogh, K. (1964). A histochemical method for the demonstration of 20α-hydroxysteroid dehydrogenase activity in rat ovaries. *J. Hist. Cytochem.*, **12**, 670–673.

Banks, J. A. and Freeman, M. E. (1978). The temporal requirement of progesterone on proestrus for extinction of the estrogen-induced daily signal controlling luteinizing hormone release in the rat. *Endocrinology*, **102**, 426–432.

Barfield, M. A. and Lisk, R. D. (1970). Advancement in behavioural estrus by subcutaneous injection of progesterone in the four-day cycle rat. *Endocrinology*, **87**, 1096–1098.

Barfield, M. A. and Lisk, R. D. (1974). Relative contributions of ovarian and adrenal progesterone to the timing of heat in the 4-day cyclic rat. *Endocrinology*, **94**, 571–575.

Barraclough, C. A. (1961). Production of anovulatory sterile rats by single injections of testosterone propionate. *Endocrinology*, **68**, 62–67.

Barraclough, C. A. and Gorski, R. A. (1961). Evidence that the hypothalamus is responsible for androgen-induced sterility in the female rat. *Endocrinology*, **68**, 68–79.

Barraclough, C. A. and Sawyer, C. H. (1955). Inhibition of release of pituitary ovulating hormone in the rat by morphine. *Endocrinology*, **57**, 329–337.

Bartke, A. and Dalterio, S. (1976). Effects of prolactin on the sensitivity of the testis to LH. *Biol. Reprod.*, **15**, 90–93.

Behrman, H. R. and Hichens. Quoted by Behrman, H. R. and Caldwell, B. V. (1974). MTP *Int. Rev. Sci.*, Vol. 8. Ed. R. O. Greep. Butterworths, pp. 63–94.

Berman, H. R., Orczyk, G. P. and Greep, R. O. (1972). Effect of synthetic gonadotrophin-releasing hormone (GnRH) on ovulation blocked by aspirin and indomethacin. *Prostaglandins*, **1**, 245–250.

Blake, C. A. (1976). A detailed characterization of the proestrous luteinizing hormone surge. *Endocrinology*, **98**, 445–450.

Blake, C. A. (1976a). Simulation of the proestrous luteinizing hormone (LH) surge after infusion of LH-releasing hormone in phenobarbital-blocked rats. *Endocrinology*, **98**, 451–460.

Blake, C. A. (1976b). Simulation of the early phase of the proestrous follicle stimulating hormone rise after infusion of luteinizing hormone-releasing hormone in phenobarbital-blocked rats. *Endocrinology*, **98**, 461–467.

Bland, K. P. and Donovan, B. T. (1966). Neural and humoral stimuli from the uterus and the control of ovarian function. *Ciba Study Group*. No. 23. Churchill: London.

Bland, K. P. and Donovan, B. T. (1966). Uterine distension and the function of the corpora lutea in the guinea-pig. *J. Physiol.*, **186,** 503–515.

Blatchley, F. R. and Donovan, B. T. (1976). Effect of intra-uterine foreign bodies and of prostaglandin administration on progesterone secretion during the oestrous cycle of the guinea-pig. *J. Endocrinol.*, **70,** 39–45.

Blatchley, F. R., Donovan, B. T., Horton, E. W. and Poyser, N. L. (1972). The release of prostaglandins and progestin into the utero-ovarian venous blood of guinea-pigs during the oestrous cycle and following oestrogen treatment. *J. Physiol.*, **223,** 69–88.

Bosley, C. G. and Leavitt, W. W. (1972). Dependence of preovulatory progesterone on critical period in the cyclic hamster. *Amer. J. Physiol.*, **222,** 129–133.

Brenner, R. M., Resko, J. A. and West, N. B. (1974). Cyclic changes in oviductal morphology and residual cytoplasmic estradiol binding capacity induced by sequential estradiol-progesterone treatment of spayed rhesus monkeys. *Endocrinology*, **95,** 1094–1104.

Brown-Grant, K. (1969). The induction of ovulation by ovarian steroids in the adult rat. *J. Endocrinol.*, **43,** 553–562.

Brown-Grant, K., Exley, D. and Naftolin, F. (1970). Peripheral plasma oestradiol and luteinizing hormone concentrations during the oestrous cycle of the rat. *J. Endocrinol.*, **48,** 295–296.

Brown-Grant, K. and Naftolin, F. (1972). Facilitation of luteinizing hormone secretion in the female rat by progesterone. *J. Endocrinol.*, **53,** 37–46.

Brownstein, M. J. *et al.* (1976). The effect of surgical isolation of the hypothalamus on its luteinizing hormone-releasing hormone content. *Endocrinology*, **98,** 662–665.

Bruce, N. W. and Hillier, K. (1974). The effect of prostaglandin $F_2\alpha$ on ovarian blood flow and corpora lutea regression in the rabbit. *Nature*, **249,** 176–177.

Bubenik, G. A., Brown, G. M. and Grota, L. J. (1975). Localization of immunoreactive androgen in testicular tissue. *Endocrinology*, **96,** 63–69.

Butcher, R. L., Collins, W. E. and Fugo, N. W. (1974). Plasma concentrations of LH, FSH, prolactin, progesterone and estradiol-17β throughout the 4-day estrous cycle of the rat. *Endocrinology*, **94,** 1704–1708.

Butler, W. R., Hotchkiss, J. and Knobil, E. (1975). Functional luteolysis in the rhesus monkey: ovarian estrogen and progesterone during the luteal phase of the menstrual cycle. *Endocrinology*, **96,** 1509–1512.

Caligaris, L., Astrada, J. J. and Taleisnik, S. (1968). Stimulating and inhibiting effects of progesterone on the release of luteinizing hormone. *Acta Endocrinol.*, **59,** 177–185.

Caligaris, L., Astrada, J. J. and Taleisnik, S. (1971). Release of luteinizing hormone induced by estrogen injection into ovariectomized rats. *Endocrinology*, **88,** 810–815.

Campbell, H. J. (1965). Effects of neonatal injections of hormone on sexual development and reproduction in the rabbit. *J. Physiol.*, **181,** 568–575.

Challis, J. R. G. *et al.* (1976). Production of prostaglandins E and Fα by corpora lutea, corpora albicantes and stroma from the human ovary. *J. Endocr.*, **68,** 401–408.

Channing, C. P. and Coudert, S. P. (1976). Contribution of granulosa cells and follicular fluid to ovarian estrogen secretion in the rhesus monkey *in vivo*. *Endocrinology*, **98,** 590–597.

Channing, C. P. and Kammerman, S. (1973). Characteristics of gonadotropin receptors of porcine granulosa cells during follicle maturation. *Endocrinology*, **92,** 531–540.

Chen, B.-L. and Ely, C. A. (1971). Immunological studies of the follicle-stimulating hormone. I. *Endocrinology*, **88**, 944–955.

Chobsieng, P. *et al.* (1975). Stimulating effect of prostaglandin E_2 on LH release in the rat. *Neuroend.*, **17**, 12–17.

Clemens, J. A., Sawyer, B. D. and Cerimele, B. (1977). Further evidence that serotonin is a neurotransmitter involved in the control of prolactin secretion. *Endocrinology*, **100**, 692–698.

Clemens, L. G. and Gladue, B. A. (1977). Effect of prostaglandin E_2 on masculine sexual behaviour in the rat. *J. Endocrinol.*, **75**, 383–389.

Cooke, B. A., van Beurdon, W. M. O., Rommerts, F. F. G. and van der Molen, H. J. (1972). Effects of trophic hormones on 3′,5′-cyclic AMP levels in rat testis interstitial tissue and seminiferous tubules. *FEBS Letters*, **25**, 83–86.

Cooper, K. J., Fawcett, C. P. and McCann, S. M. (1973). Variations in pituitary responsiveness to luteinizing hormone release factor during the rat oestrous cycle. *J. Endocrinol.*, **57**, 187–188.

Corbin, A. (1966). Pituitary and plasma LH of ovariectomized rats with median eminence implants of LH. *Endocrinology*, **78**, 893–896.

Corbin, A. and Cohen, A. I. (1966). Effect of median eminence implants of LH on pituitary LH of female rats. *Endocrinology*, **78**, 41–46.

Corbin, A., Daniels, E. L. and Milmore, J. E. (1970). An "internal" feedback mechanism controlling follicle stimulating hormone releasing factor. *Endocrinology*, **86**, 735–743.

Corbin, A. and Story, J. C. (1967). "Internal" feedback mechanism: response of pituitary FSH and of stalk-median eminence follicle stimulating hormone-releasing factor to median eminence implants of FSH. *Endocrinology*, **80**, 1006–1012.

Cramer, O. M. and Barraclough, C. A. (1971). Effect of electrical stimulation of the preoptic area on plasma LH concentrations in proestrous rats. *Endocrinology*, **88**, 1175–1183.

Critchlow, V. (1958). Ovulation induced by hypothalamic stimulation in the anesthetized rat. *Amer. J. Physiol.*, **195**, 171–174.

Cross, B. A. (1973). Unit responses in the hypothalamus. In *Frontiers in Neuroendocrinology*. Eds. W. F. Ganong and L. Martini, O.U.P., N.Y., pp. 133–171.

Cruz, A. de la, Arimura, A., Cruz, K. G. de la and Schally, A. V. (1976). Effect of administration of anti-serum to luteinizing hormone releasing hormone on gonadal function during the estrous cycle of the hamster. *Endocrinology*, **98**, 490–497.

Dang, B. T. and Voogt, J. L. (1977). Termination of pseudopregnancy following hypothalamic implantation of prolactin. *Endocrinology*, **100**, 873–880.

D'Angelo, S. A. and Kravatz, A. S. (1960). Gonadotrophic hormone function in persistent estrus rats with hypothalamic lesions. *Proc. Soc. exp. Biol. Med. N.Y.*, **104**, 130–133.

Davidson, J. M. (1969). Feedback control of gonadotropin secretion. In *Frontiers in Neuroendocrinology*. Eds. W. F. Ganong and L. Martini, pp. 343–388. O.U.P. London and Toronto.

Davidson, J. M. and Ganong, W. F. (1960). The effect of hypothalamic lesions on the testes and prostate of male dogs. *Endocrinology*, **66**, 480–488.

Davidson, J. M., Rodgers, C. H., Smith, E. R. and Bloch, G. J. (1968). Stimulation of female sex behaviour in adrenelectomized rats with estrogen alone. *Endocrinology*, **82**, 193–195.

Davidson, J. M. and Sawyer, C. H. (1961). Evidence for a hypothalamic focus of inhibition of gonadotropin by androgen in the male. *Proc. Soc. exp. Biol. Med.*, *N.Y.*, **107**, 4–7.

Davidson, J. M., Smith, E. R. and Bowers, C. Y. (1973). Effects of mating on gonadotropin release in the female rat. *Endocrinology*, **93**, 1185–1192.

Davies, L. J. and Ryan, K. J. (1972). Comparative endocrinology of gestation. *Vitam. & Horm.*, **30**, 223–279.

Debeljuk, L., Arimura, A. and Schally, A. V. (1972). Effect of estradiol and progesterone on the LH release induced by LH-releasing hormone (LH-RH) in intact diestrous rats and anestrous ewes. *Proc. Soc. exp. Biol. Med.*, *N.Y.*, **139**, 774–777.

De Kretser, D. M., Catt, K, J. and Paulsen, C. A. (1971). Studies on the *in vitro* testicular binding of iodinated luteinizing hormone in rats. *Endocrinology*, **88**, 332–337.

Desjardins, C., Ewing, L. L. and Irby, D. C. (1973). Response of the rabbit seminiferous epithelium to testosterone administered via polydimethylsiloxane capsules. *Endocrinology*, **93**, 450–460.

Dey, F. L. (1941). Changes in ovaries and uteri in guinea-pigs with hypothalamic lesions. *Amer. J. Anat.*, **69**, 61–87.

Dey, F. L. (1943). Evidence of hypothalamic control of hypophyseal gonadotropic functions in the female guinea-pig. *Endocrinology*, **33**, 75–82.

Dierschke, D. J. *et al.* (1973). Blockade by progesterone of estrogen-induced LH and FSH release in the rhesus monkey. *Endocrinology*, **92**, 1496–1501.

Döhler, K. D. and Wuttke, W. (1974). Total blockade of phasic pituitary prolactin release in rats: Effect on serum LH and progesterone during the estrous cycle and pregnancy. *Endocrinology*, **94**, 1595–1600.

Donoso, A. O. and Santolaya, R. C. (1969). Depletion of pituitary LH induced by coitus in the male hamster. *Acta physiol. Latin-Amer.*, **19**, 70–71.

Donovan, B. T. and Traczyk, W. (1962). The effect of uterine distension on the oestrous cycle of the guinea-pig. *J. Physiol.*, **161**, 227–236.

Dorrington, J. E. and Armstrong, D. J. (1975). Follicle-stimulating hormone stimulates estradiol-17β synthesis in cultured Sertoli cells. *Proc. Nat. Acad. Sci. Wash.*, **72**, 2677–2681.

Dorrington, J. H. and Fritz, I, B. (1974). Effects of gonadotropins on cyclic AMP production by isolated seminiferous tubule and interstitial cell preparations. *Endocrinology*, **94**, 395–403.

Dorrington, J. H., Moon, Y. S. and Armstrong, D. T. (1975). Estradiol-17β biosynthesis in cultured granulosa cells from hypophysectomized immature rats; stimulation by follicle-stimulating hormone. *Endocrinology*, **97**, 1328–1331.

Drouin, J., Lagacé, L. and Labrie, F. (1976). Estradiol induced increases of the LH responsiveness to LH releasing hormone (LHRH) in rat anterior pituitary cells in culture. *Endocrinology*, **99**, 1477–1481.

Dudley, C. A. and Moss, R. L. (1976). Facilitation of lordosis in the rat by prostaglandin E_2. *J. Endocrinol*, **71**, 457–458.

Dufau, M. L., Catt, K. J. and Tsuruhara, T. (1972). A sensitive gonadotropin responsive system: radioimmunoassay of testosterone production by the rat testis *in vitro*. *Endocrinology*, **90**, 1032–1040.

Dufy-Barbe, L., Franchimont, P. and Faure, J. M. A. (1973). Time-courses of LH and FSH release after mating in the female rabbit. *Endocrinology*, **92**, 1318–1321.

Dyer, R. G. (1973). An electrophysiological dissection of the hypothalamic regions which regulate the pre-ovulatory secretion of luteinizing hormone in the rat. *J. Physiol.*, **234**, 421–442.

Dyer, R. G. (1978). "Electrochemical stimulation" of the hypothalamus—a demonstration that the technique does not cause direct excitation of peptidergic neurones. *J. Physiol.*, **284**, 1–2P.

Dyer, R. G. and Burnet, F. (1976). Effects of ferrous ions on preoptic area neurones and luteinizing hormone secretion in the rat. *J. Endocrinol*, **69**, 247–254.

Dyer, R. G. and Cross, B. A. (1972). Antidromic identification of units in the preoptic and anterior hypothalamic areas projecting directly to the ventromedial and arcuate nuclei. *Brain Res.*, **43**, 254–257.

Dyer, R. G. and Dyball, R. E. J. (1974). Evidence for a direct effect of LRF and TRF on single unit activity in rostral hypothalamus. *Nature, Lond.*, **252**, 486–488.

Dyer, R. G., Pritchett, C. J. and Cross, B. A. (1972). Unit activity in the diencephalon of female rats during the oestrous cycle. *J. Endocrinol.*, **53**, 151–160.

Eckstein, B., Golan, R. and Shani, J. (1973). Onset of puberty in the immature female rat induced by 5α-androstane-3β, 17β diol. *Endocrinology*, **92**, 941–945.

Elger, W. (1966). Die Rolle der fetalen Androgene in der sexual differenzierung des Kaninchens und ihre Abgrenzung gegen andere hormonale und somatische Faktoren durch Anwendung eines starken Antiandrogens. *Arch. d'Anat. micr. Morph. exp.*, **55**, 658–743.

El Safoury, S. and Bartke, A. (1974). Effects of follicle-stimulating hormone and luteinizing hormone levels in hypophysectomized and in intact immature adult male rats. *J. Endocrinol.*, **61**, 193–198.

Eskay, R. L., Warberg, J., Mical, R. S. and Porter, J. C. (1975). Prostaglandin E$_2$-induced release of LHRH into hypophyseal portal blood. *Endocrinology*, **97**, 817–824.

Espey, L. L. and Rondell, P. (1968). Collagenolytic activity in the rabbit and sow Graafian follicle during ovulation. *Amer. J. Physiol.*, **214**, 326–329.

Everett, J. W. (1948). Progesterone and estrogen in the experimental control of ovulation time and other features of the estrous cycle in the rat. *Endocrinology*, **43**, 389–405.

Everett, J. W. and Radford, H. M. (1961). Irritative deposits from stainless steel electrodes in the preoptic rat brain causing release of pituitary gonadotropin. *Proc. Soc. exp. Biol. Med., N.Y.*, **108**, 604–609.

Everett, J. W. and Sawyer, C. H. (1950). A 24-hour periodicity in the "LH-release apparatus" of female rats, disclosed by barbiturate sedation. *Endocrinology*, **47**, 198–218.

Feder, H. H., Brown-Grant, K. and Corker, C. S. (1971). Pre-ovulatory progesterone, the adrenal cortex and the critical period for luteinizing hormone release in rats. *J. Endocrinol.*, **50**, 29–39.

Feder, H. H., Naftolin, F. and Ryan, K. J. (1974). Male and female sexual responses in male rats given estradiol benzoate and 5α-androsten-17β-ol-3-one propionate. *Endocrinology*, **94**, 136–141.

Feder, H. A., Resko, J. A. and Goy, R. W. (1968). Progesterone levels in the arterial plasma of pre-ovulatory and ovariectomized rats. *J. Endocrinol.*, **41**, 563–568.

Feder, H. H. and Ruf, K. B. (1969). Stimulation of progesterone release and estrous behaviour by ACTH in ovariectomized rodents. *Endocrinology*, **84**, 171–174.

Ferin, M. *et al.* (1974). Active immunization to 17-β-estradiol and its effects upon the reproductive cycle of the rhesus monkey. *Endocrinology*, **94**, 765–776.

Ferin, M., Tempone, A., Zimmering, P. E. and Vande Wiele, M. L. (1969). Effect of antibodies to 17β-estradiol and progesterone on the estrous cycle of the rat. *Endocrinology*, **85**, 1070–1078.

Fink, G. and Aiyer, M. S. (1974). Gonadotrophin secretion after electrical stimulation of the preoptic area during oestrous cycle of the rat. *J. Endocrinol.*, **62**, 589–604.

Fink, G., Chiappa, S. A. and Aiyer, M. S. (1976). Priming effect of luteinizing hormone releasing factor elicited by preoptic stimulation and by intravenous infusion and multiple injections of the synthetic decapeptide. *J. Endocrinol.*, **69**, 359–372.

Fink, G. and Jamieson, M. G. (1976). Immunoreactive luteinizing hormone releasing factor in rat pituitary stalk blood: effects of electrical stimulation of the medial preoptic area. *J. Endocrinol.*, **68**, 71–87.

Flerkó, B. (1970). Control of follicle-stimulating hormone and luteinizing hormone secretion. In *The Hypothalamus*. Eds. L. Martini, M. Motta and F. Fraschini., pp. 351–363. Academic Press, New York and London.

Flerkó, B. and Bárdos, V. (1961). Absence of compensatory ovarian hypertrophy in rats with anterior hypothalamic lesions. *Acta. Endocrinol.*, **36**, 180–184.

Ford, J. J. and Yoshinaga, K. (1975). The role of LH in the luteotropic process of lactating rats. *Endocrinology*, **96**, 329–334.

Franchimont, P., Chari, S. and Demoulin, A. (1975). Hypothalamus-pituitary-testis interaction. *J. Reprod. and Fert.*, **44**, 335–350.

Fraser, I. S. *et al.* (1973). Cyclical ovarian function in women with congenital absence of uterus and vagina. *J. clin. Endocrinol.*, **36**, 634–637.

Freeman, M. C., Dupke, K. C. and Croteau, C. M. (1976). Extinction of the estrogen-induced daily signal for LH release in the rat: a role for the proestrous surge of progesterone. *Endocrinology*, **99**, 223–229.

Freeman, H. E., Reichert, L. E. and Neill, J. D. (1972). Regulation of the proestrous surge of prolactin secretion by gonadotropin and estrogens in the rat. *Endocrinology*, **90**, 232–238.

Freeman, H. E., Smith, M. S., Nazian, S. J. and Neill, J. D. (1974). Ovarian and hypothalamic control of the daily surges of prolactin secretion during pseudopregnancy in the rat. *Endocrinology*, **94**, 875–882.

French, F. S. and Ritzén, E. M. (1973). Androgen-binding protein in efferent duct fluid of rat testis. *J. Reprod. Fert.*, **32**, 479–483.

Fritz, I. B., Rommerts, F. G., Louis, B. G. and Dorrington, J. H. (1976). Regulation by FSH and dibutyryl cyclic AMP of the formation of androgen binding protein in Sertoli cell-enriched cultures. *J. Reprod. Fert.*, **46**, 17–24.

Fuchs, A.-R. and Mok, E. (1974). Histochemical study of the effects of prostaglandins $F_2\alpha$ and E_2 on the corpus luteum of pregnant rats. *Biol. Reprod.*, **10**, 23–38.

Gallo, R. V. and Osland, R. B. (1976). Electrical stimulation of the arcuate nucleus in ovariectomized rats inhibits episodic luteinizing hormone (LH) release but excites LH release after estrogen priming. *Endocrinology*, **99**, 659–668.

Gay, V. L. and Midgley, A. R. (1969). Response of the adult rat to orchidectomy and ovariectomy as determined by LH-radioimmunoassay. *Endocrinology*, **84**, 1359–1364.

Gay, V. L. and Denvers, N. W. (1971). Effects of testosterone propionate and estradiol benzoate—alone or in combination—on serum LH and FSH in ovariectomized rats. *Endocrinology*, **89**, 161–168.

Gleeson, A. R., Thorburn, G. D. and Cox, R. L. (1974). Prostaglandin F concentrations in the utero-ovarian venous plasma of the sow during the late luteal phase of the oestrous cycle. *Prostaglandins*, **5**, 521–529.

Goding, J. R., McCracken, J. A. and Baird, D. T. (1967). The study of ovarian function in the ewe by means of a vascular autotransplantation technique. *J. Endocrinol.*, **39**, 37–52.

Goldman, A. S. and Baker, M. K. (1971). Androgenicity in the rat fetus of metabolites of testosterone and antagonism by cyproterone acetate. *Endocrinology*, **89**, 276–280.

Goldman, B. D., Kamberi, I. A., Siiteri, P. K. and Porter, J. C. (1969). Temporal relationship of progestin secretion, LH release and ovulation in rats. *Endocrinology*, **85**, 1137–1143.

Goodman, A. L. and Neill, J. D. (1976). Ovarian regulation of postcoital gonadotropin release in the rabbit: re-examination of a functional role for 20α dihydroprogesterone. *Endocrinology*, **99**, 852–860.

Goodman, R. L. (1978). The site of positive feedback action of estradiol in the rat. *Endocrinology*, **102**, 151–159.

Gorski, R. A. (1963). Modification of ovulatory mechanisms by postnatal administration of estrogen to the rat. *Amer. J. Physiol.*, **205**, 842–844.

Gorski, R. A. and Barraclough, C. A. (1963). Effects of low dosages of androgen on the differentiation of hypothalamic regulatory control of ovulation in the rat. *Endocrinology*, **73**, 210–216.

Gorski, R. A. and Wagner, J. W. (1965). Gonadal activity and sexual differentiation of the hypothalamus, *Endocrinology*, **76**, 226–239.

Goy, R. W. and Resko, J. A. (1972). Gonadal hormone and behavior of normal and pseudohermaphroditic nonhuman female primates. *Recent. Prog. Horm. Res.*, **28**, 707–733.

Grady, K. L. and Greenwald, G. S. (1968). Studies on interactions between the ovary and pituitary follicle-stimulating hormone in the golden hamster. *J. Endocrinol.*, **40**, 85–90.

Greep, R. O., Van Dyke, H. B. and Chow, B. F. (1942). Gonadotropins of the swine pituitary. *Endocrinology*, **30**, 635–649.

Grinwich, D. L., Hichens, M. and Behrman, H. R. (1976). Control of the LH re-receptor by prolactin and prostaglandin F₂α in rat corpora lutea. *Biol. Reprod.*, **14**, 212–218.

Grinwich, D. L., Kennedy, T. G. and Armstrong, D. T. (1972). Dissociation of ovulatory and steroidogenic actions of luteinizing hormone in rabbits with indomethacin, an inhibitor of prostaglandin biosynthesis. *Prostaglandins*, **1**, 89–94.

Hafiez, A. A., Lloyd, C. W. and Bartke, A. (1972). The role of prolactin in the regulation of testis function: the effects of prolactin and luteinizing hormone on the plasma levels of testosterone and androstenedione in hypophysectomized rats. *J. Endocrinol.*, **52**, 237–332.

Hagenäs, L. *et al.* (1975). Sertoli cell origin of testicular androgen-binding protein (ABP). *Mol. cell. Endocrinol.*, **2**, 339–350.

Halász, B. and Gorski, R. A. (1967). Gonadotrophic hormone secretion in female rats after partial or total interruption of neural afferents to the medial basal hypothalamus. *Endocrinology*, **80**, 608–622.

Hall, P. F. and Eik-Nes, K. B. (1962). The action of gonadotropic hormones upon rabbit testes *in vitro*. *Biochim. biophys. Acta*, **63**, 411–422.

Hall, P. F., Irby, D. C. and de Kretser, D. M. (1969). Conversion of cholesterol to androgens by rat testes: comparison of interstitial cells and seminiferous tubules. *Endocrinology*, **88**, 488–496.

Hansel, W. (1971). Discussion to McCracken *et al.* (1971), p. 633.

Hansson, V. *et al.* (1974). Androgen transport and receptor mechanisms in testis and epididymis. *Nature*, Lond., **250**, 387–391.

Hansson, V. *et al.* (1975). Regulation of seminiferous tubular function by FSH and androgen. *J. Reprod. Fert.*, **44**, 363–375.

Hansson, V. *et al.* (1976). Secretion and role of androgen-binding proteins in the testis and epididymis. *J. Reprod. Fert.*, Suppl. **24**, 17–33.

Hansson, V. and Tveter, K. J. (1971). Uptake and binding *in vivo* of ^{3}H-labelled androgen in the rat epididymis and ductus deferens. *Acta Endocrinol*, **66**, 745–755.

Harrington, F. E., Bex, F. J., Elton, R. L. and Roach, J. B. (1970). The ovulatory effects of follicle stimulating hormone treated with chymotrypsin in chlorpromazine blocked rats. *Acta Endocrinol*, **65**, 222–228.

Harrington, F. E. and Elton, R. L. (1969). Induction of ovulation in adult rats with follicle stimulating hormone. *Proc. Soc. exp. Biol. Med. N.Y.*, **132**, 841–844.

Harris, G. W. (1948). Electrical stimulation of the hypothalamus and the mechanism of neural control of the adenohypophysis. *J. Physiol.*, **107**, 418–429.

Harris, G. W. (1964). Sex hormones, brain development and brain function. *Endocrinology*, **75**, 627–648.

Harris, G. W. and Levine, S. (1965). Sexual differentiation of the brain and its experimental control. *J. Physiol.*, **181**, 379–400.

Harris, M. C., Makara, G. B. and Spyer, K. M. (1971). Electrophysiological identification of neurones of the tubero-infundibular system. *J. Physiol.*, **218**, 86–87P.

Henderson, K. M. and McNatty, K. P. (1977). A possible interrelationship between gonadotrophin stimulation and prostaglandin $F_2\alpha$ inhibition of steroidogenesis by granulosa-luteal cells *in vitro*. *J. Endocrinol*, **73**, 71–78.

Henderson, K. M., Scaramuzzi, R. J. and Baird, D. T. (1977). Simultaneous infusion of prostaglandin E_2 antagonizes the luteolytic action of prostaglandin $F_2\alpha$ *in vivo*. *J. Endocrinol*, **72**, 379–383.

Hess, L. L. *et al.* (1977). Estrogen-induced gonadotropin surges in decerebrated female rhesus monkeys with medial basal hypothalamic peninsulae. *Endocrinology*, **101**, 1264–1271.

Hillarp, N.-Å. (1949). Studies on the localization of hypothalamic centres controlling the gonadotrophic function of the hypophysis. *Acta Endocrinol.*, **2**, 11–23.

Hilliard, J., Penardi, R. and Sawyer, C. H. (1967). A functional role for 20-α-hydroxypregn-4-en-3-one in the rabbit. *Endocrinology*, **80**, 901–909.

Hilliard, J., Schally, A. V. and Sawyer, C. H. (1971). Progesterone blockade of the ovulatory response to intrapituitary infusion of LH-RH in rabbits. *Endocrinology*, **88**, 730–736.

Hodgen *et al.* (1974). Specific radioimmunoassay of chorionic gonadotropin during implantation in rhesus monkeys. *J. clin. Endocr. Metab.*, **39**, 457–464.

Hoffmann, P. (1960). Untersuchungen uber die hormonaler Beeinflussung der Lebensdauer des Corpus luteum im Zyklus der Frau. *Geburtshilfe u. Frauenheilk.*, **20**, 1153–1159.

Holt, J. A., Richards, J. S., Midgley, A. R. and Reichert, L. E. (1976). Effect of prolactin on LH receptor in rat luteal cells. *Endocrinology*, **98**, 1005–1013.

Hyppo, M., Motta, M. and Martini, L. (1971). "Ultrashort" feedback control of follicle stimulating hormone-releasing factor secretion. *Neuroendocrinol.*, **7**, 227–235.

Jackson, I. M. D. and Reichlin, S. (1974). Thyrotropin-releasing hormone (TRH), distribution in hypothalamic and extrahypothalamic brain tissues of mammalian and submammalian chordates. *Endocrinology*, **95**, 854–862.

Jaffe, R. B. and Keye, W. R. (1974). Estradiol augmentation of pituitary responsiveness to gonadotrophin-releasing hormone in woman. *J. clin. Endocr. Metab.*, **39**, 850–855.

de Jong, F. H., Baird, D. T. and van der Molen, H. J. (1974). Ovarian secretion rates of oestrogens, androgens and progesterone in normal women and in women with persistent ovarian follicles. *Acta. Endocrinol.*, **77**, 575–587.

Jones, G. E. and Boyns, A. R. (1974). Effect of gonadal steroids on the pituitary responsiveness to synthetic luteinizing hormone releasing hormone in the male dog. *J. Endocrinol*, **61**, 123–131.

Joslyn, W. D., Wallen, K. and Goy, R. W. (1976). Advancement of ovulation in the guinea-pig with exogenous progesterone and related effects on length of the oestrous cycle and life span of the corpus luteum. *J. Endocrinol*, **70**, 275–283.

Jost, A. (1954). Hormonal factors in the development of the fetus. *C.S.H. Symp. Quant. Biol.*, **19**, 167–181.

Jost, A. and Bergerard, Y. (1949). Culture *in vitro* d'ébauches du tractus génital du foetus de rat. *C.r. Soc. Biol.*, **143**, 608–609.

Jost, A., Vigier, B., Prépin, J. and Perchellet, J. P. (1973). Studies on sex differentiation in mammals. *Rec. Progr. Horm. Res.*, **29**, 1–35.

Kalra, S. P. (1975). Observations on facilitation of the preovulatory rise of LH by estrogen. *Endocrinology*, **96**, 23–28.

Kalra, S. P. (1976). Tissue levels of luteinizing hormone-releasing hormone in the preoptic area and hypothalamus and serum concentrations of gonadotropins following anterior hypothalamic deafferentation and estrogen treatment in the female rat. *Endocrinology*, **99**, 101–107.

Kalra, S. P. *et al.* (1971). Effects of hypothalamic and preoptic electrochemical stimulation on gonadotropin and prolactin release in prestrous rats. *Endocrinology*, **88**, 1150–1158.

Kalra, S. P. and Kalra, P. S. (1974). Effects of circulating estradiol during rat estrous cycle on LH release following electrochemical stimulation of preoptic brain or administration of synthetic LRF. *Endocrinology*, **94**, 845–851.

Kalra, S. P. and McCann, S. M. (1973). Variations in the release of LH in response to electrochemical stimulation of preoptic area and of medial basal hypothalamus during the estrous cycle of the rat. *Endocrinology*, **93**, 665–669.

Kamel, F., Wright, W. W., Mock, E. J. and Frankel, A. I. (1977). The influence of mating and related stimuli on plasma levels of luteinizing hormone, follicle stimulating hormone, prolactin, and testosterone in the male rat. *Endocrinology*, **101**, 421–429.

Kammerman, S., Canfield, R. E., Kolena, J. and Channing, C. P. (1972). The binding of iodinated hCG to porcine granulosa cells. *Endocrinology*, **91**, 65–74.

Kanematsu, S., Scaramuzzi, J., Hilliard, J. and Sawyer, C. H. (1974). Patterns of ovulation-inducing LH release following coitus, electrical stimulation and exogenous LH-RH in the rabbit. *Endocrinology*, **95**, 247–252.

Karsch, F. J. *et al.* (1973). Positive and negative feedback control by estrogen of luteinizing hormone secretion in the rhesus monkey. *Endocrinology*, **92**, 799–804.

Karsch, F. J. *et al.* (1973). An analysis of the negative feedback control of gonadotropin secretion utilizing chronic implantation of ovarian steroids in ovariectomized rhesus monkeys. *Endocrinology*, **93**, 478–486.

Karsch, F. J. and Sutton, G. P. (1976). An intra-ovarian site for the luteolytic action of estrogen in the rhesus monkey. *Endocrinology*, **98**, 553–561.

170   INTRODUCTION TO PHYSIOLOGY

Katongole, C. B., Naftolin, F. and Short, R. V. (1971). Relationship between blood levels of luteinizing hormone and testosterone in bulls, and the effects of sexual stimulation. *J. Endocrinol.*, **50**, 457–466.

Kelch, R. P. *et al.* (1972). Estradiol and testosterone secretion by human, simian, and canine testes. *J. clin. Invest.*, **51**, 824–830.

Keogh, E. J. *et al.* (1976). Selective suppression of FSH by testicular extracts. *Endocrinology*, **98**, 997–1004.

Keye, W. R. and Jaffe, R. B. (1974). Modulation of pituitary gonadotropin response to gonadotropin-releasing hormone by estradiol. *J. clin. Endocr. Metab.*, **38**, 805–810.

Kirton, K. T. *et al.* (1970). Serum luteinizing hormone and progesterone concentration during the menstrual cycle of the rhesus monkey. *J. clin. Endocrinol.*, **30**, 105–110.

Kingsley, T. R. and Bogdanove, E. M. (1973). Direct feedback of androgens: localized effects of intrapituitary implants of androgens on gonadotrophic cells and hormone stores. *Endocrinology*, **93**, 1398–1409.

Kizer, J. S. *et al.* (1976). Distribution of releasing factors, biogenic amines, and related enzymes in the bovine median eminence. *Endocrinology*, **98**, 685–695.

Kobayashi, R. M., Lu, K. H., Moor, R. Y. and Yen, S. S. C. (1978). Regional distribution of hypothalamic luteinizing hormone-releasing hormone in pro-oestrous rats: effects of ovariectomy and estrogen replacement. *Endocrinology*, **102**, 98–105.

Kledzik, G. S., Marshall, S., Campbell, G. A., Gelato, M. and Meites, J. (1976). Effects of castration, testosterone, estradiol, and prolactin on specific prolactin-binding activity in ventral prostate of male rats. *Endocrinology*, **98**, 373–379.

Korda, A. R. *et al.* (1975). Assessment of possible luteolytic effect of intra-ovarian injection of prostaglandin F$_2\alpha$ in the human. *Prostaglandins*, **9**, 443–449.

Köves, K. and Halasz, B. (1970). Location of neural structures triggering ovulation in the rat. *Neuroend.*, **6**, 180–193.

Krey, L. C., Butler, W. R. and Knobil, E. (1975). Surgical disconnection of the medial basal hypothalamus and pituitary function in the rhesus monkey I. Gonadotropin secretion. *Endocrinology*, **96**, 1073–1089.

Kubo, K., Mennin, S. P. and Gorski, R. A. (1975). Similarity of plasma LH release in androgenized and normal rats following electrochemical stimulation of the basal forebrain. *Endocrinology*, **96**, 492–500.

Kuehl, F. A. *et al.* (1970). Prostaglandin receptor site: evidence for an essential role in the action of luteinizing hormone. *Science*, **169**, 883–886.

Ledwitz-Rigby, F. *et al.* (1977). Inhibitory action of porcine follicular fluid upon granulosa cell luteinization *in vitro*: assay and influence of follicular maturation. *J. Endocrinol.*, **74**, 175–184.

Lee, V. W. K., Keogh, E. J., Burger, H. G., Hudson, B. and de Kretser, D. M. (1976). Studies on the relationship between FSH and germ cells. Evidence for selective suppression of FSH by testicular extracts. *J. Reprod. and Fert.*, Suppl. **24**, 1–15.

Legan, S. J., Coon, G. A. and Karsch, F. J. (1975). Role of estrogen as initiator of daily LH surges in the ovariectomized rat. *Endocrinology*, **96**, 50–56.

Legan, S. J. and Karsch, F. J. (1975). A daily signal for the LH surge in the rat. *Endocrinology*, **96**, 57–62.

Legan, S. J. and Karsch, F. J. (1975). Modulation of pituitary responsiveness to luteinizing hormone-releasing factor during the estrous cycle of the rat. *Endocrinology*, **96**, 571–575.

Le Maire, W. J., Rice, B. F. and Savard, K. (1968). Steroid hormone formation of the human ovary. *J. clin. Endocrinol.*, **28**, 1249–1256.

Libertun, C., Orias, R. and McCann, S. M. (1974). Biphasic effect of estrogen on the sensitivity of the pituitary to luteinizing hormone-releasing factor (LRF). *Endocrinology*, **94**, 1094–1100.

Lindner, H. R. *et al.* (1974). Gonadotropin action on cultured Graafian follicles: induction of maturation division of the mammalian oocyte and differentiation of the luteal cell. *Recent. Prog. Horm. Res.*, **30**, 79–127.

Lisk, R. D. and Reuter, L. A. (1976). Dexamethasone: increased weights and decreased [^{3}H]estradiol retention of uterus, vagina and pituitary in the ovariectomized rat. *Endocrinology*, **99**, 1063–1070.

Loeb, L. (1923). The effect of extirpation of the uterus on the life and function of the corpus luteum in the guinea-pig. *Proc. Soc. exp. Biol. Med.*, **20**, 441–443.

Lostroh, A. J. and Johnson, R. E. (1966). Amounts of interstitial cell-stimulating hormone and follicle-stimulating hormone required for follicular development, uterine growth and ovulation in the hypophysectomized rat. *Endocrinology*, **79**, 991–996.

Louvet, J. P. and Vaitukaitis, L. J. (1976). Induction of follicle-stimulating hormone (FSH) receptors in rat ovaries by estrogen priming. *Endocrinology*, **99**, 758–764.

Madhwa Raj, H. G. and Dym, M. (1976). The effects of selective withdrawal of FSH or LH on spermatogenesis in the immature rat. *Biol. Reprod.*, **14**, 489–494.

Makepeace, A. W., Weinstein, G. L. and Friedman, M. H. (1937). The effect of progestin and progesterone on ovulation in the rabbit. *Amer. J. Physiol.*, **119**, 512–516.

Manaugh, L. C. and Novy, M. J. (1976). Effects of indomethacin on corpus luteum function and pregnancy in rhesus monkeys. *Fert. Steril.*, **27**, 588–598.

Mann, D. R. and Barraclough, C. A. (1973). Changes in peripheral plasma progesterone during the rat 4-day estrous cycle: an adrenal diurnal rhythm. *Proc. Soc. exp. Biol. Med. N.Y.*, **142**, 1226–1229.

Mann, D. R. and Barraclough, C. A. (1973). Role of estrogen and progesterone in facilitating LH release in 4-day cyclic rats. *Endocrinology*, **93**, 694–699.

Marsh, J. M. and Le Maire, W. J. (1974). Cyclic AMP accumulation and steroidogenesis in the human corpus luteum: effect of gonadotropins and prostaglandins. *J. clin. Endocr. Metab.*, **38**, 99–106.

Mayer, G. and Thevenot-Duluc, A. J. (1962). Répercussion génitale chez le rat de l'administration post-natale d'androgène et d'oestrogène. *C.r. Soc. Biol.*, *Paris*, **156**, 1377–1379.

McCann, S. M. and Ramirez, V. D. (1964). The neuroendocrine regulation of hypophyseal luteinizing hormone secretion. *Rec. Progr. Horm. Res.*, **20**, 131–181.

McCracken, J. A., Baird, D. T. and Goding, J. R. (1971). Factors affecting the secretion of steroids from the transplanted ovary in the sheep. *Rec. Progr. Horm., Res.* **27**, 537–582.

McCracken, J. A. and Caldwell, B. V. (1969). Corpus luteum maintenance in a ewe with one congenitally absent uterine horn. *J. Reprod. Fert.*, **20**, 139–141.

McCullagh, D. R. (1932). Dual endocrine activity of the testes. *Science, N.Y.*, **76**, 19–20.

McCullagh, D. R. and Schneider, I. (1940). The effect of a non-androgenic testis extract on the estrous cycle in rats. *Endocrinology*, **27**, 899–902.

McDonald, P. G. and Doughty, C. (1974). Effect of neonatal administration of different androgens in the female rat: correlation between aromatization and the induction of sterilization. *J. Endocrinol.*, **61**, 95–103.

McLean, B. K. and Nikitovitch-Winer, M. B. (1973). Corpus luteum function in the rat: a critical period for luteal activation and the control of luteal maintenance. *Endocrinology*, **93**, 316–323.

McNatty, K. P., Henderson, K. M. and Sawers, R. S. (1975). Effects of prostaglandin $F_2\alpha$ and E_2 on the production of progesterone by human granulosa cells in tissue culture. *J. Endocrinol*, **67**, 231–240.

McNatty, K. P., Sawers, R. S. and McNeilly, A. S. (1974). A possible role for prolactin in control of steroid secretion by the human Graafian follicle. *Nature*, **250**, 653–655.

Means, A. R. *et al.* (1976). Follicle-stimulating hormone, the Sertoli cell, and spermatogenesis. *Recent. Prog. Horm. Res.*, **32**, 477–527.

Means, A. R. and Vaitukaitis, J. (1972). Peptide hormone "receptors": specific binding of ^{3}H-FSH to testes. *Endocrinology*, **90**, 39–46.

Midgley, A. R. and Jaffe, R. B. (1968). Regulation of human gonadotrophins. IV. *J. clin. Endocrinol.*, **28**, 1699–1703.

Midgley, A. R. and Jaffe, R. B. (1971). Regulation of human gonadotrophins. X. Episodic fluctuation of LH during the menstrual cycle. *J. clin. Endocrinol.*, **33**, 962–969.

Mikhail, G. (1970). Hormone secretion by the human ovaries. *Gynec. Invest.*, **1**, 5–20.

Mittler, J. C. (1972). Androgen effect on gonadotropin secretion by organ-cultured anterior pituitary. *Proc. Soc. exp. Biol. Med.*, *N.Y.*, **140**, 1140–1142.

Moor, R. M. (1974). The ovarian follicle of the sheep: inhibition of oestrogen secretion by luteinizing hormone. *J. Endocrinol.*, **61**, 455–463.

Moor, R. M. and Rowson, L. E. A. (1964). Influence of the embryo and uterus on luteal function in the sheep. *Nature*, **201**, 522–523.

Moss, R. L. and McCann, S. M. (1973). Induction of mating behavior in rats by luteinizing hormone releasing factor. *Science*, **181**, 177–179.

Moss, R. L. and McCann, S. M. (1975). Action of luteinizing hormone-releasing factor (LRF) in the initiation of lordosis behaviour in the estrone primed ovariectomized female rat. *Neuroend.*, **17**, 309–318.

Naftolin, F. *et al.* (1975). The formation of estrogens by central neuroendocrine tissues. *Recent Prog. Horm. Res.*, **31**, 295–315.

Naftolin, F., Ryan, K. J. and Petro, Z. (1971). Aromatization of androstenedione by the diencephalon. *J. Clin. Endocr. Metab.*, **33**, 368–370.

Negro-Vilar, A., Orias, R. and McCann, S. M. (1973). Evidence for a pituitary site of action for the acute inhibition of LH release by estrogen in the rat. *Endocrinology*, **92**, 1680–1684.

Nequin, L. G. and Schwartz, N. B. (1971). Adrenal participation in the timing of mating and LH release in the cyclic rat. *Endocrinology*, **88**, 325–331.

Nimrod, A., Erickson, G. F. and Ryan, K. J. (1976). A specific FSH receptor in rat granulosa cells: properties of binding *in vitro*. *Endocrinology*, **98**, 56–64.

Norman, R. L., Resko, J. A. and Spies, H. G. (1976). The anterior hypothalamus: how it affects gonadotropin secretion in the rhesus monkey. *Endocrinology*, **99**, 59–71.

Nunez, E. *et al.* (1971a). Mise en évidence d'une fraction protéique liant les oestrogènes dans le serum de rats impubères. *C.r. Acad. Sci. Paris.*, **272**, 2396–2399.

Nunez, E. *et al.* (1971b). Origine embryonnaire de la protéine sérique fixant l'oestrone et l'oestradiol chez la ratte impubère. *C.r. Acad. Sci. Paris*, **273**, 242–245.

Odell, W. D. and Swerdloff, R. S. (1968). Progestogen-induced luteinizing and follicle-stimulating hormone surge in postmenopausal women: a simulated ovulatory peak. *Proc. Nat. Acad. Sci.*, *Wash.*, **61**, 529–536.

O'Grady, J. P., Caldwell, B. V., Auletta, F. J. and Speroff, L. (1972). The effects of an inhibition of prostaglandin synthesis (indomethacin) on ovulation, pregnancy, and pseudopregnancy in the rabbit. *Prostaglandins*, **1**, 97–106.

Ojeda, S. R., Harris, P. G. and McCann, S. M. (1975). Effect of inhibitors of prostaglandin synthesis on gonadotropin release in the rat. *J. Endocrinol*, **97**, 843–854.

Ojeda, S. R., Jameson, H. E. and McCann, S. M. (1976). Further studies on prostaglandin E_2(PGE$_2$)-induced gonadotropin release: comparison with the effect of 15-methyl PGE$_2$, PGAs, PGBs and prostaglandin endoperoxide analogs. *Prostaglandins*, **12**, 281–301.

Ojeda, S. R., Jameson, H. E. and McCann, S. M. (1977). Hypothalamic areas involved in prostaglandin (PG)-induced gonadotropin release. I and II. *Endocrinology*, **100**, 1585–1594; 1595–1603.

Orias, R., Negro-Vilar, A., Libertun, C. and McCann, S. M. (1974). Inhibitory effect on LH release of estradiol injected into the third ventricle. *Endocrinology*, **94**, 852–855.

Orth, J. and Christensen, A. K. (1977). Localization of ^{125}I-labeled FSH in the testes of hypophysectomized rats by autoradiography at the light and electron microscope levels. *Endocrinology*, **101**, 262–278.

Pang, C. N., Zimmermann, E. and Sawyer, C. H. (1977). Morphine inhibition of the preovulatory surges of plasma luteinizing hormone and follicle stimulating hormone in the rat. *Endocrinology*, **101**, 1726–1732.

Pant, H. C., Hopkinson, C. R. N. and Fitzpatrick, R. J. (1977). Concentration of oestradiol, progesterone, luteinizing hormone and follicle-stimulating hormone in the jugular venous plasma of ewes during the oestrous cycle. *J. Endocrinol*, **73**, 247–255.

Parlow, F. N. (1961). Bioassay of pituitary luteinizing hormone by depletion of ovarian ascorbic acid. In *Human Pituitary Gonadotrophins*. Ed. A. Albert, pp. 300–310. Thomas, Springfield.

Parlow, A. F. and Reichert, L. E. (1964). Influence of follicle-stimulating hormone on the prostate assay of luteinizing hormone (LH, ICSH). *Endocrinology*, **73**, 377–385.

Patek, C. E. and Watson, J. (1976). Prostaglandin F and progesterone secreted by porcine endometrium and corpus luteum *in vitro*. *Prostaglandins*, **12**, 97–111.

Pfaff, D. W. (1973). Luteinizing hormone-releasing factor potentiates lordosis behaviour in hypophysectomized ovariectomized female rats. *Science, N.Y.*, **182**, 1148–1149.

Pfeiffer, C. A. (1936). Sexual differentiation of the hypophyses and their determination by the gonads. *Amer. J. Anat.*, **58**, 195–225.

Pharriss, B. B. (1970). The possible vascular regulation of luteal function. *Persp. Biol. Med.*, **13**, 434–444.

Pickering, A. J. M. C. and Fink, G. (1976). Priming effect of luteinizing hormone releasing factor: *in vitro* and *in vivo* evidence consistent with its dependence on protein and RNA synthesis. *J. Endocrinol*, **69**, 373–379.

Pincus, G. and Enzmann, E. V. (1935). The comparative behavior of mammalian eggs *in vivo* and *in vitro*. *J. exp. Med.*, **62**, 665–675.

Plant, T. M. *et al.* (1978). The arcuate nucleus and the control of gonadotropin and prolactin secretion in the female rhesus monkey (*Macaca mulatta*). *Endocrinology*, **102**, 52–62.

Powers, J. B. (1970). Hormonal control of sexual receptivity during the estrous cycle of the rat. *Physiol. Behav.*, **5**, 831–835.

Quadri, S. K., Norman, R. L. and Spies, H. G. (1977). Prolactin release following electrical stimulation of the brain in ovariectomized and ovariectomized-estrogen-treated rhesus monkeys. *Endocrinology*, **100**, 325–330.

Ramirez, V. D., Abrams, R. M. and McCann, S. M. (1964). Effect of estradiol implants in the hypothalamo-hypophysial region of the rat on the secretion of luteinizing hormone. *Endocrinology*, **75**, 243–248.

Ramirez, V. D. and McCann, S. M. (1965). Inhibitory effect of testosterone on luteinizing hormone secretion in immature and adult rats. *Endocrinology*, **76**, 214–217.

Rani, C. S. S. and Moudgal, N. R. (1977). Role of the proestrous surge of gonado-tropins in the initiation of follicular maturation in the cyclic hamster. *Endocrinology*, **101**, 1484–1495.

Rao, M. C., Richards, J. S., Midgley, A. R. and Reichert, L. E. (1977). Regulation of gonadotrophin receptors by luteinizing hormone in granulosa cells. *Endocrinology*, **101**, 512–524.

Reddy, V. V. R., Naftolin, F. and Ryan, K. J. (1973). Aromatization in the central nervous system of rabbits: effects of castration and hormone treatment. *Endocrinology*, **92**, 589–594.

Reddy, V. V. R., Naftolin, F. and Ryan, K. J. (1974). Conversion of androstenedione to estrone by neural tissues from fetal and neonatal rats. *Endocrinology*, **94**, 117–121.

Renaud, L. P. (1976). Tuberoinfundibular neurons in the basomedial hypothalamus of the rat; electrophysiological evidence for axon collaterals to hypothalamic and extrahypothalamic areas. *Brain Res.*, **105**, 59–72.

Renaud, L. P., Martin, J. B. and Brazeau, P. (1975). Depressant action of TRH, LH-RH and somatostatin on activity of central neurones. *Nature, Lond.*, **255**, 233–235.

Resko, J. A., Feder, H. H. and Goy, R. W. (1968). Androgen concentrations in plasma and testes of developing rats. *J. Endocrinol.*, **40**, 485–491.

Rich, K. A. and de Kretser, D. M. (1977). Effect of differing degrees of destruction of the rat seminiferous epithelium on levels of serum follicle stimulating hormone and androgen binding protein. *Endocrinology*, **101**, 959–968.

Rivarola, M. A., Podestà, E. J., Chemes, H. E. and Aguilar, D. (1973). *In vitro* metabolism of testosterone by whole human testis, isolated seminiferous tubules and interstitial tissue. *J. clin. Endocr. Metab.*, **37**, 454–460.

Rolland, R. *et al.* (1976). Prolactin and ovarian function. In *The Endocrine Function of the Human Ovary*, Eds. V. H. T. James, M. Serio and G. Giusti, pp. 305–321. Academic Press, London and N.Y.

Rowson, L. E. A., and Moor, R. M. (1967). The influence of embryonic tissue homogenate infused into the uterus, on the life-span of the corpus luteum in the sheep. *J. Reprod. Fert.*, **13**, 511–516.

Sanghero, M., Harris, M. C. and Morgan, R. A. (1978). Effects of microiontophoretic and intravenous application of gonadotrophic hormones on the discharge of medial-basal hypothalamic neurones in rats. *Brain Res.*, **140**, 63–74.

Sarkar, D. K., Chiappa, S. A. and Fink, G. (1976). Gonadotropin-releasing hormone surge in pro-oestrous rats. *Nature*, **264**, 461–463.

Sato, T. *et al.* (1974). Follicle stimulating hormone and prolactin release by prosta-glandins in rat. *Prostaglandins*, **5**, 483–490.

Sawaki, Y. and Yagi, K. (1976). Inhibition and facilitation of antidromically identi-fied tubero-infundibular neurones following stimulation of the median eminence in the rat. *J. Physiol.*, **260**, 447–460.

Scaramuzzi, R. J. and Baird, D. T. (1976). The oestrous cycle of the ewe after active immunization against prostaglandin F-2α. *J. Reprod. Fert.*, **46**, 39–47.

Schally, A. V., Redding, T. W. and Arimura, A. (1973). Effect of sex steroids on pituitary response to LH- and FSH-releasing hormone *in vitro*. *Endocrinology*, **93**, 893–902.

Schwartz, N. B. (1964). Acute effects of ovariectomy on pituitary LH, uterine weight, and vaginal cornification. *Amer. J. Physiol.*, **207**, 1251–1259.

Schwartz, N. B. (1974). The role of FSH and LH and of their antibodies on follicle growth and on ovulation. *Biol. Reprod.*, **10**, 236–272.

Schwartz, N. B., Krone, K., Talley, W. L. and Ely, C. A. (1973). Administration of antiserum to ovine FSH in the female rat: failure to influence immediate events of cycle. *Endocrinology*, **92**, 1165–1744.

Scott, R. S. and Rennie, P. I. C. (1970). Factors controlling the life-span of the corpora lutea in the pseudo-pregnant rabbit. *J. Reprod. Fesr.*, **23**, 415–422.

Setchell, B. P. and Sirinathsinghji, D. J. (1972). Antigonadotrophic activity in rete testis fluid, a possible "inhibin". *J. Endocrinol.*, **53**, lx-lxi.

Sherins, R. J. and Loriaux, D. L. (1973). Studies on the role of sex steroids in the feedback control of FSH concentrations in men. *J. clin. Endocrinol*, **36**, 886–893.

Sherman, B. M. and Korenman, S. G. (1975). Hormonal characteristics of the human menstrual cycle throughout reproductive life. *J. clin. Endocrinol.*, **55**, 699–706.

Sherwood, W. M., Chiappa, S. A. and Fink, G. (1976). Immunoreactive luteinizing hormone releasing factor in pituitary stalk blood from female rats: sex steroid modulation of response to electrical stimulation of preoptic area or median eminence. *J. Endocrinol*, **70**, 501–511.

Smith, M. S., Freeman, M. E. and Neill, J. D. (1975). The control of progesterone secretion during the estrous cycle and early pseudopregnancy in the rat: prolactin, gonodotropin and steroid levels associated with rescue of the corpus luteum of pseudopregnancy. *Endocrinology*, **96**, 219–226.

Södersten, P. and Hansen, S. (1977). Effects of oestradiol and progesterone on the induction and duration of sexual receptivity in cyclic female rats. *J. Endocrinol.*, **74**, 477–485.

Soloff, M. S., Morrison, M. J. and Swartz, T. L. (1972). A comparison of the estrone-estradiol binding proteins in the plasmas of prepubertal and pregnant rats. *Steroids*, **20**, 597–608.

Spies, H. G., Frantz, R. C. and Niswender, G. D. (1972). Patterns of luteinizing hormone in serum following administration of stalk-median eminence extracts to rhesus monkeys. *Proc. Soc. exp. Biol. Med.*, **140**, 161–166.

Spies, H. G. and Niswender, G. (1972). Effect of progesterone and estradiol on LH release and ovulation in rhesus monkey. *Endocrinology*, **90**, 257–261.

Spies, H. G., Norman, R. L., Quadri, S. K. and Clifton, D. K. (1977). Effects of estradiol-17β on the induction of gonadotropin release by electrical stimulation of the hypothalamus in rhesus monkeys. *Endocrinology*, **100**, 314–324.

Steelman, S. L. and Pohley, F. M. (1953). Assay of follicle stimulating hormone based on the augmentation with human chorionic gonadotropin. *Endocrinology*, **53**, 604–616.

Steinberger, E. and Choudhury, M. (1974). Control of pituitary FSH in male rats. *Acta Endocrinol.*, **76**, 235–241.

Steinberger, A. and Steinberger, E. (1976). Secretion of an FSH-inhibiting factor by cultured Sertoli cells. *Endocrinology*, **99**, 918–921.

Stewart-Bentley, M., Odell, W. and Horton, R. (1974). The feedback control of luteinizing hormone in normal adult men. *J. clin. Endocrinol*, **38**, 545–553.

Stouffer, R. L., Nixon, W. E. and Hodgen, G. D. (1977). Estrogen inhibition of basal and gonadotropin-stimulated progesterone production by rhesus monkey luteal cells *in vitro*. *Endocrinology*, **101**, 1157–1163.

Sundaram, K., Tsong, Y. Y., Hood, W. and Brinson, A. (1973). Effect of immunization with estrone-protein conjugate in rhesus monkey. *Endocrinology*, **93**, 843–847.

Swanston, I. A., McNatty, K. P. and Baird, D. T. (1977). Concentration of prostaglandin $F_2\alpha$ and steroids in the human corpus luteum. *J. Endocrinol.*, **73**, 115–122.

Swerdloff, R. S., Walsh, P. G. and Odell, W. D. (1972). Control of LH and FSH secretion in the male: evidence that aromatization of androgens to estradiol is not required for inhibition of gonadotropin secretion. *Steroids*, **20**, 13–22.

Talaat, M. and Lawrence, K. A. (1969). Effects of active immunization with ovine FSH on the reproductive capacity of female rats and rabbits. *Endocrinology*, **84**, 185–191.

Taleisnik, S., Caligaris, L. and Astrada, J. J. (1966). Effect of copulation on the release of pituitary gonadotropins in male and female rats. *Endocrinology*, **79**, 49–54.

Taleisnik, S., Velasco, M. E. and Astrada, J. J. (1970). Effect of hypothalamic deafferentation on the control of luteinizing hormone secretion. *J. Endocrinol.*, **46**, 1–7.

Tapper, C. M., Naftolin, F. and Brown-Grant, K. (1972). Influence of the reproductive state at the time of operation on the early response to ovariectomy in the rat. *J. Endocrinol*, **53**, 47–57.

Tejasen, T. and Everett, J. W. (1967). Surgical analysis of the preopticotuberal pathway controlling ovulatory release of gonadotropins in the rat. *Endocrinology*, **81**, 1387–1396.

Teresawa, E. and Sawyer, C. H. (1969a). Electric and electrochemical stimulation of the hypothalamo-adenohypophysial system with stainless steel electrodes. *Endocrinology*, **84**, 918–925.

Teresawa, E. and Sawyer, C. H. (1969b). Changes in electrical activity in the rat hypothalamus related to electrochemical stimulation of adenohypophyseal function. *Endocrinology*, **85**, 143–149.

Tindall, D. J. and Means, A. R. (1976). Concerning the hormonal regulation of androgen binding protein in rat testis. *Endocrinology.*, **99**, 809–818.

Treloar, O. L., Wolf, R. C. and Meyer, R. K. (1972). Failure of a single neonatal dose of testosterone to alter ovarian function in the rhesus monkey. *Endocrinology*, **90**, 281–284.

Tsafriri, A., Channing, C. P., Pomerantz, S. H. and Lindner, H. R. (1977). Inhibition of maturation of isolated rat oocytes by porcine follicular fluid. *J. Endocrinol*, **75**, 285–291.

Tsafriri, A. *et al.* (1976). Capacity of immunologically purified FSH to stimulate cyclic AMP accumulation and steroidogenesis in Graafian follicles and to induce ovum maturation and ovulation in the rat. *Endocrinology*, **98**, 655–661.

Tsou, R. C. *et al.* (1977). Luteinizing hormone releasing hormone (LHRH) levels in pituitary stalk plasma during the preovulatory gonadotropin surge of rabbits. *Endocrinology*, **101**, 534–539.

Uchida, K., Kadowaki, M. and Miyake, T. (1969). Ovarian secretion of progesterone and 20-α-hydroxypregn-4-en-3-one during rat estrous cycle in chronological relation to pituitary release of luteinizing hormone. *Endocr. Jap.*, **16**, 227–237.

Uchida, K., Kadowaki, M., Miyake, T. and Wakabayashi, K. (1972). Effects of exogenous progesterone on the ovarian progestin secretion and plasma LH and prolactin levels in cyclic rats. *Endocr. Jap.*, **19**, 323–333.

Umo, I. (1975). Effect of prostaglandin $F_2\alpha$ on the ultrastructure and function of sheep corpora lutea. *J. Reprod. Fert.*, **43**, 287–292.

Vilchez-Martinez, J. A., Arimura, A., Debeljuk, L. and Schally, A. V. (1974). Biphasic effect of estradiol benzoate on the pituitary responsiveness to LH-RH. *Endocrinology*, **94**, 1300–1303.

Vitale-Calpe, R. and Burgos, M. H. (1970a). The mechanism of spermiation in the hamster. I. Ultrastructure of spontaneous spermiation. *J. Ultrastr. Res.*, **31**, 381–393.

Vitale-Calpe, R. and Burgos, M. H. (1970b). The ultrastructural effects of coitus and of LH administration. *J. Ultrastr. Res.*, **31**, 394–406.

Wakabayashi, K. and Tamaoki, B.-I. (1966). Influence of immunization with luteinizing hormone upon the anterior pituitary-gonad system of rats and rabbits. *Endocrinology*, **79**, 477–485.

Wieck, R. F. *et al.* (1973). Periovulatory time course of circulating gonadotropic and ovarian hormones in the rhesus monkey. *Endocrinology*, **93**, 1140–1147.

Wiest, W. G., Kidwell, W. R. and Balogh, K. (1968). Progesterone catabolism in the rat ovary: a regulatory mechanism for progestational potency during pregnancy. *Endocrinology*, **82**, 844–859.

Wilber, J. F. *et al.* (1976). Gonadotropin-releasing hormone and thyrotropin-releasing hormone: distribution and effects in the central nervous system. *Rec. Progr. Horm. Res.*, **32**, 117–153.

Wiltbank, J. N. and Casida, L. E. (1956). *J. Animal. Sci.*, **15**, 134. (Quoted by Moore and Rowson, 1964.)

Winters, S. J. and Loriaux, D. L. (1978). Suppression of plasma luteinizing hormone by prolactin in the male rat. *Endocrinology*, **102**, 864–868.

Witorsch, R. J. and Smith, J. P. (1977). Evidence for androgen-dependent intracellular binding of prolactin in rat ventral prostate gland. *Endocrinology*, **101**, 929–938.

Wollman, A. L. and Hamilton, J. B. (1967). Prevention by cyproterone acetate of androgenic, but not of gonadotrophic, elicitation of persistent estrous in rats. *Endocrinology*, **81**, 350–356.

Woods, M. C. and Simpson, M. E. (1961). Pituitary control of the testis of the hypophysectomized rat. *Endocrinology*, **69**, 91–125.

Yagi, K. and Sawaki, Y. (1973). Feedback of estrogen in the hypothalamic control of gonadotrophin secretion. In *Neuroendocrine Control*. Eds. K. Yagi and S. Yoshida. John Wiley, New York, pp. 297–325.

Yen, S. S. C., Vanden Berg, G., Rebar, R. and Ehara, Y. (1972). Variation of pituitary responsiveness to synthetic LRF during different phases of the menstrual cycle. *J. clin. Endocr. Metab.*, **35**, 931–934.

Younglai, E. V., Moor, B. C. and Dimond, P. (1976). Effects of sexual activity on luteinizing hormone and testosterone levels in the adult male rabbit. *J. Endocrinol*, **69**, 183–191.

Zeballos, G. and McCann, S. M. (1975). Alterations during the estrous cycle in the responsiveness of the pituitary to subcutaneous administration of synthetic LH-releasing hormone (LHRH). *Endocrinology*, **96**, 1377–1385.

Zor, U. *et al.* (1972). Functional relations between cyclic AMP, prostaglandins, and luteinizing hormone in rat pituitary and ovary. *Adv. Cyc. Nucleot. Res.*, **1**, 503–520.

CHAPTER 3

Some Special Aspects of Hormone Receptors

Before passing to the consideration of the control of events during and after pregnancy we may pause for a brief description of some features of the binding of the hormones to their receptors with special reference to the control mechanisms, since there is a fair amount of evidence indicating that a changed sensitivity of a target organ to a sex hormone can be the result of a changed concentration of receptor material in the cell, or a changed affinity of the receptor material.

Pitfalls in Interpretation

The observation of a reduced uptake of labelled hormone by the cell, or of a reduction in the amount of labelled hormone attached to the cytoplasmic receptor material is not necessarily evidence for reduced receptor activity. Two obscuring factors are a change in the concentration of the unlabelled hormone in the blood or tissue since this will compete with the labelled hormone for binding with the receptor. A second factor is the translocation of the labelled hormone from cytoplasm to nucleus; if this is occurring at a rapid rate the labelled material will not be found in so high a concentration in the cytoplasmic material as would otherwise have been the case. This *translocation* is followed later by what has been called "replenishment", namely synthesis of new receptor material in the cytoplasm, a process that can be inhibited by protein synthesis inhibitors.

Translocation and Replenishment

An example is provided by a study by Cidlowski and Muldoon (1974) describing a fall in concentration of oestrogen receptors in uterus, pituitary and hypothalamus following injections of oestrogen, the fall amounting of 50 per cent within 10 hr; replenishment was synchronous in all three tissues and was diminished by the inhibitor of protein synthesis, cycloheximide. In general, the most reliable studies involving changes in receptor activity are carried out by isolation of the receptor material from the cytosol or nucleus and relating the quantity to the cell population, e.g. to the DNA concentration. By the so-called Scatchard plots (see, for example, Fig. 3.2), the affinity and concentration of receptors may be computed.

Pituitary and Hypothalamic Binding

Basis for Feedback

The existence of sites for specific uptake of steroid in the pituitary and hypothalamus doubtless represents the mechanism for the feedback phenomena described in Chapter 2; and variations in sensitivity during, for example, the ovarian cycle, or resulting from various treatments, might be revealed in changes in concentration of receptor material.

Male and Female

However, when cytoplasmic binding material, specific for 17-β-oestradiol, was isolated from male and female pituitaries, Korach and Muldoon (1973) found no difference in affinity of the male and female material, nor yet was the concentration of material, expressed in terms of unit weight of total protein, different; the same was true of the castrated animal, and finally, in the female, there was no alteration in the concentration of material during the phases of the oestrous cycle. In a similar type of study, Korach and Muldoon (1974) extracted receptor material from the hypothalamus of the male and female rat, and once again there was no difference in the binding characteristics or concentration between the sexes, thus confirming the observation of Whalen and Mauer (1969) that the accumulation of injected oestradiol by the hypothalamus of intact female rats did not vary with the oestrous cycle. The total binding capacity was only about one-tenth that in the pituitary. Binding of androgen occurred, but this was apparently non-specific, and this suggested that the gonadotrophic secretions of the male as well as the female are influenced by oestrogens only.*

Basis of Cyclicity

In general, it would seem from these studies on pituitary and hypothalamus that the basis of the cyclical secretion of gonadotrophic hormones is not to be sought in the character of the pituitary or hypothalamic receptors, but more in the nature of the steroidal feedback control over the liberation of releasing factors by the hypothalamus, as suggested originally by Everett (1964).

* We must note that Kato *et al.* (1969) found a pronounced fall in the degree of binding of labelled oestradiol by the anterior hypothalamus at pro-oestrus, with a return to the original value at oestrus. This might well represent increased translocation of receptors to the cell nucleus, due to the high level of oestrogen, followed by replenishment.

Effects of Oestrogen Injections

Injections of oestrogen into the rat decreased the concentration of specific receptors for this hormone, not only in the uterus, but also in the pituitary and hypothalamus; this reduction occurred in the cytoplasmic material and probably represented the migration of the receptor-oestrogen complex to the nuclei since the material was subsequently replenished, a process inhibited by the protein-synthesis inhibitor, cycloheximide.

When they compared male and female rat pituitary and hypo-thalamus, Cidlowski and Muldoon (1976) found that that replenish-ment was more rapid in the male than in the female, and they found that cycloheximide only partially blocked the replenishment. With immature rats, on the other hand, the pattern of behaviour was the same in both sexes. Thus, although the *concentrations* of receptors in the anterior pituitary and hypothalamus of rats were the same in both sexes, the dynamics of response to oestrogen were different.

Progesterone

Evidence for specific binding sites for progesterone in hypothalamus and pituitary is contradictory and has been summarized by Kato and Onouchi (1977); these authors have employed a synthetic progestin—R 5020—which binds in a highly specific manner to progesterone receptors (Philibert and Reynaud, 1974; Philibert *et al.*, 1977); receptors specific for this progestin are different from corticosterone-binding-globulin (CBG) and the receptors for glucocorticoids and oestrogens. Kato and Onouchi found that they could extract from homogenates of the hypothalamus and pituitary of oestrogen-primed immature and mature female rats binding material, sedimenting at 7 S; the K_d was 10^{-9} M, similar to that for the binding material derived from uterus; 5a-dihydro-progesterone competed with R 5020 for binding. With non-primed animals, on the other hand, there was little or no binding protein. The failure of many experimenters to find specific progesterone receptors in these tissues might be due to the metabolic conversion of progesterone to 5-a-dihydroprogesterone, which was the active material, a situation that would be comparable with the conversion of testosterone to 5a-dihydrotestosterone. In this case we should expect accumulation of 5a-dihydroprogesterone in pituitary and hypothalamus; in fact Karavolas *et al.* (1976) found accumulation of tritium-labelled material in these tissues.*

---

* It must be emphasized that progesterone is active on the uterus without conversion to 5a-DHP; the latter does have actions, affecting lordosis and ovulation.

Androgens

Korach and Muldoon (1975) found that testosterone and its reduced derivative, 5-dihydrotestosterone (5-DHT) were very poor competitors for oestradiol at pituitary binding sites; moreover they failed to find a specific receptor for testosterone. This is surprising in view of the powerful effects of testosterone and dihydrotestosterone on pituitary secretion of gonadotrophic hormones, revealed experimentally by the suppression of gonadotrophin secretion by injections of testosterone and the increase in secretion by castration (p. 129), effects that are mediated by a feedback on the hypothalamus.

Naess *et al.* (1975) have re-investigated the problem and have isolated protein material from hypothalamus, preoptic area anterior pituitary and cortex with strong affinity for testosterone and dihydrotestosterone, an affinity that was greater than that for $17\text{-}\beta\text{-}$oestradiol; in fact, when equal concentrations of labelled testosterone and unlabelled oestradiol were applied, the amount of bound label was unaffected; by contrast, cyproterone acetate competed for the labelled testosterone as effectively as unlabelled testosterone. The molecular weight of the material was about 130,000 and the reductions of binding by high temperatures and sulphydryl-agents were characteristic of receptor material isolated from primary target organs, such as ventral prostate and epididymis. From this work it seems then that the anterior pituitary and brain structures contain an androgen-receptor different from the oestrogen receptor.

Androgenization

We have seen how female rats may be made sterile if, when young, they are given doses of androgen or oestrogen; this change in their hypothalamic activity, which converts them from cycling to non-cycling animals, is reflected in a change in uptake of ^{3}H-labelled oestradiol, not only by target-tissues but by the pituitary and hypothalamus. Whereas oestrogen-induced sterility is associated with disruption of uptake by both pituitary and hypothalamus, that caused by androgen disrupts mainly pituitary uptake (McGuire and Lisk, 1969).* Biochemical studies on binding by homogenates of tissues

---

* In the study of Tuohimas and Johansson (1971) the pituitary of the androgenized rat actually took up more labelled oestradiol than the normal; it was in the uterus and in the hypothalamus that uptake was significantly decreased. According to Korach and Muldoon (1975), although testosterone and its reduced derivatives α-DHT are very poor competitors for binding at pituitary oestradiol sites under equilibrium conditions, 5α-DHT does inhibit the initial rate of formation of the oestradiol complex, an effect it does not share with progesterone and so is steroid-specific.

treated either *in vivo* or *in vitro* with ³H-oestradiol showed that androgens decreased the uptake in the nuclear fractions of the anterior hypothalamus and pituitary, whilst it *increased* the uptake in the supernatant cytosol.

Induced Ovulation

When ovulation is induced in immature female rats by injection of pregnant mare's serum, this is accompanied by the typical surges of FSH, LH, oestrogen and progesterone; these changes are accompanied by a reduction in the concentration of oestrogen-receptor material in the cytosol of both pituitary and hypothalamus (Parker *et al.*, 1976), the reduction preceding the pre-ovulatory surges of FSH and LH. During the following oestrus and dioestrus, the concentration of receptor material tended to return to its original value, and it seems that the fall was due to the translocation of receptor-oestrogen complex to the nuclei of the sensitive cells.

Changed Sensitivity to Releasing Hormone

Greely *et al.* (1974) showed that the decrease and replenishment of oestradiol receptor population in the hypothalamus and pituitary at pro-oestrus-oestrus correlated very well with sensitivity of the animal to the gonadotrophic releasing hormone, LHRH, and this suggested a genuine fluctuation in receptor concentration, which could have been due to translocation to the nucleus.

Ontogeny

The age—be it pre- or post-natal—at which the hypothalamus exhibits a specific binding capacity for steroid hormones is of great interest, since it gives a clue to the possibilities of steroid feedback on this part of the brain. Thus, studies on the male rat suggest the possibility of a negative feedback from the testis to the hypothalamus-pituitary within the first few days of life, as manifest by an increased pituitary secretion of gonadotrophin following castration; in the female the negative feedback from the ovary is manifest in the first 10–12 days. This difference may account for the fact that the serum concentrations of FSH and LH are greater in the neonatal female than in the male. Using biochemical techniques for estimation of binding of oestradiol by the hypothalamus, Kulin and Reiter (1972) found that binding by the hypothalamus was greater than that by the cerebral cortex from Day 1 to adulthood in the rat; only by Day 5, however, was the binding affected by adding non-radioactive oestradiol, indicating that only by then could the binding be considered specific. Plapinger and McEwen

(1973) measured the uptake of oestradiol by hypothalamic homogenates; nuclear preparations showed little binding at birth and this rose slightly during the first three weeks to show a sharp increase between Days 24–25 and 25–26, and this high level of nuclear binding persisted in all subsequent age-groups. The cytoplasmic preparation did not show this sharp increase so that it appears that this sharp rise in oestradiol binding capacity in the hypothalamus, occurring at the approach to puberty (p. 464), is confined to the nuclear acceptors. This nuclear accumulation of oestradiol was demonstrated auto-radiographically by Sheridan et al. (1974) in the immature rat; it certainly occurred during the "critical period" of perinatal sexual differentiation, so that the ultrastructural basis for interaction between steroid hormones and the hypothalamus is present at a stage when the physiological experiments demand it.*

In human foetal pituitary and brain tissues, Davies et al. (1975) found specific binding of oestradiol to material in the 100,000 g supernatant of cytoplasmic extracts; both male and female tissues had this binding capacity; testosterone, didydrotestosterone and cortisol did not bind; in this and other respects the material was different from sex hormone binding globulin (SHBG).

Uterine and Oviductal Receptors

Progesterone

Milgrom et al. (1972) have prepared, from the guinea-pig uterus, receptor material that specifically binds progesterone; † as Fig. 3.1 shows, the number of computed receptor sites per cell shows a sig-nificant peak at the time of the periovulatory progesterone peak;

* Plapinger et al. (1973) have examined the uptake of labelled oestradiol by the cytosol fraction of hypothalamic homogenates; the binding capacity, expressed as pmols/g. cytoso protein, is some 200 times greater in the foetal and perinatal brain than in the adult; in general it is present in high concentration in foetal life and the first perinatal week, gradually declining to undetectable levels by the end of the third week and at the time when the adult cytosol and nuclear binding sites begin to appear. This foetal binding material is different qualitatively from the adult in that it is not affected by stilboestrol and is not bound by DNA in 0·14 M NaCl; and the authors surmise that it is identical with the oestrogen-binding protein (EBP, see p. 274) present in foetal and neonatal rat plasma described by Nunez et al. (1971a, b) and Raynaud et al. (1971). Since this foetal protein can be extracted quite easily by simple washing of the brain, and since it is present in foetal cerebrospinal fluid, it is likely that it is present in the extracellular fluid of the brain, being derived in the first place from the foetal blood plasma.

† According to a recent study of Gueriguian et al. (1974), the physical characteristics of the progesterone complexing protein extracted from the human uterus are identical with those of the plasma carrier-protein which is, in fact, corticosteroid binding globulin (CBG). The amounts derived from the uterus are far too large to be accounted for by contamination by blood. However, Rao et al. (1974) have shown that their 4–5 S material in the human endo-metrium was different from CBG, so that cortisol would not compete for binding.

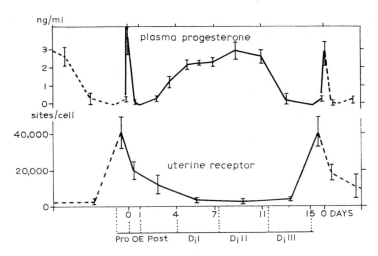

Fig. 3.1. Comparison of plasma progesterone levels and receptor sites during the oestrous cycle of the guinea pig. Note the peaks of progesterone and receptor sites around pro-oestrus and oestrus. (Milgrom *et al.*, *Endocrinology.*)

however, the subsequent more gentle rise in plasma progesterone, which follows ovulation and is due to secretion by the newly developing corpora lutea of this cycle, is not accompanied by a change in receptor activity in the uterus. If we are correct in assuming that changes in receptor activity of the uterus are bound up with the uterine changes preparatory to implantation, then it seems likely that it is the *periovulatory* rise in progesterone concentration that is important for implantation (Chapter 4). Support for this view is provided by Deanesly (1960) who showed that ovariectomy, and thus deprivation of the animal of its principal supply of progesterone, carried out from Days 3 to 7 of pregnancy, did not prevent implantation; thus, after Day 2, the period when the concentration of receptor sites is low, progesterone is no longer necessary for implantation to take place on Day 6 or 7.

Effect of Oestrogen. Corool *et al.* (1972), also working on the guinea-pig uterus, found that the specific binding material extracted from the tissue was associated with 7 S and 3·5–4 S macromolecules; the binding was considerably enhanced by oestrogen treatment, thus explaining the co-operative action of the two steroids, oestrogen and progesterone. In general, the progesterone receptors seem to be under dual control, being increased by oestrogens and decreased by progesterone; in this way we may explain the peak at pro-oestrus and oestrus and the decline at di-oestrus (Warembourg and Milgrom, 1977).

Hamster

The results of a similar type of study by Leavitt *et al.* (1974) on the hamster uterus are illustrated in Figs 3.2–3.4. In the hamster the oestrous cycle lasts four days and it is possible to predict the events taking place with considerable accuracy, the critical period for the gonadotrophin surge being 13.00 to 14.00 on Day 4 (pro-oestrus) of the cycle. By carrying out so-called Scatchard plots of the binding, as in Fig. 3.2, estimates of the number of binding sites are obtained by the

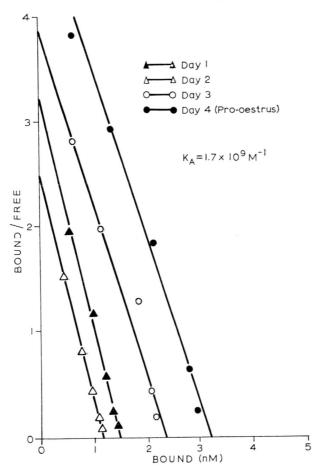

Fig. 3.2. "Scatchard-plots" of progesterone binding in uterine cytosol fractions at different stages of the oestrous cycle. From the slope of the Scatchard-plots, the affinity, K_a, is computed as $1.7.10^9 M^{-1}$. The number of binding sites relative to tissue weight is indicated by the intercept on the abscissa. (Leavitt *et al., Endocrinology.*)

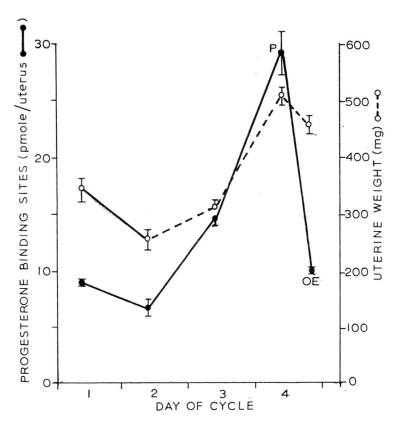

Fig. 3.3. Uterine progesterone binding sites and uterine weight during the oestrous cycle. Pro-oestrus (P) and oestrus (OE) are indicated on Day 4 of the cycle. (Leavitt *et al.*, *Endocrinology*.)

intercept on the abscissa. It will be seen that the maximum concentration of sites occurs on Day 4 (pro-oestrus); this is better demonstrated by Fig. 3.3 which shows the parallelism between binding sites and uterine weight, whilst Fig. 3.4 shows the rapid fall taking place after ovariectomy, and its restoration on treatment of the animal with oestradiol.

Gonadectomy

Ovariectomy. According to Falk and Bardin (1970) the progesterone-binding by the uterus of the ovariectomized guinea-pig can be strikingly increased by oestrogen priming. The same was found for the rabbit by Rao *et al.* (1973), the number of binding sites per cell

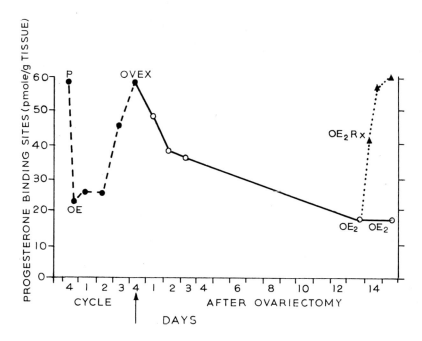

Fig. 3.4. Illustrating the fall in receptor sites following ovariectomy at pro-oestrus and its restoration by an injection of oestradiol. (Leavitt *et al.*, *Endocrinology*.)

increasing from 550 in the atrophic uterus to 3500 in the oestrogen-treated organ. Interestingly, the 8 S material, which was absent in the spayed animal, reappeared in the treated animal.

Castration. In the normal male rat, Sullivan and Strott (1973) found no specific binding of ³H-dihydrotestosterone in the cytosol obtained from prostate, epididymis and testis; immediately after castration, however, the activity of an 8–10 S fraction, derived from the accessory organs was very high, possibly as a result of removing normal androgens from the circulation by castration—androgens that tended to block uptake of the tritiated dihydrotestosterone. This level of binding activity declined, so that it fell to zero after 4 days, and this paralleled the decline in weight of the prostate, the characteristic response to castration (Fig. 3.5). However, if the binding per milligramme of tissue is plotted against time, it is seen that there is a large rise in binding activity after Day 4, suggesting that there is an androgen-independent regulation of the concentration of receptors in the tissue.

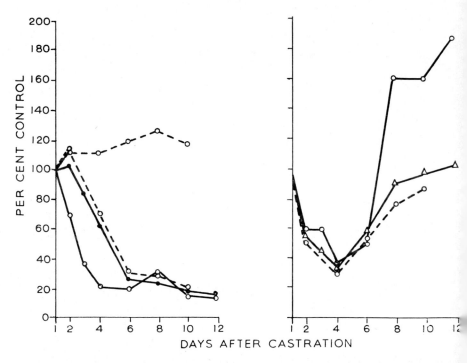

Fig. 3.5. Effects of castration on several aspects of ventral prostate atrophy and on receptor activity in rats. Results are expressed as percentage of 1-day castrate controls. *Left-hand Graph:* (●——●), prostate weight; (○ – – – ○), DNA per prostate; (□ – – – – □), DNA per mg tissue; (○——○), 8S ^{3}H-dihydrotestosterone binding per prostate. *Right-hand Graph:* (△——△), 8S ^{3}H-dihydrotestosterone binding per mg tissue; (□——□) per mg soluble protein; (○ – – – ○) per μg DNA. (Sullivan and Strott, *J. biol. Chem.*)

Positive and Negative Control

Milgrom *et al.* (1972) observed that the type of the uterine progesterone receptor as well as the absolute quantity varied with the phases of the oestrous cycle; thus, at pro-oestrus, the material of 6·7 S predominated; at oestrus 6·7 S and 4·5 S were in about equal parts, and at di-oestrus the 4·5 S material predominated. These changes are apparently induced by the changes in concentration of oestrogen in the blood during the cycle. According to Milgrom *et al.* (1973) the oestrogen-induced change in receptor activity depends on synthesis of RNA and protein, since treatment with actinomycin D or cycloheximide prevented the increase. These workers found that progesterone tended to promote the inactivation of its own carriers, so that in the absence of progesterone the receptors are stable, with a half-life of about 5 days; in the presence of progesterone the complex is transferred to the nucleus where it is rapidly inactivated. Thus there seems to be a dual control over the formation of receptors for progesterone, a positive control through oestrogens and a negative control through progesterone. So far as oestradiol receptors are concerned, it would seem from Sarff and Gorski's (1971) study that oestrogen *provokes* the synthesis of ts own receptor.

Progesterone-Suppression of Oestradiol Receptors

In many aspects progesterone may be considered to be "anti-oestrogenic", for example in its blocking of ovulation when administered at a critical point in the ovulatory cycle. An excellent experimental example is provided by the oviduct of the ovariectomized animal; unless oestrogen is administered, the oviduct atrophies, but even if this has been allowed to proceed the process may be reversed by oestradiol treatment. When reversal has been achieved, treatment with progesterone—while continuing oestradiol treatment—leads to atrophy as if the oestrogen treatment had been discontinued (Brenner *et al.*, 1974). These phenomena may be demonstrated in both rhesus monkey and the cat (West *et al.*, 1976), and in both species the atrophy and reversal of oviducal structure are accompanied by large changes in receptor concentration in cytosol and nuclear fractions of the oviducal homogenates. Thus, progesterone treatment reduced these concentrations to the level in the untreated spayed animal, whilst treatment of the spayed animal with oestradiol produced large increases. These changes are illustrated in Fig. 3.6 for both oviduct and uterus. So far as the oviduct is concerned, morphological changes strictly paralleled the changes in receptor concentration, and West *et al.* suggested that the oviduct was being "starved" of oestrogen, in that its complement of receptor material was insufficient to make use of the available amounts of oestrogen. In the uterus the situation was more complex, in that progesterone treatment, although reducing receptor concentrations, did not lead to atrophy, but rather to hypertrophy, so that progesterone stimulated growth and secretion in the oestradiol-primed animal in spite of a severe depression in oestradiol receptor levels.

Chicken Oviduct

The oviduct of the chicken synthesizes *avidin*, and this is controlled by progesterone; priming of the animal with oestrogen causes a large rise in avidin activity, and according to Toft and O'Malley (1972), priming of the animal with oestrogen over a two-week period caused a

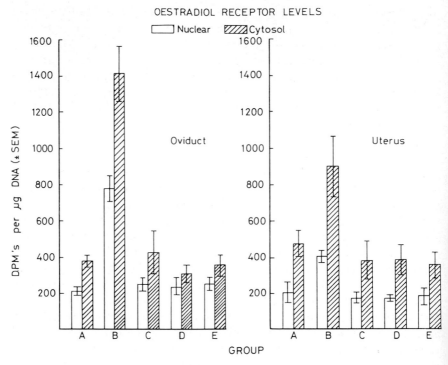

Fig. 3.6. Levels of oestradiol receptors in cat oviduct and uterine nuclear and cytosol fractions after various hormonal treatments. A. Spayed. Hypophysectomized 6–7 weeks after ovariectomy. B. Continuous E_2. Implant on Day 0. Hypophysectomy Day 14 or Day 25. C. E_2 implant Day 0. Removed Day 14. Hypophysectomy Day 28. D. E_2 implanted Day 0. Removed Day 14 and progesterone implanted instead. Hypophysectomy Day 28. E. E_2 implanted Day 0. Day 14 progesterone implanted. Day 28 hypophysectomy. (West *et al. Endocrinology.*)

large increase in the number of progesterone receptors; an interesting feature of this work was the change in the sedimentation characteristics of the receptor material from a normal 4–5 S to one of 6–8 S in the oestrogen-treated animals.

Gonadotrophin Receptor Sites

Graafian Follicle

Luteinizing hormone, LH, and follicle stimulating hormone, FSH, are bound to specific sites on their target organs, sites that are specific so that LH, for example, will not appreciably displace FSH from its binding. Human chorionic gonadotrophin (hCG) binds to the same sites as LH and so may be used as a label for these sites. Maturation of the follicle is associated with formation of the antrum and finally with transformation of granulosa into luteal cells. FSH is required for antrum formation and, in association with LH, for corpora lutea formation in hypophysectomized rats. Midgley (1973) observed that the granulosa cells of some large follicles could bind hCG (LH) and FSH whereas others could only bind FSH; in general, cells from large porcine follicles could bind more LH than small less developed ones, suggesting that maturation was accompanied by a production of LH-binding sites. Channing and Kammerman (1973) found a very strong correlation between size of preovulatory follicle and LH-binding, and Zeleznick et al. (1974) showed autoradiographically that treatment of a rat with FSH caused an increased binding of LH in granulosa cells of stimulated follicles; with isolated granulosa cells they found that pretreatment of the animal with FSH increased the binding of LH by fivefold.

Autoregulation

The reciprocal relations between the hormones and induction of receptor sites were emphasized by Rao et al. (1977), who pointed out that hormones can not only regulate receptor content for other hormones but also for themselves, thus exhibiting an autoregulation. Using granulosa cells as the target, they showed that LH, unlike FSH, decreased the number of sites for LH and FSH, thus exhibiting a negative feedback; the number of prolactin receptor sites increased strongly, by contrast; these changes are shown in Fig. 3.7; it seems that the granulosa cell, after undergoing luteinization, must become insensitive to LH but sensitive to prolactin in the rat; and this agrees with the supposed role of prolactin in promoting secretion of progesterone from the developed corpora lutea.*

Hormone-Induced Pseudopregnancy

Rajaniemi et al. (1977) have described a hormonal regimen that induces a characteristic pseudopregnancy in rats, and, using this model, have followed the changes in LH-binding sites in the maturing ovary. The treatment consisted in repeated doses of FSH from Day 25 of birth to Day 27 followed by a single dose of LH. Uterine and ovarian weights increased during the FSH injections, presumably because of the oestrogen secreted; LH injection induced a large further increase in

* Lee (1977) has demonstrated a striking reduction in the number of FSH-binding sites in the rat's ovary as luteinization was induced by successive injections of ovarian FSH and human chorionic gonadotrophin (equivalent to LH); they argued that this could not have been due to occupancy of receptor sites by the priming FSH, since after only three injections there was no change in binding. A change in affinity was also ruled out.

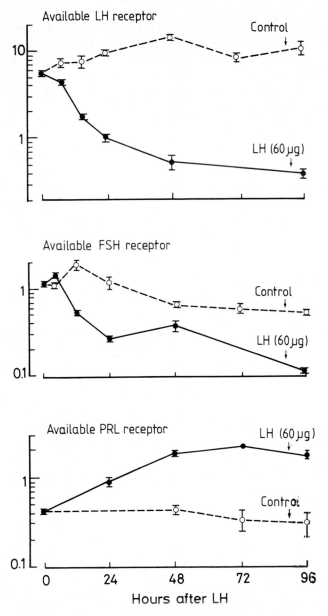

Fig. 3.7. Effect of ovine luteinizing hormone, LH, on available receptors for LH, FSH and prolactin (PRL) in granulosa cells. The abscissa represents hours after subcutaneous administration of ovine LH. The ordinate represents counts per minute from ^{125}I-labelled hormone specifically bound per μg DNA. (Rao *et al. Endocrinology.*)

ovarian weight, and this was associated with ovulation of most of the follicles and formation of corpora lutea. Corresponding with this luteal development there was an increase in the binding capacity for LH, as measured in ovarian homogenates, or radioautographically in ovarian sections; the luteal cells were especially heavily labelled and thus it appears that the secretion of progesterone, which paralleled very strikingly the increase and subsequent fall in binding by the ovaries, is governed by development of LH-receptors.

Blood-Flow to Ovary

Niswender *et al.* (1976) have suggested that the great effectiveness of LH during the luteal phase of the cycle in promoting progesterone secretion could be due, besides to increased concentration of receptors on the luteal cells, to an increased ovarian blood-flow, making a given blood-level of LH more effective. In fact, blood-flow to the ovary increases 3–7-fold during the luteal phase of the sheep's cycle, the increase being specific to the ovary containing the functional corpus luteum. An anti-LH reduced blood-flow to the corpus luteum-containing ovary.

Control Through Receptor Formation

Follicular and Luteal Development

The concept of control through formation of receptors has been applied in a very detailed fashion by Richards and Midgley (1976) to the elucidation of the various factors that control the follicular and luteal phases of ovarian development. According to their somewhat speculative model, as illustrated by Fig. 3.8, follicular development would involve the sequential actions of hormones on granulosa cells such that the action of one hormone might determine the cellular content of its own receptor and, by acting synergistically with a second hormone, might determine appearance of a receptor for a third hormone. As an example, oestradiol increases the concentration of oestradiol receptors in granulosa cells; FSH increases receptor concentration in the same cells, and oestradiol and FSH combine to increase LH receptor in these cells. Atresia is associated with loss of receptors for oestradiol, FSH and LH. Luteinization involves termination of follicular growth, and this involves loss of FSH receptors, a decrease in receptors for oestradiol and LH and an increase in receptors for prolactin. The subsequent increase in luteal cell receptor for LH and the production of progesterone will only occur if prolactin is present during the luteinization process. Apparently development of LH receptor and progesterone secretion are independent events each controlled by prolactin.

Protein Synthesis and Function

In Vol. 2, where we discussed the basic mechanisms of hormone action, it appeared that some hormones acted through inducing the synthesis of specific proteins within the target cell; the predominantly nuclear accumulation of the gonadal hormones, and their biochemical

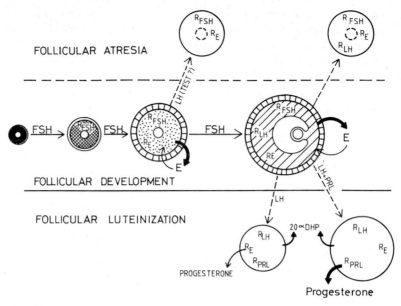

Fig. 3.8. Speculative model describing control of follicular and luteal development through production of receptors. (Richards and Midgley, *Biol. Reprod.*)

association with the chromatin of the nucleus, suggest that this is the mode of their action, and studies on RNA synthetic capacity of uterine nuclei, when treated with the oestradiol-receptor complex *in vitro*, have shown that this is comparable with that observed when the hormone is administered *in vivo*.

Uterine-Induced Protein (UIP)

Notides and Gorski (1966) measured the effects of oestrogen on the incorporation of labelled amino acids into soluble uterine protein and found that measurable incorporation, and presumably synthesis of a new protein, occurred within 30 minutes of oestrogen treatment. The effect was thus manifest well before the main increase in production of uterine proteins induced by the hormone treatment. The incorporation was blocked by cycloheximide. Barnes and Gorski (1970) later called this *induced protein*, and attributed the lag of some 40 minutes between injection of oestrogen and the beginning of incorporation of labelled amino acids into the soluble uterine protein fraction to the time required for synthesis, transport to the cytoplasm, and translation of a specific mRNA to direct the synthesis of the induced protein. Actinomycin D, which inhibits the production of the mRNA, was effective in blocking synthesis only if given before the oestrogen. Thus, an earlier step, measured as an increase in incorporation of nucleoside into uterine RNA, not blocked by puromycin and cycloheximide, could be identified within 10 minutes of treatment of the animal with oestrogen (DeAngelo and Gorski, 1970). *In vitro* studies on immature rat uteri showed that the amount of UIP formed correlated well with the amount of ³H-labelled oestradiol bound to the nuclei; the levels of oestradiol that were effective were between 2–3.10^{-8}M for maximal effect and 2.10^{-10}M for minimal effect, levels that fall well within the physiological range of effective concentrations; moreover, there was strict hormonal specificity amongst the ocstrogens, the order being: 17-β-oestradiol > diethylstilboestrol > oestriol > 17-a-oestradiol (Katzenellenbogen and Gorski, 1972).

Synthetic Rate and Ovarian Cycle

Fig. 3.9 shows the cyclicity in the rate of synthesis of UIP during the rat's oestrous cycle, the maximum occurring at pro-oestrus and the minimum at oestrus. It will be seen that exogenous oestrogen increases the ratc of synthesis at all stages. In ovariectomized animals the

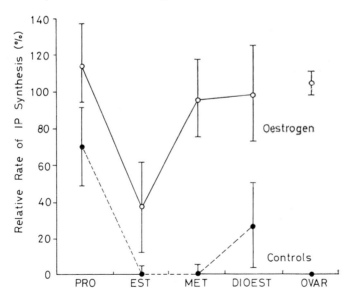

Fig. 3.9. Illustrating cyclicity in rate of synthesis of induced protein (IP) by adult rat uteri (broken line). Treatment with oestrogen increases rate of synthesis. The relative rate of synthesis in the ovariectomized rat uterus is taken as zero (OVAR on abscissa); the rate 1 hr after injection of 10 μg of oestradiol into the ovariectomized rat is taken as 100% (OVAR, open circle). (Katzenellenbogen, *Endocrinology*.)

synthesis was abolished. These results suggest that the induction of this protein plays a role—oestrogen-regulated—in the functioning of the mature uterus.*

Androgenization

When Kobayashi and Gorski (1970) studied the effects of injections of actinomycin D into 5-day-old female rats at different periods before androgenization with testosterone, they found that actinomycin D was effective in inhibiting androgenization if given six or four hours before the testosterone injection; given simultaneously, or 2 hours before, there was no effect. Puromycin was effective only when given 4 hours after the androgen, but by the time the animals were 90 days old there was no difference in the incidence of sterility in these animals compared with those treated with testosterone alone. Again, direct injection of actinomycin D into the preoptic area of the adult ovariectomized female rat suppressed the sexual behaviour that would otherwise have been induced by oestrogen-progesterone treatment (Quadagno, Shryne and Gorski, 1970).†

* An enzymatic role of UIP is suggested by the finding of phosphoprotein phosphatase activity associated with it (Vokaer *et al.*, 1974); however, Mairesse and Galand (1977) have shown that this is due to a contaminant. We may note that oestrone is quite effective in inducing the synthesis of the protein; hitherto it has been considered that oestrone acted in the rat after conversion to oestradiol (Jensen and Jacob, 1962); however, Ruh *et al.* (1973), showed that its effectiveness, which correlated well with uptake by the cell nucleus, could not have been due to prior conversion to oestradiol.
    † Quoted by Gorski (1970).

REFERENCES

Barnes, A. and Gorski, J. (1970). Estrogen-induced protein. Time course of synthesis. *Biochem.*, **9**, 1899–1904.

Baulieu, E.-E. *et al.* (1975). Steroid hormone receptors. *Vit. & Horm.*, **33**, 649–736.

Brenner, R. M., Resko, J. A. and West, N. B. (1974). Cyclic changes in oviductal morphology and residual cytoplasmic estradiol binding capacity induced by sequential estradiol-progesterone treatment of spayed rhesus monkeys. *Endocrinology*, **95**, 1094–1104.

Channing, C. P. and Kammerman, S. (1973). Characteristics of gonadotropin receptors of porcine granulosa cells during follicle maturation. *Endocrinology*, **92**, 531–540.

Cidlowski, J. A. and Muldoon, T. G. (1974). Estrogenic regulation of cytoplasmic receptor populations in estrogen-responsive tissues of the rat. *Endocrinology*, **95**, 1621–1629.

Cidlowski, J. A. and Muldoon, T. G. (1976). Sex-related differences in the regulation of cytoplasmic estrogen receptor levels in responsive tissues of the rat. *Endocrinology*, **98**, 833–841.

Corool, P., Falk, R., Freifeld, M. and Bardin, C. W. (1972). *In vitro* studies of progesterone binding proteins in guinea-pig uterus. *Endocrinology*, **90**, 1464–1469.

Davies, L. J., Naftolin, F., Ryan, K. J. and Siu, J. (1975). A specific, high-affinity, limited capacity estrogen binding component in the cystosol of human fetal pituitary and brain tissues. *J. clin. Endocr. Metab.*, **40**, 909–912.

Deanesley, R. (1960). Implantation and early pregnancy in ovariectomized guinea-pigs. *J. Reprod. Fert.*, **1**, 242–248.

De Angelo, A. B. and Gorski, J. (1970). Role of RNA synthesis in the estrogen induction of a specific uterine protein. *Proc. Nat. Acad. Sci. Wash.*, **66**, 693–700.

Everett, J. W. (1964). Central neural control of reproductive functions of the adenohypophysis. *Physiol. Rev.*, **44**, 373–431.

Falk, R. and Bardin, C. W. (1970). Uptake of tritiated progesterone by the uterus of the ovariectomized guinea-pig. *Endocrinology*, **86**, 1059–1063.

Gorski, R. A. (1970). Localization of hypothalamic regulation of anterior pituitary function. *Amer. J. Anat.*, **129**, 219–222.

Greeley, G. H., Muldoon, T. G., Allen, M. B. and Mahesh, V. B. (1974). Pituitary sensitivity to LRF and estradiol-17β receptor population during the rat estrous cycle. *J. Steroid Biochem.*, **5**, 388.

Gueriguian, J. L., Sawyer, M. E. and Pearlman, W. H. (1974). A comparative study of progesterone- and cortisol-binding activity in the uterus and serum of pregnant and non-pregnant women. *J. Endocrinol.*, **61**, 331–345.

Hansson, V. and Tveter, K. J. (1971). Uptake and binding *in vivo* of ^{3}H-labelled androgen in the rat epididymis and ductus deferens. *Acta endocrinol.*, **66**, 745–755.

Jensen, E. V. and Jacobson, H. I. (1962). Basic guides to the mechanism of estrogen action. *Rec. Progr. Horm. Res.*, **18**, 387–414.

Karavolas, H. J., Hodges, D. and O'Brien, D. (1976). Uptake of ^{3}H-progesterone and ^{3}H5a-dihydroprogesterone by rat tissues *in vivo*. *Endocrinology*, **98**, 164–175.

Kato, J. and Onouchi, T. (1977). Specific progesterone receptors in the hypothalamus and anterior hypophysis of the rat. *Endocrinology*, **101**, 920–928.

Kato, J., Inaba, M. and Kobayashi, T. (1969). Variable uptake of tritiated oestradiol by the anterior hypothalamus in the postpubertal female rat. *Acta. endocr.*, **61**, 585–591.

Katzenellenbogen, B. S. (1975). Synthesis and inducibility of the uterine estrogen-induced protein, IP, during the rat estrous cycle: clues to uterine estrogen sensitivity. *Endocrinology*, **96**, 289–297.

Katzenellenbogen, B. S. and Gorski, J. (1972). Estrogen action *in vitro*. Induction of the synthesis of a specific uterine protein. *J. biol. Chem.*, **247**, 1299–1305.

Kobayashi, F. and Gorski, R. A. (1970). Effects of antibiotics on androgenization of the neonatal female rat. *Endocrinology*, **86**, 285–289.

Korach, K. S. and Muldoon, T. G. (1973). Comparison of specific 17-β-estradiol-receptor interactions in the anterior pituitary of male and female rats. *Endocrinology*, **92**, 322–326.

Korach, K. S. and Muldoon, T. G. (1974). Studies on the nature of the hypothalamic estradiol-concentrating mechanism in the male and female rat. *Endocrinology*, **94**, 785–793.

Korach, K. S. and Muldoon, T. G. (1975). Inhibition of anterior pituitary estrogen-receptor complex formation by low-affinity interaction with 5a-dihydrotestosterone. *Endocrinology*, **97**, 231–236.

Kulin, H. E. and Reiter, E. O. (1972). Ontogeny of the *in vitro* uptake of tritiated estradiol by the hypothalamus of the female rat. *Endocrinology*, **90**, 1371–1374.

Leavitt, W. W., Toft, D. O., Strott, C. A. and O'Malley, B. W. (1974). A specific progesterone receptor in the hamster uterus: physiologic properties and regulation during the oestrous cycle. *Endocrinology*, **94,** 1041–1053.

Lee, C. Y. and Takahashi, H. (1977). Follicle-stimulating hormone receptors in rat ovaries: decrease in numbers of binding sites associated with luteinization. *Endocrinology*, **101,** 869–875.

Mairesse, N. and Galand, P. (1977). Oestrogen-induced uterine protein: presence of associated phosphoprotein phosphatase activity before oestrogen treatment. *J. Endocrinol*, **72,** 81–85.

McEwen, B. S. and Pfaff, D. W. (1970). Factors influencing sex hormone uptake by rat brain regions I. *Brain Res.*, **21,** 1–16.

McEwen, B. S., Pfaff, D. W. and Zigmond, R. E. (1970). Factors influencing sex hormone uptake by rat brain regions. II. and III. *Brain Res.*, **21,** 17–28; 29–38.

McGuire, J. L. and Lisk, R. D. (1969). Oestrogen receptors in androgen or oestrogen stimulated sterilized female rats. *Nature*, **221,** 1068–1069.

Midgley, A. R. (1973). Autoradiographic analysis of gonadotropin binding to rat ovarian tissue sections, *Adv. exp. Med. Biol.*, **36,** 365–378.

Milgrom, E., Alger, M., Perrot, M. and Baulieu, E.-E. (1972). Uterine progesterone receptors during the estrous cycle and implantation in the guinea-pig. *Endocrinology*, **90,** 1071–1078.

Milgrom, E., Allouch, P., Atger, M. and Baulieu, E.-E. (1973). Progesterone-binding plasma protein of pregnant guine-pig. *J. biol. Chem.*, **248,** 1106–1114.

Naess, O., Attramadal, A. and Aakvaag, A. (1975). Androgen binding proteins in the anterior pituitary, hypothalamus, preoptic area and brain cortex of the rat. *Endocrinology*, **96,** 1–9.

Niswender, G. D., Reimers, T. J., Dickman, M. A. and Nett, T. M. (1976). Blood flow: a mediator of ovarian function. *Biol. Reprod.*, **14,** 64–81.

Notides, A. and Gorski, J. (1966). Estrogen-induced synthesis of a specific uterine protein. *Proc. Nat. Acad. Sci. Wash.*, **56,** 230–235.

Nunez, E. *et al.* (1971a). Mise en évidence d'une fraction protéique liant les oestrogènes dans le sérum de rats impubères. *C.r. Acad. Aci. Paris*, **272,** 2396–2399.

Nunez, E. *et al.* (1971b). Origine embryonnaire de la protéine sérique fixant l'oestrone et l'oestradiol chez la ratte impubère. *C.r. Acad. Sci. Paris*, **273,** 242–245.

Parker, C. R., Costoff, A., Muldoon, T. G. and Mahesh, V. B. (1976). Actions of pregnant mare serum gonadotropin in the immature female rat. *Endocrinology*, **98,** 129–138.

Philibert, D. and Raynaud, J. P. (1974). Binding of progesterone and R 5020, a highly potent progestin, to human endometrium and myometrium. *Contraception*, **10,** 457.

Philibert, D., Ojasoo, T. and Raynaud, J.-P. (1977). Properties of the cytoplasmic progestin-binding protein in rabbit uterus. *Endocrinology*, **101,** 1850–1861.

Plapinger, L. and McEwen, B. S. (1973). Ontogeny of estradiol-binding sites in rat brain. I. *Endocrinology*, **93,** 1119–1128.

Plapinger, L., McEwen, B. S. and Clemens, L. E. (1973). Ontogeny of estradiol-binding sites in rat brain, II. *Endocrinology*, **93,** 1129–1139.

Rajaniemi, H. J., Midgley, A. R., Duncan J. A. and Reichert, L. E. (1977). Gonadotropin receptors in rat ovarian tissue: III. Binding sites for luteinizing hormone and differentiation of granulosa cells to luteal cells. *Endocrinology*, **101,** 898–910.

Rao, M. C., Richards, J. S., Midgley, A. R. and Reichert, L. E. (1977). Regulation of gonadotrophin receptors by luteinizing hormone in granulosa cells. *Endocrinology*, **101,** 512–524.

Rao, B. R., Wiest, W. G. and Allen, W. M. (1973). Progesterone "receptor" in rabbit uterus. I. *Endocrinology*, **92**, 1229–1240.

Rao, B. R., Wiest, W. G. and Allen, W. M. (1974). Progesterone receptor in human endometrium. *Endocrinology*, **95**, 1275–1281.

Raynaud, J.-P., Mercier-Bodard, C. and Baulieu, E. E. (1971). Rat estradiol binding plasma protein (EBP). *Steroids*, **18**, 767–788.

Richards, J. S. and Midgley, A. R. (1976). Protein hormone action: a key to understanding ovarian follicular and luteal development. *Biol. Reprod.*, **14**, 82–98.

Ruh, T. S., Katzenellenbogen, B. S., Katzenellenbogen, J. A. and Gorski, J. (1973). Estrone interaction with the rat uterus: *in vitro* response and nuclear uptake. *Endocrinology*, **92**, 125–134.

Sarff, M. and Gorski, J. (1971). Control of estrogen-binding protein concentration under basal conditions and after estrogen administration. *Biochemistry*, **10**, 2557–2563.

Sheridan, P. J., Sar, M. and Stumpf, W. E. (1974). Autradiographic localization of ^{3}H-estradiol or its metabolites in the central nervous system of the developing rat. *Endocrinology*, **94**, 1386–1390.

Sullivan, J. N. and Strott, C. A. (1973). Evidence for an androgen-independent mechanism regulating the levels of receptor in target tissue. *J. biol. Chem.*, **248**, 3202–3208.

Toft, D. O. and O'Malley, B. W. (1972). Target tissue receptors for progesterone: the influence of estrogen treatment. *Endocrinology*, **90**, 1041–1045.

Tuohimas, P. and Johansson, R. (1971). Decreased estradiol binding in the uterus and anterior hypothalamus of androgenized female rats. *Endocrinology*, **88**, 1159–1164.

Vokaer, A., Iacobelli, S. and Kram, R. (1974). Phosphoprotein phosphatase activity associated with estrogen-induced protein in rat uterus. *Proc. Nat. Acad. Sci. Wash.*, **71**, 4482–4486.

Warembourg, M. and Milgrom, E. (1977). Radioautography of the uterus and vagina after [^{3}H] progesterone injection into guinea-pigs at various periods of the estrous cycle. *Endocrinology*, **100**, 175–181.

West, N. B., Verhage, H. G. and Brenner, R. M. (1976). Suppression of the estradiol receptor system by progesterone in the oviduct and uterus of the cat. *Endocrinology*, **99**, 1010–1016.

Whalen, R. E. and Maurer, R. A. (1969). Estrogen "receptors" in brain; an unsolved problem. *Proc. Nat. Acad. Sci. Wash.*, **63**, 681–685.

Zelesnik, A. J., Midgley, A. R. and Reichert, L. E. (1974). Granulosa cell maturation in the rat: increased binding of human chorionic gonadotropin following treatment with follicle-stimulating hormone *in vivo*. *Endocrinology*, **95**, 818–825.

CHAPTER 4

Pregnancy : Implantation

"RESCUE" OF THE CORPUS LUTEUM

Progesterone and Oestrogen

Progesterone

Successful fertilization of the ovum leads to implantation of the zygote which, on its way to the uterus, has multiplied by mitotic division to become a blastocyst; during this period—the *progravid phase* of the normal ovarian cycle—the corpus luteum has been developing and secreting its hormones—mainly progesterone—controlling changes in the uterus that prepare it for implantation. These changes would have taken place whether fertilization occurred or not, but regression of the corporus luteum—*leuteolysis*—is inhibited by the implantation of the blastocyst in the uterus. Thus, progesterone secretion is sustained and, as Fig. 4.1 shows for the monkey, the concentration in blood-plasma rises to higher levels than during the ovulatory peak, with subsequent variations. Progesterone serves to reduce motility of the uterus and it also, by its feedback action on the hypothalamico-pituitary axis, suppresses ovulation.

Oestrogen

Although progesterone is commonly called the steroid hormone of pregnancy, oestrogen continues to exert its trophic action on the enlarging uterus, and on the mammary glands. The changes in blood oestrogen concentration of the monkey during the early stages of pregnancy are included in Fig. 4.1. Some idea of the increasing amounts of oestrogen synthesis taking place within the pregnant woman is given by the estimates of Oakey (1970); thus one week before menstruation a non-pregnant woman excretes about 20 μg of oestriol, 10 μg of oestrone, and 4 μg of oestradiol-17β daily. If pregnancy occurs, the quantity of oestrogen excreted increases at once, and progressively

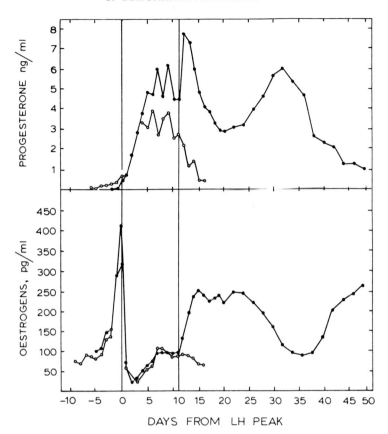

Fig. 4.1. Mean serum progesterone and oestrogen concentrations during the fertile and nonfertile cycle in rhesus monkeys. The estimations are normalized to the day of the LH-peak (Day 0); the vertical line drawn at Day 11 after the LH-peak indicates the mean day of the rescue of the corpus luteum. (Atkinson *et al.*, *Biol. Reprod.*)

so that at term the daily excretion may be 40 mg oestriol, 2 mg of oestrone and 1 mg of oestradiol-17β. These figures by no means represent the total oestrogen produced, which can be assessed by the arterio-venous differences in concentration of oestrogen precursors in the umbilical artery and vein; on this basis the amount of oestrogen secreted by the placenta of a foetus near term is probably as high as 330 mg per day.

Pseudopregnancy

This condition, occurring in mammals with relatively short gestational periods, is typically induced by copulation of a rabbit with a non-fertile male (p. 57); in this species, the act of copulation induces ovulation, and the hormonal secretions that promote receptivity of the uterus for the fertilized ovum take place and are sustained in spite of the absence of a conceptus. The ferret, an extreme example, gives a period of pseudopregnancy of the same duration as that of a normal pregnancy, and this suggests that, in this species at any rate, the whole process of pregnancy is governed by the neuroendocrine system of the mother, being determined by the act of copulation and the consequent endocrine secretions.

Placental Gonadotrophin

In many species, however, an important change in the pituitary-ovarian relationship soon takes place; thus, in species including the rat, sheep, cow and the primates the dependence of the ovaries on pituitary gonadotrophic hormones for synthesis and secretion of their steroid hormones becomes less, whilst the placenta produces a gonadotrophin that promotes ovarian secretion. In man this is secreted by the chorion, and the gonadotrophin is called *human chorionic gonadotrophin*—hCG—with mainly luteotrophic activity and thus serving to maintain progesterone secretion.* In addition, the placenta acquires the power to synthesize steroid hormones (Zander and Münstermann, 1956), so that after quite a short period, in the primate (20–30 days, Amoroso and Porter, 1966) pregnancy is not terminated either by hypophysectomy or by ovariectomy, the quantities of steroids, produced from this non-ovarian source, being adequate to sustain pregnancy. Thus the rescue of the corpus luteum in species that produce a placental gonadotrophin could be the consequence of the luteotrophic hormone, CG, secreted by the placenta. To this extent the conceptus is contributing to the control of the gestational process. The secretion of steroid hormones by the placenta, moreover, reduces the importance of the ovarian source so that in the primates, for example, the ovaries can be dispensed with early in pregnancy, the placental source of steroids during the last two-thirds of pregnancy being adequate to maintain pregnancy. The importance of the conceptus extends to this

* In the mare, the gonadotrophin is found mainly in the blood, and the material isolated from this is called pregnant mare's serum gonadotrophin, or PMS; whereas hCG has predominantly LH activity, PMS has both LH and FSH activity. Secretion of hCG in the urine is an early sign of pregnancy and formed the basis of the Ascheim-Zondek urinary test; it appears a few days prior to the expected menstrual period, and within a few days o ⇁implantation. Modern techniques of radioimmunoassay have permitted its assay in the bl^{oo}d distinct from the pituitary hormones. In the pregnant woman it is first detected 8–12 da^{ys} after the presumed date of ovulation, i.e. a day or so after implantation; at this time the blastocyst is only 1 mm^3 in volume, and it is remarkable that the syncytiotrophoblast can synthesize sufficient hormone to become detectable in the maternal systemic blood (Vaitukaitis et al., 1976).

aspect of control since the foetal adrenal supplies the placenta with precursors of the progesterone and oestrogen that it synthesizes and subsequently secretes.

Placental Synthesis of Progesterone

The synthesis of progesterone by the human placenta was demonstrated by Zander and Münstermann (1956) who found progesterone in the umbilical artery and vein and in placental tissue. In the guinea-pig, pregnancy may continue after ovariectomy as early as Day 20–27 (gestation period some 67 days) and Fig. 4.2 shows the progesterone

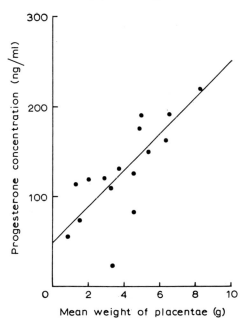

Fig. 4.2. Plasma progesterone concentration in systemic blood of ovariectomized pregnant guinea pigs plotted against placental weight. (Heap and Deanesly, *J. Endocrinol.*)

concentration in systemic blood of animals ovariectomized on Day 28 post-conception, as a function of placental weight. The correlation with weight suggests that secretion is governed by the size of the placenta rather than by any specific trophic mechanism (Davies and Ryan, 1972). In the monkey, ovariectomy after Day 22 to 24 after mating does not promote abortion, indicating the dominant role of the placenta in producing the steroid at an early stage. A similar situation pertains in the human, the corpus luteum apparently serving no useful purpose after the second month of pregnancy (Csapo *et al.*, 1972, 1973). Thus luteotomy at 49 ± 2 days caused abortion with a large fall in plasma progesterone; at 61 ± 4 days there was no abortion and plasma progesterone fell only slightly. Abortion due to early luteotomy could be prevented by progesterone injections.

Species Variations

Before proceeding with any detailed consideration of the control mechanisms in the maintenance and termination (parturition) of pregnancy we might consider the variations in patterns of control that have developed in evolution. This is the more important because so much experimental work has been carried out on non-primate species; moreover, amongst the non-primate mammals there are striking differences, and these are essentially related to the relative degrees to which the control mechanisms operate through the ovary and placenta. As we shall see, the prime factor governing the maintenance of pregnancy is the level of progesterone available to the myometrium, which determines the level of its excitability, a high level of the steroid maintaining a state of quiescence, or "progesterone block", favourable to retention of the conceptus. The study of the levels of progesterone in the blood during pregnancy might therefore indicate the state of hormonal control, whilst ovariectomy might indicate the relative importance of the ovarian and placental sources. The rapid metabolism of this hormone ($\frac{1}{2}$-life of 6 minutes) makes it an especially useful index to the control being exerted, any change in secretion being rapidly reflected in a change in blood- and tissue-levels.

Effects of Ovariectomy

Tables I and II classify the mammalian species on the basis of ovariectomy-intolerance and ovariectomy-tolerance, in the sense that in the opossum, for example, abortion invariably follows ovariectomy whereas in the monkey, an example of the tolerant species, ovariectomy after Day 25 of pregnancy does not cause abortion, and delivery occurs at the normal time.

Spectrum of Behaviour. As Davies and Ryan (1972) emphasize, however, the division shown by the Tables is somewhat arbitrary, and it is best to think of a spectrum of dependence on the ovary from absolute, as in the goat where placental progesterone synthesis is absent, through partial, as in the rat and guinea-pig, to total independence as in primates. Thus the rat, although classed as intolerant, will carry 90 per cent of its foetuses to full term if ovariectomy is postponed to Day 18, i.e. four days before term. Again, the guinea-pig,

4. PREGNANCY: IMPLANTATION

TABLE I

Ovariectomy-intolerant mammals. (Davies and Ryan, 1972)

Species	Approximate length of gestation (days)	Regression of corpora lutea before term	Key references
Opossum	12·5	No	Amoroso and Finn (1962)
Hamster	16	No	Amoroso and Finn (1962)
Mouse	20	No	Amoroso and Finn (1962)
Rat	22	No	Amoroso and Finn (1962)
Rabbit	31	No	Amoroso and Finn (1962)
Ferret	42	No	Hammond and Marshall (1930)
Pig	114	No	Amoroso and Finn (1962)
Armadillo	120*	No	Amoroso and Finn (1962)
Goat	151	No	Amoroso and Finn (1962)

* 3·5–4 months of delayed implantation; 4 months of postimplantation gestation.

TABLE II

Ovariectomy-tolerant mammals. (Davies and Ryan, 1972)

Species	Approximate length of gestation (days)	Earliest successful oophorectomy	Regression of corpora lutea before term	Key references
Cat	63	49	Yes	Amoroso and Finn (1962)
Dog	63	Late	Yes	Courrier (1945)
Guinea-pig	68	40	No	Amoroso and Finn (1962)
Sheep	144	55	Yes	Amoroso and Finn (1962)
Macaca mulatta	166	25	—	Amoroso and Finn (1962)
Baboon	175	—	Yes	Kriewaldt and Hendricks (1968)
Human	267	40	Yes	Amoroso and Finn (1962)
Cow	280	207	No	Amoroso and Finn (1962) Courrier (1945)
Horse	330	170	Yes	Amoroso and Finn (1962)

classed as tolerant, will only carry its pregnancy to term if it is carrying 1–2 foetuses, indicating a partial dependence on the ovary when the number of placentae available is too small. Similarly, in sheep, although they are tolerant, the ovaries continue to secrete large quantities of

progesterone until quite late in pregnancy (Linzell and Heap, 1968);*
the same is true for the cow (Erb et al., 1967).

Uterine Quiescence. The cause of abortion is presumably the
failure to maintain the uterus quiescent. Thus, Courrier (1941) slit
the uterine horn of a 19-day pregnant rabbit and delivered one foetus
in its amniotic sac into the peritoneal cavity, retaining its placental
attachment.† The two remaining foetuses of the same horn were left
intact, and the rabbit was ovariectomized. On Day 28 it was found
that the foetus in the abdominal cavity was alive and growing normally,
whereas the other two had died shortly after the operation and had
been extruded through a rupture in the uterine wall. Finally, Haterius
(1936) found that whereas ovariectomy of the pregnant rat before
Day 14 always leads to abortion, if all the foetuses except one were
removed, leaving all the placentas intact, the surviving foetus developed
to term. Thus, in so far as maintaining the uterus quiescent is concerned,
even in intolerant species, provided there is some placental secretion of
progesterone, tolerance may be manifest under conditions where a
reduced supply of the steroid will suffice. In the case of Haterius's
experiment it appears that the presence of several foetuses in the uterus
requires the secretion of more progesterone to maintain the uterus
quiescent than the presence of a single one.‡

Local Action of Placental Progesterone

Supply of progesterone from the ovaries to the uterus must be through
the general circulation, so that where an animal is dependent on this
source, a reduced plasma level will indicate reduced ability to sustain
the quiescent state, other things being equal. However, when progesterone is secreted by the placenta such a close relation between plasma
level and uterine function need not be expected since it is possible that
the secreted steroid is carried directly to the myometrium by way of
the uterine lymphatics and interstitial fluid. Thus the lymphatic bed
of the human uterus is extensive and the vessels increase in size in
association with pregnancy (Maurizio and Ottaviani, 1934). Evidence
bearing on this local action will be considered later; for the moment we
must note that the failure to influence gestation in some species by

* The relative contributions of ovary and placenta to progesterone production have been
assessed by Linzell and Heap (1968) by measuring arterio-venous differences in concentration. Between 119–126 days of pregnancy in *goats* the ovaries produced 10 mg/day whilst the
placenta produced none; by contrast in the *sheep* the placenta produced 14 mg/day and the
ovaries only about 2 mg/day; the adrenal production was less than 2 per cent of the ovarian.

† In this "ectopic" pregnancy the placenta became attached to the intestinal wall.

‡ Haterius noticed that the birth mechanism, as opposed to gestation, was impaired in the
overiectomized rat with a single foetus. Presumably because this requires the secretion of
ovarian oestrogen (p. 343).

administration of exogenous progesterone need not necessarily mean that the uterine state is not under the influence of this steroid, but simply that the plasma level does not reflect the effective level in the myometrium.

Hypophysectomy

Dependence on the hypophysis for secretion of progesterone should vary amongst species, according to the extent to which placental gonadotrophin can substitute for the pituitary gonadotrophin. Table III compares the effects of both ovariectomy and hypophysectomy in

TABLE III

Comparison of the effects of hypophysectomy and ovariectomy in different mammalian species. (Van Tienhoven in *Reproductive Physiology of Vertebrates*, 1968, Saunders, Philadelphia)

Species	Effect*		Earliest date†	
	HypoX	OvX	HypoX	OvX
Rabbit	Aborted	Aborted	Near term	Near term
Dog	Aborted	Continued	Near term	Day 30 (?)
Guinea-pig	Continued	Continued	Day 40–41	Day 20
Mouse	Continued	Absorbed, aborted	Day 10m	—
Rat	Continued	Absorbed, aborted	Day 11	—
Sheep	Continued	Continued	Day 50	Day 54
Rhesus monkey	Continued	Continued	Day 27	Day 25
Woman	Continued	Continued	Week 12	Day 30

* Effect is indicated as either *aborted* or *absorbed* if the operation caused these effects at any time during pregnancy except close to term, or as *continued* if pregnancy was not interrupted.
† Earliest date at which the operation was compatible with the maintenance of pregnancy.

different mammalian species. Usually, but not always, ovariectomy and hypophysectomy have the same effects, but in the rat, for example, postponement of hypophysectomy until after Day 11 permits continuation of pregnancy, indicating the operation of a placental gonadotrophin, since ovariectomy at this date causes abortion.

Length of Gestation. The transition from maternal to embryonic and foetal control over the gestational process is not abrupt, as suggested by Tables I and II so that it is not possible to correlate it with a single feature of the gestational process. However, the control over the duration of pregnancy when this is long, as in primates, will clearly be easier if the conceptus is able to provide the signal that it is ready to be born than if the period is built into the mother's hypothamico-pituitary system, and, as Davies and Ryan emphasize, the development of the foeto-placental unit whereby the foetus takes over control both of its development and final emergence from the mother may be the answer to the problems emerging as gestational periods become lengthy. Thus the condition of pseudopregnancy, indicating maternal control, is a feature of short-gestation animals, whilst complete foetal control is characteristic of primates; inspection of Tables I and II shows that, with some exceptions, the intolerant mammals are those with short gestation periods.

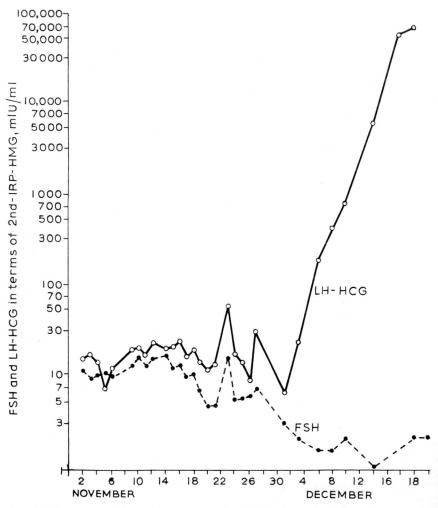

Fig. 4.3. Serum gonadotrophin concentrations during a normal human menstrual cycle and the early period of the ensuing pregnancy. Note log. scale of ordinates. (Jaffe *et al.*, *J. clin. Endocrinol.*)

Bearing in mind these variations in dependence on ovary and hypophysis, let us return to the study of the control mechanisms in the initiation and maintenance of pregnancy.

Plasma Gonadotrophin Levels

Human Subjects

The development of radioimmunoassays for the pituitary and ovarian hormones has permitted accurate assessments of the changes in plasma levels taking place in the early stages of pregnancy. Figure 4.3 from Jaffe *et al.* and Fig. 4.4 from Parlow *et al.* (1971) on human

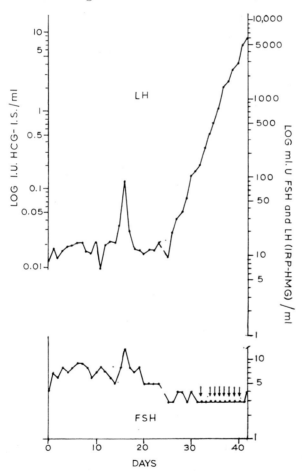

Fig. 4.4. Serum FSH and LH during early human pregnancy. Arrows indicate that the levels are below the sensitivity of the test (3 milli-international units/ml). (Parlow *et al., J. clin. Endocrinol.*)

subjects show that the characteristic feature is the suppression of FSH secretion to barely detectable or undetectable levels; within 26 days of the mid-cycle peak of LH, the levels of LH had risen enormously and were doubtless due to secretion of human chorionic gonadotrophin (HCG), which reacted similarly to LH in the radioimmunoassay and has essentially luteotrophic activity, promoting the secretion of progesterone.

Mouse

In the mouse, an "intolerant animal", Murr *et al.* (1974) found a rise in the plasma LH-level on Day 3, reaching a peak at Days 7–12 (Fig. 4.5); the rise begins on Day 3 and is presumably related to implantation; there is definitely no corresponding rise in FSH (Fig. 4.6).

Dominance of Luteotrophic Hormone

Thus, in both the tolerant and intolerant species, it is essentially the luteotrophic hormone, either of pituitary or of placental origin, that dominates behaviour during pregnancy, maintaining progesterone secretion, whilst follicular development is inhibited through suppression of secretion of FSH.

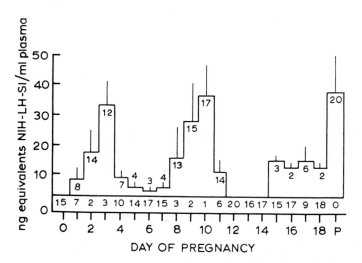

Fig. 4.5. Concentrations of LH in plasma throughout pregnancy and at parturition (P) in the mouse. The numbers below the line at 3 ng/ml represent the numbers of animals with plasma levels below the minimum detectable limit of the assay. The numbers within the bars represent animals with detectable levels. (Murr *et al.*, *Endocrinology*.)

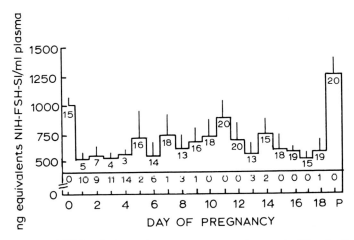

Fig. 4.6. Plasma FSH levels throughout pregnancy and at parturition in the mouse. Conventions as for Fig. 4.5. (Murr *et al.*, *Endocrinology*.)

Relation to Implantation

The time-relations between the hormonal secretions and the early events in pregnancy in the monkey are indicated schematically by Fig. 4.7 from Meyer (1972). The relation of gonadotrophin secretion to progesterone secretion has been examined with considerable accuracy by Hodgen *et al.* (1974), and Atkinson *et al.* (1975), employing assays for the placental hormone that distinguish it from the pituitary gonadotroph, LH. Figure 4.8 from Hodgen *et al.* shows that the "rescue" of the corpus luteum, manifest as a steep rise in progesterone secretion, occurred 24 to 48 hr after chorionic gonadotrophin was initially detected in blood.*

Figure 4.9 from Atkinson *et al.* (1975) shows the actual concentrations of chorionic and pituitary gonadotrophs from Days 7 to 60 of a fertile cycle. The peaks at Day 10 after the ovulatory peak may be compared with the "rescue" of the corpus luteum shown in Fig. 4.8 by the rise in

---

* The delay between first detection of mCG in blood and the rise in blood progesterone has been discussed by Atkinson *et al.* (1975); they have suggested a direct vascular drainage from the uterus to the ovary, similar to that described for the sheep, but such a portal system has not been described in the primate. However, Meyer (1972) detected mCG in ovarian venous blood three days before its detection in the peripheral circulation, whilst the experiments of Riesen *et al.* (1970) are highly suggestive. They found that the concentration of progesterone in the ovarian vein on the side of the corpus luteum was 407 ng/ml, comparing with 83 ng/ml for the opposite side; however, these concentrations were considerably higher than the peripheral level, namely less than 7 ng/ml, and this suggested some crossing over by way of the uterus. When a haemostat was applied to prevent flow of blood from the ovary, containing the corpus luteum, to the uterus, the concentration in the ovarian blood from the side without the corpus luteum dropped to nearly that of the peripheral blood.

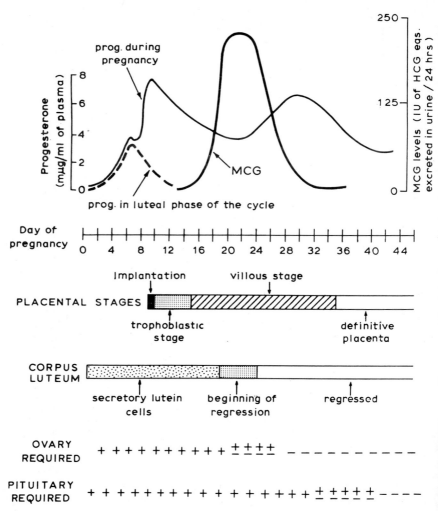

Fig. 4.7. Time-relations between progesterone and chorionic gonadotrophin (mCG) secretion and stages in early pregnancy in the monkey, *M. mulatta.* (Meyer, *Acta endocrinol.*)

progesterone concentration in the blood, and a causal relation is suggested between the two. The secondary rise in blood progesterone corresponds with the larger peak in mCG secretion; however, when the steroid hormone levels are plotted on the same graph as those of mCG, as in Fig. 4.10, where Day 0 is the day of corpus luteal rescue, it will be seen that the large rise in secretion of mCG is not accompanied by a sustained rise in progesterone; and this may well be due to the luteo-

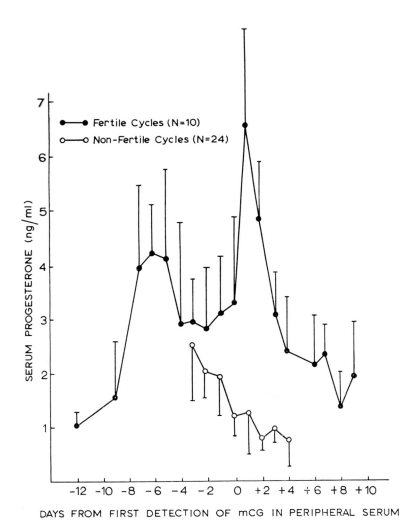

Fig. 4.8. Mean serum concentrations of progesterone synchronized around the initial detection of chorionic gonadotrophin (mCG) in serum during fertile menstrual cycles of rhesus monkeys. The progesterone values of nonpregnant monkeys are plotted around Day 23 of the menstrual cycle for comparative purposes. (Hodgen *et al.*, *J. clin. Endocrinol.*)

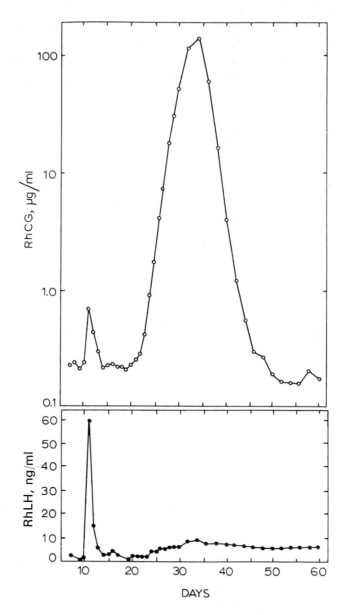

Fig. 4.9. Relation between secretions of LH (RhLH) and chorionic gonadotrophin (RhCG) in the blood during a fertile cycle in the monkey from Days 7 to 60. (Atkinson *et al.*, *Biol. Reprod.*)

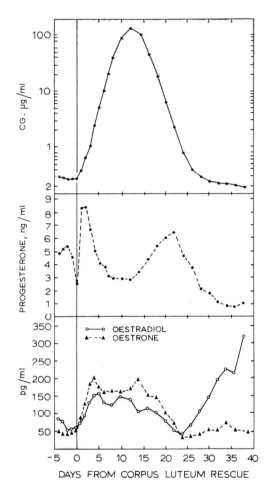

Fig. 4.10. Chorionic gonadotrophic hormone (CG), progesterone, and oestrogen concentration in serum during early pregnancy in rhesus monkeys. The values are normalized to the day of corpus luteum rescue. (Atkinson, *et al., Biol. Reprod.*)

lytic action of oestrogen, whose secretion is increased (bottom graph) during this phase. Thus administration of oestrogen to monkeys, or women, during the early luteal phase shortens the menstrual cycle and curtails progesterone secretion.

Termination of mCG Secretion

By Day 40 of pregnancy, secretion of chorionic gonadotrophin ceases in the rhesus monkey (Hodgen *et al.*, 1974); this is not true of other primates studied, which continue to secrete chorionic gonadotrophin throughout pregnancy (see, for example, Varma *et al.*, 1971, and Vaitukaitis, 1974, for women, and Hobson, 1971, for the baboon). The termination in the rhesus monkey is understandable as the placenta quite early in pregnancy becomes the main source of progesterone, and its secretion presumably does not require a trophic hormone.

Changes in Character of Chorionic Gonadotrophin

During the course of human pregnancy, the concentrations of hCG in blood, urine and placenta vary in a parallel manner, indicating that the fall in plasma levels in later pregnancy are probably due to reduced synthesis, rather than increased metabolism. Vaitukaitis (1974) observed changes in the character of the extracted hormone during pregnancy, and these were probably related to the completeness, or otherwise, of synthesis. Thus, at a late stage the proportion of free a-subunits increased, and two varieties of these could be differentiated from the "authentic" subunit obtained from highly purified hCG. In addition, diffuse immunological peaks of the hormone preparation were obtained, suggesting incomplete forms of hCG with varying amino acid and carbohydrate contents; and this might account for variations in the ratio of biological and immunological activities observed during pregnancy (Wide, 1962; Morgan and Canfield, 1971); thus removal of only a few sialic acid residues would decrease the biological activity of the hormone, and explain the lowered biological activity after the first trimester.

Effect of Ovariectomy

The importance of the placental supply of progesterone early in pregnancy in the primate is shown by the effects of ovariectomy on Day 23; as Fig. 4.11 shows, progesterone secretion is not affected whereas that of oestrogen is; by Day 40, however, oestrogen secretion is resumed at the time corresponding to the normal secondary gestational rise shown in Fig. 4.1, which thus has a placental origin.

Recrudescence of Corpus Luteum

After Day 35, the secretion of trophic hormone by the monkey's placenta ceases, and the progesterone of pregnancy is largely derived from the placenta. However, the corpus luteum does not become completely non-functional during late pregnancy. According to Treloar *et al.* (1972), from Days 20 to 100 of pregnancy there is a regression in size of the corpus luteum, but from Days 100 to 155 there is a "rejuvenation" in size and vascularity. Moreover, on Day 155 the concentration of progesterone in ovarian vein plasma is 14-times that in

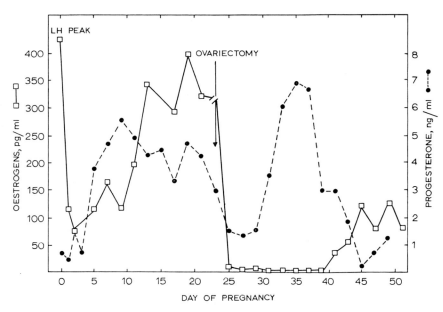

Fig. 4.11. The effect of bilateral ovariectomy on Day 23 of pregnancy of the monkey on circulating levels of progesterone and oestrogens in the rhesus monkey. The secretion of progesterone is not affected whereas that of oestrogen is reduced. (Atkinson *et al.*, *Biol. Reprod.*)

peripheral plasma and 8-times that in uterine plasma. Morphologically, the cells of the 155-day pregnant corpus luteum reveal the capacity for synthesis, and in fact the corpus luteum at term seems more active than that at Day 2 (Kocring *et al.*, 1973). Since the chorionic gonadotrophin does not maintain function of the corpus luteum in late pregnancy, the question arises as to what is the luteotroph.

Role of Chorionic Gonadotrophin

We may regard the appearance of this hormone in evolution as the first step in permitting the extension of the pregnancy beyond the duration of pseudopregnancy, the latter being governed by the maternal pituitary-hypothalamus. In this way we can account for the presence of chorionic gonadotrophins in species like the rat that are intolerant to ovariectomy and hypophysectomy. In the second place its function could be that of assisting to maintain the corpus luteum functional in the face of diminishing pituitary secretion of LH. Thus, during pregnancy, the level of gonadotrophins in the pituitary is low, perhaps due to negative feedback from the oestrogens secreted at progressively higher rates during pregnancy. By Day 21 of human pregnancy, however, there are degenerative signs in the corpus luteum, and after Day 25 the corpus luteum can be dispensed with,* so that the placental hormone is not required to provoke synthesis and release of *ovarian* steroids. The placenta acquires the ability to synthesize these steroids, so that it is possible that the chorionic gonadotrophin acts as a trophic hormone on the placental steroid-secreting cells. However, it must be emphasized that during later stages of pregnancy the high levels of gonadotrophins in the blood are not associated with high levels of progesterone, in fact, it can happen that the peak gonadotrophin level is associated with a nadir in progesterone concentration (Neill *et al.*, 1969).

The low plasma levels of progesterone in the later stages of pregnancy in many species do not necessarily indicate an absence of dependence on this steroid in the maintenance of uterine quiescence; as we have indicated earlier, it will be the concentration of progesterone in the tissue that is important, and this need not be the same as that in the general circulation. We may note, also, that in the monkey, its characteristic chorionic gonadotrophin—mCG—has dropped to about zero concentration by days 35–38 of pregnancy, the gestation time being 168 days (Tullner and Hertz, 1966).

IMPLANTATION AND EARLY EMBRYONIC DEVELOPMENT

Passage of the Ovum Along Oviduct

The passage of the ovum from the ovary into the oviduct and thence into the uterus has been described in Volume 2; it is essential that the ovum should remain within the duct for an appropriate time in order to increase the chances of a successful meeting with a sperm, and this retention within the duct is achieved by what has been called *isthmic block*, a closure of the isthmus preventing passage from the ampulla to the uterus. Thus, transport of the egg through the ampulla to the isthmus takes only a few minutes whereas transport from the isthmus to the uterus requires 3–4 days.

Failure of the isthmic block leads to premature escape of the ovum,

---

* In the rat, the pregnancy corpus luteum is maintained by the pituitary for the first half of pregnancy, and only then does the placental gonadotrophin take over (Pencharz and Long, 1933); in man, hypophysectomy does not terminate pregnancy if carried out after 12 weeks, and in the monkey 5 weeks (Kaplan, 1961). Morishige and Rothchild (1974) have discussed the identity of the rat's luteotroph of pregnancy; the results of treatment with anti-LH and ergocomine (ECO), which blocks prolactin release, suggest that before Day 8, progesterone secretion is governed by a luteotrophic complex of prolactin plus LH, yielding to one of LH plus placental luteotrophin—rPL—from Day 8 until Day 12; i.e. rPL replaces prolactin as pregnancy proceeds.

and it may be that one of the actions of contraceptive pills is to cause abnormal relaxation of the oviducal musculature leading to premature escape into the uterus (Croxatto *et al.* 1969) or alternatively to induce tubal arrest or "tube locking" (Greenwald, 1963). During its stay in the oviduct the ovum relies on the secretions of the duct for its sustenance, and the formation of these, as well as the motility of the duct, are under hormonal and nervous control.

Oviducal Motility

By inserting a cannula, containing at its end a pressure-measuring device, into the Fallopian tube of women during laparotomy, Maia and Coutinho (1968) were able to record contractile activity during normal cycling and under other conditions. Activity consisted in small contractions manifest as rises in pressure of 5–10 mm Hg, interspersed at regular intervals by outbursts of increased activity leading to pressures as high as 20 mm Hg. Periods of quiescence lasting several minutes could follow an outburst of activity. The oviduct showed greater activity during menstruation and the early progestational phase of the cycle. Just after ovulation activity became strong, and subsequently it diminished. The pattern of behaviour during normal cycling suggests that oestrogen promotes muscular activity whilst progesterone favours the relaxed condition. The enhanced activity due to oestrogen need not necessarily promote onward movement but rather, according to Coutinho (1974), it might cause isthmic block through powerful localized constriction of the duct, similar to the contraction of the cervical muscle in rats which permits the retention of uterine fluid during oestrus, whilst the subsequent influence of progesterone leads to the relaxation that allows the fluid to escape in di-oestrus (Blandau, 1945).

Steroid Hormone Effects. Direct experiments assessing the roles of oestrogens and progestogens on transport along the oviduct have been carried out by Harper (1966), who transferred newly ovulated eggs, stained with toluidine blue, into the oviduct of a recipient female rabbit. Ovariectomy decreased the rate of transport to 4.5 ± 0.8 mm/min and treatment with progesterone failed to decrease the rate further. Administration of 2–4 μg of oestradiol benzoate restored the rate to 8.4 ± 0.9 mm/min, the rate seen in the oestrous animal; progesterone did not depress this rate. When progesterone was given to the intact oestrous animal the rate was reduced to that seen in pseudopregnancy, namely 5.7 ± 0.4 mm/min. The results are summarized in Fig. 4.12; the author concluded that the maximal rate was achieved by a suitable combination of oestrogen and progesterone.

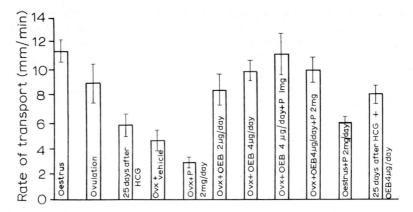

Fig. 4.12. Rate of transport of eggs through the ampulla of the rabbit in relation to hormonal status. P = progesterone; OEB = oestradiol benzoate. (Harper, *Endrocrinology.*)

Other studies, summarized by Chang and Harper (1966), have shown that the picture is complex, depending on the dosage of the hormone and the time of its administration; moreover, the actual effect of an increased motility, induced by oestrogen, may vary, sometimes leading to isthmic block and therefore favouring retention of the ovum, and sometimes promoting passage through the isthmus to the uterus, the effect perhaps being critically dependent on the level of progesterone. Furthermore, the study of De Mattos and Coutinho (1971) has emphasized that the ampulla and isthmus can behave independently (Fig. 4.13) so far as the muscular responses to the hormonal environment are concerned. In general, they found that pregnancy depressed motility whilst ovariectomy caused a progressive increase in the amplitude of pressure-changes some 3–5 days later. Oestrogen produced rapid, small amplitude changes in the normal and ovariectomized animals, whilst progesterone tended to reduce activity.

Innervation of Oviduct. The site of most dense adrenergic innervation of the oviduct is the distal isthmus, i.e. that half adjacent to the ampullary-isthmic junction (Owman and Sjöberg, 1966) it is here, also, that the concentration of noradrenaline is highest (Holst *et al.*, 1974; Bodtke and Harper, 1972). Moreover, by combined electrical and mechanical recordings from the oviduct, Talo and Brundin (1971) showed that there was a pacemaker centre in a discrete area around the ampullary-isthmic junction, and from here electrical activity spreads through the entire isthmus, preceding the rise in pressure. Thus it is this region, subject to adrenergic control, that is most concerned with

mm Hg

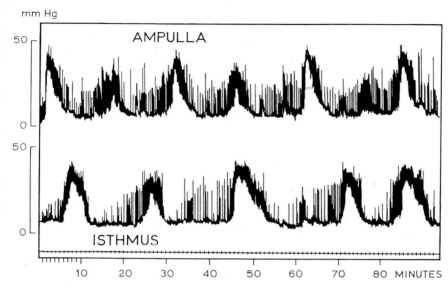

Fig. 4.13. Illustrating asynchrony in outbursts of increased motility of rabbit oviduct between ampulla and isthmus. (De Mattos and Coutinho, *Endocrinology*.)

isthmic block and hence inhibition of implantation. In the hypophysectomized immature rat the superovulation induced by treatment with pregnant mare's serum (PMS) followed by chorionic gonadotrophin (hCG), described by Hopkins and Pincus (1963), could be reduced by reserpine; and the same is true of the rabbit (Bodtke and Harper, 1972).

Relation to Steroid Secretion. Pauerstein *et al.* (1970) showed that oestrogen tends to block passage of the fertilized ovum along the tube, causing an isthmic block lasting for some 60 hr or longer; the adrenergic α-blocker, phenoxybenzamine, antagonized this tube-locking although it had no effect of its own, so that it was concluded that oestrogen exerted its action through adrenergic nerves. When Coutinho *et al.* (1970) catheterized human oviducts, they measured bursts of activity after intravenous injections of adrenaline and noradrenaline except during the luteal phase of the ovarian cycle. Isoproterenol markedly inhibited activity, an effect that was blocked by propanolol, which therefore increased the effectiveness of adrenaline (Coutinho *et al.*, 1971).

Alpha–Beta Balance. In their study of the isolated Fallopian tube *in vitro*, Seitchik *et al.* (1968) showed that α-adrenergic drugs caused constriction and β-adrenergic drugs relaxation, and they suggested that the steroid hormones affected the balance between α- and β-adrenergic activity. These authors have discussed the conflicting results on the effects of ovarian hormones on the oviduct; they emphasized that the isthmus and ampulla must be considered as separate systems, responding independently to nervous and hormonal influences. Moreover, as indicated above, a mere increase in motility of the musculature need not increase rate of movement of the ovum, in fact the increase induced by oestrogen is more often associated with isthmic block due to a spasm of the duct.*

Prostaglandins

According to Spilman and Harper (1973), prostaglandins of the E-class suppress oviducal motility, as measured by a balloon-tipped catheter in the rabbit's oviduct; $PGF_{1\alpha}$ and $PGF_{2\alpha}$ increased activity to give a sustained spasmodic contraction. Thus, PGF tends to retain the embryo by occlusion, whilst PGE allows it to pass to the uterus by abolishing this occlusion. Indomethacin, the inhibitor of prostaglandin synthesis, did not interfere with the process of egg transport, however (El-Banna *et al.*, 1976), but since it is able to disturb early pregnancy it may be that prostaglandins are necessary to support the newly implanted embryo.

The Tubal or Oviducal Fluid

Mastroianni *et al.* (1969) placed a cannula in the monkey's fimbria and led it out on to the surface of the abdomen so that continuous collection of fluid could be made. In a similar manner Restall and Wales (1966) were able to collect fluid from the ewe. Figure 4.14 shows the variation in rate of fluid formation during the oestrous cycle

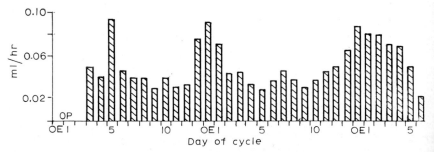

Fig. 4.14. Variation in rate of formation of oviducal fluid in the sheep during the oestrous cycle. (Black *et al.*, *J. Reprod. Fert.*)

---

* Johns and Paton (1976) found the circular muscle of the ampulla more sensitive to nor-adrenaline than the longitudinal, presumably because of the much greater density of innervation of the circular muscle.

in the sheep; at oestrus it reaches 1·4 ml/day and subsides to 0·4 ml/day during the luteal phase (Black *et al.*, 1963).† The composition of the fluid, determined in these two species, is indicated in Table IV. A striking feature is the low concentration of glucose (0·11–0·44 meq/litre

TABLE IV

Composition (meq/litre) of oviducal fluid from sheep (Restall and Wales, 1966) and monkey (Mastroianni *et al.*, 1969)

Species	Na$^+$	K$^+$	Ca^{++}	Mg^{++}	PO$_4$$^{3-}$	Cl$^-$	HCO$_3$$^-$	Lactate	Glucose
Sheep	137	8·2	2·8	0·75	1·1	121	19·7	3·7	—
Monkey	150–170	4–6	2·2–2·6 —	—		—	—	2–3·6	0·11–0·44

or 2–4 mg/100 ml) and the high concentration of lactate; this is consistent with the known metabolic requirement of the mammalian ovum for lactate and the failure to make use of glucose or substrates of the citric acid cycle (Brinster, 1965). Moghissi (1970) found a protein concentration in human fluid of 3·26 g/100 ml; when examined electrophoretically the individual proteins were the same as those in blood plasma.

Specific Protein. However, a more elaborate study on the rabbit's tubal fluid by Kay and Feigelson (1972) showed that there were some glycoprotein fractions in the tubular fluid that were peculiar to it, whilst others that were found in the blood serum were absent from tubular fluid so that the fluid is presumably a secretion by the cellular lining of the tube rather than a simple exudate from the blood plasma. Kay and Feigelson extracted a protein with molecular weight 16,000; after ovariectomy the protein was absent, and could be restored with parenteral *oestrogen* rather than progesterone. They called this "oestrogen-modulated protein". Later Goswami and Feigelson (1974) found the same protein in uterine fluid of intact oestrous rabbits; this disappeared with ovariectomy and was induced by oestrogen, but now progesterone induced this uterine preparation far more effectively than oestrogen. Thus, the protein seems to be secreted separately by uterine and oviducal cells, being induced to different degrees by oestrogen and progesterone according to the site of secretion. Whether or not

† The accumulation of fluid in the oviduct bears some analogy with the accumulation of fluid in the uterus of some rodents, such as the rat, during the oestrous cycle. This requires closure of the cervix during oestrus, and subsequent relaxation after ovulation, the closure being mediated through oestrogen and the subsequent relaxation through progesterone. We may note that prolonged treatment with oestrogen prevents fluid accumulation, but this is due to the stimulation of prolactin secretion which promotes relaxation of the constricted cervix (Kennedy and Armstrong, 1972).

this protein is identical with blastokinin and uteroglobin (p. 245) remains to be seen.

Control of Secretion. So far as control over the secretion is concerned, Mastroianni and Wallach (1961) found that during oestrus in the rabbit and for a day after mating (ovulation), the rate of secretion was about 1·5 ml/24 hr, and that during the course of pregnancy this decreased to 0·8 ml/24 hr. There was no obvious change in lactate concentration with pregnancy, whilst progesterone decreased the rate of secretion in oestrogen-primed castrates.

Implantation

The fertilized ovum undergoes a certain degree of development, through mitotic division, to reach the blastocyst stage before it is ready for implantation—or nidation—in the uterus. At the same time histological changes occur in the uterus that bring it from a "neutral" to a "receptive" state.

Receptive State

Thus we may remove unimplanted blastocysts from a pregnant rat and try to cause them to develop in a recipient pregnant rat; if this was attempted on, say, Day 1 after impregnation of the recipient, there was a delay in implantation, revealed at autopsy by the embryos being smaller than their conceptual age would have demanded. If transfer of the blastocysts was delayed for four to five days after initiation of pregnancy in the recipient, there was no delay in implantation, as indicated by the size of embryos (Psychoyos, 1966).

Three Uterine Stages

If transfer was delayed after the fifth day of pregnancy, no implantation occurred, the blastocysts being rejected from the uterus.

Thus, we may characterize, in all, three stages of the uterus. An initial peroid when it is not receptive, in the sense that implantation does not occur immediately a blastocyst appears in its lumen; during this *neutral* period the blastocyst must await morphological changes taking place in the uterus before it reaches the *receptive stage*, when implantation can occur. This receptive stage soon changes to one in which the blastocyst is actively expelled, called the *non-receptive stage*. In the normal reproductive cycle of the rat, the blastocysts do not enter the uterus while it is in the neutral stage since they are passing down the oviducts during this period of about 5 days. As we shall see, however, the neutral state can be prolonged, both naturally—during lactation—and experimentally through various manoeuvres, in which case the blastocyst remains in the uterus awaiting the change to the receptive state.

Decidualization

This is the name given to the changed histology of the myometrium leading up to and following implantation of the fertilized zygote; before this occurs, the myometrium must be prepared, a process that is obvious in the rapid multiplication by mitosis of the epithelial cells. Decidualization consists in the differentiation of connective-tissue cells into characteristic entities called *decidual cells*; these appear very soon after the blastocyst has made contact, causing a swelling on the outside of the uterus at the site of implantation. Each swelling is an organized structure—the *decidua**—in the centre of which is the *implantation chamber*. Thus the endometrium has been changed into the decidua. A prominent feature, revealed by the electron microscope, is the large number of tight junctions apparently sealing the intercellular gaps; if, as has been suggested, these represent a means of ensuring intercellular communication, the junctions are probably of the gap-type (Volume 1). As indicated above, however, before this decidualization takes place other changes must occur, so that ultimately the process of decidualization is triggered off, apparently by contact of the blastocyst with the altered endometrium.

Histological Changes in Uterus. To describe these in the mouse (Finn, 1971), there is a large increase in the number of mitoses in the epithelial cells of lumen and glands on the 2nd and 3rd days after insemination. On the fourth day, the day before implantation, an abrupt change takes place, the stromal cells now undergoing mitosis and not the epithelial cells. The significance of this multiplication of stromal cells is not clear—perhaps it de-differentiates them, preparatory to transformation into decidual cells (Finn, 1971).

At the time of copulation the lumen of the uterus is open, its epithelial surfaces are irregular, and it contains a large volume of fluid; just before implantation, the lumen appears as a slit in a mesometrial-antimesometrial direction, and immediately before implantation the lumen closes by close apposition of the two surfaces; as Martin *et al.* (1970) have described it, the epithelium shows a wavy corrugation, the one surface appearing as the miror-image of the other suggesting that, before fixation, the surfaces had been closely apposed; and in fact electron-microscopy shows the formation of many microvillous projections of the epithelial cells that interdigitate. According to Nilsson

* As defined, this is "the gestational endometrium that is shed at parturition".

(1966), this interlocking is a feature of the final attachment stage of the implantation process. Thus, the whole process of implantation requires secretion of both progesterone and oestrogen; when the ovariectomized animal was treated with progesterone alone the apposing surfaces of the epithelium were always separated by a space, whereas, if oestrogen was given as well, the cells showed close contacts over wide areas with gaps of only 150Å.; Nilsson suggested that oestrogen induced an increase in intercellular adhesiveness; such adhesiveness would be critical for forming attachment with the blastocyst.*

Decidual Response. The stimulus for decidualization and subsequest implantation seems to be the contact of the blastocyst with the uterine wall; experimentally, decidualization can be provoked by mechanical, electrical or chemical (e.g. pyrathiazine), stimulation of the uterine wall; thus injection of arachis oil into one horn of the uterus of a pseudopregnant rat or mouse will produce a deciduoma, manifest as a gross enlargement by comparison with the other, control, horn (Fig. 4.15). This enlargement, which may amount to a fourfold increase in weight in 72 hours, is the result of cellular proliferation as well as of increase in cell size (Finn and Keen, 1963). Mechanical injury is just as effective (Finn and Keen, 1963). The receptivity of the uterus is estimated experimentally by the magnitude of the decidual response, and this can be estimated either by comparing horn weights in bicorned uteri, or by taking advantage of the pronounced increase in capillary permeability that takes place in response to the stimulus; this can be measured by escape of intravenously injected labelled protein into the uterine tissue. In the rat Kraicer and Shelasnyak (1959) obtained an optimal response between 2 a.m. and noon of the fifth day of pregnancy; at this time the blastocyst was pressed between the enlarged cells of the epithelium of the antimesometrial wall; implantation had not occurred, but was clearly about to.†

---

* Finn and Martin (1967) have described three features of the uterus during implantation that are useful signs of the process for the experimenter. These are :
(1) an increased local capillary permeability as revealed by escape of intravenous Pontamine sky blue forming a band of colour across the uterus when the blastocyst is present.
(2) Passage of primary invasive cells from the blastocyst through the uterine epithelium. Alkaline phosphatase increases in the endometrial stroma around the implanting blastocyst.
(3) The uterine stroma shows oedema around the implanting blastocyst. The primary decidual zone, consisting of closely packed cells around the uterine lumen, appears later.

† Shelesnyak (1962) has reviewed the techniques for inducing decidualization; although he produces convincing evidence that histamine is involved in the process, subsequent studies have rendered this unlikely (Finn, 1971). Of the chemical inducers, Finn found that, in addition to arachis oil, carrageenin—a sulphated mucopolysaccharide—was also highly effective. This substance typically stimulates the formation of granulation tissue and its activity in decidualization supports the analogy between the decidua and the body's response to injury.

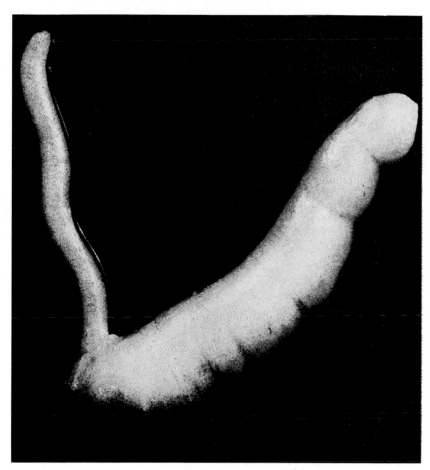

Fig. 4.15. Deciduoma in uterus of pseudopregnant rat following injection of arachis oil into this horn; the other horn, which was not traumatized, is of normal size. (Wiest, *Ciba Study Group*.)

Physiological Significance. The decidua is essentially the initial nesting place for the blastocyst, whilst later it is replaced as a lining to the uterus by the uterine epithelium. Thus temporarily it acts as a separating wall, perhaps defending the underlying uterine tissue from the powerful invasive action of the trophoblast. The high concentration of glycogen in the decidua cells also suggests a nutritive role for the blastocyst until vascularization develops (Finn, 1971).

Non-Receptive State. According to Bitton *et al.* (1965) the failure of deciduogenesis and the rejection of the blastocyst, occurring some five days after fertilization, are related to a pronounced fall in capillary permeability of the endometrium; thus the escape of labelled albumin from blood into the tissue, in response to an experimental trauma, occurs on the fifth day, i.e. when receptivity occurs, but if the trauma is made later than this there is an abrupt fall in the escape of labelled albumin.

Contraceptive Polypeptide. Kent (1975) suspected that the failure of a hamster to ovulate four days after coitus must be due to the presence of an inhibitory factor secreted within the oviduct possibly by the embryo. In fact 2-cell embryos of the progravid hamster do contain a tetrapeptide—threonyl-prolyl-arginyl-lysine—which, when injected subcutaneously into non-bred hamsters, prevents ovulation although the animals exhibit the psychic manifestation of oestrus.

Delayed Implantation

Species Variations

There are very pronounced species variations in the time elapsing between fertilization of the ovum and the final implantation of the blastocyst (Table V).* In the rat, the delay is only some 6 days, just long enough for the uterus to become receptive; in the mouse, implanta-

TABLE V

Graphical illustration of the reproductive cycles of species showing delayed implantation (Hammett, 1935)

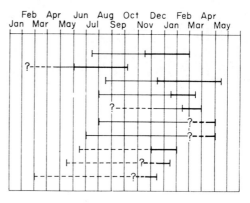

The first vertical mark indicates the average time when the sexes pair; this represents the time of ovulation, except possibly in the bears. The light horizontal line indicates the duration of the quiescent period (unimplanted blastocysts), the heavy line that of active development. The vertical line separating these indicates the time of implantation, the last vertical mark that of parturition.

* In most bats of the families Vespertilionidae and Rhinolophidae there is a delay between copulation and implantation, but this is because the sperm remain in the uterus throughout the winter, whilst *ovulation* only occurs in the spring.

tion occurs a day earlier, but this varies with the strain (Bindon, 1969); in monkeys it occurs on Days 8 to 9 and in the human and rabbit on Day 7. In some species there are very long delays of many weeks or or even several months, notably in the reindeer, armadillo, mink and bear (Hammett, 1935).

Badger. The situation, with regard to the badger, has been summarized by Canivenc (1966); most females mate soon after giving birth in February, whilst implantation takes place in the following December, thus giving a progestational phase of as long as ten months. During this time the corpora lutea pass through several phases suggestive of variable secretory activity; this is indicated in Table VI, and it is seen that cell-size and progesterone content attain a maximum during

TABLE VI

Changes in cell size and progesterone concentration in the corpus luteum during pregnancy (Canivenc, 1966)

	Reproductive phase					
	Metaplasia (Jan.-Feb.)		Resting (Feb.-Dec.)		Preimplantation (Nov.-Dec.)	Implantation (Dec.-Feb.)
Cell size (μ)	13·75	11	9·28	10·95	19·10	26·6
Progesterone (μg/g)	35·8	27·8	19·5	46·5	54·1	76·5

the implantation period; during the resting phase the cell size is small, and vascularization, prominent in actively secreting glands, is notably absent. It would seem, therefore, that delayed implantation in this species is the result of suppression of progesterone secretion; since thyroid activity showed a corresponding cycle, Canivenc suggested that implantation resulted from a general hypothalamic stimulation, involving release of gonadotrophic and thyrotrophic hormones.

Opportunistic Breeding. In the marsupial, the ability to delay implantation is exploited to permit rapid production of new embryos when there is a sudden improvement in the availability of vegetation for food. It is well known, of course, that in the kangaroo only 33 days or so after implantation the foetus is delivered, and its subsequent development depends on suckling in the pouch. Under good conditions most females have simultaneously one offspring running at heel, suckling from an elongated teat outside the pouch, and a very much smaller suckling attached to another teat inside the pouch; moreover, a blastocyst, in what has been called diapause, can be flushed out of the uterus. As indicated in Fig. 4.16, the processes of oestrus with egg fertilization, embryonic diapause followed by gestation, pouch suckling, running at heel and final independence take place cyclically, so that under good conditions breeding is continuous.

When the conditions deteriorate, the first effect is a heavier mortality of the young just emerging from the pouch, and this is revealed by the presence of an elongated teat that is dry due to loss of the young at heel. However, the females still carry a pouch-young. Under severe conditions of drought, the pouch-young may die but the female has in its uterus an embryo that is advanced in development, its development being no longer inhibitied by the suckling of the pouch sib. When born, this young replaces the now moribund young in the pouch and may complete the first two months of life before succumbing to starvation, since at this age the young does not make great demands on the mother. On succumbing it is replaced by a new young that has developed in the uterus, and this renewal of the dying pouch-young may be repeated several times. In effect then, the embryonic diapause, or delayed implantation is reduced to meet the requirement for a new pouch-young.

Thus, breeding can continue well into a drought, and the mother can profit rapidly by an improvement in conditions, so that shortly after

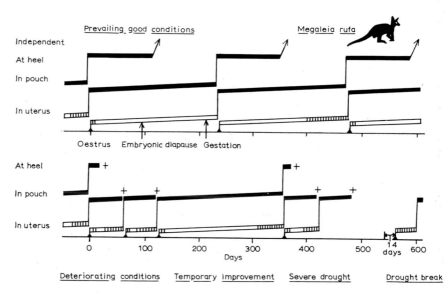

Fig. 4.16. Diagram to illustrate the opportunistic breeding pattern of the red kangaroo, which responds to prevailing drought and rainfall. For details see text. (Tyndal-Biscoe, *Life of Marsupials.*)

a drought all females are carrying pouch-young, an observation that led to the erroneous supposition that the mothers were able to predict the onset of rain. In fact, of course, the pouch-young would have died if the drought had continued. When drought is severe with no rain for two or three years the females eventually enter anoestrus so that all breeding ceases. With the onset of rain the maturation of follicles apparently begins immediately, since after fourteen days some 65 per cent of non-lactating females were in, or near to oestrus. Since it takes some 10 days for follicles to mature this means that the animals are responding to rain rather than the growth of new vegetation.

Black Bear. This animal has been studied because its seasonal mating is combined with delayed implantation and nursing in a den to ensure that the young are fit to face life in the spring. Thus, gestation lasts 7 months but breeding occurs in June to early July; delayed implantation means that embryonic growth does not begin until five months after implantation and the young are born in a very immature state weighing only 500 g in late January; however, they are protected from the cold by segregation in the winter den. The male shows a seasonal variation in size of its gonads and this is associated with a seasonal variation in the testosterone level in its blood, being some 153 ng/100 ml in mid-March to mid-June and 47 ng/100 ml in mid-July to November (McMillin et al., 1976). This circannual rhythm in testosterone level is of special interest since the peak occurs when the male is segregated from the female in its own den, so that exposure to the receptive female, which is a common cause of elevated testosterone levels in monkeys (Rose et al., 1972) and rat (Purvis et al., 1974) may be eliminated, and we must seek the rhythm in the light-dark cycle, working through the pineal gland.*

Hormonal Control of Implantation

Steroids

It is generally agreed that secretion of both oestrogen and progesterone is necessary for the earliest stages of pregnancy, both to cause development of the ovum and implantation of the blastocyst. Thus, ovariectomy on Day 1 of pregnancy leads to failure of egg development and implantation; however, by injection of progesterone daily after ovariectomy the viability of the zygote may be maintained for a long time, although implantation is delayed. This was shown by Cochrane and Meyer (1957) who ovariectomized rats between the 45th and 60th hours after appearance of semen in the oviducts. Treatment with

* It has been argued that the annual rhythm in the monkey is governed by an annual rhythm in the female (Gordon and Bernstein, 1973).

progesterone delayed implantation for as long as the daily injections were maintained; at any time during this period, which was extended to 45 days, injections of oestrogen together with the progesterone brought about implantation within five days of beginning the treatment. Thus, the blastocysts had remained quiescent but viable and only awaited the oestrogen, imposed on a background of progesterone, for the development of the receptive state of the uterus. Histological examination showed that in these ovariectomized animals treated with progesterone, mitosis had not been stimulated, the hormone inducing this being oestrogen; this does not mean that no uterine changes were taking place since Martin et al. (1970) showed that lining up of the epithelial surfaces, leading to an apparent closure of the lumen, occurred in pregnant ovariectomized mice treated only with progesterone; thus, the other important features of the implantation complex, decidual reaction and implantation, require oestrogen.

Oestrogen Secretion

It was considered by Psychoyos (1966) that, *in the natural state* the oestrogen necessary for implantation is secreted by the ovary late on the evening of the fourth day of pregnancy, since, when ovariectomy is performed late on the evening of this day, implantation occurs normally, whereas if ovariectomy occurs earlier, exogenous oestrogen is required (Psychoyos, 1966).

The occurrence of a critical period is illustrated by Fig. 4.17 from Zeilmaker (1963), where it will be seen that, if the operation of ovariectomy was delayed until 01.00 hours of Day 4, implantation was normal, provided progesterone treatment was maintained.

Oestrogen-Surge. The concept of an "oestrogen surge", imposed on a background of continuous secretion of progesterone, as the trigger for implantation, has been questioned by, among others, Finn and Martin (1969; 1974), who pointed out that Zeilmaker (1963) had found that implantation occurred in ovariectomized mice when given *repeated small doses* of oestrogen. Their experiments on the decidual response to arachis oil showed that maximal responsiveness could be achieved on Day 5 post-copulation (equivalent to Day 5 after oestrus in the mouse) by daily injections of low doses of oestrogen together with progesterone (Table VII).

Similarly Psychoyos (1969) showed that a state of receptivity for implantation could be induced in castrated rats by a regimen of progesterone plus oestrogen injections. As in the natural condition, this period of receptivity was followed by a period during which implantation of blastocysts, artificially introduced, was inhibited.

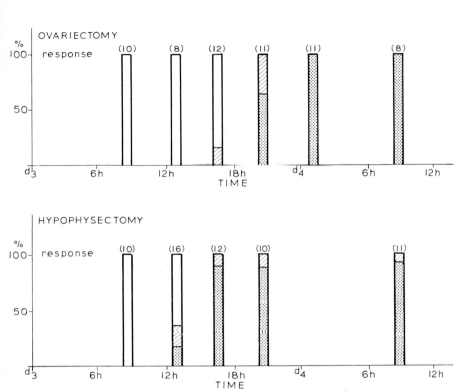

Fig. 4.17. Effects of ovariectomy and hypophysectomy at various times on the third and fourth days of pregnancy on blastocyst implantation in the rat. Black: implantation sites only; shaded: incomplete implantation; white: blastocysts only. Number under each bar is number of rats of that particular time-group. (Zeilmaker, *Acta endocrinol.*)

Priming. Thus, as Finn and Martin (1974) emphasize, along with many other aspects of progesterone action, a preliminary "priming" of the animal with oestrogen, in this case presumably secreted during pro-oestrus, is necessary. On this view, the secretion of both ovarian hormones, oestrogen and progesterone, is patterned to produce maximum uterine sensitivity at the time when the blastocyst is ready to attach. Nevertheless, determinations of oestradiol-17β in the blood-plasma of pregnant rats by Watson *et al.* (1975) have shown a steep rise in level late on Day 3, a rise that can be suppressed by an anti-oestrogen such as Tamoxifen (p. 280), which likewise suppresses implantation. The "surge" is illustrated by Fig. 4.18, which also shows the effects of Tamoxifen.

TABLE VII

Decidual response after injection of arachis oil into left horn of uterus of ovariectomized mice treated with hormones as indicated (Finn and Martin, 1969)

The maximal decidual response, measured as an increase in weight of the treated horn, occurs in Groups 2 and 4 when the animals were given daily doses of oestrogen up to Day 5, and killed on Day 6. (The mice were primed with 100 ng oestradiol for 3 days after ovariectomy and then left for 1 day)

| Group | Treatment | | | | | | No of mice | Decidual response Mean (mg) ± S.E. |
	Day 2	Day 3	Day 4	Day 5	Day 6	Day 7		
1	7 ng Oe ⎱ — ⎰	P + 7 ng Oe	P + 7 ng Oe	K	—	—	9	6·3 ± 0·9
2	7 ng Oe ⎱ — ⎰	P + 7 ng Oe	P + 7 ng Oe	P + 7 ng Oe	killed	—	10	40·4 ± 4·4
3	7 ng Oe ⎱ — ⎰	P + 7 ng Oe	P + 7 ng Oe	P + 7 ng Oe	P + 7 ng Oe	killed	10	29·3 ± 4·8
4	7 ng Oe	P + 10 ng Oe	P + 15 ng Oe	P + 15 ng Oe	killed	—	5	51·6 ± 6·8

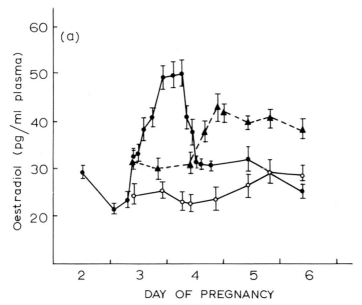

Fig. 4.18. Illustrating the steep rise in plasma level of oestradiol in blood of rats during early pregnancy. (●——●), normal rats; (▲——▲) rats receiving 0·1 mg Tamoxifen/kg; (○——○), rats receiving 0·2 mg Taxomifen/kg. (Watson *et al.*, *J. Endocrinol.*)

Oestrogen Receptors

Evidence supporting the "oestrogen surge" concept is given by Fig. 4.19 which shows that the concentration of oestrogen receptors in the uterine tissue of pregnant rats passed through a maximum at Day 5, the receptive period for implantation of the blastocyst in the rat; the Figure shows that it is the endometrial tissue that is important.*

Contraceptive Activity of Steroids

Since successful implantation depends on an adequate balance of oestrogen and progesterone secretions during the critical period, it is not surprising that steroid injections can prevent implantation. The subject will be discussed more fully later; here we may note that injections of oestrogen will prevent implantation; especially potent is oestradiol-17-β when compared with oestriol, epi-oestriol and oestradiol-16α; and Müller and Wotiz (1977) consider that the modes of action of oestradiol-17β, on the one hand, and of the remaining phenolic steroids on the other, are fundamentally different. Thus oestradiol-17β acts by virtue of its powerful oestrogenic

* Oestrogen, while it promotes implantation of the delayed blastocyst also, of necessity, brings about an increased metabolic activity (Prasad *et al.*, 1968). However there is no evidence of the presence of oestradiol-binding receptors in the blastocyst, so that Prasad *et al.* (1974) suggest that oestradiol acts through synthesis of cAMP or in some other indirect fashion.

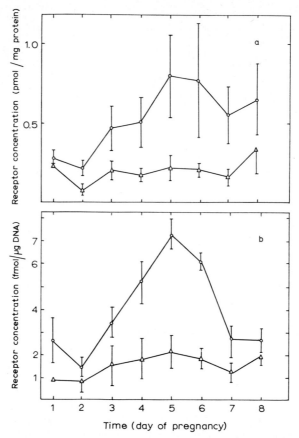

Fig. 4.19. Concentrations of endometrial (○———○) and myometrial (△———△) oestrogen receptors in the rat during early pregnancy. The results are shown in relation to total soluble protein (*a*) or DNA content (*b*). (Mester *et al.*, *Nature*.)

activity, manifest in uterine growth, whereas the others, with only some 8–0·8 per cent of the anti-implantation activity, act by competing for oestradiol's receptors. Thus the anti-implantation activity of oestriol can be prevented by giving oestradiol.

Suckling Animals

Delayed Implantation

Cochrane and Meyer (1957) were able to delay implantation for as long as 45 days in their experiments on ovariectomized rats; this compares with a maximum delay of some 28 days described by Weichert (1942) in suckling pregnant rats. In general, when a suckling rat becomes pregnant the gestation period is prolonged to some 27 to 40 days, with most periods falling between 32 and 35 days. This prolongation is due entirely to a delay in implantation, the blastocysts lying free in the lumen of the uterus.

Role of Oestrogen. Weichert showed that daily injections of oestrogen into pregnant suckling rats from Day 3 up to Days 11–14 caused blastocyst implantation and subsequent development, similar to that of non-suckling rats in some 68 per cent of the treated animals. He considered that the oestrogen required for implantation was being lost in the milk, since there was good evidence that the delay in implantation increased with the number of suckling young; thus, Enzmann *et al.* (1932) found that each suckling mouse delayed implantation by 21 hr so that implantation could occur on Days 7–16 according to the number of sucklings, compared with Day 6 for the nullipar.

Duration of Suckling. The duration of suckling is also important; thus Zeilmaker (1964) showed that, with a standard litter of 10 suckling rats, lactation until the 13th hour of the third day of pregnancy did not delay implantation; each day of suckling after this caused a day's delay, and spontaneous implantation only occurred from Day 8 onwards. If the litter was removed on any of the first five days following the third day of lactation-pregnancy, implantation followed.

Pituitary Secretions

Psychoyos' studies on ovariectomy in the rat and mouse suggest that the maintenance of the ovum and subsequent implantation of the blastocyst depend on a background secretion of progesterone together with a "surge" of oestrogen imposed on this. More precisely, the progesterone must be secreted at an adequate level for at least forty-eight hours; and the secretion of oestrogen at the end of this period is the basic hormonal sequence for implantation. As soon as this sequence is accomplished, the uterus undergoes specific modifications leading, within twenty-four hours, to an optimum endometrial sensitivity to allow of implantation, a sensitivity that lasts some twelve hours (Psychoyos, 1966). The pituitary secretions should reflect these requirements, and a great many studies have been carried out on the effects of hypophysectomy,* and hypophysectomy plus ovariectomy, in order to

---

* There are wide variations respecting the effects of hypophysectomy (and ovariectomy) on the course of pregnancy and these have been summarized in Table I, p. 205. So far as the rat and mouse are concerned, hypophysectomy between Days 7 and 10, i.e. after implantation, causes resorption of the foetuses; if delayed till Day 11, the foetuses may be unaffected and carried to term. It must be noted, however, that although rats may survive *in utero* up to term, delivery is seriously impaired so that Pencharz and Long (1933) found that, usually, the mother died after a prolonged gestation or else gave birth to both dead and living progeny. In the guinea-pig the *early* pregnancy is independent of the pituitary so that hypophysectomy during the first 1–3 days allows embryonic development for at least 26 days (Bland and Donovan, 1969).

elucidate the role of the gonadotrophins, including prolactin, in permitting implantation and sustaining subsequent growth of the embryo. Zeilmaker's study on the rat (Fig. 4.17, bottom, p. 233) shows that hypophysectomy delayed implantation, the difference between this and ovariectomy consisting in the shorter period during which hypophysectomy could be carried out to prevent implantation; thus hypophysectomy on Day 3 + 17 hr had no effect whereas ovariectomy was still effective.

Luteinizing Hormone

The pituitary gonadotrophin necessary for implantation seems to be LH in the rat. Thus, Macdonald et al. (1967) were able to induce implantation in hypophysectomized rats by administration of LH at any time up to 12 days after copulation, whereas FSH and prolactin had no effect. Under these conditions the ovaries are luteinized, so that the oestrogen necessary for implantation must have come from corpora lutea, follicles, or both.

Critical Period

In mice, Bindon (1969) showed that hypophysectomy on Day 1 of pregnancy completely prevented implantation, in fact development and viability of the zygote were prejudiced, entry into the uterus and loss of the zona pellucida were delayed, and by Day 8 there was no sign of blastocysts in the uterus. Injections of progestogens preserved the blastocysts, and implantation could be achieved by a subsequent injection of oestrogen. By changing the day of hypophysectomy, Bindon estimated the time when a pituitary-induced secretion of oestrogen required for implantation took place. As Table VIII shows, the critical period is between 20 and 24 hours on Day 3.

TABLE VIII

Effects of hypophysectomy on the number of embryos in pregnant mice given a single injection of a long-acting progestogen, Depo-Provira (After Bindon, 1969)

Time of hypophysectomy		Proportion with embryos
Day 3	16·00	0/15
	20·00	0/13
	24·00	9/15
Day 4	08·00	13/15
	12·00	14/17
	16·00	14/16

Exogenous Gonadotrophins and Antisera

PMSG and hCG

Bindon (1971) studied the effects of exogenous gonadotrophins on implantation in either hypophysectomized pregnant mice or pregnant suckling mothers. Pregnant mare's serum (PMSG), with both FSH- and LH-activities, was effective in both types of mouse, whereas human chorionic gonadotrophin, which has mainly LH-activity, was ineffective.* When purified ovine gonadotrophins were used, a mixture of FSH and LH was required. Munshi *et al.* (1972) prepared an antiserum to ovine LH; this was effective in preventing implantation in mice when administered on any of Days 2–7 of pregnancy, an effect that could be inhibited by 1 mg of progesterone given with the LH-antiserum on Day 4, and then daily from Day 4 to Day 18, when implantation was 100 per cent. When given on Day 1 of pregnancy, the antiserum was ineffective, presumably because subsequent secretion of LH was adequate to provide the background of progesterone of duration necessary for the subsequent secretion of oestrogen.

Rat Anti-LH

In the rat, Madhwa Raj *et al.* (1968) found that an antiserum to LH, administered on Day 4 after fertilization, inhibited implantation; the expected surge of oestrogen was likewise inhibited. An anti-FSH was ineffective, so that it would seem that LH is the pituitary hormone predominantly in control of implantation in the rat. After Day 4, the anti-LH had no effect; thus on the morning of Day 4 inhibition of implantation occured whereas, if treatment was delayed till 6 p.m. of the same day, implantation was normal. The effects of the anti-LH could be reversed by oestradiol-17β. Thus, it would appear that, in the rat, the pituitary secretion of LH on Day 4 is the important event for implantation, triggering the release of oestrogen.

Anti-LHRH

Arimura *et al.* (1976) also concluded that Day 4 was critical for implantation in the rat, so that there were no effects of an anti-LHRH if it was administered on Days 3 or 5; given on Day 4, it caused a three-day delay in implantation. The effects could be counteracted by LHRH or oestrogen. The fact that administration of the anti-LHRH

* According to Choudary and Greenwald (1969) a combination of LH and FSH is necessary for maintenance of corpora luteal function in the mouse.

was ineffective on Day 3 means that the antiserum to the releasing hormone is rapidly inactivated, by contrast with the anti-LH.*

The effects of oestrogen, in counteracting the effects of antisera, confirm the involvement of this steroid in nidation, but the role of progesterone is not so clear; the peak in the plasma level on Days 3–5 of the rat suggests that it is necessary; moreover, the anti-releasing hormone decreased the plasma progesterone.

Pituitary FSH-Content

Suppression of FSH Secretion

Bindon (1970) controlled the day of implantation in mice by employing pregnant suckling animals and simply removing the sucklings from the mother, which is sufficient to induce implantation (McLaren 1968). Thus litters were removed from suckling pregnant mice on Day 5, the mothers were killed at intervals and their pituitaries estimated for FSH and compared with controls with litters not removed. The FSH content of the pituitaries of mothers with suckling litters was high, suggesting suppression of secretion. On removing the litters the content was reduced, suggesting secretion, and by 32 hr the FSH contents had returned to the control values; by this time there was evidence of some implantation. The rate of incorporation of radioactive uridine into the uterus is a measure of FSH-induced oestrogen secretion, and this was increased by removal of the litters. Thus, the prime event in delayed implantation in the mouse seems to be a suppression of the FSH secretion that produces the oestrogen, necessary to be imposed on the background of progesterone that takes place after ovulation.

Shedding the Zona Pellucida

Steroid Requirements

In the rat, the blastocyst sheds its zona pellucida, preparatory to implantation, during the afternoon of Day 5 of pregnancy; during this time both progesterone and oestrogen are being secreted. Dickmann (1968) injected blastocysts from pregnant rats into the uteri of ovariectomized recipient rats, and killed them some 10, 24 and 72 hr later, the uteri being flushed out to examine the state of the blastocysts. Shedding of the zona pellucida certainly occurred in the ovariectomized recipients, but it was considerably delayed by comparison with control, pseudopregnant, recipient rats. Earlier it was shown that delaying also occurred under the influence of progesterone, so that it

---

* Watson *et al.* (1975) have discussed the role of gonadotrophins in the implantation stage of pregnancy; they conclude that FSH is responsible for the oestrogen surge, acting synergistically with LH and prolactin.

would appear that oestrogen is necessary, perhaps in cooperation with progesterone, for shedding.

Ovariectomy. McLaren (1971) noted that the oestrogen secretion necessary for successful implantation in the mouse apparently occurred on Day 4 post-coitum (82 hr p.c.), so she carried out ovariectomy on Day 3, before this secretion, and collected blastocysts from the uterus on Day 7·5; many of the blastocysts had not lost their zonae pellucidae, whereas if ovariectomy was postponed till Day 4 there were no persistent zonae. Therefore it seemed that oestrogen was responsible for the uterine secretion of a lysin.

Lytic Factor

In a valuable discussion of the problem of the zona pellucida in mice, McLaren concluded that both a lytic factor and the natural tendency of the blastocyst to expand were the two factors involved in the shedding process. Thus, in the case of delayed implantation, McLaren (1968) found that growth, and therefore expansion of the non-implanted blastocyst, did not cease abruptly at the date of normal implantation (4 days p.c. in the mouse); instead, there was expansion until Day 7·5, when the blastocyst ceased mitosis and went into "developmental arrest", lying free but closely apposed to the uterus. In the lactating or ovariectomized animal the lysing was absent, but the blastocyst eventually burst out, leaving intact zonae which could be recovered free in the uterus (McLaren, 1970). The unfertilized egg, placed in the uterus of a pseudopregnant animal, does not expand, but its zona is lysed. During the normal implantation process, the blastocyst expands, but the thinned zona resulting from this expansion is usually lysed before emergence of the blastocyst. Thus the theories of zone loss, according to one of which the blastocyst escapes by a hatching process and according to the other, by a lytic removal of the zona, are reconciled by McLaren's studies.

Effect of Suckling

As with implantation itself, the preliminary shedding of the zona is delayed if the animal is suckling. This is demonstrated by the following:

	Day 5		Day 6	
	Non-suckling	Suckling	Non-suckling	Suckling
No. of blastocysts in zona	22	86	0	1
No. of blastocysts zona-free	62	67	12	47

Blocking by Drugs

Ovulation may be blocked by injections of tranquillizers, such as chlorpromazine, by barbiturates, and by reserpine, presumably acting through the hypothalamus. Blockage of implantation differs in that barbiturates and reserpine have no effect; a tranquillizer, trifluoperazine blocks both ovulation and implantation.

Exteroceptively Induced Block

Alien Males

Bruce (1960) observed that implantation was blocked in some 80 per cent of newly mated female mice if they were exposed for 3 days to alien males i.e. males of a different strain from that of the stud-male. Contact with the males was not necessary, so that either the sight or smell was the important factor. Females rendered anosmic by removal of the olfactory bulbs were not susceptible (Bruce and Parrott, 1960), and it was sufficient simply to place straw from the cages of strange males in the cage of the pregnant female to delay implantation (Parkes and Bruce, 1962), so that there is no doubt that it is the smell of the foreign male that governs the delay.

Since the hypothalamus has a strong afferent input from the olfactory tract, it is likely that the effect is mediated through this part of the brain, influencing the gonadotrophin release from the anterior pituitary and thus the level of ovarian steroid hormones during the critical implantation period. In fact Parkes and Bruce (1961) have shown that the block may be overcome by an injection of prolactin, which is luteotrophic in the mouse. Examination of the ovaries showed a variable luteal response; in 9 out of 17 mice there were enlarged corpora lutea typical of pseudopregnancy, but in the remainder the appearance was typical of pro-oestrus (Bruce, 1960).*

Pheromones

The substance in the urine responsible for these effects falls into the class of pheromones, a term commonly used for externally voided substances that convey information between members of the same species; strictly speaking, the mode of transmission is via the olfactory sense, but it is usual to include other modes of transmission (Bronson, 1971).

* Repeated blockages were without effect on the subsequent fertility of females (Bruce, 1962); there was, however, a tendency to spontaneous pseudopregnancy, a rare condition in the rat; this may be due to storage of LH in the pituitary as a result of suppressed secretion, with its subsequent release, pseudopregnancy being essentially an expression of sustained luteotrophic activity (p. 57).

Signalling Pheromone. In general they have been classified as signalling and priming pheromones as indicated in Table IX; the signalling agent, typically seen as a sex attractant, may be demonstrated by presenting an animal with two tunnels in a two-choice experiment, the frequency with which it chooses that containing the scent of the animal at the end of the tube being an indication of the presence of an attractant.

TABLE IX

Postulated mouse pheromones (Bronson, 1971)

Signalling pheromones:	Priming pheromones:
1. Fear substance	1. Oestrus-inducer
2. ♂ sex attractant	2. Oestrus-inhibitor
3. ♀ sex attractant	3. Adrenocortical activator
4. Aggression-inducer	
5. Aggression-inhibitor	

Priming Pheromone. Priming pheromones are typically those that induce gonadotrophin release. So far as reproductive physiology is concerned, the pheromones can induce or inhibit oestrus and ovulation; they can accelerate the onset of puberty and block implantation. Their urinary secretion depends on the androgen state of the male so that castration abolishes the liberation completely within 10–15 days (Lombardi *et al.*, 1976), an effect that is reversed by testosterone injection. Interestingly, the female urine becomes active if she is injected with testosterone.

Sex Attractant. A pheromone voided by a rodent that attracts members of the opposite sex was established by Carr *et al.* (1965); Gawienowski *et al.* (1975) made ether extracts of various tissues of the rat and tested male and females for their attraction to these, presenting them with two tunnels at the end of which one of the extracts was placed, and measuring the frequency with which they moved into each tunnel. The extract that proved to be most potent was one from the preputial gland, that from the male attracting the female and *vice versa*. If the male was castrated, the extract ceased to be effective; similarly, the male was only attracted to extracts from receptive females. Male extracts were not attractive to males, and female extracts were not attractive to females. The volatile factor in the extracts included methyl-substituted C7 and C8 aliphatic alcohols.

Mouse Oestrous Cycle. As Bronson (1971) has emphasized, it is common to think of the mouse cycle as lasting 4–5 days, whereas under appropriate olfactory environmental conditions cycles as long as 11–12 days can be considered normal. Apparently two factors are operative; one that is produced by females tending to suppress oestrous cycling, and another, in male urine, that accelerates cycling. Thus isolation from other females, or exposure to males, shortens the cycle. When females are housed together, instead of singly, oestrus is partly suppressed and 25 per cent may become pseudopregnant, a rare condition for a mouse (Parkes and Bruce, 1961). Place this pseudo-pregnant mouse with a male, and oestrus returns immediately. That a priming pheromone is involved, is indicated by the relief of oestrus-suppression by olfactory bulbectomy, and its independence of vision and contact.

The male pheromone is better understood; it occurs in the urine, whether this is voided in the usual manner or taken from the bladder, so that the preputial gland, a source of the sex attractant pheromone (Gawienowski *et al.*, 1975), is apparently not involved.

Female Pheromone on Male Mouse. Maruniak and Bronson (1976) exposed male mice to the smell of female mouse urine and found elevated levels of LH and FSH in their blood; the potency of the female urine was independent of her ovarian state. These authors have discussed the two-way nature of pheromonal communication between the sexes, which results in altered gonadatrophic secretion, the most prominent in the female being increased LH secretion with concomitant oestradiol secretion, and in the male a release of LH and concomitant rise in testosterone secretion.

The action of pheromones in accelerating the onset of puberty will be described later (p. 438).

Spacing of the Embryos

It is generally believed that in the polytocous mammal there is a regular spacing of the implantation sites determined, perhaps, by the development of local refractory zones around each implantation site that inhibit implantation in these zones. However, McLaren and Michie's (1959) experiments suggest that spacing is, in fact, randomly achieved and is not the consequence of "serial implantation" whereby the first embryo to arrive implants in the uterus nearest to the oviduct, the second a little farther down, and so on. Rather, the eggs may be mixed by uterine contractions before implantation and adopt essentially random positions at implantation.

Uterine Component in Control of Implantation

State of the Blastocyst

Blastocysts exist in either a proliferative or quiescent state; this latter quiescent, or diapausal state, is characteristic of the blastocyst in delayed implantation, whether this has been induced experimentally, as in the ovariectomized rat, or facultatively, as during suckling, or as a compulsory condition, as in the lengthy delays in such species as the roe deer. In this quiescent state, embryonic mitotic activity ceases and metabolic activity is minimal. All the evidence indicates that the quiescent state is governed by the uterus, possibly by the secretion of a specific protein; thus blastocysts flushed out of the uterus during delayed implantation will grow, *in vitro*, in the same way as blastocysts, collected just prior to normal implantation, will (Gwatkin, 1969); furthermore, if the quiescent blastocysts are transplanted to an ectopic site, e.g. under the kidney capsule, they will grow, although those in the uterus of the same animal will not (Kirby, 1967). Again, tumour cells (Walker carcinoma) will invade the uterine wall of the rat when introduced into the cavity provided the uterus will allow implantation of a blastocyst; if the hormonal condition of the animal is unsuitable, as in ovariectomy, this invasion fails (Short and Yoshinaga, 1967). Gwatkin, in his studies on *in vitro* growth of blastocysts, found that, as with many other cultured cells, amino acids and a high-molecular weight component of blood serum were necessary to sustain growth; and the suggestion arises that the uterus secretes specific products necessary to permit blastocyst development, or alternatively, it secretes an inhibitor that maintains the quiescent state of the blastocyst in delayed implantation.

Uteroglobin or Blastokinin

Krishnan and Daniel (1967) separated a protein from rabbit uterus in early pregnancy that promoted development of morulae and blastocysts *in vitro*, and described the protein as *blastokinin*. Later, Beier (1968) described a protein, molecular weight about 30,000, in the endometrial secretions of the pregnant rabbit, the amount extracted being influenced by treatment of the animal with oestrogen and progesterone; he called this *uteroglobin*. It seems likely that these proteins are the same, so we may refer to them as blastokinin.

Fig. 4.20 shows the changes in the amount of blastokinin, expressed as a percentage of the total extractable protein, following normal pregnancy (a), treatment with progesterone (b) and treatment with progesterone in ovariectomy (c). In these ovariectomized, progesterone-treated animals, blastocysts could be implanted successfully, and it was suggested that blastokinin acted as a steroid carrier supplying progesterone to the embryo in a usable and non-toxic form (Arthur and Daniel, 1972).* When an anti-blastokinin was administered to

---

* The conflicting claims regarding the binding of progesterone and oestradiol to uteroglobin have been discussed by Fridlansky and Milgrom (1976); progesterone is, indeed, bound, with a relatively high affinity ($K_D = 4 \cdot 10.^{-7}M$), but 5α pregnane-3,20-dione is bound more avidly. Oestradiol is not bound specifically. The binding of progesterone depends on the mode of preparation of the uteroglobin.

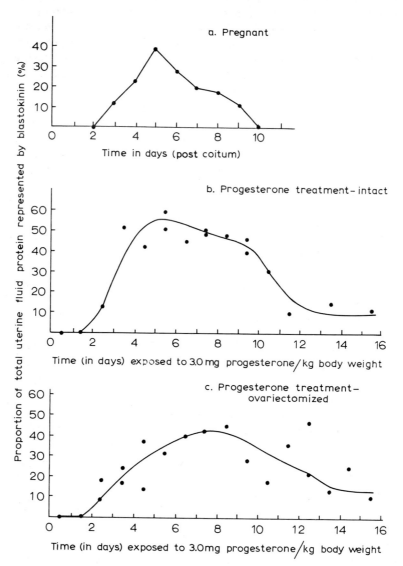

Fig. 4.20. The proportion of the total uterine fluid protein represented by blasto-kinin recovered from (a) uteri of pregnant rabbits, (b) progesterone-treated intact rabbits, and (c) progesterone-treated ovariectomized rabbits. (Arthur and Daniel, *Fert. Ster.*)

pregnant rabbits, implantation, which normally occurs on Day 7 post-coitum, was usually prevented (Krishnan, 1971).

Blood Levels. With the development of a radioimmunoassay for blastokinin, its appearance in the uterine fluid could be assessed quantitatively; thus Mayol and Longenecker (1974) found a concentration of 20,000 ng/ml on Day 3 increasing to 370,000 ng/ml on Day 6, a level some 2000 times that on Day 0. By Days 10–12 the level had fallen to 5–10,000 ng/ml. According to Bullock and Connell (1973), the molecular weight is 10,000–14,500, so that the earlier estimate in the region of 27,000 might have been due to dimer formation. These authors found some blastokinin in non-pregnant rabbit uterine fluid, whilst Noske and Feigelson (1976) produced immunological evidence for the presence of uteroglobin (blastokinin) in the male reproductive tract, and in non-reproductive tract-tissues, such as the digestive tract.

Other Species than Rabbit

The appearance of large amounts of protein, including the specific blastokinin, seems to be peculiar to the rabbit, so that in many of the other species examined the quantities of a blastokinin-like protein identified in the uterine fluid have been much smaller.

Human Uterine Proteins. Roberts *et al.* (1976) have examined the uterine flushings of non-pregnant women; the main components were serum proteins but some uterus-specific proteins were identified electrophoretically, but, in contrast to Daniel (1968), a band corresponding to blastokinin was not identified. Only one of the bands found by Roberts *et al.* varied in intensity with the menstrual cycle.

Rat. Experimentally, the changes in uterine fluid associated with implantation can be studied by ovariectomy, which prevents implantation. The capacity to implant is restored by a regimen of exogenous progesterone and oestradiol (Psychoyos, 1969). Surani (1975, 1976) identified a protein of molecular weight 70,000 which appeared when both progesterone and oestradiol were administered to the ovariectomized rat within one hour of the oestradiol injection. This protein failed to appear if only progesterone was given, and thus may be related to implantation. The protein profile at 13–20 hr in these non-pregnant animals was similar to that in the fluid obtained from pregnant animals on Day 5, the time of implantation.

Cow. In the cow the embryo lies free in the uterus for about 30 days, during which time extensive embryonic differentiation occurs before firm attachment to the uterine wall. This species therefore allows the examination of the uterine fluid during a considerable period between conception and implantation. Roberts and Parker (1974) have described the electrophoretic identification of several proteins, specific for the uterus. The appearance of two cathode-migrating proteins on Day 12 may well coincide with the period of rapid elongation of the blastocyst occurring at about this time. They cannot be embryonic products since they occur in unmated cows at about the same time of the cycle. A slower cathode-migrating protein, appearing between Days 16 and 35 of pregnancy is probably produced by the conceptus and Roberts and Parker suggest that it is connected with the "rescue" of the corpus luteum through some anti-luteolytic factor. Thus, in the non-pregnant cow, the level of blood progesterone falls most rapidly on Day 18; it is about this time that the conceptus has an influence on progesterone levels, so that it could be that an anti-luteolytic factor—in this case an anti-prostaglandin $F_{2\alpha}$—is produced by the uterus or conceptus.

Pig. Murray *et al.* (1972) showed that the total quantity of protein flushed from the pig's uterus increased strikingly between Days 12 to 15 and showed an abrupt fall on Day 16, which was synchronous with the fall in progesterone level accompanying luteolysis. Fractionation of the proteins revealed the presence of material specific for the uterine fluid, and it was two of these fractions—IV and V—that showed the large fall associated with luteolysis. The secretory pattern of these two proteins may be associated with implantation, which occurs on about Day 11 of pregnancy.* In a later study, Squire *et al.* (1972) examined the proteins electrophoretically and found the appearance of specific fractions, not present in the early stage of the oestrous cycle, from Day 12 onwards. Whether or not one or more of the protein fractions is an anti-luteolysin, as suggested for the bovine fluid, has been discussed by Murray *et al.* (1972).

Roe Deer. This artiodactyl exhibits a very prolonged period of delayed implantation, starting when the blastocyst loses its zona pellucida a few days after ovulation and continuing for 5 months until December–January; during the first two weeks of January the blastocyst shows rapid elongation and implantation, followed by a gestation period of five months until May. Aitken (1974) combined analysis of the uterine fluid with electron microscopy of the blastocyst and uterine glands. In mid-January there was a steep rise in amino acids, proteins and other constituents of the uterine fluid, and this was associated with changes in the endometrial glands suggesting secretion of a mucoid fluid.

---

* Krishnan and Daniel (1967) suggested that blastokinin was primarily concerned with blastulation rather than with the later stages of blastocyst enlargement and implantation. Murray *et al.* emphasize that the secretory pattern of their proteins corresponds more with the period of implantation than blastulation, which occurs, in the pig, between Days 6 and 8.

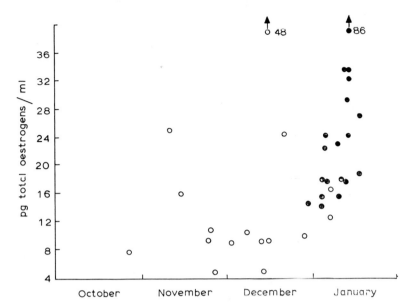

Fig. 4.21. Concentrations of total oestrogens in roe deer blood plasma during delayed implantation. ○, Blastocyst in uterus; ●, embryo in uterus. (Aitken, *J. Reprod. Fert.*)

These changes might well be controlled by the level of oestrogen in the maternal blood (Fig. 4.21).*

Activator or Inhibitor

It can be argued that the blastocyst is under the control of an inhibitory factor secreted by the uterus, and in this way it is easy to account for the rapid increase in metabolic activity of the quiescent blastocyst on removal from the uterus (Kirby, 1967). Alternatively it can be argued that control is exerted by an excitatory substance that is withheld during delayed implantation and secreted when hormonal conditions favour implantation, for example in the ovariectomized rat treated with progesterone and oestrogen. Weitlauf (1974) concludes that an inhibitory factor is important in determining the state of quiescence or diapause, a view shared by Bitton-Casimiri *et al.* (1976).

These authors showed that incorporation of ³H-uridine into RNA by

---

* Daniel (1968) was unable to find traces of blastokinin in the uterine washings of the northern fur seal, and armadillo; both animals with lengthy delays in implantation; small amounts were found in the milk, with a short and variable delay of about 14–17 days.

the quiescent blastocyst *in vitro* was initially much less than that of normal blastocysts, but this steadily increased to reach the normal value in 24 hours (Table X). It is possible, therefore, that an inhibitory substance maintains metabolism low in the diapausing blastocyst;

TABLE X

Incorporation of ^{3}H-uridine into RNA of blastocysts
(Bitton-Casimiri *et al.*, 1976)

Incubation period (hr)	Normal (ct/min/blastocyst)	Dormant
0	42 $\pm$ 5	20 $\pm$ 2
3	49 $\pm$ 8	25 $\pm$ 4
6	57 $\pm$ 6	34 $\pm$ 3
12	81 $\pm$ 13	32 $\pm$ 10
18	136 $\pm$ 3	105 $\pm$ 12
24	154 $\pm$ 5	139 $\pm$ 19

removal from this influence allows the synthesis of RNA which then governs the synthesis of specific proteins that permit the rapid development and implantation of the blastocyst.

Mode of Action. Surani has discussed the possible mode of action of high-molecular weight compounds on blastocyst development and implantation. Serum proteins alone will promote growth of mammalian cells *in vitro* and the same occurs for normal embryos in diapause, and it has been suggested that this action is brought about by a coordination of unrelated biochemical events and has been called a "pleiotopic response" (Hershko *et al.*, 1971) and this can be achieved not only by serum albumin but also by insulin and cAMP. It could be, then, that the specific proteins identified in uterine washings exert some sort of pleiotopic control over the embryo.

REFERENCES

Aitken, R. J. (1974). Delayed implantation in roe deer (*Capreolus capreolus*). *J. Reprod. Fert.*, **39,** 225–233.

Amoroso, E. C. and Finn, C. A. (1962). Ovarian activity during gestation, ovum transport and implantation. In *The Ovary*, Vol. 1. Ed. S. Zuckerman. Academic Press, N.Y. and London, pp. 451–537.

Amoroso, E. C. and Porter, D. G. (1966). Anterior pituitary function in pregnancy. In *The Pituitary Gland*, Vol. 2. Eds. G. W. Harris and B. T. Donovan. Butterworth, London, pp. 364–411.

Aref, I., Hefnawi, F., Kandil, O. and Aziz, M. T. A. (1973). Effect of minipills on physiologic responses of human cervical mucus, endometrium, and ovary. *Fert. Steril.*, **24,** 578–583.

Arimura, A., Nishi, N. and Schally, A. V. (1976). Delayed implantation caused by administration of sheep immunogamma globulin against LHRH in the rat. *Proc. Soc. exp. Biol. Med.*, **152,** 71–75.

Arthur, A. T. and Daniel, J. C. (1972). Progesterone regulation of blastokinin production and maintenance of rabbit blastocysts transferred into uteri of castrate recipients. *Fert. Steril.*, **23,** 115–122.

Atkinson, L. E. *et al.* (1975). Circulating levels of steroids and chorionic gonodotropin during pregnancy in the rhesus monkey, with special attention to the rescue of the corpus luteum in early pregnancy. *Biol. Reprod.*, **12,** 335–345.

Beier, H. M. (1968). Uteroglobin: a hormone-sensitive endometrial protein involved in blastocyst development. *Biochim. biophys. Acta*, **160,** 289–291.

Bindon, B. M. (1969). The role of the pituitary gland in implantation in the mouse: delay of implantation by hypophysectomy and neurodepressive drugs. *J. Endocrinol.*, **43,** 225–235.

Bindon, B. M. (1969). Follicle-stimulating hormone content of the pituitary gland before implantation in the mouse and rat. *J. Endocrinol.*, **44,** 349–356.

Bindon, B. M. (1970). Preimplantation changes after litter removal from suckling mice. *J. Endocrinol.*, **46,** 511–516.

Bindon, B. M. (1971). Gonadotrophin requirements for implantation in the mouse. *J. Endocrinol.*, **50,** 19–27.

Bitton, V., Vassent, G. and Psychoyos, A. (1965). Réponse vasculaire de l'uterus au traumatisme, au cours de la pseudogestation chez la ratte. *C. r. Acad. Sci. Paris*, **261,** 3474–3477.

Bitton-Casimiri, V., Brun, J. L. and Psychoyos, A. (1976). Uptake and incorporation of ^{3}H-uridine by normal or diapausing rat blastocysts after various periods of culture. *J. Reprod. Fert.*, **46,** 447–448.

Black, D. L., Duby, R. T. and Riesen, J. (1963). Apparatus for the continuous collection of sheep oviduct fluid. *J. Reprod. Fert.*, **6,** 257–260.

Bland, K. P. and Donovan, B. T. (1969). Control of luteal function during early pregnancy in the guinea-pig. *J. Reprod. Fert.*, **20,** 491–501.

Bodtke, R. R. and Harper, M. J. K. (1972). Changes in the amount of adrenergic neurotransmitter in the genital tract of untreated rabbits, and rabbits given reserpine or iproniazid during the time of egg transport. *Biol. Reprod.*, **6,** 288–299.

Brinster, R. L. (1965). Studies on the development of mouse embryos *in vitro*. II. The effect of energy source. *J. exp. Zool.*, **158,** 59–68.

Bronson, F. H. (1971). Rodent pheromones. *Biol. Reprod.*, **4,** 344–357.

Bruce, H. M. (1960). A block to pregnancy in the mouse caused by proximity of strange males. *J. Reprod. Fert.*, **1,** 96–103.

Bruce, H. M. (1962). Continued suppression of pituitary luteotrophic activity and fertility in the female mouse. *J. Reprod. Fert.*, **4,** 313–318.

Bruce, H. M. and Parrott, D. M. V. (1960). Role of olfactory sense in pregnancy block by strange males. *Science*, **131,** 1526.

Bullock, D. W. and Connell, K. M. (1973). Occurrence and molecular weight of rabbit uterine "blastokinin". *Biol. Reprod.*, **9,** 125–132.

Carr, W. J., Loeb, L. S. and Dissinger, M. L. (1965). Responses of rats to sex odors. *J. comp. physiol. Psychol.*, **59,** 370–377.

Canivenc, R. (1966). Progestational activity in the badger (*Meles meles* L). Ciba Study Group No. 23. Churchill, 1966, pp. 46–58.

Chang, M. C. and Harper, M. J. K. (1966). Effects of ethinyl estradiol on egg transport and development in the rabbit. *Endocrinology*, **78**, 860–872.

Choudary, J. B. and Greenwald, G. S. (1969). Luteotropic complex of the mouse. *Anat. Rec.*, **163**, 373–388.

Cochrane, R. L. and Meyer, R. K. (1957). Delayed nidation in the rat induced by progesterone. *Proc. Soc. exp. Biol. Med. N.Y.*, **96**, 155–159.

Courrier, R. (1941). Evolution de la grossesse extra-utérine chez la lapine castrée. *C. r. Soc. Biol.*, **135**, 820–822.

Courrier, R. (1945). *Endocrinologie de la Gestation*. Masson et Cie. (Quoted by Davies and Ryan, 1972.)

Coutinho, E. M. (1974). Hormonal control of oviductal motility and secretory functions. In *Reproductive Physiology*. Ed. R. O. Greep. Butterworth, London. University Park Press, Baltimore, pp. 133–153.

Coutinho, E. M., Maia, H. and Filho, J. A. (1970). Response of the human Fallopian tube to adrenergic stimulation. *Fert. Ster.*, **21**, 590–594.

Coutinho, E. M., de Mattos, C. E. R. and Rita da Silva, A. (1971). The effect of ovarian hormones on the adrenergic stimulation of the rabbit Fallopian tube. *Fert. Ster.*, **22**, 311–317.

Croxatto, H. *et al.* (1969). Fertility control in women with a progestogen released in microquantities from subcutaneous capsules. *Amer. J. Obstet. Gynec.*, **105**, 1135–1138.

Csapo, A. I. *et al.* (1972). The significance of the human corpus luteum in pregnancy maintenance. *Amer. J. Obstet. Gynec.*, **112**, 1061–1067.

Csapo, A. I., Pulkinnen, M. O. and Wiest, W. G. (1973). Effects of luteotomy and progesterone replacement therapy in early pregnant patients. *Amer. J. Obstet. Gynec.*, **115**, 759–765.

Csapo, A. I. and Wiest, W. G. (1969). An examination of the quantitative relationship between progesterone and the maintenance of pregnancy. *Endocrinology*, **85**, 735–746.

Daniel, J. C. (1968). Comparison of electrophoretic patterns of uterine fluid from rabbits and mammals having delayed implantation. *Comp. Biochem. Physiol.*, **24**, 297–299.

Davies, L. J. and Ryan, K. J. (1972). The uptake of progesterone by the uterus of the pregnant rat *in vivo* and its relationship to cytoplasmic progesterone-binding protein. *Endocrinology*, **90**, 507–515.

Davies, L. J. and Ryan, K. J. (1972). Comparative endocrinology of gestation. *Vitam. & Horm.*, **30**, 223–279.

De Mattos, C. E. R. and Coutinho, E. M. (1971) Effects of the ovarian hormones on tubal motility of the rabbit. *Endocrinology*, **89**, 912–917.

Dickmann, Z. (1968). Does shedding of the zona pellucida by the rat blastocyst depend on stimulation by the ovarian hormones. *J. Endocrinol.*, **40**, 393–394.

El-Banna, A. A., Sacher, B. and Schilling, E. (1976). Effect of indomethacin on egg transport and pregnancy in the rabbit. *J. Reprod. Fert.*, **46**, 375–378.

Enzmann, E. V., Saphir, N. R. and Pincus, G. (1932). Delayed pregnancy in mice. *Anat. Rec.*, **54**, 325–341.

Erb, R. E. *et al.* (1967). *J. Dairy Sci.*, **51**, 420. (Quoted by Davies and Ryan, 1972.)

Finn, C. A. (1971). The biology of decidual cells. *Adv. Reprod. Physiol.*, **5**, 1–26.

Finn, C. A. and Keen, P. M. (1963). The induction of deciduomata in the rat. *J. Embryol. exp. Morph.*, **11**, 673–682.

Finn, C. A. and McLaren, A. (1967). A study of the early stages of implantation in mice. *J. Reprod. Fert.*, **13**, 259–267.

Finn, C. A. and Martin, L. (1969). Hormone secretion during early pregnancy in the mouse. *J. Endocrinol.*, **45**, 57–65.

Finn, C. A. and Martin, L. (1974). The control of implantation. *J. Reprod. Fert.*, **39**, 195–206.

Fridlansky, F. and Milgrom, E. (1976). Interaction of uteroglobin with progesterone, 5α pregnane-3, 20-dione and estrogens. *Endocrinology*, **99**, 1244–1251.

Gawienowski, A. M., Orsulak, P. J., Stellwicz-Sapuntzakis, M. and Joseph, B. M. (1975). Presence of sex pheromone in preputial glands of male rats. *J. Endocrinol.*, **67**, 283–288.

Gordon, T. P. and Bernstein, I. S. (1973). Seasonal variation in sexual behavior of all-male rhesus troops. *Amer. J. phys. Anthropol.*, **38**, 221–226.

Goswami, A. and Feigelson, M. (1974). Differential regulation of a low-molecular weight protein in oviductal and uterine fluids by ovarian hormone. *Endocrinology*, **95**, 669–675.

Greenwald, G. S. (1963). *In vivo* recording of intraluminal pressure changes in the rabbit oviduct. *Fert. Ster.*, **14**, 666–674.

Gwatkin, R. B. L. (1969). Nutritional requirements for post-blastocyst development in the mouse. *Internat. J. Fert.*, **14**, 101–105.

Hammett, G. W. D. (1935). Delayed implantation and discontinuous development in mammals. *Quart. Rev. Biol.*, **10**, 432–447.

Hammond, J. and Marshall, F. H. A. (1930). Oestrus and pseudo-pregnancy in the ferret. *Proc. Roy. Soc. B*, **105**, 607–630.

Harper, M. J. K. (1966). Hormonal control of transport of eggs in cumulus through the ampulla of the rabbit oviduct. *Endocrinology*, **78**, 568–574.

Haterius, H. O. (1936). Reduction of litter size and maintenance of pregnancy in the oophorectomized rat: evidence concerning the endocrine role of the placenta. *Amer. J. Physiol.*, **114**, 399–406.

Heap, R. B. and Deanesley, R. (1966). Progesterone in systemic blood and placentae of intact and ovariectomized pregnant guinea-pigs. *J. Endocrinol.*, **34**, 417–423.

Hershko, A. *et al.* (1971). Pleiotypic responses. *Nature New Biol.*, **232**, 206–211.

Hobson, B. M. (1971). Production of gonadotrophin, oestrogens and progesterone by the primate placenta. *Adv. Reprod. Physiol.*, **5**, 67–102.

Hodgen, G. D. *et al.* (1974). Specific radioimmunoassay of chorionic gonadotropin during implantation in rhesus monkey. *J. clin. Endocrinol.*, **49**, 457–464.

Holst, P. J., Cox, R. I. and Braden, A. W. H. (1974). The distribution of noradrenaline in the sheep oviduct. *Austral. J. Exp. Biol. Med. Sci.*, **48**, 563–565.

Hopkins, T. F. and Pincus, G. (1963). Effects of reserpine on gonadotropin-induced ovulation in immature rats. *Endocrinology*, **73**, 775–780.

Jaffe, R. B., Lee, P. A. and Midgley, A. R. (1969). Serum gonadotropins before, at the inception of, and following human pregnancy. *J. clin. Endocrinol.*, **29**, 1281–1283.

Johns, A. and Paton, D. M. (1976). The effect of cocaine, desipramine, 6-hydroxy-dopamine and indomethacin on the estrogen-dominated ampulla of rabbit oviduct to (-)-norepinephrine. *Biol. Reprod.*, **14**, 248–252.

Kaplan, N. M. (1961). Successful pregnancy following hypophysectomy during the twelfth week of gestation. *J. clin. Endocrinol.*, **21**, 1139–1145.

Kay, E. and Feigelson, M. (1972). An estrogen modulated protein in rabbit oviducal fluid. *Biochim. biophys. Acta.*, **271**, 436–441.

Kennedy, T. G. and Armstrong, D. T. (1972). Extra-ovarian action of prolactin in the regulation of uterine lumen fluid accumulation in rats. *Endocrinology*, **90,** 1503–1509.

Kent, H. A. (1975). Contraceptive polypeptide from hamster embryos. Sequence of amino acids in the compound. *Biol. Reprod.*, **12,** 504–507.

Kirby, D. R. S. (1967). Ectopic autografts of blastocysts in mice maintained in delayed implantation. *J. Reprod. Fert.*, **14,** 515–517.

Koering, M. J., Wolf, R. C. and Meyer, R. K. (1973). Morphological and functional evidence for corpus luteum activity during late pregnancy in the rhesus monkey. *Endocrinology*, **93,** 686–693.

Kraicer, P. F. and Shelesnyak, M. C. (1959). Détermination de la période de sensibilité maximale de l'endomètre à la décidualisation au moyen de déciduomes provoqués par un traitement empruntant la voie vasculaire. *C.r. Acad. Sci. Paris,* **248,** 3213–3215.

Kriewaldt, F. H. and Hendricks, A. C. (1968). *Lab. Anim. Care*, **18,** 361. (Quoted by Davies and Ryan, 1972.)

Krishnan, R. S. (1971). Effect of passive administration of antiblastokinin on blastocyst development and maintenance of pregnancy in rabbits. *Experientia*, **15,** 955–956.

Krishnan, R. S. and Daniel, J. C. (1967). "Blastokinin": inducer and regulator of blastocyst development in the rabbit uterus. *Science, N.Y.*, **158,** 490–492.

Linzell, J. L. and Heap, R. B. (1968). A comparison of progesterone metabolism in the pregnant sheep and goat: sources of production and an estimation of uptake by some target organs. *J. Endocrinol.*, **41,** 433–438.

Lombardi, J. R., Vandenbergh, J. G. and Mal Whitsett, J. (1976). Androgen control of the sexual maturation pheromone in house mouse urine. *Biol. Reprod.*, **15,** 179–186.

Macdonald, G. J., Armstrong, D. T. and Greep, R. O. (1967). Initiation of blastocyst implantation by luteinizing hormone. *Endocrinology*, **80,** 172–176.

Madhwa Raj, H. G., Sairam, M. R. and Moudgal, N. R. (1968). Involvement of luteinizing hormone in the implantation process of the rat. *J. Reprod. Fert.*, **17,** 335–341.

Maia, H. and Coutinho, E. M. (1968). A new technique for recording human tubal activity *in vivo*. *Amer. J. Obstet. Gyn.*, **102,** 1043–1047.

Martin, L., Finn, C. A. and Carter, J. (1970). Effects of progesterone and oestradiol -17β on the luminal epithelium of the mouse uterus. *J. Reprod. Fert.*, **21,** 461–469.

Maruniak, J. A. and Bronson, F. H. (1976). Gonadotropic responses of male mice to female urine. *Endocrinology*, **99,** 963–969.

Mastroianni, L. and Wallach, R. C. (1961). Effects of ovulation and early gestation on oviduct secretions in the rabbit. *Amer. J. Physiol.*, **200,** 815–818.

Mastroianni, L. *et al.* (1969). Some observations on Fallopian tube fluid in the monkey. *Amer. J. Obstet. Gyn.*, **103,** 703–709.

Maurizio, E. and Ottaviani, G. (1934). Comportamento delle reti lifatiche sotto-sierose e muscolari dell' utero della donna durante la gravidanza. *Ann. Ost. Ginecol.*, **56,** 1251–1277.

Mayol, R. F. and Longenecker, D. E. (1974). Development of a radioimmunoassay for blastokinin. *Endocrinology*, **95,** 1534–1542.

McLaren, A. (1968). A study of blastocysts during delay and subsequent implantation in lactating mice. *J. Endocrinol.*, **42,** 453–463.

McLaren, A. (1970). The fate of the zona pellucida in mice. *J. Embryol. exp. Morph.*, **23,** 1–19.

McLaren, A. (1971). Blastocysts in the mouse uterus and the effect of ovariectomy, progesterone and oestrogen. *J. Endocrinol*, **50**, 515–526.

McLaren, A. and Michie, D. (1959). The spacing of implantations in the mouse uterus. *Mem. Soc. Endocrin.* No. 6. C.U.P., pp. 65–75.

McMillin, J. M., Seal, U. S., Rogers, L. and Erickson, A. W. (1976). Annual testosterone rhythm in the black bear (*Ursus americanus*). *Biol. Reprod.*, **15**, 163–164.

Mester, I., Martel, D., Psychoyos, A. and Baulieu, E. E. (1974). Hormonal control of oestrogen receptor in uterus and receptivity for ovoimplantation in the rat. *Nature*, **250**, 776–778.

Meyer, R. K. (1972). Chorionic gonadotrophin, corpus luteum function and embryo implantation in the rhesus monkey. *Acta endocrinol. Suppl.*, **166**, 214–217.

Moghissi, K. S. (1970). Human Fallopian tube fluid. I. Protein composition. *Fert. Ster.*, **21**, 821–829.

Morgan, F. J. and Canfield, R. E. (1971). Nature of the subunits of human chorionic gonadotropin. *Endocrinology*, **88**, 1045–1053.

Morishige, W. K. and Rothchild, I. (1974). Temporal aspects of the regulation of corpus luteum function by luteinizing hormone, prolactin and placental luteotrophin during the first half of pregnancy in the rat. *Endocrinology*, **95**, 260–274.

Müller, R. E. and Wotiz, H. H. (1977). Post-coital contraceptive activity and estrogen receptor binding affinity of phenolic steroids. *Endocrinology*, **100**, 513–519.

Munshi, S. R., Purandare, T. V. and Rao, S. S. (1972). Effect of antiserum to ovine luteinizing hormone in corpus luteum function in mice. *J. Reprod. Fert.*, **30**, 7–12.

Murr, S. M., Bradford, G. E. and Geschwind, I. I. (1974). Plasma luteinizing hormone, follicle-stimulating hormone and prolactin during pregnancy in the mouse. *Endocrinology*, **94**, 112–121.

Murray, F. A. *et al.* (1972). Quantitative and qualitative variation in the secretion of protein by the porcine uterus during the estrous cycle. *Biol. Reprod.*, **7**, 314–320.

Neill, J. D., Johansson, E. D. B. and Knobil, E. (1969). Patterns of circulating progesterone concentrations during the fertile menstrual cycle and the remainder of gestation in the rhesus monkey. *Endocrinology*, **84**, 45–48.

Nilsson, O. (1966). Estrogen-induced increase of adhesiveness in uterine epithelium of mouse and rat. *Exp. Cell Res.*, **43**, 239–241.

Noske, I. G. and Feigelson, M. (1976). Immunological evidence of uteroglobulin (blastokinin) in the male reproductive tract and in nonreproductive ductal tissues and their secretions. *Biol. Reprod.*, **15**, 704–713.

Oakey, R. E. (1970). The progressive increase in estrogen production in human pregnancy; an appraisal of the factors responsible. *Vitam. & Horm.*, **28**, 1–36.

Owman, C. and Sjöberg, N.-O. (1966). Adrenergic nerves in the female genital tract of the rabbit. *Z. Zellforsch. mikrosk. Anat.*, **74**, 182–197.

Owman, C., Sjöberg, N.-O., Svensson, K.-G. and Walles, B. (1975). Autonomic nerves mediating contractility in the human Graafian follicle. *J. Reprod. Fert.*, **45**, 553–556.

Parkes, A. S. and Bruce, H. M. (1961). Olfactory stimuli in mammalian reproduction. *Science*, **134**, 1049–1054.

Parkes, A. S. and Bruce, H. M. (1962). Pregnancy-block in female mice placed in boxes soiled by males. *J. Reprod. Fert.*, **4**, 303–308.

Parlow, A. F., Daane, T. A. and Dignam, W. J. (1971). On the concentrations of radio-immunassayable FSH circulating in blood throughout human pregnancy. *J. clin. Endocrinol.*, **31,** 213–214.

Pauerstein, C. J., Fleming, B. D. and Martin, J. E. (1970). Estrogen-induced tubal arrest of ovum. *Obst. Gyn.*, **35,** 671–675.

Pencharz, R. I. and Long, J. A. (1933). Hypophysectomy in the pregnant rat. *Amer. J. Anat.*, **53,** 117–139.

Prasad, M. R. N., Dass, C. M. S. and Mohla, S. (1968). Action of oestrogen on the blastocyst and uterus in delayed implantation—an autoradiographic study. *J. Reprod. Fert.*, **16,** 97–104.

Prasad, M. R. N., Sar M. and Stumpf, W. E. (1974). Autoradiographic studies on [³H]oestradiol localization in the blastocysts and uterus of rats during delayed implantation. *J. Reprod. Fert.*, **36,** 75–81.

Psychoyos, A. (1966). Recent researches on egg implantation. Ciba Study Group No. 23, Churchill, pp. 4–15.

Psychoyos, A. (1969). Hormonal requirements for egg implantation. *Adv. Biosci.*, **4,** 275–290.

Purvis, K. and Haynes, N. B. (1974). Short term effects of copulation, human chorionic gonadotrophin-injection and non-tactile association with a female on testosterone levels in the male rat. *J. Endocrinol.*, **60,** 429–439.

Restall, B. J. and Wales, R. G. (1966). The chemical composition of the fluid from the Fallopian tube. *Austral. J. Biol. Sci.*, **19,** 687–698.

Riesen, J. W., Koening, M. J., Meyer, R. K. and Wolf, R. C. (1970). Origin of ovarian venous progesterone in the rhesus monkey. *Endocrinology*, **86,** 1212–1214.

Roberts, G. P. and Parker, J. M. (1974). Macromolecular components of the luminal fluid from the bovine uterus. *J. Reprod. Fert.*, **40,** 291–303.

Roberts, G. P., Parker, J. M. and Henderson, S. R. (1976). Proteins in human uterine fluid. *J. Reprod. Fert.*, **48,** 153–157.

Rose, R. M., Gordon, T. P. and Berstein, I. S. (1972). Plasma testosterone levels in male rhesus: influence of sexual and social stimuli. *Science*, **177,** 643–644.

Seitchik, J., Goldberg, E., Goldsmith, J. P. and Pauerstein, C. (1968). Pharmacodynamic studies of the human Fallopian tube *in vitro*. *Amer. J. Obstet. Gyn.*, **102,** 727–738.

Shelesnyak, M. C. (1962). Decidualization: the decidua and the deciduoma. *Persp. Biol. Med.*, **5,** 503–518.

Short, R. V. and Yoshinaga, K. (1967). Hormonal influences on tumour growth in the uterus of the rat. *J. Reprod. Fert.*, **14,** 287–293.

Spilman, C. H. and Harper, M. J. K. (1973). Effect of prostaglandins on oviduct motility in estrous rabbits. *Biol. Reprod.*, **9,** 36–45.

Squire, G. D., Bazer, F. W. and Murray, F. A. (1972). Electrophoretic patterns of porcine uterine protein secretions during the estrous cycle. *Biol. Reprod.*, **7,** 321–325.

Surani, M. A. H. (1975). Hormonal regulation of proteins in the uterine secretion of ovariectomized rats and the implications for implantation and embryonic diapause. *J. Reprod. Fert.*, **43,** 411–417.

Surani, M. A. H. (1976). Uterine luminal proteins at the time of implantation in rats. *J. Reprod. Fert.*, **48,** 141–145.

Talo, A. and Brundin, J. (1971). Muscular activity in the rabbit oviduct: a combination of electric and mechanic recordings. *Biol. Reprod.*, **5,** 67–77.

Treloar, O. L., Wolf, R. C. and Meyer, R. K. (1972). The corpus luteum of the rhesus monkey during late pregnancy. *Endocrinology*, **91,** 665–668.

Tullner, W. W. and Hertz, R. (1966). Chorionic gonadotrophin levels in rhesus monkey during early pregnancy. *Endocrinology*, **78,** 204–207.

Tyndal-Biscoe. (1973). *Life of Marsupials.*, Arnold, London.

Vaitukaitis, J. L. (1974). Changing placental concentrations of human chorionic gonadotrophin and its subunits during gestation. *J. clin. Endocrinol.*, **38,** 755–760.

Vaitukaitis, J. L., Ross, G. T., Braunstein, G. D. and Rayford, P. L. (1976). Gonadotropins and their subunits: basic and clinical studies. *Rec. Prog. Horm. Res.*, **32,** 289–321.

Van Tienhoven, A. (1968). *Reproductive Physiology of Vertebrates.* Saunders, Philadelphia.

Varma, K., Larraga, L. and Selenkov, H. A. (1971). Radioimmunoassay of serum human chorionic gonadotrophin during normal pregnancy. *Obstét. Gynéc.*, **37,** 10–18.

Watson, J. *et al.* (1975). Plasma hormones and pituitary luteinizing hormone in the rat during the early stages of pregnancy and after post-coital treatment with Tamoxifen. (ICI 46,474). *J. Endocrinol.*, **65,** 7–17.

Weichert, C. K. (1942). The experimental control of prolonged pregnancy in the lactating rat by means of estrogen. *Anat. Rec.*, **83,** 1–17.

Weitlauf, H. M. (1974). Metabolic changes in the blastocysts of mice and rats during delayed implantation. *J. Reprod. Fert.*, **39,** 213–224.

Wide, L. (1962). An immunological method for the assay of human chorionic gonadotrophin. *Acta Endocrinol.* Suppl. **70,** pp. 1–111.

Wiest, W. G. (1969). Progesterone interactions in the rat uterus. *Ciba Study Group No. 34*, pp. 57–72.

Zander, J. and von Münstermann, A.-M. (1956). Progesterone in menschlichem Blut und Geweben. III. *Klin. Wchschr.*, **34,** 944–953.

Zeilmaker, G. H. (1963). Experimental studies on the effects of ovariectomy and hypophysectomy on blastocyst implantation in the rat. *Acta. Endocrinol.*, **44,** 355–366.

Zeilmaker, G. H. (1964). Quantitative studies on the effect of the suckling stimulus on blastocyst implantation in the rat. *Acta Endocrinol.*, **46,** 483–492.

CHAPTER 5

Pregnancy : Maintenance and Prevention

MAINTENANCE OF PREGNANCY

The maintenance of pregnancy demands, among other things, that the uterus should grow to adapt its capacity to the size of the growing conceptus, and that in spite of the stretch that will clearly be imposed on its smooth muscle, spontaneous contractions should be suppressed. At the same time, the mammary glands must develop in preparation for lactation.

Contractile State of the Myometrium

The uterine muscle, because of the large size of its cells during pregnancy, was one of the first to be studied by impalement with an intracellular electrode, and it was observed that its excitability and resting potential depended on its "hormonal state"; thus, when removed during pregnancy, when it was under the dominant influence of progesterone (p. 200), it showed less spontaneous activity and had a higher resting potential than when taken after delivery, when it was presumably more under the influence of oestrogens (Marshall and Csapo, 1961).

Hormonal State

The steroid hormone that seems to be vital for the maintenance of pregnancy is progesterone, secreted by the corpus luteum in the early stages* and, in many species including primates, by the placenta. Thus, in species that maintain and are dependent on their corpora lutea during pregnancy, such as the goat, removal of the ovary or simply of the active corpora lutea, brings pregnancy to an end, pre-

---

* It was Allen and Corner (1929) who showed that implantation and maintenance of pregnancy in ovariectomized rabbits could be sustained by injections of extracts of corpora lutea; the principle—progesterone—was later purified and its structure was determined by Butenandt *et al.* (1934).

sumably due to the fall in blood progesterone since pregnancy can be maintained by administration of exogenous progesterone.* The effect is undoubtedly on myometrial activity since it has been found that the dose of progesterone required to maintain pregnancy in the ovariectomized rabbit, for example, is some five times that required to maintain endometrial proliferation (Heckel and Allen, 1938). Again Adams (1965) showed that, in pseudopregnant rabbits, one or two corpora lutea were adequate to maintain endometrial proliferation but not to allow implantation of blastocysts which were expelled from the uterus; with four corpora lutea half the blastocysts were retained and with eight corpora lutea some 90 per cent were retained.

Tolerant Species. In species that tolerate ovariectomy the progesterone comes mainly from the placenta, so that the demonstration that progesterone is likewise important for maintenance of uterine quiescence is not so simple. De Snoo (1919) observed that in women with complete accidental separation of the placenta, the consequence was delivery within 24–48 hr, and he suggested that the placenta had an inhibiting influence on uterine contractions. Since then a number of experimental studies have demonstrated the local effects of the placenta on uterine activity; for example Gillette (1966) implanted microballoons in the cow's myometrium in a row from a cotyledonary site to an intercotyledonary area; uterine activity, as measured by the pressure in these balloons, decreased progressively with distance from the cotyledonary site. Again, Daniel and Renner (1960) found that the myometrium of the cat's uterus at the placental site lacked spontaneous activity and the ability to propagate an action potential, by contrast with tissue taken from interplacental regions. At term these differences disappeared.† Finally, placental separation results in abortion in a number of species including guinea-pigs (Porter, 1969).

Blood Progesterone

Corpus Luteum Dependent

The changes in blood progesterone during the early stages of pregnancy have been considered earlier. The changes in the rat throughout pregnancy are shown in Figure 5.1; in this and other corpus luteum-

---

* Although ovariectomy in the monkey after Days 22 to 24 does not promote abortion, the monkey ovary does continue to secrete progesterone throughout pregnancy although the functional importance of this supply is not clear (Walsh *et al.*, 1974).

† Kuriyama and Csapo (1961) found differences in the responses of the rabbit's uterus to mechanical stimulation according to the state of gestation. Post-partum, a stimulus was propagated along the whole uterine strip; in early pregnancy the effect was restricted to a local action; in late pregnancy the "uterine block" was no longer uniform and was restricted to the placental portion of the uterine wall.

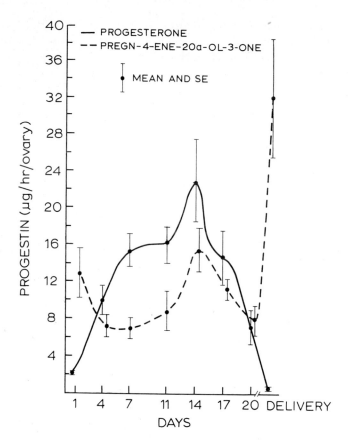

Fig. 5.1. Progesterone and pregn-4-ene-20a-ol-3-one secretory rates into ovarian venous blood of the rat during pregnancy and within 6 hr after parturition. (Hashimoto *et al.*, *Endocrinology.*)

dependent species, notably the rabbit, sow and goat, there is a progressive rise to a peak around mid-pregnancy, followed by a slow decline to low levels. The levels represent corpus luteum function throughout pregnancy.

Part Dependent

In species that are only partly dependent on the ovaries for continued pregnancy, such as the sheep, the pattern is basically similar, the peak occurring late at 130–140 days and being followed by a sharp decline. Rather similar concentrations are found in the cow (Erb *et al.*, 1967). Ovariectomy in these species does not cause abortion, but it does reduce the level of progesterone in the blood, indicating continued function of the corpora lutea. That the plasma level does not reflect total synthesis is well illustrated by this species; thus in the ovariectomized sheep the plasma concentration is about half normal (Fylling, 1970), suggesting equal rates of production by placenta and ovary; however, estimates of production by the placenta by analysis of uterine venous blood indicated that the placenta was producing progesterone at some five times the rate of the ovary (Linzell and Heap, 1968).

Placenta Dependent

In completely placenta-dependent species such as the monkey and human the secretion of the metabolic product of progesterone, namely pregnanediol, increases steadily throughout pregnancy (Fig. 5.2), suggesting an increasing secretion of progesterone; the plasma levels of progesterone remain fairly steady, however, as Figure 5.3 illustrates for the rhesus monkey; an initial peak is due to corpus luteal function and a second peak is attributable to placental secretion. In the human, a plateau is reached with concentrations considerably higher than those in the monkey (Fig. 5.4).

Myometrial Concentrations

It seems very likely that steroids synthesized by the placenta are carried directly to the uterine muscle, so that peripheral plasma levels

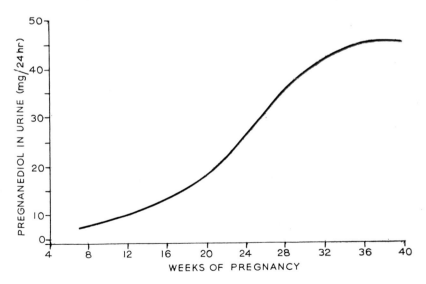

Fig. 5.2. Excretion of pregnanediol during normal pregnancy in the human. (Hobson, *Adv. Reprod. Physiol.*)

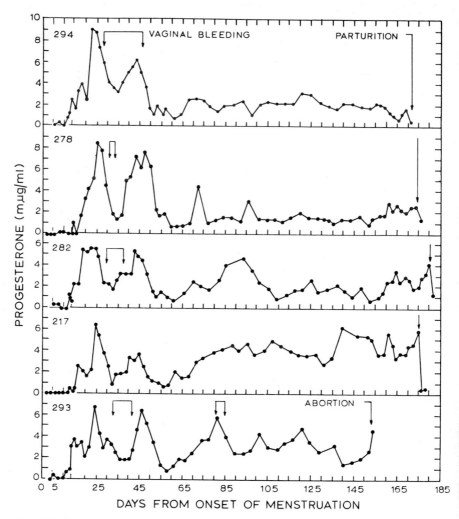

Fig. 5.3. Progesterone concentrations in the peripheral plasma of five individual pregnant monkeys. Days from the onset of menstruation can be converted to days of pregnancy by subtracting 13. (Neill *et al., Endocrinology.*)

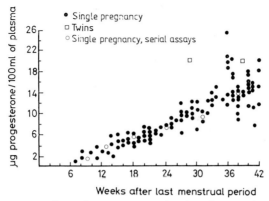

Fig. 5.4. The concentration of progesterone in the plasma of normal pregnant women. (Yannone *et al.*, *Amer. J. Obstr. Gynecol.*)

of progesterone may not have much relevance. Kumar and Barnes (1965) measured progesterone in the human placenta, myometrial tissue, and blood from the periphery, a uterine vein and the retro-placental pool. The results are shown in Table I; the high concentrations in the retroplacental pool and uterine vein by comparison with peripheral blood are obvious. Also the myometrial tissue overlying the placenta had a higher concentration than that at an antiplacental site. Towards term the ratio of these concentrations tends to unity.

Metabolism of Progesterone

The placenta and uterus are capable of metabolizing progesterone, the nature of the metabolic products and the extent of the metabolism varying with species. In the ovary of the rat there is no doubt from the work of Wiest (p. 71) that this metabolism is related to control over secretion, so that there is an inverse relation between blood progesterone and blood-20a-hydroxyprogesterone (20a-OHP), but this is not true of placental metabolism. Since the metabolic products are much less active than progesterone, the metabolism may facilitate control of uterine function by removing the active progestogen rapidly.*

Metabolic Clearance. Changes in the plasma level of progesterone could be at least partly due to altered rates of metabolism, e.g. by the liver, as opposed to altered rates of secretion. Illingworth *et al.* (1970) found a metabolic clearance rate (MCR) of 112 ± 7 litres plasma/day/kg

---

* The relative contributions of ovary and placenta to progesterone production have been assessed by Linzell and Heap (1968) by measuring arterio-venous differences in concentration. Between 119–126 days of pregnancy in goats the ovaries produced 10 mg/day whilst the placenta produced none; by contrast, in the sheep the placenta produced 14 mg/day and the ovaries only about 2 mg/day; the adrenal production was less than 2 per cent of the ovarian.

TABLE I

Progesterone concentrations in blood and tissue of pregnant women. (Kumar and Barnes, 1976)

| Stage | Placenta (μg/g) | Plasma (μg/100 ml) | | | Myometrical tissue (μg/g) | | P/A |
		Retroplacental pool	Uterine vein	Peripheral vein	Overlying placenta (P)	Anti-placenta (A)	
10–12 weeks	3·5	24·6	10·2	4·0	0·08	0·05	1·6
38 weeks	2·5	65·4	18·5	6·4	—	—	—
40 weeks	2·0	41·8	13·7	6·1	0·05	0·04	1·2

ın the normally cycling guinea-pig; this was reduced to $8·3 ±0·8$ by Days 15–20 of pregnancy and continued at this low rate throughout gestation, and only slowly returned to higher values post-partum. The decreased clearance coincided with the rise in plasma progesterone, and might have been related to the appearance in guinea-pig plasma of a binding protein with high affinity for progesterone. In women, clearance of progesterone remains high during pregnancy, so that weight-for-weight the woman produces some four times as much progesterone during pregnancy as the guinea-pig; thus the reduced clearance in the latter animal represents an economy in synthesis.

Tissue Concentrations

As indicated earlier, it is the concentration of progesterone in the myometrium at any moment that is likely to be the determining factor so far as uterine motility is concerned, and where this concentration has been studied a better correlation with this and function exists than that between plasma-level and function. Thus, Csapo's studies on the rat have shown that a myometrial concentration of $2 \ \mu g/100$ g is necessary to maintain uterine quiescence consistent with pregnancy, values below this leading to abortion (Csapo and Wiest, 1969; Wiest, 1970). In humans, the concentration is higher than in systemic blood, but the concentration in the retroplacental pool or uterine vein is actually greater, indicating the importance of the local concentrations by comparison with that in systemic blood (Kumar and Barnes, 1965; Zander and Von Münstermann, 1956).

Progesterone Receptor

If the uterus is a target organ for progesterone we must expect to be able to isolate a receptor protein from its cytoplasm and nucleus; Davies and Ryan (1972b) have reviewed the contradictory experimental findings in this respect; in their own study of homogenized uteri of pregnant rats they were able to isolate a high-affinity carrier like that described by Milgrom and Baulieu (p. 183) for non-pregnant uteri and similar to CBG (corticotrophin binding globulin). Uteri taken from late pregnancy gave a preparation that bound less strongly, whilst *in vivo* the ability to take ^{3}H-labelled progesterone into the tissue also fell; thus, at 11 days concentration-ratios (Uterus/Plasma) of $2·1± 0·1$ were obtained but at 21 days they had fallen to $0·57± 0·05$.*

---

* A more specific binding protein for progesterone has been extracted from rat and rabbit uteri by McGuire and Bariso (1972); it has a molecular weight of 85,000 and is different from that described by Milgrom and Baulieu (1970) which also binds cortisol. Again Rao *et al.* (1974) have extracted a 4–5S complex from human endometrium with high affinity for progesterone ($Ka = 1·2$–$3·4.10^9 M^{-1}$); this could be distinguished from CBG by the negligible displacement of progesterone by cortisol.

Oestrogen

Placental and Ovarian Synthesis

In considering the secretion of oestrogens during pregnancy we must once again bear in mind the difference between the tolerant and intolerant species. In the intolerant (short-gestation) species oestrogen synthesis is carried out by the follicles from acetate without the need for preformed steroids. The corpora lutea, on the other hand, make progesterone but do not synthesize androgens and subsequently convert these to oestrogens. The ability of the placenta to synthesize oestrogens correlates very well with gestational length; thus, as Davies and Ryan emphasize, there is no known species with a gestational length of greater than 70 days in which placental oestrogen synthesis is absent. Thus formation of oestrone and oestradiol-17β by aromatization of androstenedione *in vitro* has been described for the placentae of primates, horse, pig, sheep, cow and goat, whilst the short-gestation animals like rat, rabbit, and guinea-pig failed to carry out this synthesis (see, for example, Ainsworth *et al.*, 1969). Thus the aromatization system plus the 17-ol dehydrogenase system are present in the placentas of these long-gestation animals. These placentas, however, could not metabolize the C_{21}-precursors, such as progesterone, to oestrogens, and therefore lacked the 17,20-desmolase required for this.*

That this *in vitro* synthetic capacity is related to physiological function is indicated by the abrupt fall in plasma oestrogens following delivery of the placenta (see, for example, Challis, 1971); whilst both hypophysectomy and ovariectomy in the pregnant sow fail to prevent the normal rise in plasma oestrogen.

Foeto-Placental Unit

According to this concept, developed largely by Diczfalusy (1964) the synthesis of oestrogen is dependent on the combined capabilities of the foetal adrenal gland and the placenta, the placenta producing C_{21} compounds (pregnenolone and progesterone) but lacking the 17,20-desmolase activity that permits their conversion to C_{19} compounds (androgens) that can act as precursors to oestrogens. The foetal adrenal possesses this activity, removing the side-chain from the C_{21} compounds or synthesizing C_{19} compounds *de novo;* these are sent to the placenta (Fig. 5.5) for conversion by the placenta to oestrogens

---

* The capacity of rat placental tissue to synthesize steroid hormones has been examined by Townsend and Ryan (1970). Androstene was not converted to oestrogen, whilst the conversion of pregnenolone to progesterone was very limited. These limited capacities fully explain the failure of the rat to maintain pregnancy after ovariectomy.

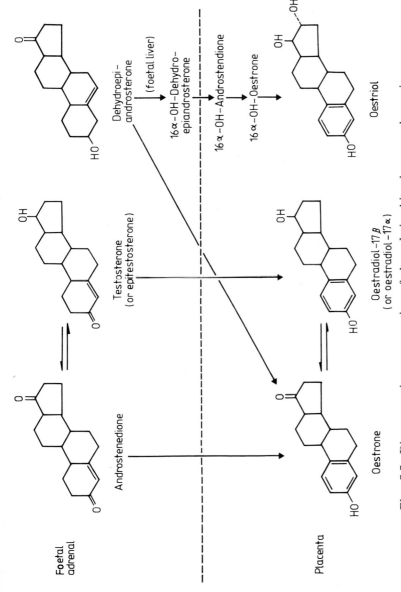

Foetal adrenal

Androstenedione

Testosterone (or epitestosterone)

Dehydroepi-androsterone

(foetal liver)

16α–OH–Dehydro-epiandrosterone

16α–OH–Androstendione

16α–OH–Oestrone

Placenta

Oestrone

Oestradiol–17β (or oestradiol–17α)

Oestriol

Fig. 5.5. Diagrammatic representation of the relationships between the major oestrogens in pregnancy urine and their adrenal androgen precursors. The pathways are well established for human beings. (Davies and Ryan, *Vitam. & Horm.*, 1972.)

(Siiteri and MacDonald, 1966; Diczfalusy, 1969). This concept, originally applied to the human, has been extended to other species; thus, Davies *et al.* (1970) made homogenates of placenta and adrenal gland from the sheep and monkey and showed that they could obtain oestrogen from pregnenolone only when they combined the extracts, foetal or maternal adrenal being adequate; none of the tissue extracts alone could carry out this synthesis.

Secretion During Pregnancy

Secretion of oestrogens continues throughout pregnancy and this is reflected in a gradual rise in their concentrations. The blood levels in the human are shown in Figure 5.6 and those of the monkey in Figure 5.7. Just before term the rise in blood level becomes steeper and, if oestrogens tend to increase excitability of uterine muscle, this rise might well be an element in the parturitional complex (p. 292).

Role of Increased Plasma Oestrogens

The role of the increased maternal oestrogens during pregnancy is not clear, but the apparently universal appearance of the foeto-placental unit in long gestation period animals, exhibiting cooperation in synthetic activity between the foetus and placenta, strongly suggests

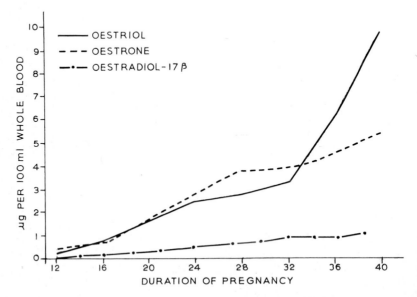

Fig. 5.6. Blood oestrogen levels during human pregnancy showing mean values for oestriol, oestrone and oestradiol-17β. (Hobson, *Adv. Reprod. Physiol.*)

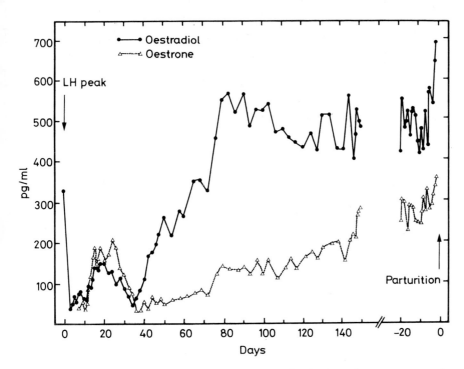

Fig. 5.7. Mean circulating oestradiol and oestrone levels throughout pregnancy in 8 rhesus monkeys. The data on the left are normalized to the day of the LH peak (Day 0 of pregnancy) while the preparturient data (6 animals) are normalized to the day preceding parturition. Note the striking rise and fall of both oestrogens, especially oestrone, during early pregnancy (Days 11–35) and the rise of oestradiol during mid- and late pregnancy. (Atkinson *et al.*, *Biol. Reprod.*)

that the increasing levels represent a control mechanism in gestational periods that last longer than some 70 days. As we shall see, parturition is preceded in many species by a sharp rise in blood oestrogen which might well be a factor in increasing uterine activity and overcoming the progesterone block necessary for uterine quiescence.

Metabolism

The large quantities of oestrogen formed during pregnancy, if they obtained access to the foetal blood, might well be toxic to the foetus unless converted to less active products. Space will not permit a description of the metabolic products excreted by the pregnant female, suffice it to say that in the human, for example, large quantities of oestriol are excreted in the urine. This probably represents a hydroxylation process carried out by the foetal adrenal on C_{19} precursors to oestrogen, i.e. they are not products of oestrogen but rather they represent the existence of a shunt that prevents the synthesis of too much oestrone and oestradiol by the placenta. Thus oestriol is far less potent as an oestrogen than oestradiol, and actually inhibits some aspects of oestradiol action (Hisaw *et al.*, 1954). A further protective mechanism is the hydroxylation of oestrogens directly by the foetus as they arrive from the placenta; again, sulpho-conjugation is extensive in the foetus.

Maternal and Foetal Blood Levels. As a result of these metabolic activities, the concentrations of oestradiol and oestrone in foetal plasma are much lower than in the maternal plasma, whilst the concentration of oestriol is some ten times higher.

Some results of analyses of oestrogens in maternal and foetal blood obtained at Caesarian section are shown in Table II. The differences in concentration of the oestrogens, oestrone and oestradiol, in maternal

TABLE II

Oestrogen levels (μg/100 ml) in amniotic fluid and foetal and maternal blood plasma (Roy 1962)

Fluid	Oestriol	Oestrone	Oestradiol
Amniotic fluid	73	0·5	0·24
Umbilical vein	55	0·59	0·34
Umbilical artery	54	0·45	0·31
Uterine vein	8·7	2·5	1·3
Arm vein	6·7	2·0	0·9

arm blood on the one hand and uterine venous blood on the other are highly significant statistically and indicate a release of the oestrogens from the placenta to the general maternal circulation. Roy (1962) computed that these differences would represent transport of 15,000, 850 and 200 μg of oestriol, oestrone and oestradiol to the maternal circulation every 24 hours.* In a similar type of study Maner *et al.* (1963) found essentially the same relations between foetal and maternal blood levels, especially the greater concentrations in uterine vein than in peripheral venous blood.

Gonadotrophins

The switch from pituitary to placental gonadotrophin secretion, characteristic of the primate and many other species, has been described earlier (p. 202). Figure 5.8 shows the gradual fall in hCG activity

---

* The high concentrations of oestriol in the cord blood are probably due to accumulation of conjugated oestriol in the foetus; this would not cross the placental barrier easily. Only 16–20 per cent of the cord oestriol was in the unconjugated form.

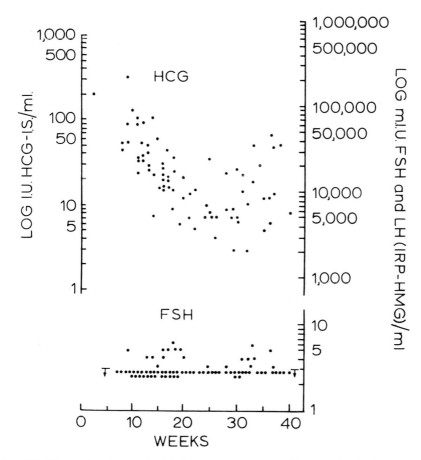

Fig. 5.8. Illustrating the gradual fall in concentration of human chorionic gonado-trophin (hCG) in the blood during pregnancy compared with the relatively constant level of FSH. (Parlow *et al.*, *J. clin. Endocrinol.*)

during the course of pregnancy; the activity is high, by comparison with the non-pregnant condition, but falls rapidly at term (Fig. 5.9). FSH levels remain fairly constant throughout pregnancy; after delivery the plasma levels rise, presumably connected with lactation.

Abortion by Anti-Serum

An anti-LH administered to rats from Days 7 to 14 of pregnancy will terminate gestation, an effect that can be overcome by treatment with progesterone (Madhwa Raj and Moudgal, 1970); this confirms the dependence of the rat on its pituitary until about mid-term. Nishi *et al.*

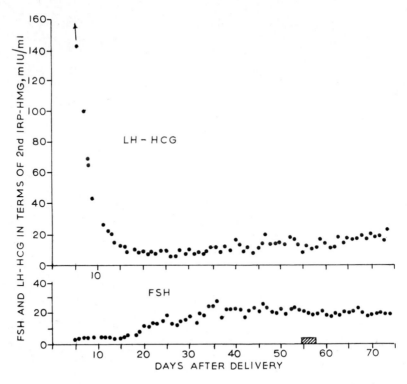

Fig. 5.9. Serum gonadotrophin concentrations in a woman from 5 to 74 days post-partum. She did not breast-feed. (Jaffe *et al.*, *J. clin. Endocrinol.*)

(1976) used an antiserum to the *releasing* hormone, LHRH, which, with its limited period of action, permits a more accurate assessment of the period of dependence on the pituitary. Given on Days 9–10 it induced resorption of the foetuses; given on Days 8 or 11 there was partial resorption whilst given on Day 12 only one out of 5 resorbed. The effects of the anti-LHRH were reflected in the changes in blood progesterone. The fact that anti-LHRH is effective after Day 7 indicates that pituitary prolactin, which is luteotrophic in the rat, plays a relatively small role in maintaining the corpora lutea, since anti-LHRH does not affect prolactin release. Thus, the placental gonado-troph starts to substitute for pituitary prolactin after Day 7.

Testosterone Stimulation

Androgens, in addition to oestrogen and progesterone, can stimulate the uterus; their effect seems to be primarily on uterine growth whereas that of oestrogen is manifest in both growth and secretion by the endometrium. Androgens do not compete with oestrogens for receptor sites in the uterus, and, in fact, the uterine cytosol contains receptor material with a high degree of specificity for testosterone (Giannopoulos, 1973). Interestingly, the ^{3}H-testosterone found in the nucleus of the uterine cells disappeared after about 2 hours, whereas the ^{3}H-oestrogen complex remained for much longer, and this may explain why 2000 times as much testosterone is required to promote uterine growth as oestrogen. The separate modes of action of the two classes of steroid are revealed by the different effects of anti-oestrogens and anti androgens; thus the anti-oestrogens inhibit the uterotrophic action of oestrogens but not that of androgen, whilst the anti-androgens inhibit the uterotrophic action of testosterone but not that of oestrogen; and this is consistent with the observation that oestrogen and testosterone do not compete for each other's sites in the uterus.*

Alpha Foetoprotein

Human Foetal Serum

This is a glycoprotein of molecular weight 70,000 that is found in appreciable concentrations in human foetal serum, increasing from 10 mg/100 ml at 6 weeks to a maximum of about 340 mg/100 ml at 14 weeks; after birth the concentration declines rapidly with a "half-life" of about 3·5 days (Gitlin and Boesman, 1967). In maternal plasma there is a peak concentration of about 200 ng/ml at 32 weeks (Leek et al., 1975). The concentration in adult serum is very small, 2–16 ng/ml (Van Furth and Adinolfi, 1969).

Amniotic Fluid

In the hope that the concentration in the amniotic fluid might be an index to gestational age, Seppälä and Ruislahti (1972) have measured this. At 32 weeks the concentration was 533 ng/ml; at 40 weeks it had fallen to 15–275 ng/ml; at 41 weeks it was 45–235 ng/ml and at 42 weeks, 23–90 ng/ml. In pregnancy complicated by anencephaly the concentration, measured at 26–38 weeks, was abnormally high (Randle and Cumberbatch, 1973).

Foetal Rat

The protein has also been identified in the foetal rat where it is mainly synthesized by the liver; transfer across the placenta takes place so that the protein is present in the maternal plasma, the concentration being about one twentieth to one thirtieth that in the foetus (Sell and Alexander, 1974; Lai et al., 1976). Peak levels are reached in foetal plasma and amniotic fluid and in maternal plasma at the same time, namely Day 19; this is in contrast to the human where the peaks in foetal plasma and amniotic fluid are reached in the second trimester whilst that in the maternal plasma in the middle third of the third trimester (Randle and Cumberbatch, 1973; Leek et al., 1975).

* The uterus differs from the prostate in that testosterone binds much more strongly to receptor sites than does 5a-dihydrotestosterone (Giannopoulos, 1973; Rennie and Bruchovsky, 1972).

Oestrogen-Binding Globulin

The adult rat serum has no specific binding capacity for oestradiol beyond that due to the serum albumin; in the pregnant rat, however, a protein binding specifically to oestradiol appears in the plasma and this is apparently derived from the foetus. Thus Nunez *et al.* prepared from foetal plasma a protein that bound with high affinity to oestradiol but not at all to progesterone or testosterone; as Table III shows, the binding was very high in the 19-day embryonic rat and

TABLE III

Binding of oestradiol, mM/mg protein, by protein extracted from foetal blood serum (Nunez *et al.*, 1971)

Foetal	$1·2.10^{-9}$
5-Day	$44·8.10^{-9}$
10-Day	$32·8.10^{-9}$
15-Day	$28·0.10^{-9}$
21-Day	$8·3.10^{-9}$
28-Day	$1·0.10^{-9}$

declined to low values during the approach to adulthood. In the pregnant mother there was some activity, which fell after delivery, and was presumably derived from the foetus. A similar protein was described by Raynaud *et al.* (1971), its sedimentation coefficient being 4·5 S. The physiological significance of the appearance of this embryonic and foetal protein is not clear; it could be that it protects the embryo and foetus from the high concentrations of oestrogen in the maternal plasma prevalent during pregnancy.

CONTRACEPTION

The voluntary prevention of pregnancy is of fundamental importance in human society so that the subject should not escape mention in a textbook of physiology. As summarized by Diczfalusy (1968), "the mode of action of contraceptive drugs poses a difficult problem mainly because the physiologic mechanisms involved in the regulation of reproductive processes in the human are incompletely understood as yet". It will be obvious, however, that there are several points of attack on the process, e.g. maturation of ova and sperm, ovulation, oviducal mobility, implantation of the blastocyst.

Anti-Fertility Pills

The use of mechanical contrivances to prevent fertilization of the ovum, or subsequent implantation, is rapidly being replaced by the so-called anti-fertility pills, the most successful of which act primarily by a block of ovulation, achieved by combined administration of oestrogen and progestogen. Additional effects of the steroids may well include the development of an unfavourable uterine condition for implantation, as well as an alteration in the oviducal fluid that impedes the passage of the sperm.

Combined and Sequential Types

The steroid hormones or their synthetic substitutes may be administered in combination—*combined type*—or in sequence—*sequential type*.

Combined Type

A mixture of oestrogen and progestogen is taken daily for 21 days beginning on the fifth day after the beginning of menstruation, i.e. before ovulation is expected. After the 21st day, administration ceases for seven days, and a new cycle of 28 days is begun. Some two days after the cessation of administration "withdrawal bleeding" begins. Figure 5.10 illustrates a typical regimen based on this 28-day cycle, the subject taking a pill every day, those marked with an X being placebos; the period of bleeding is shaded.

Mode of Action. It is generally agreed that the mode of action is through blockage of ovulation, the pre-ovulatory peak of LH being suppressed; as to the precise mode of suppression of ovulation the

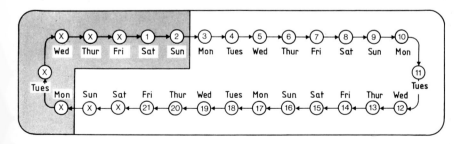

Fig. 5.10. Typical regimen for taking the combined type of pill; the subject takes a pill every day, those marked with an X being placebos. The period of bleeding is shaded. (Brotherton, *Sex Hormone Pharmacology*.)

situation is vague. According to Eisenfeld (1974) the oestrogenic component prevents pregnancy through its negative feedback on the FSH secretion by the pituitary, thereby preventing maturation of the ovum; whilst the progestogen acts by suppressing LH-secretion and thence the "ovulatory surge". The block of ovulation leaves the ovary quiescent with low secretion of ovarian steroids. Histologically this quiescence is manifest as an arrest at an early stage in follicular development, the ovary being reduced in size with a smooth white juvenile appearance with no evidence of fresh corpora lutea. As we might expect, the normal cycle of changes in the endometrium associated with the menstrual cycle is disrupted. During the first few cycles the glandular proliferation is at first diminished and then abolished; as treatment continues the glands become uniformly small with inactive epithelium, so that secretions are minimal whilst the stroma becomes dense and inactive. Abnormal thin-walled sinusoids appear from which the "withdrawal bleeding" presumably occurs.

Withdrawal Bleeding. As indicated above, this occurs as a result of cessation of treatment, some two to four days after the last active pill has been taken, so that it occurs more regularly and predictably than does normal menstrual bleeding.*

Sequential Pill

In this treatment the natural sequence of hormone secretion during a normal ovarian cycle is mimicked; thus oestrogen is given for 14–16 days of the cycle, beginning on Day 5 and this is followed by a mixture of oestrogen plus progestogen for the following 5–7 days. The theory behind the treatment is, essentially, to establish negative feedbacks on the pituitary-hypothalamus system, so as to prevent the secretion of the gonadotrophic hormones responsible for follicular development and ovulation. According to Balin et al. (1969), the oestrogens cause suppression of FSH secretion, whilst the LH-peaks occur at random; thus ovulation does not occur largely because of the absence of mature follicles. The changes produced in the endometrium are different from those described for the combined pill (Diczfalusy, 1968) the proliferative phase being slower to develop and passing straight into the decidual phase. The use of this treatment has been given up in favour of the combined pill.

The Mini-Pill

The treatment consists in a daily dose of a progestogen; under this treatment a woman will settle down to a regular bleeding which seems

---

* "Breakthrough bleeding" may occur while the pills are being taken.

to be neither menstrual nor "withdrawal" in character, and probably results from the continuous stimulation of the uterus by the progestogen; at intervals, the dose being insufficient to support further growth of the tissue, the latter then breaks down. It is thus best described as a "breakthrough bleeding". Under this treatment ovulation is not always inhibited, whilst the level of FSH in the blood is unaffected and so is that of oestrogen.

Cervical Mucus. The anti-fertility effect seems to depend on changes in the physical character of the cervical mucus and in the histology of the endometrium that interfere with sperm penetration and blastocyst implantation respectively (Aref et al., 1973). Thus, prior to ovulation a considerable mucorrhea occurs, the cervical fluid being thin and watery with low viscosity and a low concentration of albumen, a condition that favours sperm penetration. In the normal luteal phase of the menstrual cycle, on the other hand, the fluid is thick and tacky and hostile to sperm penetration, presumably as a result of the secretion of progesterone by the corpus luteum. In the sequential treatment there is no change in the cervical mucus until progesterone is administered. With continuous progestogen contraception, as with chlornadinone acetate, Martinez-Manautou et al. (1967) found that the mucus acquired the high viscosity characteristic of the luteal phase of the menstrual cycle.

Post-Coital Pills

Implantation may be interfered with by large doses of a progestogen or oestrogen; but the anti-fertility reliability is not so great as with the combined or sequential treatment. So far as oestrogens are concerned, according to Blye (1973), diethylstilboestrol or ethinyl oestradiol will prevent pregnancy if given within 72 hr of unprotected mid-cycle coital exposure. The author has discussed the possible luteolytic mechanism, but emphasizes that once nidation has occurred, oestrogens do not affect the course of pregnancy. Therefore, the treatment must start as soon as possible after coitus, within 24 hr preferably but not later than 72 hr, and the treatment must be continued for 5 days.

Anti-Oestrogens

Interest in the anti-oestrogen stems from the possibility of interfering with the normal series of physiological changes that constitute the menstrual or ovarian cycle. Thus, of the natural steroids, progesterone may be regarded as an anti-oestrogen possibly exerting its anti-fertility effect through altering the character of the mid-cycle cervical mucus.

Types of Anti-Hormone

As Figure 5.11 shows, in general, a steroid hormone may have an *inactive* compound of very similar structure, an *antagonist* with a steroid-related structure, and an *antagonist* with a non-steroidal structure.

Mode of Action

It was originally considered that the action was due to a simple competition for oestrogen at cytoplasmic receptor sites (Korenman, 1970; Rochefort and Capony, 1972); however, although this competition undoubtedly occurs, the situation is complex and involves interference with the interaction of the oestrogen with its nuclear acceptor. Thus, Katzenellenbogen and Ferguson (1975) studied nafoxidine and C–I–628; these are anti-oestrogens in that they block oestrogen-induced synthesis of uterus-specific protein (IP), and in doing so they reduce the oestrogen-binding capacity of the uterine cytoplasmic receptors, and are carried to the nucleus. The return of uterine synthetic activity ran parallel with the return of receptor sites in the cytoplasm so that the anti-oestrogen was, in fact, removing these sites by translocation to the nucleus. Treatment of the tissue with excess of ^{3}H-oestradiol resulted in a considerable uptake of the hormone by the nuclei, but this did not release the block so that in some way the translocation of the anti-oestrogen to the nucleus prevents oestrogen from exerting its own action. Cidlowski and Muldoon (1976) examined the effects of 10^{-11} molar concentrations of 17-β-oestradiol, MER–25, CI–628 and dimethylstilboestrol on binding of ^{3}H-oestradiol in anterior pituitary, hypothalamus and uterus *in vitro*; as Table IV shows, the relative inhibitions were roughly the same for all tissues, but when the

TABLE IV

In vitro equilibrium competition of anti-oestrogens for oestrogen receptor binding sites in anterior pituitary, hypothalamus and pituitary (Cidlowski and Muldoon, 1976)

Inhibitor	Per cent inhibition		
	Anterior pituitary	Hypo-thalamus	Uterus
17β-Oestradiol	99·6	98·0	99·3
MER–25	20·9	7·1	11·4
CI–628	50·0	29·0	26·8
Dimethylstilboestrol	98·4	97·0	97·2

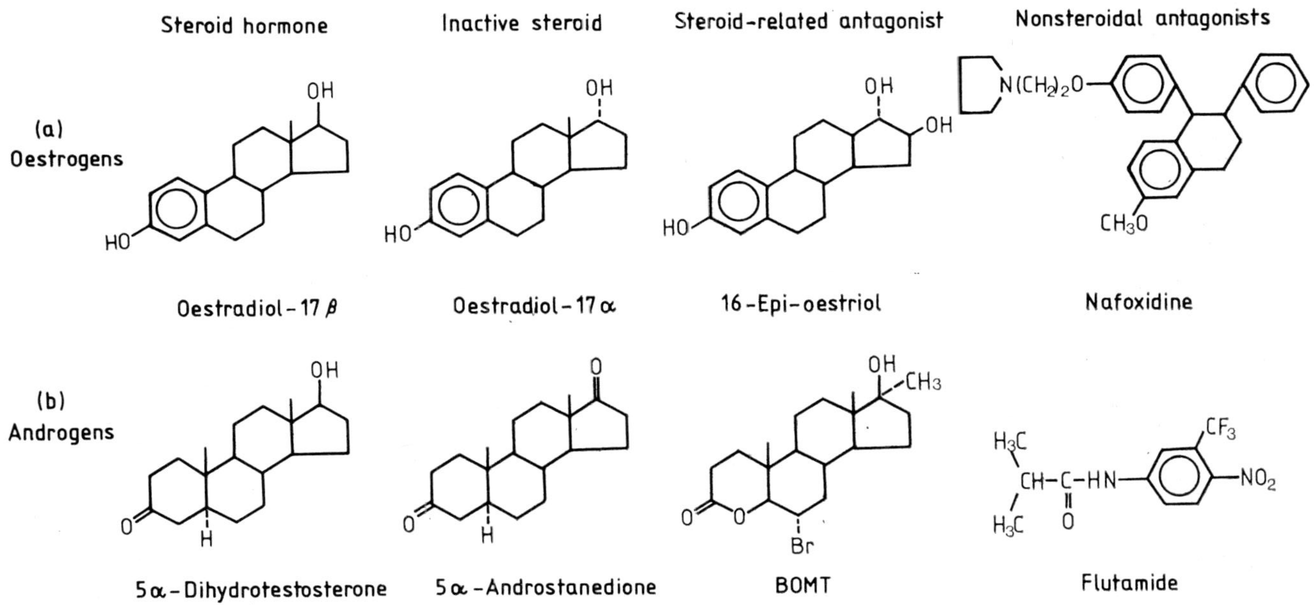

Fig. 5.11. Structures of selected steroid hormones and their antagonists. BOMT, 6α-Bromo-17β-hydroxy-17α-methyl-4-oxa-5α-androstane-3-one. (Mainwaring, *Vitam. & Horm.*, 1975.)

in vivo effectiveness of the anti-oestrogens was compared, marked differences were found according to the particular tissue studied, thereby emphasizing the implication of other factors besides *in vitro* binding in determining anti-oestrogenic activity. Thus, Figure 5.12 shows the changes in binding after *in vivo* administration of CI–628 and MER–25; in uterus and anterior pituitary binding was reduced, with very little replenishment following CI–628; in the hypothalamus binding was hardly affected. With dimethylstilboestrol binding was strongly inhibited but replenishment was fairly rapid; the hypothalamus behaved similarly to the other tissues, by contrast with CI–628, so that it is important, when discussing anti-oestrogen action, to specify the tissue studied.*

Tamoxifen

This triphenylethylene derivative (I.C.I. 46,474) is an anti-oestrogen that initially stimulates uterine growth, but later suppresses it; it

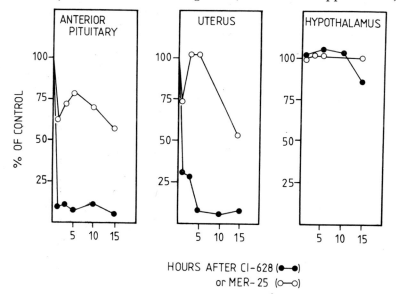

Fig. 5.12. Depletion and replenishment of cytosol oestradiol receptors following CI-628 or MER-25. At time zero, groups of castrate female rats were administered a single dose of either CI-628 or MER-25. Groups of animals were killed at the designated intervals and the specific binding capacity of the extracted cytosol material was determined. Results are presented as percentages of the vehicle-treated control animals' receptor binding concentrations. (Cidlowski and Muldoon, *Biol. Reprod.*)

* Using much higher concentrations of CI–628, Whalen *et al.* (1975) have found a slow depletion of hypothalamic receptors with subsequent slow replenishment.

prevents implantation and vaginal cornification (Harper and Walpole, 1967), and thus has great potentialities as a post-coital anti-fertility agent. Its action on the uterus has been studied by Koseki *et al.* (1977) in the hope of elucidating the mechanism of its action in terms of receptor binding and translocation. The phase of decreasing uterine weight that follows the initial period of stimulation was not associated with decrease in nuclear or cytoplasmic receptors, and the authors suggested that the later-emerging failure to promote uterine growth was due to a failure to "process" the nuclear-oestradiol receptor complex. Thus, the oestradiol-acceptor complex of the nucleus must be disposed of if nuclear function is to continue, and this disposal does not consist in a simple loss to the cytoplasm but some more active process of removal.*

Alternatively, there may be a more direct influence on the synthesis of oestrogen; thus, Watson *et al.* (1975) employed doses of Tamoxifen that delayed implantation in rats for 20–24 hr. In this event, the "oestrogen surge" that occurs at midnight on Day 3 of pregnancy was delayed, whilst doses that prevented implantation completely eliminated the surge, and it was suggested that Tamoxifen exerted a part of its effect by inhibiting synthesis of oestradiol from progesterone. So far as the mechanism of a failure to implant is concerned, it has been suggested that Tamoxifen increases speed of migration through the oviduct and its expulsion by uterine contractions; however, this is unlikely, so that an interference with uterine metabolism is probably the main factor in the anti-fertility effect (Major and Heald, 1974).

Clomiphene

This drug has been used to induce ovulation in anovulatory women, i.e. it is the reverse of a contraceptive agent; yet in rats it blocks ovulation, probably through an anti-oestrogenic effect, lowering the sensitivity of the pituitary to LHRH. Under appropriate conditions, however, it will induce ovulation in the rat when this has been blocked, e.g. when they are in constant oestrus in consequence of suprachiasmatic lesions (Döcke, 1971).

* Terenius (1970) points out that there are two major classes of non-steroidal anti-oestrogens. (a) Amino-ether derivatives of polycyclic phenols. (b) Diphenolic compoundst Class (a), typified by U-111000A acts systemically whereas Class (b), typified by meso-butoestrol only antagonizes oestrogen when this and the anti-oestrogen are applied together locally. Terenius studied the affinities of the two types to uterine receptors. Mesobutoestrol (Type b) had a very high affinity but this was easily reversed and so it only acts under special conditions. U-111000A had a lower affinity but attachment was much less reversible.

Anti-Androgens

Interest in these compounds is sustained by the hope that they will prove a valuable means of contraception by inhibiting sperm maturation or otherwise interfering with male reproductive physiology. A limitation on their use will clearly be the extent to which they affect the male libido.

Competition for Receptors

Typical anti-androgens are cyproterone and cyproterone acetate (Fig. 5.13); they antagonize the action of androgens such as testosterone or 5α-dihydrotestosterone by competing for the hormones at the site of the receptor protein (Stern and Eisenfeld, 1969). They do not exert their effects by inhibiting the conversion of testosterone to dihydrotestosterone (Stern and Eisenfeld, 1971). They are able to compete at the receptor sites by virtue of their similar structure to that of the androgens, and the same may be said of 17-methyl-β-nortestosterone (Tveter, 1971) and BOMT (Fig. 5.13). Flutamide is not a steroid* but it, likewise, acts as an anti-androgen through competitive antagonism at the target organ (Peats et al., 1974).

Biological Effects

The anti-androgens have acquired interest clinically both in pathology and in the attempt to employ them in male contraception, in the hope that compounds might be found that would, say, inhibit spermatogenesis or sperm maturation, leaving libido and the accessory organs, such as the prostate, functional. Thus, in rats and guinea-pigs, cyproterone acetate in high doses led to regression of accessory glands and

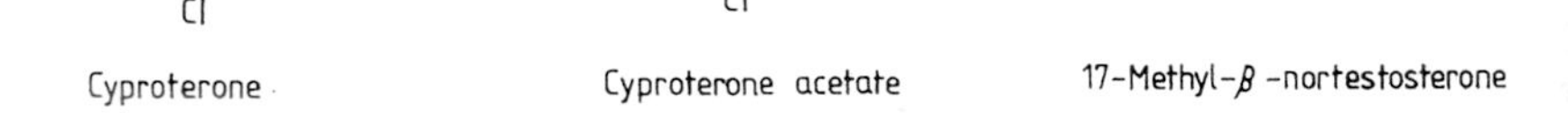

Cyproterone   Cyproterone acetate   17-Methyl-β-nortestosterone

Fig. 5.13. Structural formulae of some anti-androgens. (Neumann and Schenk, *J. Reprod. Fert.*)

* According to Tymoczko and Liao (1976) the two molecules, dihydrotestosterone and flutamide may have a similar gross geometrical structure.

sterility without loss of libido, but in other species, including man, inhibition of libido is prominent, the sterility being due to inhibition of spermatogenesis. Moreover, much larger doses are required to inhibit spermatogenesis than to inhibit the secretions of prostate and seminal vesicles. Figure 5.14 illustrates the results of an experiment; in (a) and (c) the effects of large and small doses of cyproterone acetate on testicular weight are shown; only the large dose causes reduction in testicular weight; by contrast, (b) and (d), the prostate weights were reduced by both doses; with the lower doses moreover fertility, although reduced, was by no means completely abolished (Neumann and Schenk, 1976).

Cyproterone and Cyproterone Acetate. Neumann and Schenk emphasize that these two anti-androgens are not to be treated as the same; cyproterone is described as a "pure" anti-androgen since it mainly attacks the site of action of the androgen at its points of negative feedback in the pituitary-hypothalamus; as a result, secretion of

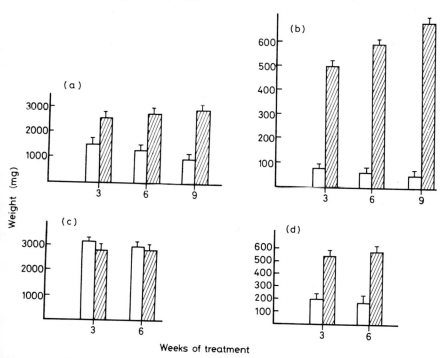

Fig. 5.14. The weights of the testes (a, c) and prostate glands (b, d) of rats after treatment (white columns) with 10 (a, b) or 2·5 (c, d) mg cyproterone acetate per animal per day compared with those of controls (cross-hatched columns). Bars represent S.E.M. (Neumann and Schenk, *J. Reprod. Fert.*)

gonadotrophins increases and this augments the secretion of testosterone and thus tends to overcome the anti-androgen effect. Thus, an initial anti-androgen effect will be manifest, say, as a reduced prostate weight, but later this will disappear. Figure 5.15 shows how cyproterone administration to male rats actually increases the concentration of testosterone in the blood above normal and even above that produced by a typical stimulant of gonadal secretion, human chorionic gonado-trophin (hCG). The removal of the negative feedback on the pituitary will produce "castration cells" similar to those produced by gonad-ectomy (Neumann, 1966). Cyproterone acetate has more complex activity including progestogenic, so that the increased gonadotrophin secretion that occurs with cyproterone is not manifest, and this means that the anti-androgen effect is much greater and more lasting.

Ovulation. The progestogenic effect of cyproterone acetate is manifest in an inhibition of ovulation in females, whereas cyproterone, the pure anti-androgen, does not affect ovulation. When ovulation is inhibited by treatment of the female with testosterone, cyproterone will counteract this since it displaces testosterone from the receptor on the hypothalamus-pituitary system that regulates gonadotrophin secretion. Thus, testosterone inhibits ovulation because it prevents, by receptor competition, the release of LH normally induced by oestrogen at the preovulatory surge. Cyproterone competes with testosterone, and

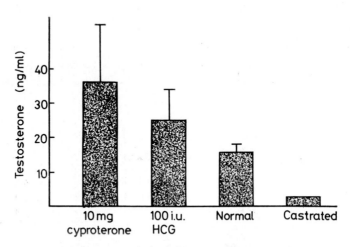

Fig. 5.15. Testosterone concentrations in the serum of 42-day-old rats after 12 days of treatment with 10 mg cyproterone/day or 100 i.u. hCG. Because of inhibition of the central effect of the androgens, cyproterone causes an increase in testosterone secretion via stimulation of gonadotrophin secretion. (Neumann and Schenk, *J. Reprod. Fert.*)

prevents it from exerting this inhibitory action on LH secretion (Neumann and Schenk, 1976).*

REFERENCES

Adams, C. E. (1965). Influence of the number of corpora lutea on endometrial proliferation and embryo development in the rabbit. *J. Endocrinol.*, **31,** xxix-xxx.

Ainsworth, L., Daenen, M. and Ryan, K. J. (1969). Steroid transformations by endocrine organs from pregnant mammals. IV. *Endocrinology*, **84,** 1421–1429.

Allen, W. M. and Corner, G. W. (1929). Normal growth and implantation of embryos after very early ablation of the ovaries, under the influence of extracts of the corpus luteum. *Amer. J. Physiol.*, **88,** 340–346.

Aref, I., Hefnani, F., Kandil, O. and Abdel Aziz, M. T. (1973). Effect of minipills on physiologic responses of human cervial mucus, endometrium and ovary. *Fert. Steril.*, **24,** 578–583.

Atkinson, L. E. *et al.* (1975). Circulating levels of steroids and chorionic gonadotrophin during pregnancy in the rhesus monkey, with special attention to the rescue of the corpus luteum in early pregnancy. *Biol. Reprod.*, **12,** 335–345.

Balin, H., Wan, L. S. and Rajam, R. (1969). Sequential approach to oral contraception therapy. *Internat. J. Fert.*, **14,** 300–308.

Blye, R. P. (1973). The use of estrogens as postcoital contraceptive agents. *Amer. J. Obstet. Gynecol.*, **116,** 1044–1050.

Brotherton, J. (1976). *Sex Hormone Pharmacology*. Academic Press, London and N.Y.

Butenandt, A., Westphal, U. and Coblen, H. (1934). Uber einen Abbau des Stigmasterins zu corpus-luteumwirksamer Stoffen, ein Beitrag zur Konstitution des Corpus-luteums Hormons. *Ber. deutsch. Chem. Ges.*, **67,** 1611–1616.

Challis, J. R. G. (1971). Sharp increase in free circulating oestrogens immediately before parturition in sheep. *Nature*, **229,** 208.

Cidlowski, J. A. and Muldoon, T. G. (1976). Dissimilar effects of antiestrogens upon estrogen receptors in responsive tissues of male and female rats. *Biol. Reprod.*, **15,** 381–389.

Csapo, A. I. and Wiest, W. G. (1969). An examination of the quantitative relationship between progesterone and the maintenance of pregnancy. *Endocrinol.*, **85,** 735–746.

Daniel, E. E. and Renner, S. A. (1960). Effect of the placenta on the electrical activity of the cat uterus *in vivo* and *in vitro*. *Amer. J. Obstet. Gynecol.*, **80,** 229–248.

Davies, L. J., Ryan, K. J. and Petro, Z. (1970). Estrogen synthesis of adrenal-placental tissues of the sheep and the iris monkey *in vitro*. *Endocrinology*, **86,** 1457–1459.

Davies, L. J. and Ryan, K. J. (1972a). Comparative endocrinology of gestation. *Vitam. & Horm.*, **30,** 223–279.

Davies, L. J. and Ryan, K. J. (1972b). The uptake of progesterone by the uterus of

---

* Progesterone is anti-androgenic, but does not compete very effectively with testosterone or dihydrotestosterone for binding at the rat's ventral prostate or seminal vesicle receptor, so that its action may well be to prevent the conversion of testosterone to dihydrotestosterone by substrate competition for the enzyme responsible for the conversion (Voight *et al.*, 1970; Stern and Eisenfeld, 1971).

the pregnant rat *in vivo* and its relationship to cytoplasmic progesterone-binding protein. *Endocrinol.*, **90**, 507–515.

De Snoo, K. (1919). *Ned. Tijdschr. Geneesk.*, **2**, 306. (Quoted by Davies and Ryan, 1972.)

Diczfalusy, E. (1964). Endocrine functions of the human fetoplacental unit. *Fed. Proc.*, **23**, 791–798.

Diczfalusy, E. (1968). Mode of action of contraceptive drugs. *Am. J. Obstet. Gynec.*, **100**, 136–163.

Diczfalusy, E. (1969). Steroid metabolism in the human foeto-placental unit. *Acta Endocrinol.*, **61**, 649–664.

Döcke, F. (1971). Studies on the anti-ovulatory and ovulatory action of clomiphene citrate in the rat. *J. Reprod. Fert.*, **24**, 45–54.

Eisenfeld, A. J. (1970). ³H-estradiol: *in vitro* binding to macromolecules from the rat hypothalamus, anterior pituitary and uterus. *Endocrinology*, **86**, 1313–1318.

Eisenfeld, A. (1974). Oral contraceptives: ethinyl estradiol binds with higher affinity than mestranol to macromolecules from the sites of anti-fertility action. *Endocrinology*, **94**, 803–807.

Erb, R. E. *et al.* (1967). *J. Dairy Sci.*, **51**, 420. (Quoted by Davies and Ryan, 1972a).

Fylling, P. (1970). The effect of pregnancy, ovariectomy and parturition on plasma progesterone level in sheep. *Acta Endocrinol.*, **65**, 273–283.

Giannopoulos, G. (1973). Binding of testosterone to uterine components of the immature rat. *J. biol. Chem.*, **248**, 1004–1010.

Gillette, D. D. (1966). Placental influence on uterine activity in the cow. *Amer. J. Physiol.*, **211**, 1095–1098.

Gitlin, D. and Boesman, M. (1967). Sites of serum α-fetoprotein synthesis in the human and in the rat. *J. clin. Invest.*, **46**, 1010–1016.

Harper, M. J. K. and Walpole, A. L. (1967). Mode of action of I.C.I. 46,474 in preventing implantation in rats. *J. Endocrinol*, **37**, 83–92.

Hashimoto, I., Henricks, D. M., Anderson, L. L. and Melampy, R. M. (1968). Progesterone and pregn-4-en-20α-ol-3-one in ovarian venous blood during various reproductive states in the rat. *Endocrinology*, **82**, 333–341.

Heckel, G. P. and Allen, W. M. (1938). Prolongation of pregnancy in the rabbit by the injection of progesterone. *Amer. J. Obstet. Gynaec.*, **35**, 131–137.

Hisaw, F. L., Velardo, J. I. and Goolsby, C. M. (1954). Interaction of estrogens on uterine growth. *J. clin. Endocr.*, **14**, 1134–1143.

Hobson, B. M. (1971). Production of gonadotrophin, oestrogens and progesterone by the primate placenta. *Adv. Reprod. Physiol.*, **5**, 67–102.

Illingworth, D. V., Heap, R. B. and Perry, J. S. (1970). Changes in the metabolic clearance rate of progesterone in the guinea-pig. *J. Endocr.*, **48**, 409–417.

Jaffe, R. B., Lee, P. A. and Midgley, A. R. (1969). Serum gonadotropins before, at the inception of, and following human pregnancy. *J. clin. Endocrinol.*, **29**, 1281–1283.

Katzenellenbogen, B. S. and Ferguson, E. R. (1975). Antiestrogen actiteon in uterus: biological ineffectiveness of nuclear bound estradiol after antiestrogen. *J. Endocrinol*, **97**, 1–12.

Korenman, S. G. (1970). Relation between estrogen inhibitory activity and binding to cytosol of rabbit and human uterus. *Endocrinology*, **87**, 1119–1123.

Koseki, Y., Zava, D. T., Chamness, G. C. and McGuire, W. L. (1977). Estrogen receptor translocation and replenishment by the antioestrogen Tamoxifen. *Endocrinology*, **101**, 1104–1110.

Kumar, D. and Barnes, A. C. (1965). Studies in human myometrium during pregnancy. *Amer. J. Obstet. Gynecol.*, **92**, 717–719.

Kuriyama, H. and Csapo, A. (1961). Placenta and myometrial block. *Amer. J. Obstet. Gynecol.*, **82**, 592–599.

Lai, P. C. W. *et al.* (1976). Pat alpha-fetoprotein: isolation, radioimmunoassay and fetal-maternal distribution during pregnancy. *J. Reprod. Fert.*, **48**, 1–8.

Leek, A. E., Ruoso, C. F., Kitau, M. J. and Chard, T. (1975). Maternal plasma alphafetoprotein levels in the second half of normal pregnancy: *J. Obstet. Gynecol. Brit. Commonw.*, **82**, 669–673.

Linzell, J. L. and Heap, R. B. (1968). A comparison of progesterone metabolism in the pregnant sheep and goat: sources of production and an estimation of uptake by some target organs. *J. Endocr.*, **41**, 433–438.

Madhwa Raj, H. G. and Moudgal, N. R. (1970). Hormonal control of gestation in the intact rat. *Endocrinology*, **86**, 874–889.

Mainwairing, W. I. P. (1975). Steroid hormone receptors: A survey. *Vitam. & Horm.*, **33**, 223–245.

Major, J. S. and Heald, P. J. (1974). The effects of I.C.I. 46,474 on ovum transport and implantation in the rat. *J. Reprod. Fert.*, **36**, 117–124.

Maner, F. D. *et al.* (1963). Interrelationship of estrogen concentrations in the maternal circulation, fetal circulation and maternal urine in late pregnancy. *J. clin. Endocr. Metab.*, **23**, 445–458.

Marshall, J. M. and Csapó, A. I. (1961). Hormonal and ionic influences on the membrane activity of uterine smooth muscle cells. *Endocrinology*, **68**, 1026–1035.

Martinez-Manautou, J., Giner-Velasquez, J. and Rudel, H. (1967). Continuous progestogen contraception: a close relationship study with chlornadinone acetate. *Fed. Proc.*, **18**, 57–62.

McGuire, J. L. and Bariso, C. D. (1972). Isolation and preliminary characterization of a progestogen specific binding macromolecule from the 273,000 g supernatant of rat and rabbit uteri. *Endocrinology*, **90**, 496–506.

Milgrom, E. and Baulieu, E. E. (1970). Progesterone in uterus and plasma. I. Binding in rat uterus 105,000 g supernatant. *Endocrinology*, **87**, 276–287.

Neill, J. D., Johansson, E. D. B. and Knobil, E. (1969). Patterns of circulating progesterone concentrations during the fertile menstrual cycle and the remainder of gestation in the rhesus monkey. *Endocrinology*, **84**, 45–48.

Neumann, F. (1966). Auftreten von Kastrationszellen im Hypophysenvorlappen männlicher Ratten nach Behandlung mit einen Antiandrogen. *Acta endocrinol.*, **53**, 53–60.

Neumann, F. and Schenck, B. (1976). New antiandrogens and their mode of action. *J. Reprod. Fert.* Suppl. **24**, 129–141.

Nishi, N., Arimura, A., de la Cruz, K. G. and Schally, A. V. (1976). Termination of pregnancy by sheep anti-LHRH gamma globulin in rats. *Endocrinology*, **98**, 1024–1030.

Nunez, E. *et al.* (1971). Mise en évidence d'une fraction protéique liant les oestrogènes dans le serum de rats impubères. *C.r. Acad. Sci. Paris*, **272**, 2396–2399.

Parlow, A. F., Daane, T. A. and Dignam, W. J. (1970). On the concentration of radioimmunoassayable FSH circulating in blood throughout pregnancy. *J. clin. Endocr.*, **31**, 213–214.

Peats, E. A., Henson, M. F. and Neri, R. (1974). On the mechanism of the antiandrogenic action of flutamide (α-a-α-trifluoro-2-methyl-4'-nitro m-propionotoluidide) in the rat. *Endocrinology*, **94**, 532–540.

Porter, D. G. (1969). Progesterone and the guinea-pig myometrium. In *Progesterone*. Ciba Study Group No. 34. pp. 79–86.

Randle, G. H. and Cumberbatch, K. N. (1973). Alpha-fetoprotein levels in amniotic fluid in normal pregnancy and in pregnancy complicated by anencephaly. *J. Obstet. Gynaec. Brit. Commonw.*, **80**, 1054–1058.

Rao, B. R., Wiest, W. G. and Allen, W. M. (1974). Progesterone "receptor" in human endometrium. *Endocrinology*, **95**, 1275–1281.

Raynaud, J.-P., Mercier-Bodard, C. and Baulieu, E. E. (1971). Rat estradiol binding plasma protein (EBP). *Steroids*, **18**, 767–788.

Rennie, P. and Bruchovsky, N. (1972). *In vitro* and *in vivo* studies on the functional significance of androgen receptors in rat prostate. *J. biol. Chem.*, **247**, 1546–1554.

Rochefort, H. and Capony, F. (1973). Étude comparée du comportement d'un anti-oestrogene et de l'oestradiol dans les cellules utérines. *C.r. Acad. Sci. Paris Ser. D*, **276**, 2321–2324.

Roy, E. J. (1962). The concentration of oestrogens in maternal and foetal blood obtained at Caesarian section. *J. Obstet. Gynaec. Brit. Commonw.*, **69**, 196–202.

Sell, S. and Alexander, D. (1974). Rat alpha fetoprotein. V. Catabolism and fetal-maternal distribution. *J. Nat. Cancer Inst.*, **52**, 1483–1489.

Seppälä, M. and Rueslahti, E. (1972). Alpha fetoprotein in amniotic fluid: an index of gestational age. *Am. J. Obsts. Gynec.*, **114**, 595–598.

Siiteri, P. K. and MacDonald, P. C. (1966). Placental estrogen biosynthesis during human pregnancy. *J. clin. Endocr. Metab.*, **26**, 751–761.

Stern, J. M. and Eisenfeld, A. J. (1969). Androgen accumulation and binding to macromolecules in seminal vesicles: inhibition by cyproterone. *Science, N.Y.*, **166**, 233–235.

Stern, J. M. and Eisenfeld, A. J. (1971). Distribution and metabolism of ^{3}H-testosterone in castrated male rats; effects of cyproterone, progesterone and unlabeled testosterone. *Endocrinology*, **88**, 1117–1125.

Terenius, L. (1970). Two modes of interaction between oestrogen and anti-oestrogen. *Acta. Endocrinol*, **64**, 47–58.

Townsend, L. and Ryan, K. J. (1970). *In vitro* metabolism of pregnenolone-7α-^{3}H,-progesterone-4-^{14}C and androstenedione-4-^{14}C by rat placental tissue. *Endocrinology*, **87**, 151–155.

Tveter, K. J. (1971). Effect of 17α-methyl-β-nortestosterone (SK & F 7690) on the binding *in vitro* of 5α-dihydrotestosterone by macromolecular components from the rat ventral prostate. *Acta Endocrinol.*, **66**, 352–356.

Tymoczko, J. L. and Liao, S. (1976). Androgen receptors and the molecular basis for the action of antiandrogens in the ventral prostate. *J. Reprod. Fert.* Suppl. **24**, 147–162.

Van Furth, R. and Adinolfi, M. (1969). *In vitro* synthesis of the foetal α-globulin in man. *Nature*, **222**, 1296–1299.

Voigt, W., Fernandez, E. P. and Hsia, S. L. (1970). Transformation of testosterone into 17β-Hydroxy-5α-androstan-3-one by microsomal preparations of human skin. *J. biol. Chem.*, **245**, 5594–5599.

Walsh, S. W., Wolf, R. C. and Meyer, R. K. (1974). Progesterone, progestins and 17-hydroxypregn-4-ene-3,20-dione in the utero-ovarian, uterine and peripheral blood of the pregnant rhesus monkey. *Endocrinology*, **95**, 1704–1710.

Watson, J. *et al.* (1975). Plasma hormones and pituitary luteinizing hormone in the rat during the early stages of pregnancy and after post-coital treatment with Tamoxifen (ICI 46,474). *J. Endocrinol.*, **65**, 7–17.

Whalen, R. E., Martin, J. V. and Olsen, K. L. (1975). Effect of oestrogen antagonist on hypothalamic oestrogen receptors. *Nature*, **258**, 742–743.

Wiest, W. G. (1970). Progesterone and 20α-hydroxypregn-4en-3-one in plasma, ovaries and uteri during pregnancy in the rat. *Endocrinology*, **87**, 43–48.

Yannone, M. E., McCurdy, J. R. and Goldfien, A. (1968). Plasma progesterone levels in normal human pregnancy, labor and the puerperium. *Amer. J. Obstet. Gynecol.*, **101**, 1058–1061.

Zander, J. and von Munstermann, A.-M. (1956). Progesterone in menschlichem Blut und Geweben. III. *Klin. Wchschr.*, **34**, 944–953.

CHAPTER 6

Pregnancy : Parturition and Labour

Gestation Period

After a period of gestation appropriate to the species, varying from some 20 days in the rat and mouse, through 170 days in the rhesus monkey to some 270 days in the human,* the foetus is ready to be expelled from the uterus. Since the processes of change that take place in the foetus are more far-reaching and rapid than those in the maternal receptacle, it is reasonable to consider the hypothesis that it is the foetus that finally gives the signal that it is ready to emerge; and the problem with which we are most concerned is the nature of this signal and the steps between this and the powerful contractions of the uterine muscle that finally expel the foetus. In this connection we must bear in mind that, within the class of mammals, there may well be and are, indeed, considerable variations in the mechanism, so that a theory based on the sheep or goat may not be applicable to primates or rats. This point is especially cogent because the pregnant ewe and goat are ideal experimental subjects for this kind of study, since it is possible by previous implantation of cannulae into the blood vessels of foetus and mother, to carry out experimental manoeuvres under conditions that approximate very closely to the normal. Because of this a great deal of our information is derived from these two species.

* The gestation period of the guinea-pig is some 67 days, far longer than that of animals of comparable size; this is due to the late stage of development at which it is born. Approximate gestation periods in days for some species are as follows:

Hamster	14	Rabbit	32	Sheep	140	Human	265
Mouse	20	Dog	61	Goat	148	Cow	285
Rat	20	Cat	65	Monkey	170	Elephant	660

These are times from mating which, in cycling animals correspond fairly closely to the day of ovulation. In the human the time of pregnancy is dated from the first day of the last menstrual period; 66 per cent of normal females give birth within two weeks before or after 280 days after this day. Since ovulation occurs on about the 15th day after beginning menstruation, this would make a gestation period of about 265 days.

Effector Mechanisms

The mechanisms available to the mother, responding to the hypothetical signal, are several. The ultimate mechanism is single, namely contraction of the uterine muscle, but this may be initiated in several ways, as Csapo (1969) has enumerated them.

Uterine Stretching

First there is the increase in size of the foetus; this causes stretch, which itself promotes uterine growth (Csapo *et al.*, 1965); this was shown by inserting balloons into the two horns of a parturient ovariectomized rabbit's uterus, the one inflated with saline and the other virtually empty; the horn containing the saline-filled balloon was heavier than the unstretched horn.

Oxytocin

A second factor is the neurohypophyseal hormone oxytocin, which in addition to promoting release of milk from the mammary glands also causes contraction of uterine smooth muscle.

Fall in Blood-Progesterone

According to Csapo, the secretion of oxytocin at term is unlikely to be the governing factor in promoting the lengthy process of parturition; it probably acts as a trigger, initiating contractions, but its subsequent action is rather to modulate activity initiated by other agencies. Thus, its action is antagonized by progesterone, so that to be an effective trigger it must await a fall in the blood-level of this hormone.

Rise in Blood-Oestrogen

In fact, Csapo lists a sudden fall in progesterone secretion as the *fourth* factor, the *third* factor being the progressive rise in oestrogen level which would increase uterine smooth muscle excitability.

Excitability of Myometrium

Thus, the excitability of the myometrium, as revealed by a variety of electrophysiological studies, is reduced by treatment of the ovariectomized animal with progesterone and increased by treatment with oestrogen*, so that in a given state of the animal, e.g. during the

---

* So far as the isolated uterine muscle is concerned, the effects of steroid hormones on excitability are not consistent with the notion that oestrogen increases excitability. Thus, a powerful oestrogen, such as the synthetic diethylstilboestrol, strongly *inhibits* the myometrium, as measured by spontaneous activity, and the responses to acetylcholine, of uterine segments (Batra and Bengtsson, 1978).

phases of oestrus, the excitability of the tissue and its tendency to spontaneous activity will be governed by the balance of progesterone and oestrogen secretions. The fall in concentration of progesterone in the blood supplying the uterus is, in Csapo's (1976; 1977) view, the key to parturition, the fall converting the previously "suppressed" muscle of the uterus to one capable of producing the co-ordinated contractions required for delivery. Thus, the suppressor, in Csapo's view, is progesterone; once its suppressing activity is removed other factors, such as oxytocin and prostaglandin synthesis and release within the myometrium and decidua, can exert their actions.

Uterine Volume

It has been suggested that a factor in determining the time of parturition is the uterine volume, so that in a polytocous species like the rat a large litter would be associated with a shorter gestation. The evidence on this subject has been summarized by Biggers et al. (1963) and Davies and Ryan (1972) and is conflicting; thus in pigs, litter-size does not affect gestational period (e.g. Cox, 1964); in guinea-pigs a mean gestational length of 70 days was found when the litter-size was one, and 65 days when it was six (Goy et al., 1957).

A similar inverse relation between litter-size and gestation period has been observed in mice, cattle, sheep, goats, rabbits and hamsters. That crowding of the uterus is not the basis of this inverse relationship was shown by Biggers et al. (1963) who found that, if mice were ovariectomized unilaterally, all ova implanted in the uterine horn of the opposite side, without any change in average litter-size. Under these conditions, then, the uterine horn becomes overcrowded, but the same inverse relation between litter-size and gestation period pertains, so that the total number of foetuses in the female apparently determines size of the litter. Biggers et al. suggested that some humoral factor was important, but this was probably not progesterone since, when one Fallopian tube was occluded, the number of foetuses was halved without affecting the number of corpora lutea, whilst there was no effect on the inverse relationship.

Primates. In singleton pregnancies in primates, if a correlation exists it is in the reverse direction so that a large infant is associated with a long gestation (Hendricks, 1964). In human twins, the unimportance of uterine size becomes clear when we appreciate that the combined foetal weight of the twins is equal to that of a singleton at 32 weeks; nevertheless, gestation proceeds to some 37 weeks when the mean combined weight is some 5 kg.

Species Variations in Control Mechanism

Short and Long Gestation

As with the maintenance of gestation, so with the act of parturition, we must bear in mind the differences in control mechanisms that exist amongst mammalian species. Thus, short-period gestational animals, such as the ferret, rat, rabbit, etc., depending on the maternal ovary for maintenance of uterine quiescence, may be expected to have their period of gestation determined largely, if not exclusively, by maternal events, whilst the long-gestational animals, typified by the primates, being dependent on the foeto-placental unit for steroid hormone synthesis, may be expected to have gestation determined by a signal from the foetus. In general, this prognostication is correct.

Maternal Secretion of Oestrogen

Thus manipulation with the rat maternal pituitary-ovarian system influences parturition; we have seen that ovariectomy in this "intolerant" species brings about abortion, and the same is true of hypophysectomy. However, if these operations are postponed till just before parturition, then they prolong pregnancy, apparently due to failure of secretion of maternal oestrogen; thus injection of oestrogen into ovariectomized rats results in parturition within 12–24 hours (Csapo, 1969). Again, Newton (1935) observed that killing mouse foetuses *in utero* did not prevent the placentae from being delivered at term; such "fetectomized" rats carried their placentae to term, however, only if their ovaries were intact, so that removal of the foetuses, accompanied by ovariectomy, led to abortion of the placentae (Newton and Lits, 1938). Thus, in these short-gestational animals, it could well be that the duration of pregnancy is programmed by the maternal pituitary-hypothalamic system.

Final Common Pathway

In long-gestational animals, such as primates, there is strong reason to believe that the foetus, operating through the foeto-placental unit, brings pregnancy to an end at the appropriate time.

In both types of mammal, however, it could be that the final common pathway for parturition was through altered steroid levels in the blood or, more probably, the myometrium. Thus, oestrogen increases myometrial excitability and progesterone opposes this; a swing of the balance in favour of oestrogen might well be the initiating event in the regular uterine contractions characteristic of labour.

Oestrogen

The importance of oestrogen in normal delivery is suggested by Figure 6.1 which describes the delays in labour of rats caused by several manipulations. The top row indicates control labours taking place between Days 22 and 23. Ovariectomy 48 hours before term delayed and arrested delivery, the majority of the foetuses eventually dying *in utero* unless rescued by hysterectomy. Treatment of these ovariectomized animals with oestrogen restored the condition to normal, whether it was administered on Day 21 or at term (arrows).

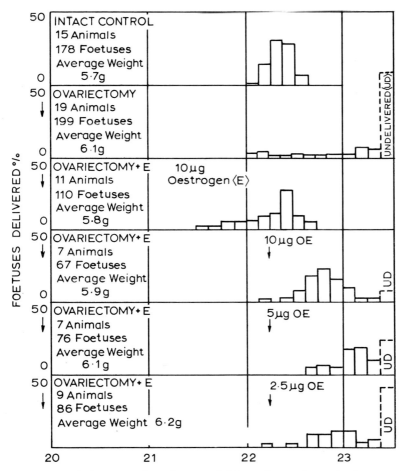

Fig. 6.1. Delayed, interrupted, and arrested labour in ovariectomized rats and its correction by treatment with oestrogen. Ordinate: per cent of foetuses delivered. Abscissa: days of pregnancy. (Csapo, *Ciba Study Group.*)

The Oestrogen Surge

We have already seen how the concentration of oestrogens in the human maternal plasma rises during pregnancy, whilst the urinary excretion of oestriol may increase by a factor of 1000 to 10,000. As Figure 6.2 shows for the pregnant ewe, the rise in oestrogens of the plasma just before term is very striking; whether this is due to the rapid availability of adrenal corticosteroids, which can act as precursors for maternal synthesis, as suggested by Ainsworth and Ryan (1966), is by no means certain. As we shall see, in this species and probably in many other long-gestation species, the secretion of adrenal corticosteroids seems, in fact, to be the trigger that determines parturition, and this could happen by increasing maternal oestrogen synthesis or by interference with progesterone synthesis, or both.

In the monkey, the principal urinary oestrogen of pregnancy is oestrone, and Fig. 6.3 shows the rapid rise in urinary excretion during the last ten pre-parturition days. In the rat, too, Yoshinaga *et al.* (1969) have described a steep pre-parturient rise in oestrogen secretion by the ovary.*

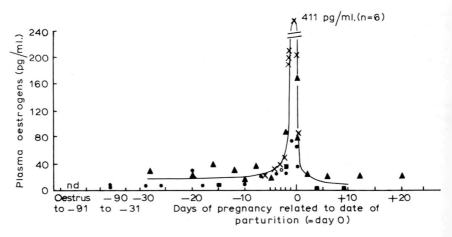

Fig. 6.2. Illustrating the rise in plasma oestrogens in the ewe before term. Different symbols refer to different ewes. (Challis, *Nature.*)

---

* Klopper (1973) has critically examined the evidence for an involvement of oestrogen in induction of labour. He concludes that a rise in oestrogen level in the blood is not a dominant or indeed an essential feature of the onset of labour, so that the hormone may well exert only a permissive action.

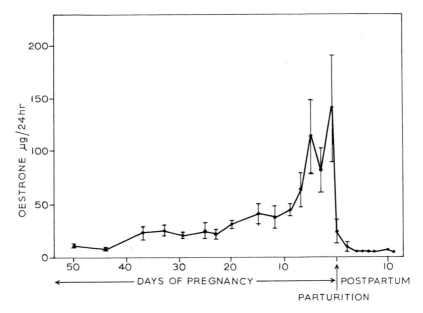

Fig. 6.3. Urinary excretion of oestrone during pregnancy and post-partum in the monkey. (Liskowski *et al.*, *Biol. Reprod.*)

Progesterone

If progesterone is the "hormone of pregnancy", maintaining the uterus quiescent, then parturition could follow from a sudden drop in the plasma level, caused either by luteolysis, when there is a functional corpus luteum, as in the rat, or by interference with placental synthesis. Fylling's (1970) study of the sheep revealed a steep fall before parturition from some 20 ng/ml to 1 to 2 ng/ml. Figure 6.4 shows a similar change associated with an increase in prostaglandin secretion, and Figure 6.15 (p. 316) shows corresponding changes in the goat.*

Uterine and Plasma Levels

As we have indicated earlier, the level of progesterone in the contractile tissue might be more relevant than that in the plasma. Wiest (1970) analysed uterus and plasma of the pregnant rat and, after implantation,

* Anderson *et al.* (1975) have measured the changes in steroid metabolism of the placentae of ewes, immediately after induction of labour with foetal injections of corticosteroid, in an endeavour to elucidate the mechanism of the reduced progesterone concentration before term. The induction of the enzyme, steroid 17α-hydroxylase, is the most prominent change; this enzyme could cause reduced placental progesterone by favouring synthesis of oestrogens from it.

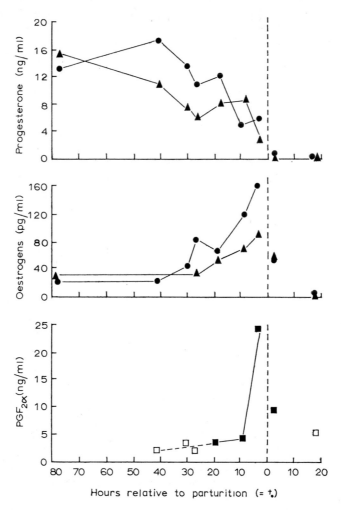

Fig. 6.4. The concentrations of progesterone, total unconjugated oestrogens, and prostaglandin $F_2\alpha$ in the uterine blood of a sheep during late pregnancy and parturition (t_0). Progesterone and oestrogens determined in samples from the right (▲) and left (●) uterine veins. $PGF_2\alpha$ (■) measured in pooled samples from the right and left uterine veins; (□), values at the limit of assay. (Challis *et al.*, *J. Reprod. Fert.*)

placental tissue. Some results are shown in Table I. In general, a concentration of 2 μg/100 g in the uterus was sufficient to prevent

TABLE I

Concentrations of progesterone (μg/100 g or ml) in uterus, placenta and blood plasma of the rat during pregnancy (Wiest, 1970)

Day	Uterus	Placenta	Plasma
2	15·6	—	5·6
4	20·9	—	7·4
7	14·8	—	8·5
11	22·5	27	11·2
14	20·2	4·2	13·2
17	14	3·5	11·9
19	11·5	4·0	11·0
20	4·6	2·3	9·9
21	2·6	2·2	2·4
22	1·7	2·1	1·6

delivery. The high levels in the uterus in the early stages of pregnancy are clearly related to placentation and indicate a capacity of the uterus to establish higher concentrations than in the plasma.

Blood Progesterone in Primates

The existence of a sudden drop in progesterone level, *before* term, has been questioned; thus, Figure 6.5 from Sommerville (1969) shows the plasma levels of progesterone and pregnanediol in a pregnant woman for four weeks before term and some hours post-partum. There is quite clearly no *sudden* change in plasma level of either of these substances, so that the post-partum fall in the plasma level of pro-gesterone could well be due to loss of the placental supply, the corpus luteal supply being small in the human (p. 204). The results of several other studies on humans,* summarized by Bengtsson (1973), have also failed to demonstrate any fall in progesterone level in the blood until the placenta has been delivered, and the same is true of the monkey (Fig. 5.3, p. 262).

Exogenous Progesterone. If "progesterone block" is a real phenomenon, then we should expect to be able to prolong gestation by increasing the level of progesterone in the maternal blood. Once again,

* That the evidence on this point is not uniform is shown by the study of Turnbull *et al.* (1974) who found a steady rise in blood oestrogens in women during the last three weeks of pregnancy; this was accompanied by a steady fall in progesterone level, but there was no sudden fall pre-partum.

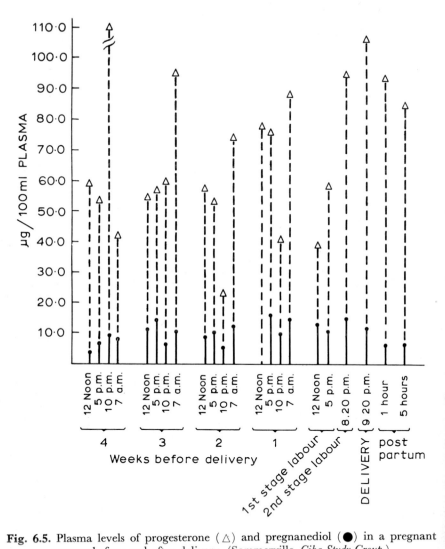

Fig. 6.5. Plasma levels of progesterone (△) and pregnanediol (●) in a pregnant woman before and after delivery. (Sommerville, *Ciba Study Group*.)

however, we meet species variations in the response to progesterone; thus, according to Schofield (1964), a daily dosage of progesterone will delay parturition in the rabbit, and the same appears to be true of the rat, whereas in the guinea-pig, human and sheep, progesterone injections have no effect.* According to Schofield, the basis of these species variations may well be the dependence, or otherwise, on placental progesterone. Thus, those species, such as human, sheep and guinea-pig, that are able to rely exclusively on the placental source towards the end of pregnancy are uninfluenced by the plasma level, and they show no sudden drop at term,† whereas the species more dependent on an ovarian source do show a prolongation of gestation with exogenous progesterone and also show the sudden drop at term. Schofield suggested that in the placenta-dependent species the progesterone synthesized by the placenta might reach the uterus directly by diffusion (Zarrow et al., 1960) whereas, of course, the ovarian supply would have to come through the general circulation. Thus, the plasma level would be relevant for the "intolerant" species but irrelevant for the "tolerant" species. This view has been supported by Davies and Ryan (1972); thus the concentration of progesterone in the retroplacental pool of blood is very high by comparison with peripheral blood, whilst local access from the placenta to the uterus is very likely by virtue of the complex subserosal and muscular lymphatic system of the human uterus, which exhibits an increase in calibre and number of its vessels during pregnancy (Maurizio and Ottaviani, 1934).

Prostaglandins

The effects of prostaglandins on the process of parturition must be considered from two aspects, namely the events triggering the process of parturition and the events during labour. The first aspect considers the role of prostaglandins in luteolysis, where the pregnancy depends throughout its length on an intact corpus luteum, e.g. goat, rabbit, and mouse, or in otherwise antagonizing or causing withdrawal of progesterone from the maternal circulation.

* Bengtsson and Schofield (1963) were unable to prolong gestation in the sheep by daily injections of progesterone a week before term. Hindson et al. (1969) recorded intra-uterine pressure in the pregnant ewe and, when uterine contractions indicative of labour had begun, they injected a large dose of progesterone; they were able to abolish the contractions, and delivery was delayed for seven days.

† The sheep seems to be an exception, however, since it does show a large drop in blood progesterone at term (Fig. 6.4, p. 296). This decrease towards term, however, is due mainly to failure of the ovarian supply; in the guinea-pig and cow the decline is due to ovarian failure, placental secretion being maintained till term.

Luteolysis

The role of $PGF_{2\alpha}$ as a luteolysin in the ovarian or menstrual cycles has been discussed earlier, and we have seen that there is little doubt that this prostaglandin can inhibit *in vitro* synthesis of progesterone by granulosa cells of the corpus luteum. This luteolytic action doubtless explains the control exerted by the uterus on luteolysis. Infusion of $PGF_{2\alpha}$ into the rabbit's aorta on Day 21 of pregnancy (normal gestation 32 days) caused parturition some 40 hours later. The uterine contractions did not occur immediately but after 15 hours, so that it may be that the first effect of the prostaglandin was exerted through its luteolytic action, and later its oxytocic (p. 330) action induced labour, the prostaglandin being effective on the uterus because of the lowered blood progesterone level (Nathanielsz *et al.*, 1972).

Effects on Corpus Luteum. The lowered blood progesterone was associated with involution of the corpus luteum, suggesting a luteolytic action of the prostaglandin. Thus, all the intracellular organelles commonly associated with secretory activity were absent. Although the blood progesterone was reduced by treatment on all days tested, the failure to find regressive changes during some parts of the cycle permits the distinction between *functional luteolysis*, when progesterone secretion alone is affected, and *morphological luteolysis* when both morphological and blood changes are found. There were no effects on interstitial cells or follicles so that, although oestrogen is the natural luteotrophin in rabbits, the luteolytic attack seems not to be on the cells of origin of oestrogen. Koering and Kirton (1973) suggest that the attack is on the oestrogen receptor; thus, according to Lee and Jacobsen (1971), the maximal amount of receptor activity occurs during the mid-luteal phase of pseudo-pregnancy, so that if the same happens during mid-pregnancy, this might account for the large effects on Day 14.

Blood PG Levels

Estimation of the levels of prostaglandins of the F series in the plasma of pregnant rabbits showed a gradual increase after midpregnancy to reach highest values immediately prior to, or coincident with, the preparturient decline in concentration of progesterone (Challis *et al.*, 1973). In rats, Strauss *et al.* (1975) found that administration of $PGF_{2\alpha}$ to pregnant animals on Days 19 and 20 caused a rapid fall in blood progesterone followed, on Day 21, by premature delivery. The same effect occurred in hypophysectomized rats, so that the action must have been a direct one on the corpora lutea. Injection of a synthetic progestin—Depoprovera—antagonized the effect of $PGF_{2\alpha}$ although the blood-progesterone fell, and delivery was normal. Thus, prostaglandin acts first as a luteolysin and later directly on the uterus, promoting contractions (oxytocic effect).

PG-Infusions in the Goat

By infusing $PGF_{2\alpha}$ into the uterine vein of goats Currie and Thorburn (1973) induced premature parturition; this was due to a luteolytic action as manifest by the fall in blood progesterone; if the rate of infusion of PG was as low as $0.94 \ \mu g/min$, there was no change in blood-progesterone, and no induction of labour. The concentrations of PG in the utero-ovarian venous plasma with infusions at the rate of 1.8 to $7.5 \ \mu g/min$, which were adequate to induce labour, were comparable with those found during normal labour.

Primates

We have seen that, although the natural luteolysin in the primate is probably not a prostaglandin, $PGF_{2\alpha}$ does have a luteolytic effect so that its secretion can interfere with luteal function. In the monkey Kirton et al. (1970) demonstrated a marked decline in progesterone secretion by the ovaries when $PGF_{2\alpha}$ was injected on Days 11, 12 or 13 postovulatory of a fertile cycle (Fig. 6.6). If it was injected earlier, there was no effect, indicating a greater vulnerability of the corpus luteum in the early stage of pregnancy when it is being stimulated by the blastocyst, i.e. while chorionic gonadotrophin is stimulating the ovary. When $PGF_{2\alpha}$ was effective, namely between Days 11 and 13, the plasma level of progesterone was high, due to the secretion of chorionic gonadotrophin; at this time, however, the placenta is not yet secreting steroid hormones in quantity so that the source of the progesterone is still the ovary, and the decline in plasma level is definitely due to luteolysis.*

Myometrial Action. During the later stages of gestation, $PGF_{2\alpha}$ administered to pregnant monkeys precipitated premature delivery; in this case the effect was not due to luteolysis but a direct action on the myometrium, and a consistent change in maternal steroid concentrations does not seem to be a prerequisite for premature delivery (Challis et al., 1974).

---

* Kirton et al. (1970) found that PGE_2 was generally more effective in inducing the uterine contractions that led to abortion than $PGF_{2\alpha}$; a similar difference was found in humans (Karim and Filshie, 1970).

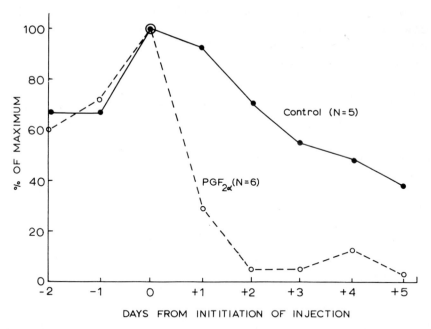

Fig. 6.6. Decline in progestin level of plasma following injection of $PGF_2\alpha$ in monkeys. (Kirton *et al.*, *Proc. Soc. exp. Biol. Med.*)

PG-Synthesis Inhibitors

We have seen (Volume 2) that aspirin and indomethacin exert their therapeutical effects by inhibiting the synthesis of prostaglandins; if synthesis and release of prostaglandin are a necessary concomitant of parturition, then these inhibitors should prolong labour. In fact, Aiken (1972) found that indomethacin increased the duration of parturition in rats and increased the proportions of pups dying during (not prior to) birth. He found that normal uteri, taken at parturition, produced *in vitro* some 20 times more PGF_{2a} than uteri taken on the 18–19th days of pregnancy, and that this difference was reduced by indomethacin treatment. In monkeys, Novy *et al.* (1974) prolonged gestation from a normal 167 ± 0.4 days to 180 ± 2 days with indomethacin. In women, too, Waltman *et al.* (1972) found that the abortion, induced by intra-amniotic injection of saline, could be delayed by treatment with indomethacin, whilst subjects taking high doses of aspirin during the last two trimesters showed a significantly increased duration of gestation and in the mean duration of spontaneous labour (Lewis and Schulman, 1973).

Role of Prostaglandins

In summary, then, the synthesis and release of prostaglandins constitute an important element in parturition. In those species like the goat, rabbit and mouse that depend on an intact corpus luteum throughout pregnancy the luteolytic action of $PGF_{2\alpha}$ is of obvious significance. In the sheep and cow, parturition is associated with definite changes in placental synthesis and secretion of steroid hormones, and these changes may be causally related to the synthesis and secretion of prostaglandin. Figure 6.4 of Challis et al. (1972) showing the changes in progesterone oestrogen and $PGF_{2\alpha}$ in uterine venous blood of a pregnant ewe during the hours leading to parturition, is highly suggestive; moreover, Caldwell et al. (1972) have shown that progesterone plus oestrogen cause a rise in $PGF_{2\alpha}$ in the jugular vein plasma of ovariectomized sheep, so that the sequence of events might be: secretion of progesterone plus oestrogen leading to uterine synthesis and release of $PGF_{2\alpha}$ which might then precipitate parturition.

Man, along with non-human primates and the guinea-pig, are placenta-dependent species so far as steroid hormone secretion during pregnancy is concerned; unlike the sheep and cow, however, labour starts without the level of hormones in the maternal circulation showing obvious evidence of altered placental hormone metabolism (Liggins et al., 1977). Nevertheless, the steadily increasing secretion of oestrogens during pregnancy in the face of low or steady levels of progesterone would lead to a progressive increase in the oestrogen: progesterone ratio in the tissue and this may well reach a point where "progesterone block" is overcome, and prostaglandin secretion is sufficient to induce the contractions of labour.

Synthesis by Uterus during Pregnancy

The ability of the rat's uterus to synthesize prostaglandin $F_{2\alpha}$ increases during the course of pregnancy. Harney et al. (1974) measured release in vitro from the uterus taken from rats at different stages in pregnancy. Maximal release occurred on Day 22 at term, and after this, 1–3 days post-partum, the release was much smaller; the critical time for the increased production seemed to be Day 18, when blood progesterone falls and oestrogen rises. Thus, it seems likely that synthesis of prostaglandins is influenced by the blood levels of the ovarian steroid hormones; however, according to Harney et al., the steep oestrogen surge before parturition, when it occurs, is unlikely to be the cause of the increased synthesis of $PGF_{2\alpha}$; if this surge were responsible for parturition, then it would exert its effect by increasing the sensitivity of the uterus to oxytocic substances, including prostaglandins.

Metabolism and Synthesis

Carminati *et al.* (1975) have studied both metabolism (during incubation) and synthesis (from arachidonic acid *in vitro*) of PGF_{2a} and PGE_2 by placenta, uterus and ovary of the rat during pregnancy. The most striking changes in uterine synthesis and metabolism occur during late pregnancy; as the authors point out, in preparation for parturition luteolysis must take place, uterine blood-flow must increase and uterine contractions must occur, all processes in which prostaglandins are involved. Placental synthesis and metabolism are very high on Day 11, that of PGE_2 being several times greater than that of PGF_{2a}; and these changes correlate with the high concentration of progesterone in the pregnant rat ovary, uterus and placenta; at this time, too, placental control of the rat's corpus luteum is initiated.

Antiluteolytic Factor

If we grant that in, say, the guinea-pig, PGF_{2a} is luteolytic, we may postulate a normal suppression of its synthesis by the uterus, perhaps by an antiluteolytic factor secreted by the conceptus (Bland and Donovan, 1969). Maule Walker and Poyser (1974) found that the guinea-pig uterus produced less F_{2a} on Day 15 of pregnancy than on Day 15 of the oestrous cycle (Fig. 6.7) and they consider that an antiluteolytic factor is secreted by the trophoblast of the conceptus preventing synthesis of PGF_{2a} and its release into the uterine vein.*

Release from Progesterone Block

Csapo (1976) has emphasized the importance of the hormonal state of the myometrium with respect to the induction of labour, insisting that the simple view of the uterus as a passive organ responding by rhythmic contractions when a certain level of prostaglandin has accumulated in its tissue is quite inadequate to describe the events leading to labour. In his studies on rabbits he showed that massive doses of PGF_{2a}, given into the uteral lumen, failed to produce the high intraluminal pressure with superimposed cyclical variations characteristic of labour unless the intraluminal injection was given after Day 29, i.e. during the parturient period; at this time, moreover, much smaller doses were just as effective. When uterine tissue was analysed at

* The value of a luteolytic mechanism is immediately evident when it is appreciated that an infertile cycle might lead to an indefinite state of pseudopregnancy; once fertilization occurs this process must be rendered inoperative, but the mechanism whereby luteolysis is inhibited is not certain; it may be that the decidual reaction is adequate to prevent the release of the factor or that there is competition with luteotrophic hormone at the receptor site in the ovary.

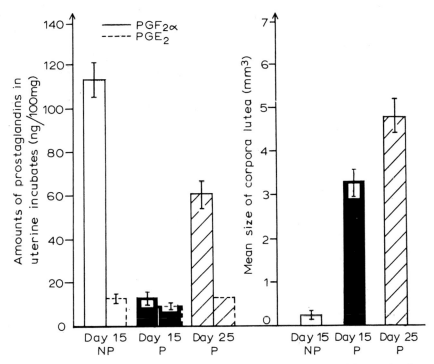

Fig. 6.7. Amounts of prostaglandins (PGF$_2$a and PGE$_2$) produced by uterine incubates (left-hand figure) compared with size of corpora lutea. NP, nonpregnant guinea pigs; P, pregnant guinea pigs. (Maule Walker and Poyser, *J. Endocrinol.*)

different stages it appeared that the period of high susceptibility to the action of prostaglandins actually occurred when the levels in the uterine tissue were on the decline (Fig. 6.8). These findings thus do not support the concept that labour is initiated by a precipitous increase in uterine PG concentration, but rather suggest that an earlier event is release of the endometrium from inhibitory block, i.e. the progesterone block postulated by Csapo. Analysis of the uterus of the pregnant rabbit showed, indeed, that the levels of progesterone (P in Fig. 6.8) fell continuously during the perinatal period. The actual levels are higher than in plasma, and since the major source of progesterone in the rabbit is the ovary whence it is carried in the general circulation to the uterus, it is clear that these high levels are maintained by the presence of a specific binding material (Davies *et al.*, 1974). Thus, at the critical period around the 31st day of pregnancy when the rabbit's uterus shows high sensitivity to PG's it is the level of progesterone that falls rather than the level of PG that rises. Hence, whilst the triggering

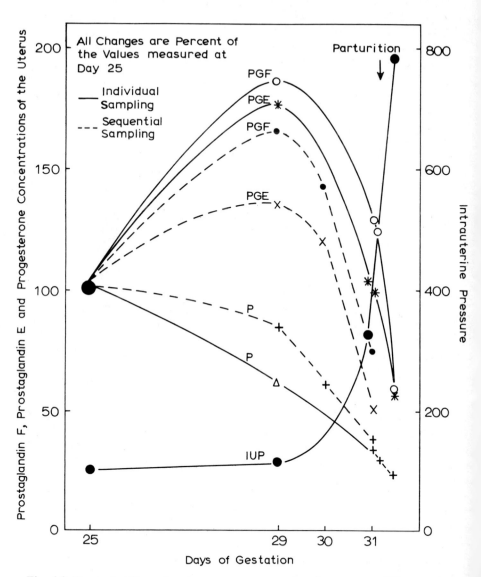

Fig. 6.8. Prostaglandin and progesterone concentrations in uterus at different stages of gestation in the rabbit. P = progesterone; IUP = cyclic IUP. Labour is clearly not precipitated by a rise in prostaglandin concentration in uterine tissue. (Csapo, *Prostaglandins.*)

event for labour is presumably the intrinsic uterine stimulant, prostaglandin, with or without the assistance of oxytocin, the decisive controlling step is the decrease in uterine progesterone below a critical value.

ROLE OF THE FOETAL PITUITARY

Clinical Evidence

A number of clinical studies on humans and farm animals have suggested that the foetus exerts some influence over the date of parturition. For example, Milic and Adamsons (1969) showed that in some 22 cases of uncomplicated human pregnancies involving anencephaly of the foetus, pregnancy was prolonged perhaps to as long as 52 weeks. This anencephaly was invariably associated with adrenal hypoplasia, and in Turnbull and Andersen's (1969) studies of premature expulsions, taking place for no known reason after the 20th week, the adrenals of the foetus were unusually large compared with those of a similar gestational age delivered as a result of therapeutic abortion or some complication of pregnancy. Again, a genetically controlled adrenal insufficiency, found in Holstein-Friesian cattle, is associated with delayed parturition, and the finding that the male contribution to the phenotype influences this defect indicates that it is a defect in the calf rather than in the mother. Holm et al. (1961) showed that in these cases of delayed pregnancy the adenohypohyphysis of the calf was smaller than normal and the animal showed signs of adrenocortical deficiency, such as a persistent hypoglycaemia, especially during fasting. Again, habitually aborting Angora goats have a genetically controlled hyperplasia of the adrenal cortex (Van Rensburg, 1963) with retarded growth of the corpora lutea.

Experimental Studies

Pituitary Lesions

Liggins et al. (1967) carried out electrocoagulation of the pituitary in foetal sheep in utero between Days 93–143, and demonstrated prolongation of pregnancy if the operation was carried out by Day 124. If operated on Day 143 delivery was spontaneous. Interestingly, if the ewe was bearing two or three lambs, all of the foetuses had to be hypophysectomized to bring about prolongation, indicating that a single pituitary was adequate to initiate the stimulus to parturition. Later, Liggens et al. (1973), showed that permanent section of the pituitary stalk was adequate to produce the same prolongation of pregnancy.

ACTH

The hormone secreted by the pituitary to give the signal is probably ACTH; thus the foetal adrenal gland undergoes very rapid growth, confined to the cortex, in the last few days of pregnancy, as revealed by the changes in weight illustrated by Figure 6.9 and it is conceivable that the secretion of foetal corticosteroids is of fundamental importance in exciting parturition, perhaps by a direct action on the maternal hypothalamus, but possibly, also by providing precursors available to the placenta for synthesis of maternal oestrogens (Ainsworth and Ryan, 1966).

Adrenalectomy

Experimentally Drost and Holm (1968) carried out adrenalectomy in sheep foetuses *in utero*, and found that, when there was no evidence of adrenal regeneration, delivery was delayed. None of the overdue ewes showed signs of impending parturition, such as mammary development, softening of the sacro-ileac ligaments and swelling and

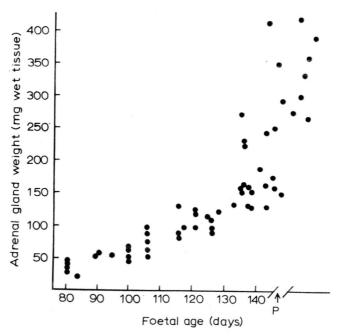

Fig. 6.9. Weight of right adrenal at various gestational ages in normal foetal lambs. P, parturition. (Liggins, *J. Endocrinol.* from Comline and Silver.)

relaxation of the vulva. Injections of dexamethasone into adrenalectomized foetuses allowed normal parturition (Liggins *et al.*, 1973).

ACTH Infusions

Again, Liggins (1968) infused foetal sheep *in utero* with ACTH
between Days 88 and 129 and after 4–7 days of infusion parturition
occurred, indicating premature delivery, since the gestation period is
normally some 150 days. The adrenals of the prematurely born lambs
were of the same average weight as those of lambs born at normal term,
and since, as Figure 6.9 shows, the foetal adrenal doubles its weight
during the last few days of pregnancy, the ACTH must have had
a powerful trophic action bringing about precocious development
and presumably secretion of corticosteroids, the principal corticosteroid
of the sheep being cortisol. Cortisol infusions were also effective, but
not into the mother.*

Concentrations of Foetal Corticosteroids

Bassett and Thorburn (1969) showed that the plasma level of
corticosteroid of the foetal lamb increased before parturition, and
similar increases have been described for the bovine foetus near term;
thus they rose from around 10 ng/ml 9 days before term to a maximum
of 60–100 ng/ml at delivery. In the rat, Dupouy *et al.* (1975) found that
the foetal corticosterone levels increased to a peak at Day 18 of pregnancy, and remained high until parturition. At every stage of gestation
the levels were higher in the foetal plasma than in the maternal, and
since maternal adrenalectomy on Day 14 did not affect the plasma
levels on Day 18 onwards it appears that the maternal corticosteroid
was largely derived from the foetuses passing across the placental
barrier. The fact that maternal levels correlated with the number of
foetuses during the last 3 days of pregnancy further supports the
foetal source.†

* A possible role of the foetal adrenal in providing steroid precursors, e.g. androstenedione,
for oestrogen synthesis by the mother has been investigated by Flint *et al.* (1976) in ewes
whose foetuses had been adrenalectomized. However, the hormonal changes in the mother
preceding parturition were the same whether she was carrying normal or adrenalectomized
lambs. Again, induction of large rises in foetal blood ACTH, by anoxia, did not increase
foetal adrenal secretion of androgens (Jones *et al.*, 1977).

† In the guinea-pig it seems unlikely that the foetal corticosteroids play any role in parturition. Dalle and DeLost (1976) found the plasma corticosteroid was much lower in the foetus
than in the mother in late pregnancy, confirming Illingworth *et al.* (1970). The plasma cortisol
of the foetus does rise in late pregnancy but this is probably derived from the mother and
may also be related to a rise in corticosteroid binding globulin which, according to Seal and
Dod (1967, quoted by Dalle and DeLost), rises to values 7–8 times those in the adult.

Dexamethasone Infusions

In a later study, Liggins (1969) infused the synthetic dexamethasone into 100–121 day-old foetal lambs and in 48 hours at all doses between 0·5 and 4·0 mg/24 hr parturition was induced; at lower doses the delay was longer, as indicated by Figure 6.10. Interestingly, the effect was not antagonized by progesterone. Since infusion of corticosteroids into the mother were ineffective we may postulate a "second messenger" that influences the maternal hypothalamus, triggering off the secretion of oxytocin. However, the failure of Liggins to accelerate parturition by maternal infusion of cortisol or dexamethasone may have been due to the small dose employed, since Fylling (1971) found that although

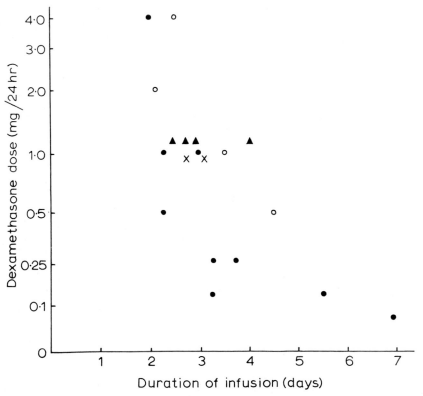

Fig. 6.10. Relation between rate of infusion of dexamethasone into foetuses and duration of infusion required to induce parturition. Dexamethasone was infused continuously either alone or mixed with deoxycorticosterone (DOC) or progesterone. Progesterone was given to ewes in some experiments. ● = dexamethasone; o = dexamethasone plus DOC; x = dexamethasone plus progesterone (foetal); ▲ = dexamethasone plus progesterone (maternal). (Liggins, *J. Endocrinol.*)

Liggins' dose of 4 mg/24 hr was ineffective, 6 mg/24 hr for 4 days did, in fact, cause premature delivery. He considered that the corticosteroid affected myometrial contractility.

Also, Adams and Wagner (1970) working on cattle with a normal gestation period of 280–290 days, caused premature birth as early as the 246th day with intramuscularly injected dexamethasone, and the same with ewes. They pointed out that habitually aborting goats have hyper-secreting adrenals, the abortion being associated with failure of the corpus luteum.

Rabbit

Prostaglandin $F_{2\alpha}$ is luteolytic in the rabbit and thus causes abortion in the pregnant animal (p. 300). In normal pregnant animals there is a steady increase in plasma level of PGF in the blood plasma which may be related to the preparturient fall in blood progesterone that takes place in this species. Since dexamethasone induces premature delivery, Challis et al. (1975) suggested that this steroid might act by stimulating synthesis of prostaglandins; however, they were unable to relate the rapid fall in blood progesterone that followed treatment with dexamethasone with an increase in blood prostaglandin. Interestingly, these authors observed that a fall in blood progesterone, which correlated with a fall in endometrial progesterone, did not of itself induce parturition, pregnancy being continued for some days after the level had fallen to levels normally associated with parturition.

Altered Foetal Pituitary Sensitivity to ACTH

Nathanielsz et al. (1977), when summarizing their studies on the role of corticotrophin, ACTH, in the foetal sheep on parturition, suggested that the increased secretion of corticoids occurring in this species before parturition might be due to an increased sensitivity of the foetal pituitary to its trophic hormone. They found a sharp increase in foetal plasma corticotrophin (ACTH) concentration 24–48 hr before delivery, this being the cause of the increased plasma cortisol concentration *during* delivery, so that care must be taken to ensure that the corticoid levels observed are concerned with the events leading to parturition rather than those occurring at and after this event.

Evidence for this increased sensitivity is provided by the study of Madill and Bassett (1973) on the response of foetal adrenal gland *in vitro* to ACTH, and it may be that this increased sensitivity, if it occurs, is brought about by the loss of an inhibiting factor.*

* According to the calculation of Nathanielsz (quoted by Nathanielsz et al., 1977) there takes place a 6·5-fold increase in the size of the sheep's adrenal cortex in the last eight days of intra-uterine life, and this, alone might account for the increased rate of secretion at this period.

Membrane ACTH receptors

An increased concentration of these corticotrophin receptors, involved in adenylate cyclase activity leading to steroid secretion, might be crucial to the increased secretion of corticoids at term. Experiments by Albano *et al.* (1976) have shown that, in the rabbit, there occurs a pronounced increase in adenylate cyclase activity that can be stimulated by ACTH between Days 21–24 of gestation. These changes can be suppressed by injections of cortisol into the mother, and since this crosses the blood-placental barrier, the effects may be interpreted as a negative feedback on the foetal pituitary, inhibiting the pituitary secretion of ACTH and in turn inhibiting the induction of receptors that would normally have been brought about by circulating ACTH.

Role of Prolactin

Winters *et al.* (1975) have suggested, on the basis of their measure-ments of foetal prolactin plasma levels in humans, that this pituitary hormone controls adrenal growth; as Fig. 6.11 shows, from Abel *et al.* (1977) there is, in the sheep foetus, a sharp increase in plasma prolactin that may well be the cause of the increased cortisol concentration, since Nathanielsz *et al.* have been unable to demonstrate that the increased cortisol concentration is preceded by an increase in ACTH.

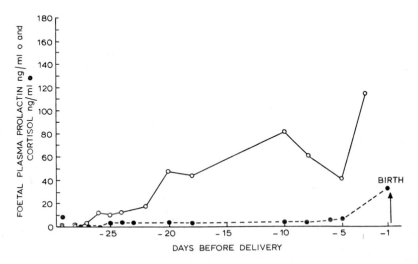

Fig. 6.11. Plasma cortisol and prolactin levels in a sheep foetus during the last 29 days of gestation. Delivery occurred on Day 144 of gestation. (Nathanielsz, *Ciba Symp.* from Abel *et al.*)

Parturition in the Goat

A recent study of Currie and Thorburn (1977) on the goat is worth considering in some detail, since it provides unequivocal evidence for the role of foetal adrenal in parturition and suggests the chain of events leading to this. Figure 6.12 shows the striking rise in foetal plasma corticosteroids during late gestation; Fig. 6.13 shows on the same graph the fall in maternal plasma progesterone twenty-four hours before parturition and the episodic rises in concentration of $PGF_{2\alpha}$ in the utero-ovarian vein ispilateral with the foetus and ovary containing the corpus luteum. When premature delivery was induced by infusion of ACTH into the foetus, delivery was associated with a precipitate fall in maternal plasma progesterone and rises in utero-ovarian $PGF_{2\alpha}$ (Fig. 6.14). According to Currie and Thorburn, the release of PGF is the penultimate event for luteal regression, the

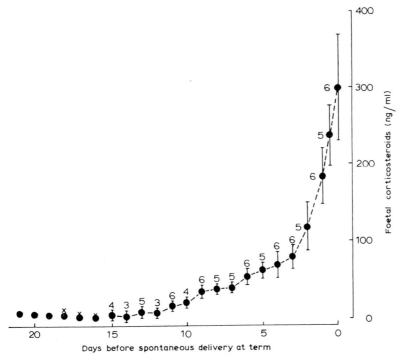

Fig. 6.12. Plasma corticosteroids in foetal goats during late gestation. Values are means $\pm$ S.E.M. for 3–6 foetuses during the last 15 days before spontaneous parturition at term. Individual values are shown before Day 15. The last point indicates samples taken during advanced labour. (Currie and Thorburn, *J. Endocrinol.*)

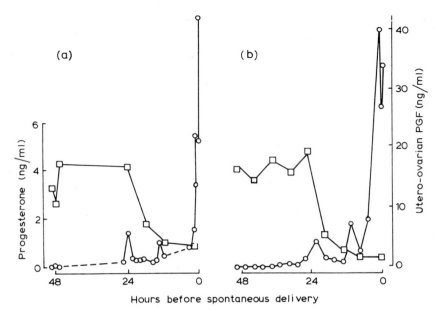

Fig. 6.13. Relationship between maternal jugular concentrations of progesterone (□) and concentrations of prostaglandin F (PGF, o) in the utero-ovarian vein ipsilateral to the foetus and ovary with corpus luteum. Observations were made in one goat before and during spontaneous parturition at term after two pregnancies (*a, b*). The last samples were collected at completion of foetal delivery. (Currie and Thorburn, *J. Endocrinol.*)

concentrations in utero-ovarian blood at the time of luteal regression, namely 5–25 ng/ml, being sufficient to act as a powerful luteolytic signal (Currie and Thorburn, 1973). When comparable utero-ovarian peaks of PGF occurred in the vein contralateral to the ovary with the corpus luteum, luteal regression did not occur, emphasizing the local control of this factor.

Source of PGF. The source of the PGF appearing in the utero-ovarian vein 24 hours before foetal delivery is probably placental, and may well be in response to the high level of oestrogen. Thus, the early releases of PGF occur on the side of the infused foetus, indicating a local mechanism, so that synthesis of oestrogen by the foetal placenta might cause high local concentrations of oestrogen in the adjacent maternal placental tissues which might promote maternal PG synthesis, perhaps by labilizing lysosomes and releasing phospholipases which would generate precursors for PG synthesis.

Pre-Partum Rise in PGF. The results obtained by Thorburn and his collaborators agree with those of Umo *et al.* (1976); however, these authors emphasize that the pre-partum rise in PGF begins 48 hr before delivery rather than 24 hr, and they found

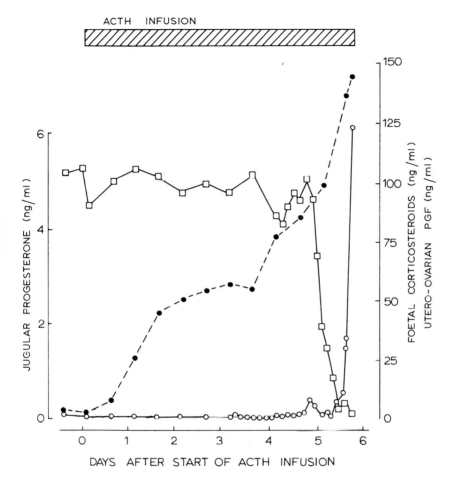

Fig. 6.14. ACTH-induced premature parturition in the goat. Relationship between maternal jugular concentrations of progesterone (□), foetal corticosteroids (○) and utero-ovarian venous prostaglandin F (PGF, ●) during infusion of ACTH into the foetus. Utero-ovarian samples were collected from the side draining the pregnant uterine horn and were ipsilateral to the ovary with the corpus luteum. The infusion lasted until completion of foetal delivery when the last samples were obtained. (Currie and Thorburn, *J. Endocrinol.*)

no evidence of episodic bursts but rather a continuous increase culminating in the peak of 60 ng/ml at delivery. The primary increase in PGF preceded the decline in blood progesterone. These relations are illustrated by Figure 6.15. That PGF was exerting a role in promoting uterine contractions was suggested by the increased motility of this organ running parallel with increased PGF concentration in the uterine vein.

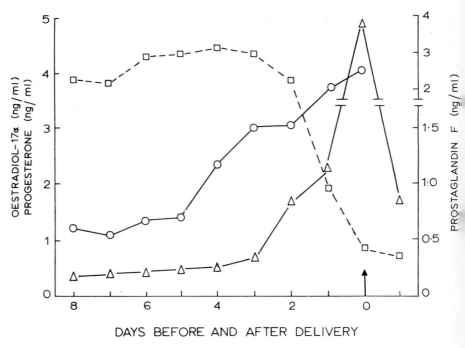

Fig. 6.15. Changes in the uterine venous plasma of oestradiol-17α (◯), progesterone (▢), and prostaglandin F (△) before and after delivery in the goat (arrow). (Umo *et al.*, *J. Endocrinol.*)

Release from Progesterone Block

In general, then, the relation between the foeto-placental unit and progesterone withdrawal, whether the progesterone is derived from the ovary or placenta or both, found in different species including the ewe and rabbit, confirms the view of Csapo that a fundamental requirement for parturition in many if not all species is the withdrawal of the inhibitory influence of progesterone on the uterus. So far as the goat is concerned, the primary event is the raised adrenal corticosteroids which induce a rise in utero-ovarian $PGF_{2\alpha}$ which acts as a luteolysin; and the studies of Hoffmann *et al.* (1977) suggest a similar mechanism, the basic feature being a sudden withdrawal of progesterone due to cessation of function of the corpus luteum in late pregnancy, a process related to the increase in foetal corticosteroids and increased production of oestrogen, especially in the foetal compartment.

The Basic Scheme

Based largely on their own studies on the ewe, Liggins *et al.* (1973) have suggested the schemes illustrated by Figure 6.16 and 6.17 for the control mechanisms of parturition in the sheep. Figure 6.16 is self-explanatory; Figure 6.17 shows the steps leading to release of cortisol into the foetal circulation; in the placenta it causes increased secretion of oestrogen which may act positively on the maternal posterior pituitary provoking secretion of oxytocin; progesterone secretion by the placenta is inhibited, and the reduced blood level might also act positively on the maternal pituitary. The increased secretion of oestrogen by the placenta provokes synthesis and release of PGF_{2a}, which acts to increase myometrial sensitivity to oxytocin as well as having a direct effect on uterine muscular activity. The lowered progesterone may alter myometrial sensitivity, removing the "progesterone block" of Csapo, but a more certain action might well be through an influence on the synthesis of PGF_{2a}.

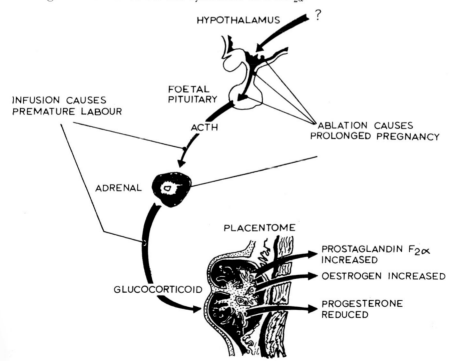

Fig. 6.16. Schematic diagram of the pathway by which the foetal lamb influences endocrine events in the ewe. Also shown are experimental procedures that have been used to modify the activity of the pathway. (Liggins *et al.*, *Rec. Progr. Horm. Res.*)

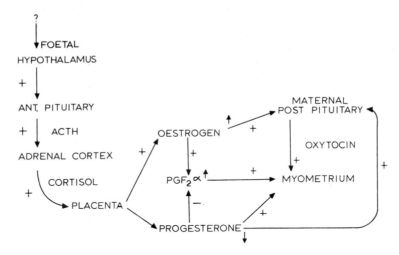

Fig. 6.17. Proposed model of the mechanism controlling the initiation of parturition in the ewe. Androgen from foetal adrenal cortex has been postulated as a major precursor of placental production of oestrogen, but this pathway is omitted from the model. (Liggins *et al.*, *Rec. Progr. Horm. Res.*)

Additional Roles of Adrenal Corticosteroids

Besides initiating parturition in such species as the goat and sheep, the foetal adrenal secretions serve other important functions, such as to stimulate the secretion of surfactants (Volume 1), and the rapid development of the mammary gland with the initiation of lactation. Thus, the viability of the newborn, lactation and parturition are apparently initiated by a single mechanism (Thorburn *et al.*, 1972).

PARTURITION IN PRIMATES

The scheme outlined above applies to animals such as the goat, sheep and cow but the situation in the human and non-human primates is clearly different since the changes in progesterone level, so obvious in the lower mammals, do not occur here.* Again, the role of the foetal pituitary-hypothalamic-adrenal system described for the sheep and goat is by no means so obvious in the primate, and may well be insignificant (Liggins *et al.*, 1977). It is difficult to escape the conclusion, however, that prostaglandins, especially F_{2a}, play an important role in both the initiation and development of labour.

* The progesterone fall before parturition occurs in guinea-pig, sheep, goat, rat and cow (Bedford *et al.*, 1972).

Monkeys

We have seen that a pre-partum fall in maternal progesterone does not occur; if the function of the fall, occurring in the sheep and goat, is to permit the action of prostaglandin on the myometrium as the final link in the parturitional process, then the absence of this fall may merely mean that in the monkey the prostaglandins can act in the presence of the maternal progesterone.

Although Challis *et al.* (1974) failed to observe a fall in maternal plasma progesterone until after parturition, the concentrations of both oestradiol-17β and ocstriol increased some 5 days before parturition. As will be discussed later, the rise in maternal blood oestrogens might well be a significant factor in triggering parturition in primates.

Role of the Foetus

The role of the foetus in determining parturition has not been established so unequivocally as in the sheep and goat; however, as summarized by Challis *et al.* (1977), the evidence does suggest that death of the foetus, or removal of the foetus leaving the placenta intact (foetectomy), causes loss of the precise control over the timing of parturition.

Foetectomy. This loss of timing control could result from two causes, namely failure of the placenta, which is damaged after foetectomy, to synthesize adequate progesterone, in which case premature delivery might be expected, or the loss of the foetal adrenal gland might remove the foetal signal for parturition. Lanman *et al.* (1975) carried out foetectomy and foetectomy combined with ovariectomy in monkeys; 5 out of 8 animals actually had prolonged gestation periods as a result of foetectomy, in the sense of retention of the placenta, whilst the maternal peripheral progesterone concentration was maintained at about the normal pregnancy value during the period of placental retention; when the placenta was expelled, the progesterone concentration fell to very low values indicating that, although damaged, it was capable of sustaining pregnancy levels. The effects of ovariectomy, however, suggested that the placenta was exerting a feedback action on the ovary, stimulating it to produce progesterone when its own production became inadequate, so that loss of the placenta caused not only loss of the placental supply but also that of the corpus luteum.

Role of Foetal Adrenal

The effects of the foetal adrenal have been studied by experimental adrenalectomy and hypophysectomy, this latter manoeuvre being achieved either by decapitation of the foetus or by implantation of a radioactive source on the sella turcica. Using this latter technique, Chez et al. (1970) found a significant lengthening of the gestation period of monkeys, whilst the foetal portion of the adrenal was involuted; four out of the five animals that were of normal growth being more than two and a half weeks late. The same operation on the maternal pituitary had no effect.

Anencephaly. In this respect, the study of Novy (1977) on experimental anencephalic foetuses is of interest; after 73–78 days' gestation the foetus was exposed and its head removed, the carotid jugular and vertebral vessels being ligated. The foetus was then returned to the uterus; 10 out of 19 foetuses were judged to have been alive at birth, and of these a much higher percentage were born after term (later than 175 days) than in a corresponding group of normals; this change was considered to be highly significant since intra-uterine surgery of any sort has a tendency to favour premature delivery.

Function of Corticosteroids

It must be appreciated that the primate placenta is highly permeable to corticosteroids (Bashore et al., 1970), by contrast with the sheep's, so that it is likely that the foetal blood corticosteroids are of mainly maternal origin, and if this is true it is most unlikely that the foetus would control parturition by a direct action of its corticosteroids on the mother; and Novy has suggested that the foetus plays a less direct role by contributing to oestrogen biosynthesis of the foeto-placental unit during late pregnancy and, through this, and the consequent increase in maternal oestrogens, induces parturition either directly or through prostaglandin.

Maternal Oestrogen Levels. Thus the rises in oestrone and oestradiol concentrations in the maternal blood plasma at late pregnancy are well established for the monkey (Fig. 6.18) and for the human (Fig. 5.6, p. 268) and abnormally high concentrations of oestradiol are associated with premature delivery (Atkinson et al., 1975). According to Novy et al. (1976) (quoted by Novy, 1977), foetal decapitation and atrophy of the foetal adrenal lower maternal concentrations of oestradiol and prolong gestation. Finally, as we have seen, foetectomy prolongs gestation and lowers the maternal concentration of *oestrogen* (Lanman et al., 1975). Thus, it seems that the foetal adrenal is a source

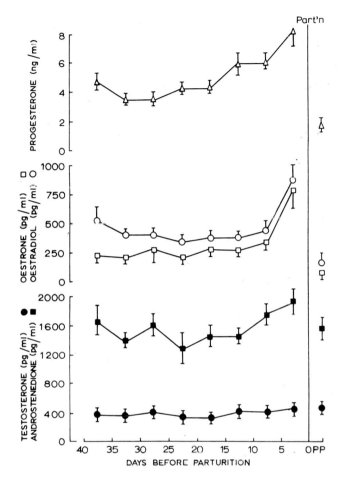

Fig. 6.18. The concentrations of progesterone ($\triangle$), oestrone ($\square$), oestradiol ($\bigcirc$), testosterone ($\bullet$) and androstenedione ($\blacksquare$) in the maternal peripheral plasma of rhesus monkeys during the last 40 days of pregnancy. Day of parturition, Day 0; PP, post-partum. (Challis *et al.*, *Ciba Symp.*)

of maternal oestrogens in late pregnancy; presumably androgens are elaborated by the foetal adrenal and employed by the foeto-placental unit in the synthesis of oestrogens.

Response to Corticotrophin

That the monkey's foetal adrenal responds to corticotrophin was shown in chronic experiments in which the foetus was infused with corticotrophin whilst the maternal response was suppressed by infusion into her of dexamethasone; the levels of foetal oestrone and maternal oestradiol rose sharply.*

Prostaglandin

That prostaglandin synthesis by the uterus-placenta is important for delivery seems very likely from the observed effects of prostaglandins in inducing delivery in monkeys and the effects of synthesis-inhibitors such as indomethacin in prolonging gestation. Thus, Novy et al. (1974) found that in monkeys the gestation length varied about a mode of 168 days, with an average of 161 days. With three indomethacin-treated monkeys the gestation times were 183, 185 and 187 days respectively. Since parturition in the primate does not depend on luteolysis the effects of prostaglandin synthesis inhibitors are to be sought at the level of direct uterine action, in fact, as Manaugh and Novy (1976) have shown, prostaglandins are not luteolytic in the monkey, and the delayed parturition caused by indomethacin was not associated with changed steroid levels in the maternal circulation.

Amniotic Fluid Concentrations. Challis et al. (1977) have analysed the amniotic fluid of pregnant monkeys; as Fig. 6.19 shows, there is a sharp rise in concentrations of PGF and of 13,14-dihydro-15-keto-$PGF_{2\alpha}$ (PGFM) during the latter part of gestation; this parallel increase suggests an increase in prostaglandin production, rather than a decrease in primary prostaglandin metabolism, around the time of parturition.

Humans

Foetal Adrenal

The hormonal control over parturition in the human is as uncertain as in the monkey; it was generally accepted that the foetal adrenal influenced the time of parturition; however, the study of Milic & Adamsons (1969) has been quoted as evidence *against* the hypothesis for a role of the foetal adrenal, since they found that many anencephalic foetuses could be born prematurely. Nevertheless, these results indicate unequivocally that anencephally is associated with defective *timing* of parturition. Thus in normals the chance of a foetus being delivered 2 weeks before or after 280 days is 66 per cent, whereas in anencephalics it was only 2 per cent. Moreover, the foetuses with shortened gestation were those that exhibited foetal abnormalities such as polyhydramnios.

---

* Mueller-Heubach et al. (1972) found no effect on gestation time of monkeys in which the foetal adrenal had been removed; however, there was evidence of residual adrenal tissue at delivery. Bosu et al. (1973) found no effect on gestation time of infusion of dexamethasone, although there was a considerable reduction in maternal blood-oestrogens.

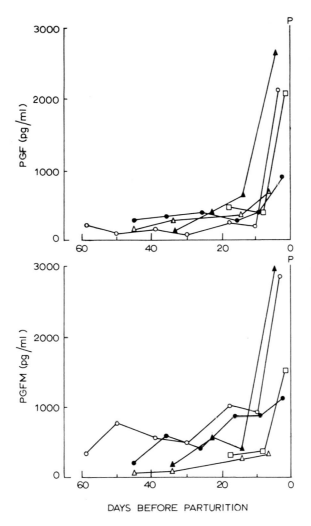

DAYS BEFORE PARTURITION

Fig. 6.19. The concentrations of PGF and 13, 14-dihydro-15-keto-PGF$_2$ (PGFM) in serial samples of amniotic fluid taken from 5 rhesus monkeys during the latter part of gestation. Parturition (P) occurred on Day 0. Lengths of gestation varied from 144 to 166 days. (Challis *et al.*, *Ciba Symp.*)

Further evidence favouring a foetal influence, operating through the adrenal, is given by Anderson *et al.* (1969) who, of eight anencephalics not showing hydramnios, found that the length of gestation was inversely related to the weight of the foetal adrenals and thickness of the foetal zone on cross-section. Thus the foetal adrenal hypoplasia, which may have been responsible for lengthened gestation, could have been due to the lack of pituitary stimulation.

Finally, Miyakama *et al.* (1976) found that the concentration of cortisol in the *cord* blood of humans with anencephalic foetuses was less than normal although the maternal peripheral blood concentration was normal. The maternal blood-oestrogen was very low, whilst the progesterone concentration was normal.*

Prostaglandins

So far as prostaglandins are concerned, there is no doubt that inhibitors of prostaglandin synthesis, such as aspirin, prolong gestation in humans as well as monkeys; moreover, Hillier *et al.* (1974) found higher concentrations of $PGF_{2\alpha}$ in amniotic fluid of mothers in whom labour occurred spontaneously than in that of mothers with induced labour (with oxytocin or PGE). Again, Willman and Collins (1976) showed that the increasing amounts of prostaglandins in the amniotic fluid as pregnancy developed were probably due to the decidua.

Steroids

So far as maternal steroid secretions are concerned, the study of Turnbull *et al.* (1974) has revealed an unequivocal rise in maternal oestrogen and fall in progesterone (Fig. 6.20), and the fall in plasma progesterone is probably not due to a failing placenta since oestrogen synthesis, which in the primate during late pregnancy is predominantly due to the utero-placental unit, is increasing. The authors agreed, however, that there was no sudden rise in oestrogen levels at parturition, since the levels in the second stage of labour were no different from those in the week preceding labour.

Lysosomal Hypothesis

Szego *et al.* (1971) pointed out that an important aspect of steroid hormone action was the changes in tissue structure that they induced, changes that occurred rapidly, e.g. in the uterus during the menstrual cycle, or in the cervix at parturition. Such changes would require the

---

* Earlier, Mikayama *et al.* (1974) had found that the ACTH concentrations in maternal and cord blood were normal, and the ACTH does not cross the human placental barrier. We may note that Cawood *et al.* (1976) found no evidence of foetal transfer of cortisone or DOC to the mother, in so far as mothers with anencephalic foetuses did not excrete smaller amounts of tetrahydro-derivatives than mothers with normal foetuses.

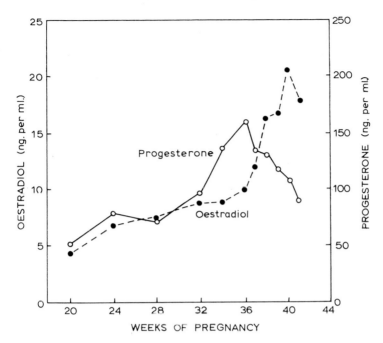

Fig. 6.20. Mean levels of plasma progesterone and oestradiol measured serially in 33 normal human primigravidae from 20 weeks up to spontaneous onset of labour. (Turnbull *et al.*, *Lancet*.)

release of cellular hydrolytic enzymes, achieved by an increased breakdown of lysosomes. They showed that sex steroids, such as oestradiol-17β and diethylstilboestrol, did, in fact, increase the fragility of lysosomes derived from the preputial gland of the rat; adrenal corticosteroids had the opposite effect, so that stability was governed by a balance between the two types of hormone. Hempel *et al.* (1970) noted that inflammatory diseases such as rheumatoid arthritis frequently subsided during the last three months of pregnancy and recurred several weeks after parturition, and suggested that, if the cause of the inflammatory disease was a pathological release of hydrolytic enzymes into the tissue by lysosomes (Volume 2), then perhaps the pregnant female's serum contained a principle that increased lysosomal stability; in fact lysosomes from leucocytes were more stable to attack by detergent when pregnant serum was added. The effect was not related to the cortisol concentration in the plasma.

Relation to Prostaglandins. A role of prostaglandins in lysosomal fragility is suggested by Weiner and Kaly's (1972) observation that incubation of liver lysosomes with $PGF_{2\alpha}$ accelerated release of the enzyme β-glucuronidase, whereas aspirin had the opposite effect. This might explain the luteolytic effect of PGF_{2a}, but more generally, according to Gustavii (1972), the release of prostaglandins by the decidua might well lead to release of hydrolytic hormones that would be the important initiating factor in labour; in fact, he suggested that parturition might be described as delayed menstruation, a situation in which large amounts of prostaglandins are produced by the endometrium. Gustavii was impressed with the abortion or induced delivery brought about by mere injections of hypertonic saline, either intra- or extra-amniotically; the damage to the placenta under these conditions was too small to reduce progesterone secretion significantly (Gustavii and Brunk, 1972), and he considered that the uterine contractions were primarily induced by damage to the decidua. Thus, examination of decidual cells taken at Caesarian section at term showed degenerative changes compared with mid-pregnancy hysterotomy controls, which were due to release of acid phosphatase into the cytoplasm (Gustavii, 1975). The chain of events, according to Gustavii, might be: damage to decidual cells, causing release of lysomal enzymes; these might include phospholipase A which could split arachidonic acid from cell lipids producing the precursors for prostaglandin synthesis (Gustavii, 1974). The scheme is illustrated in Fig. 6.21.

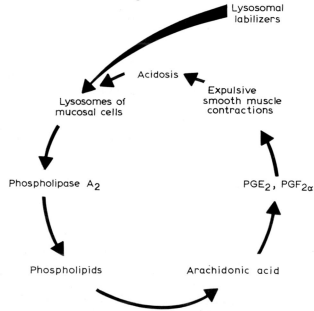

Fig. 6.21. Hypothetical scheme of reactions for triggering the onset of expulsive contractions in mucosa-lined muscles. (Gustavii, *Ciba Symp.*)

FURTHER ASPECTS OF SPECIES VARIATIONS

Pig and Sheep

As summarized by Heap *et al.* (1977) there are some interesting variations amongst species that rely on the foetal message to induce labour. Thus the pig and sheep rely on the foetal message, manifest as a rise in foetal cortisol. However, the pig relies on *ovarian* progesterone production, and the preparturient fall in this steroid is due to luteolysis, probably induced by PGF_{2a}; in the sheep, the release of PGF_{2a} is effective on the myometrium but is not the cause of the decline of progesterone, which is due to the action of foetal cortisol on the placental steroid metabolism.

Rabbit

In the rabbit, the foetus apparently does not provide the message; foetal decapitation tends rather to curtail gestation than prolong it.

Carnivores

Among the carnivores, the life-span and functional activity of the corpus luteum in the ferret, blue fox, and mink are similar whether the animal is pregnant or pseudopregnant, indicating the absence of a placental or foetal signal for luteolysis.

Dog. In the dog, on the other hand, the corpus luteum of pseudopregnancy lasts longer than that of pregnancy, whereas in the cat it is shorter. Furthermore, high doses of an analogue of PGF_{2a}, sufficient to induce parturition in a sow, are ineffective in inducing parturition in the ferret. Thus, although all these species rely on a fall in plasma progesterone for parturition, the mechanisms for producing this fall must be different, so that PGF_{2a} is not a universal common pathway in the sequence of changes of steroid hormone secretion.

Guinea-pig

The guinea-pig is interesting since there are high plasma progesterone concentrations at term, and this is consistent with the absence of an inhibitory action of this steroid on the guinea-pig's myometrium, which may rely on relaxin for this (Porter, 1972).

Hierarchical Theory

As developed by Heap *et al.* (1977), this theory suggests that there is a hierarchy in control over the process of parturition, but that the order of the hierarchy may change with species. Thus in the sheep the foetal unit, comprising its hypothalamus, pituitary and adrenal, occupies a dominant place in the hierarchy; the placenta occupies a lower place in the sequence, and its steroid hormonal syntheses can be modified by the foetal message, the foetal cortisol transforming placental steroidogenesis from a pathway directed mainly to progesterone synthesis to one directed to oestrogen synthesis (Anderson *et al.*, 1974). The maternal hypothalamus, pituitary, adrenal and ovary have an inferior position since all can be removed without imperilling gestation or delivery in the second half of gestation. The final stages, myometrial contraction and cervical relaxation rely on higher control, namely placental $PGF_{2\alpha}$ production and posterior pituitary oxytocin secretion. Figure 6.22 illustrates the schemes for sheep and pig; in the latter, as with the goat, the maternal ovary is involved. In the rabbit, the foetus is apparently unimportant, so that the maternal hypothalamus may well be number one in the hierarchy.

LABOUR

Oxytocin

The foetus is expelled as a result of the powerful contractions of the uterine musculature, and it is likely that, when this happens, the fall in blood or myometrial progesterone that takes place at parturition is an important factor in permitting these contractions, as suggested by Csapo.

Vaginal Distension

The posterior pituitary secretes the hormone oxytocin in response to vaginal distension; thus, Ferguson (1941) stretched one of the two cervices of the post-partum rabbit and this caused contraction of the other uterus after a latent period suggesting action of a blood-borne agent. Again Andersson (1951) found that distension of the cervix caused milk-ejection in goats, suggesting the secretion of oxytocin; and Roberts and Share (1968) found that the reflex release, in response to vaginal distension, measured by oxytocin activity in the jugular venous blood, occurred in normally cycling ewes as well as in pregnant animals.

Plasma Level Pre-Partum

Thus, it is reasonable to assume that secretion of oxytocin, which causes strong contractions of the isolated uterus as well as of mammary tissue, plays its role during parturition. This is shown by Figure 6.23 which illustrates the striking increase in oxytocin concentration—measured by its milk-ejection potency—in the blood plasma of cows just a few minutes pre-partum.

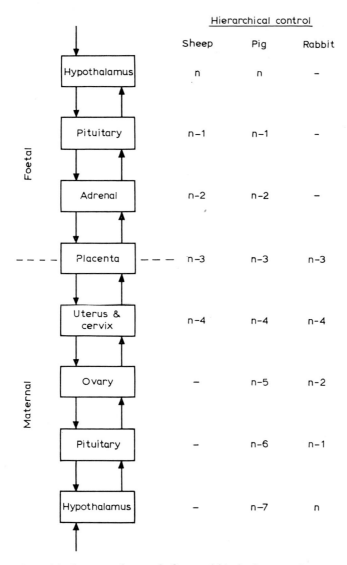

Fig. 6.22. Hierarchical system of control of parturition in three species. The dominant level of the hierarchy denoted as letter n is indicated at the foetal hypothalamus in sheep and pig, and the maternal hypothalamus in the rabbit. A dash indicates a level with negligible role in the initiation of parturition. (Heap *et al.*, *Ciba Symp.*)

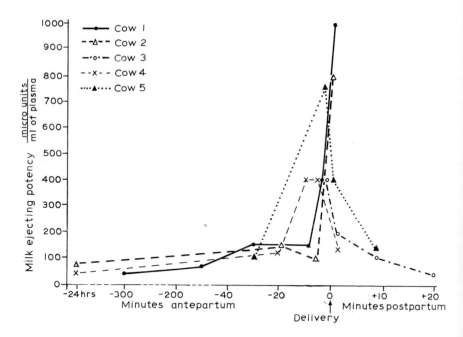

Fig. 6.23. Milk-ejecting activity in blood plasma of five cows estimated as oxytocin. Note large increase during last few minutes before delivery and a very rapid decline in the first few minutes post-partum. (Fitzpatrick, *Ciba Study Group.*)

Exogenous Oxytocin

Injection into animals before term, however, does not usually initiate labour, presumably because the action is opposed by progesterone. In women, nevertheless, the infusion of oxytocin is a standard clinical practice for the induction of labour although the concentrations developed are probably considerably greater than any that might occur normally during labour (Chard, 1973).*

Prostaglandins

Plasma Levels

Since Csapo's (1969) analysis, the importance of the prostaglandins in inducing labour has been recognized; the prostaglandin most

---

* The human foetal pituitary secretes oxytocin since the concentration in the umbilical artery is higher than in the umbilical vein at delivery (Chard, 1973), the highest concentrations being reached during the expulsive phase of delivery. The significance of the secretion has been discussed by Chard (1972); it must be emphasized that in the human, maternal levels of oxytocin show no significant or consistent change up to and during labour (Chard, 1972).

intimately concerned being $PGF_{2\alpha}$. Thus, according to Karim (1968), this may be detected in maternal blood during labour, apparently preceding the onset of uterine contractions.

Again, Currie (1975) showed that during induced labour in the goat such large amounts of $PGF_{2\alpha}$ were synthesized and released by the uterus and/or placenta that, during advanced labour, a concentration of about 3 ng/ml in the arterial plasma was established. When it is appreciated how rapidly the prostaglandins are inactivated in the general circulation, largely through the lungs, the appearance of detectable concentrations indicates considerable local liberation.

Amniotic Fluid

Analysis of the tissues of the foeto-placental unit, and of amniotic fluid might therefore give a better picture of prostaglandin synthesis during pregnancy. In general $PGF_{2\alpha}$ and PGE_2 are present in measurable concentrations in the tissues of the foeto-placental unit (Willman and Collins, 1976) whilst the concentrations in amniotic fluid increase from early pregnancy to term and from term to late labour (Karim and Devlin, 1967; Karim, 1971; Keirse et al., 1974).

Thus, Karim and Devlin (1967) found only PGE_1 in women not in labour whilst in strong labour prostaglandins E_1, E_2, $F_{1\alpha}$ and $F_{2\alpha}$ were all present in the amniotic fluid; if labour was induced by artificial rupture of the membranes only PGE_1 and PGE_2 were present. In a similar way, Keirse et al. (1974) found that oxytocin-induced labour caused a smaller rise in amniotic $PGF_{2\alpha}$ than that during spontaneous labour. Thus, the contractions of the uterus, which can cause release of prostaglandins (p. 333), are not the sole cause of the large rise in amniotic fluid prostaglandins during severe labour. The relation between prostaglandin synthesis and labour is well illustrated by Figure 6.24 which shows the concentration of PGF in the human amniotic fluid as a function of cervical dilatation. Keirse et al. considered that the fall in progesterone secretion by the placenta, and the increase in oestrogen secretion, that take place during late pregnancy tend to cause a gradual increase in prostaglandin synthesis leading finally to the onset of labour.*

---

* Prostaglandin is doubtless concerned in the uterine contractions associated with menstrual bleeding; in fact Pickles (1957) described an oxytocic factor in menstrual fluid that was subsequently identified as a mixture of prostaglandins $F_{2\alpha}$ and E_2 (Eglinton et al., 1963). Karim (1971) has used PGE_2 or $PGF_{2\alpha}$ as a once-a-month contraceptive, inducing menstrual bleeding.

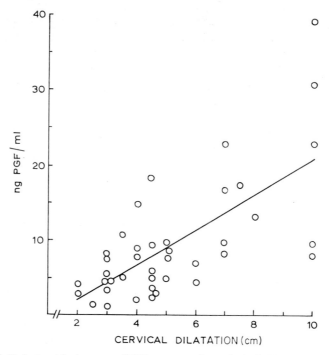

Fig. 6.24. Relationship between PGF content of amniotic fluid and degree of spontaneous cervical dilatation during spontaneous labour in women. (Keirse *et al.*, *J. Obstet. Gynaec. Br. Commonw.*)

Fluctuations in Plasma PG during Labour

According to Sharma *et al.* (1973) rapid fluctuations in plasma concentration are manifest during labour, each uterine contraction being associated with an increased level. It was difficult to determine whether the rise in concentration was the consequence of the contraction, or whether it was the stimulus, since the peak in any rise occurred some 15–45 sec after the uterine contraction; and this is close to the circulation-time from uterus to periphery (30–55 sec). If the circulation-time were greater than 60 sec the result would suggest that the prostaglandin induced the contraction, whereas if it were less than 60 sec the result would suggest that it was released into the circulation at the time of contraction. In view of the uncertainty in assessing the true circulation-time, the situation must remain unclear.*

* According to Currie (1975) the reduction in blood-flow through the uterus in the advanced stages of labour is adequate to cause release of prostaglandins.

PG-Induced Labour

That PGF_{2a} can induce labour has been amply proved; thus Karim *et al.* (1968) reported successful induction in ten out of ten cases; the uterus behaved normally, relaxing completely between each wave of contraction, so that there was no evidence of spasm. Karim and Filshie (1970) also found that therapeutic abortion could be induced during the first and second trimesters, but whereas an infusion at the rate of $4 \mu g/min$ would induce labour at normal term, some $50 \mu g/min$ were necessary during the first and second trimesters.

Monkey. Figures 6.25 and 6.26 show the increase in uterine contractions induced by PGF_{2a} and PGE_2 in the pregnant monkey; as with the human, PGE_2 is more effective than PGF_{2a}. In the sheep, Liggins and Grieves (1971) established that the large increase in PGF_{2a} found in the maternal cotyledon preceded labour by at least 24 hours. When they induced labour artificially by injection of dexamethasone (p. 310), and suppressed this labour by large doses of progesterone, the rise in $F_{2}a$ still occurred, so that the accumulation and synthesis were not due to labour itself, but rather the prostaglandin seemed to be either a "second messenger" or the active—oxytocic—principal. Certainly it has a powerful contractile effect on isolated uterine strips.

Release due to Distension of Uterus

Distension of the uterus will also induce the liberation of PGF_{2a}; this was shown by Poyser *et al.* (1970) who removed the guinea-pig's uterus on Day 3 of oestrus and incubated it in Tyrode solution; one

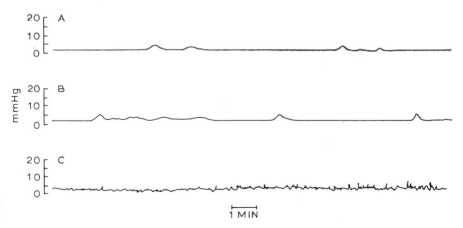

6.25. Normal uterine contractions recorded in tranquillized monkeys at: A, 40 days; B, 140 days; C, 75 days of pregnancy. (Kirton *et al.*, *Biol. Reprod.*)

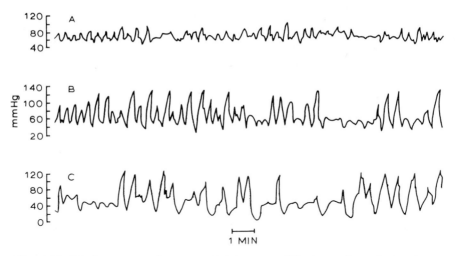

Fig. 6.26. Uterine contractions recorded in tranquillized monkeys during intra-venous infusions of PGE_2 or PGF_2a at Day 35 of pregnancy. Letters on tracing are 10 min after initiation of each infusion. Doses infused were: A, PGF_2a 60 $\mu g/min$; B, PGE_2 8 $\mu g/min$; C, PGF_2a 8 $\mu g/min$. (Kirton *et al.*, *Biol. Reprod.*)

horn was distended whilst the other acted as control. The test horn contained 1 μg of F_{2a} compared with 0·1 μg in the control horn. We have already seen that distension of the uterus causes luteolysis in guinea-pigs (p. 90) and this is presumably mediated by PGF_{2a}.

Oxytocin-Induced Release of PGF_{2a}

Oxytocin causes an *in vivo* release of prostaglandins (Sharma and Fitzpatrick, 1974; Mitchell *et al.*, 1975) which might be attributed to its action on the uterine myometrium since we have seen that uterine contractions and vaginal distension themselves can cause release of the prostaglandin. However, infusion of oxytocin into the uterine artery of pregnant ewes caused release of prostaglandins without affecting myometrial activity, so that a direct action on synthesis is likely (Mitchell *et al.*, 1975).

Incubated Myometrium. Proof of this action was provided by Roberts *et al.* (1976) who incubated endometrium from ewes at different stages of the oestrous cycle and observed an enhanced production of PGF_{2a} on addition of oxytocin, the enhancement being greatest on the day of oestrus. Binding receptor material was isolated from homogenates of both endometrium and myometrium with high specific affinity ($K_a = 5$ to 7. $10^{-10}M$), the number of sites rising to a peak at oestrus. Endometrium is the principal source of uterine prostaglandins (Pickles *et al.*, 1965) so that the finding of high-affinity binding material for oxytocin in this tissue is significant, the more so as the binding capacity at oestrus was twice that of myometrium. Thus ovarian steroids could exert an influence on $PGF_{2\alpha}$ synthesis by regulating the binding of oxytocin to the endometrium.

Role of Oestrogens

Effects on Myometrial Excitability

The steep rise in maternal plasma oestrogen is a prominent feature of late pregnancy; this could be of significance if oestrogen directly increased myometrial excitability and thus opposed the hypothetical "progesterone block" of Csapo. *In vitro*, oestradiol and progesterone both *inhibited* spontaneous electrical and muscular activity of rat's myometrium (Saldivar and Melton, 1966) and progesterone and oestradiol also inhibited the contractile response to oxytocin (Barnafi and Croxatto, 1963), whereas *in vivo* there is little doubt that oestradiol exerts a powerful stimulatory action. Once again, therefore, we are presented with the possibility of an indirect action of the steroid hormone.

Effects on PG Secretion. As Fig. 6.27 shows, a maternal injection of stilboestrol causes a rise in elimination of $PGF_{2\alpha}$ into the uterine vein, an effect that is inhibited by simultaneous administration of progesterone. Thus, a rise in secretion of oestrogens accompanied by a fall in progesterone might well cause release of sufficient prostaglandin to

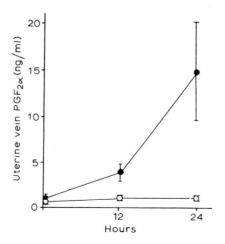

Fig. 6.27. The concentration of prostaglandin F in uterine vein blood after a maternal injection of stilboestrol administered alone (●), or after a continuous intravenous infusion of progesterone (○). (Liggins *et al.*, *Rec. Progr. Horm. Res.*)

increase uterine contractions. On this basis, the role of progesterone is to inhibit secretion of prostaglandins whilst that of oestrogen is to promote this; in this case progesterone should not inhibit the effects of infused prostaglandin; and this is true, raising the concentration of progesterone to unphysiological levels being without effect on the prostaglandin-induced changes of myometrial sensitivity to oxytocin (Fig. 6.28).

Inhibition of Oxytocin Release

The role of progesterone, *in vivo*, may be exerted through an inhibition of synthesis of PGF_{2a} (Liggins *et al.*, 1973) but also it might reflexly inhibit the release of oxytocin.

Evidence favouring this action was provided by Roberts and Share (1968–1970) who showed, first, that in normal cycling and lactating ewes vaginal distension, by inflation of a balloon, raised the concentration of oxytocin in the blood; injections of progesterone prevented this rise, whilst oestradiol propionate facilitated it; thus, before treatment with oestrogen, the rise in plasma level was fourfold whilst after 1 week of oestrogen treatment the rise was 25-fold.

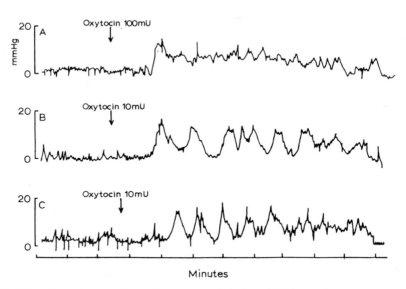

6.28. The effect of a continuous intravenous infusion of 200 mg of progesterone per 24 hours on prostaglandin F_2a (PGF_2a)-induced changes in myometrial sensitivity to oxytocin. Recording A: Threshold response to oxytocin before PGF_2a. Recording B: Threshold response to oxytocin during intra-aortic infusion of PGF_2a at a rate of 10 μg per minute. Recording C: Threshold response to oxytocin during simultaneous infusion of PGF_2a and progesterone. (Liggins *et al.*, *Rec. Progr. Horm. Res.*)

Effects of Steroid Hormones on Sensitivity to PG's and Oxytocin

Although the steroid hormone secretions exert an influence on the myometrium indirectly, by modifying the synthesis of prostaglandins and the reflex release of oxytocin, we must not ignore their direct action on the myometrium, first, in modifying the spontaneous activity, and secondly by modifying the sensitivity of the muscle to oxytocin and prostaglandins. This aspect has been studied by Kuriyama and Suzuki (1976) who examined the electrical activity of rat's myometrium taken from rats at different stages of pregnancy and from non-pregnant ovariectomized animals previously subjected to different regimens of oestrogen and progesterone. They found an increased sensitivity to oxytocin during the last stage of pregnancy that could have been due to the increased levels of oestrogen and lowered levels of progesterone pertaining at this stage. Sensitivity to PGE_2 began to increase earlier in pregnancy (middle) and continued to do so until term. The difference in sensitivity to PG may be illustrated by Fig. 6.29. (A) in the record, shows the spike-activity in control non-pregnant muscle with bursts occurring about every 2–3 min. PGE_2 at 10^{-7} g/ml had a threshold effect, (B), on increasing frequency of spike discharge and of the individual bursts; 10^{-9} g/ml produced a first-grade response* (C). Essentially the same effects of the prostaglandin were achieved by 10^{-9} and 10^{-8} g/ml PGE_2 in the post-partum myometrium (E and F). The results on tissue from steroid-treated ovariectomized rats showed unequivocally that the sensitivity to oxytocin and PGE_2 was influenced by progesterone and oestrogen, although the effects were complex. Thus oestrogen increased, and progesterone suppressed, activity in response to oxytocin and PGE_2. If the two hormones were administered in succession, the sensitivity was determined by the second hormone. With simultaneous administration, the sensitivity was largely that of an oestradiol-treated myometrium. So far as mimicking the effects of pregnancy, this could be done by steroid treatment, but the high sensitivity at parturition and post-partum could not be simulated, i.e. the "progesterone dominated" myometrium can be simulated but not the "oestrogen dominated" myometrium.

---

* The effects on the myometrium were classified into three grades according to effects on the membrane potential; in grade 1 this remained unaltered; in grade 2 depolarization led to the elimination of the quiescent periods; in grade 3 depolarization led to the blocking of spike generation.

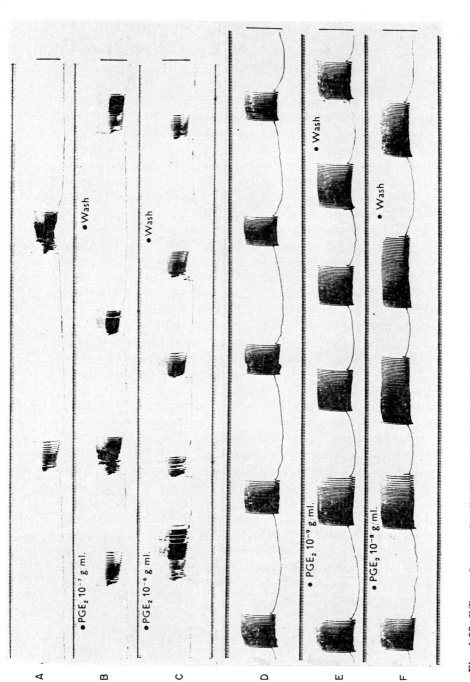

Fig. 6.29. Effects of prostaglandin E_2 on the electrical activity of rat's myometrium in non-pregnant and pregnant animals. A and D non-pregnant and pregnant controls. B and C non-pregnant animals treated with 10^{-7} (B) and 10^{-6} g/ml PGE_2. E and F, post-partum myometrium, just after delivery treated with 10^{-9} g/ml PGE_2 and 10^{-8} g/ml PGE_2. (Kuriyama and

Adrenergic Influence

It is often desirable, clinically, to delay labour in order to allow the action of injected corticosteroids on the foetus, the steroids being given to alleviate the ante-partum respiratory stresses on the immature infant. Liggins and Vaughan (1973) showed that salbutamol, a β-adrenergic stimulant, given to women in premature labour, postponed labour for greater than 24 hr in 85 per cent of cases. Similarly vitodrine, a sympathomimetic amine, reduces the contractions of labour (Landesman *et al..* 1971).

WIDENING OF THE BIRTH CANAL

Birth of the foetus at term involves its passage through the constricted portion of the uterus where it debouches into the vagina—the *cervix uteri* or *ostium uteri*. Distension of this is ultimately limited by the dimensions of the pelvic girdle, so that if these dimensions are inadequate to permit passage normal birth is impossible.

Pelvic Girdle

This girdle is composed of separate bones that come into close approximation at the symphysis pubis, ischiopubic symphysis and sacro-ileac symphysis; and it was early recognized that, in the human, there was a "pubic relaxation" or separation of the bones of the symphysis pubis before and during labour that permitted the egress of the foetus.

Pelvic Relaxation

Chamberlain (1930, 1937), using X-ray methods of examination, showed that pelvic relaxation occurred by the fourth week of pregnancy; the greatest degree of separation of the pubic bones occurs between the 5th and 7th months of pregnancy and achieves an average of some 1·2 cm. When different species are compared some notable differences are found; thus the guinea-pig, with its exceptionally large foetuses—its gestation period is 67 days—shows the relaxation in a marked form, and this is absolutely necessary for parturition. In the male, the symphysis passes from an indifferent embryonic stage of hyaline cartilage into an osteochondral stage, and proliferation of cartilage and ossification eventually lead to complete ossification. In the female, the symphysis passes from an indifferent to a chondrofibrous stage, and then develops a fissure containing fibroblasts and connective tissue leading to a separation of the bony elements of the symphysis. In pregnancy, there is a progressive vascularization of the symphyseal ligament with little or no resorption of bone, but the connective tissue is greatly increased in quantity. After parturition regressional changes comparable to scar formation take place in the ligaments in addition to bone regeneration.

Measurement. The onset of relaxation can be recognized by manual palpation, and the time required for its onset has been used as a measure of the effects of hormones on the process. Alternatively, the separation of the pubic bones can be estimated from X-ray photographs (Hall, 1948), or, in the albino mouse, by transillumination and direct observation in the dissecting microscope (Steinetz *et al.*, 1960), in which case a fairly accurate quantitative assay of relaxin (p. 342) has been made.

Pelvic Relaxation in the Mouse

The guinea-pig has been studied in detail because of the magnitude of the effects; another experimental animal studied in some detail is the mouse. During pregnancy, the separation of the pubic bones may amount to as much as 5–6 mm, the bones beginning to separate on the 13th day of pregnancy, and the gap widening at a rate of about 1 mm per day until parturition on the night of Day 19 (Fig. 6.30). By the 3rd or 4th day postpartum the gap has narrowed to less than 2 mm.

Histology. The histological changes taking place at the symphysis have been described by Hall (1947). By Day 12 there was some proliferation of the articular cartilage bordering the joint cavity so

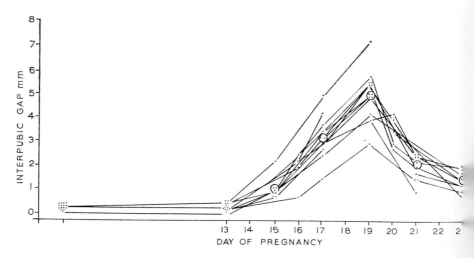

Fig. 6.30. Illustrating the course of separation of the symphysis pubis during the course of pregnancy in the mouse. (Hall and Newton, *J. Physiol.*)

that the separation of the bones was slightly increased; there was little or no evidence of bone resorption. By Day 15 the gap now measured 1 mm and the bones were connected by a ligament composed mainly of richly cellular chondroid tissue with areas of criss-crossing collagenous fibres. During the last five days of pregnancy an intensive process of bone resorption takes place, so that as the medial ends of the bones are eaten away the gap widens and the ligament lengthens. In these regions of bone resorption there is intense proliferation of collagen producing cells, their fibres contributing to the ligament.

Free-Tailed Bat

In order that the pregnant bat may deliver her single foetus there must be an enormous expansion of the bony birth canal, as illustrated schematically in Figure 6.31. The pubic joint in the male is a typical symphysis whereas in the female there is an interpubic ligament interconnecting the coxal bones; this contains an abundance of elastic fibres, and during parturition this ligament stretches to some fifteen times its original length.

Sexual Dimorphism

As Crelin (1969) has pointed out, the difference between the male and female pelvis has been recognized since antiquity; he quotes Celsus as stating that the shape of the female pelvis was such that it did not hinder birth. However, it was not appreciated until much later that this sexual dimorphism was not present at birth. Thus, Morton (1942) found no difference between the male and female pelvic inlet in foetal and pre-pubertal humans; in the female, at puberty, the pelvic inlet

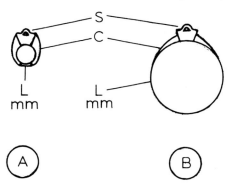

Fig. 6.31. A diagram of the actual size of the bony birth canal of a non-pregnant free-tailed bat is shown in A. The increase in length of the interpubic ligament necessary for the birth canal to accommodate a full-term foetus is shown in B. S, sacrum; C, coxal bone; L, interpubic ligament. (Crelin and Newton, *Anat. Rec.*)

showed a tendency to flattening. Thus, the ratio of the antero-posterior diameter to the greatest transverse diameter was less than 0·9 in foetuses, whereas in the female at puberty this ratio increased to 1·0–1·2, and the pelvic inlet was described as dolichopellic. This sexual dimorphism thus probably depends on gonadal secretions, and this was proved by Crelin (1958). In the mouse he found that the dimorphism develops by about 3 weeks after birth; gonadectomized animals all had female-type pelvises, so here again we find that the female type is the "neutral type" whilst maleness, consisting in a narrowing of the pelvis, requires the secretion of male steroid. When male or female mice were gonadectomized at birth and treated with testosterone they all developed male-type pelvises.

The Connective Tissue Progenitor

The remarkable changes in tissue structure taking place at the symphysis pubis suggest that the process is not one of simple growth, but rather of a transformation of cell-types, so that cells that were apparently differentiated cartilage cells are transformed into fibroblasts producing the collagen that appears in large quantities. Thus, Crelin concluded that, during pregnancy, the first effects of the altered hormonal environment was to cause the release of enzymes that converted the tissue from a relatively water-poor to a water-rich consistency, thereby permitting the release of the chondrocytes from their lacunae and allowing a re-orientation of the collagen fibres permitting separation of the pubic bones. The released chondrocytes arrange themselves along the transversely orientated collagen fibres and become fibroblasts, and after parturition revert to their original chondrocyte state. Thus, the chondrocytes are not morphologically fixed but can transform under the influence of hormones. Similarly, with the oesteocytes of the medial ends of the coxal bones at the symphysis; these are probably released from their matrix and dedifferentiate into progenitor cells that either become fibroblasts or osteoblasts, or merge with other progenitor cells to become typical multinucleate osteoclasts—the cells involved in bone resorption.

Hormonal Control

Relaxin

Hisaw (1926) was the first to postulate a hormonal control over the pubic relaxation process; he showed that, in virgin guinea-pigs, subcutaneous injections of extracts of pregnant rabbit serum produced sufficient pubic relaxation to be manifest by manual manipulation of the pelvis within 8 hours; the treatment had to be given in early post-oestrus, suggesting that "oestrogen-priming" was necessary. Extracts of rabbit's placenta, or amniotic liquor, were also effective. Since later workers were able to induce some relaxation of the symphysis pubis with oestrogen alone, or with combinations of oestrogen and progesterone, the identity of the principle in the pregnant rabbit's serum as a specific hormone, later called by Hisaw *relaxin*, was called into question, but the development of improved extraction procedures and quantitative assay methods has established relaxin as a hormone in its own right, being a protein of molecular weight in the region of 8–10,000 containing a high proportion of basic amino acids (Cohen, 1963).

Blood Levels. Relaxin has been extracted from the blood of pregnant guinea-pigs, cats, dogs as well as of rabbits, whilst the corpora lutea of the sow are a rich source, where it may be identified by immunofluorescence (Larkin *et al.*, 1977). The material appears in the blood of the rabbit during the 6th to 7th days of pregnancy, and disappears within 12–24 hours of parturition.

Role of Steroid Hormones

Castrated females do not respond to treatment with relaxin, but only after a preliminary treatment with oestrogen has brought them into an artificial state of oestrus. Castrated males could not respond, owing to the masculinizing of their pelvises at puberty; however, if they were first feminized by ovarian grafts they responded readily. Hisaw *et al.* (1944) showed that the relaxation in the guinea-pig caused by steroid hormone treatment, e.g. progesterone after oestrogen priming, required days to bring about, compared with 8 hr for relaxin, and he considered that the steroids had to induce the formation of relaxin before achieving their effects. Thus progesterone induced the appearance of relaxin in normal blood but not in that of hysterectomized females.*

Oestrogen

When relaxation is brought about by prolonged treatment with oestrogen alone, the histology of the process is different from the normal, or relaxin-induced relaxation; thus oestrogen promotes resorption of bone and proliferation of loose connective tissue whereas in the normal, or relaxin-induced situation, the prominent feature was the break-up and dissolution of collagen fibres (Talmage, 1947). The resorption of bone that occurs in gonadectomized mice on treatment with oestrogen is striking and may result eventually in resorption of the whole pubic

---

* Zarrow (1948) listed three treatments that would produce relaxation of the symphysis in guinea-pigs, namely lengthy treatment with oestrogen alone; treatment with progesterone after oestrogen priming; and treatment with relaxin after oestrogen priming.

ramus (Gardner, 1936); this may be regarded as a "programming" of the osteocytes during development to respond to oestrogen by dissolution of their matrix, thus differentiating them from the rest of the bones of the body which, in the mouse, respond by increased ossification to oestrogen treatment (Gardner and Pfeiffer, 1948). Thus pubic bones, autotransplanted to the tibia of oestrogen-treated mice, underwent resorption whilst the tibia bone, continuous with it, underwent increased ossification (Pinnell and Crelin, 1963).

Role of Progesterone

Although progesterone, given to the oestrogen-primed guinea-pig, induced pelvic relaxation, other evidence, derived from the mouse, indicates an inhibitory role for this, possibly analogous with the "progesterone block" of uterine contractions. Thus, Hall and Newton (1947), in their study of the effects of steroid hormones and relaxin on the pubic gap in ovariectomized mice, showed that progesterone did not improve the effects of oestrogen alone, nor yet of oestrogen plus relaxin. During normal pregnancy, moreover, daily administration of progesterone after ovariectomy caused an actual diminution in the pubic gap, which was prevented by giving oestrogen as well. Again, Hall (1949) inhibited the pelvic response in ovariectomized mice to oestrogen plus relaxin with progesterone. If, as has been claimed (p. 295) the decline in blood progesterone at the approach of term is the signal for the various events involved in parturition, then an inhibitory role of progesterone on pelvic relaxation fits into the picture.

Guinea-Pig. The effects of progesterone in this species may be different; thus, Figure 6.32 shows the relative lengths of the pubic symphysis ligament in guinea-pigs treated with oestrogen, oestrogen plus relaxin, and oestrogen plus relaxin plus progesterone, and it will be seen that progesterone improves activity. However, progesterone undoubtedly has an inhibitory effect, so that the sharp fall in progesterone level during days 51–60 of gestation just prior to the rapid elongation, by contrast with the continued rise in oestrogen secretion, is probably an important factor in inducing ligament growth. According to Wahl *et al.*, who measured the collagenase activity in the ligament, oestrogen plus low levels of progesterone *prime* the tissue to respond to relaxin secretion. Post-partum, the rapid falls in concentration of all three hormones cause an enhanced secretion of collagenase which brings about a rapid degradation of the ligament.

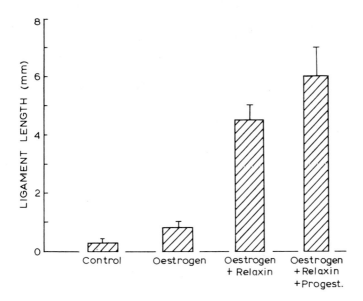

Fig. 6.32. Effects of different hormonal treatments on length of pubic symphysis ligament in non-pregnant guinea-pigs. (Wahl *et al.*, *Endocrinology.*)

Secretion of Relaxin

The hormone is extracted from both placenta and ovary but the situation varies with the species; thus in man it is produced only by the placenta and the same seems to be true of the rat.* In the rat, the cells producing the secretion, recognized by their granular nature, and the acidophilia of their granules (it will be recalled that the hormone contains mainly basic amino acids) are largely concentrated in what Selye and McKeown (1935) called the *metrial gland*; this was recognized during involution of experimental deciduomata as an organ in the mesometrial side of the uterus (Fig. 6.33).

Metrial Gland. This organ appears normally during pregnancy in the rat and clearly belongs to the maternal rather than the foetal tissue.† In ovariectomized rats Selye *et al.* (1942) caused formation of

---

* Anderson *et al.* (1975) found relaxin in luteal cells of the rat's ovary and in only one experiment out of six did they find the hormone in the metrial gland; similarly Zarrow and O'Connor (1966) found relaxin in the luteal cells of the rabbit but none in the uterus. Both these studies employed immunofluorescence to identify the hormone, and it is possible that the weak antigenicty of the hormone, whose molecular weight is only 6000 (Sherwood and O'Byrne, 1974), accounts for failure to identify it. In the sow and sheep, Bryant (1972) identified relaxin in uterus and ovary.

† Dickson and Bulmer (1961) consider that at least some of the cells in the rat's metrial gland are derived from the embryo.

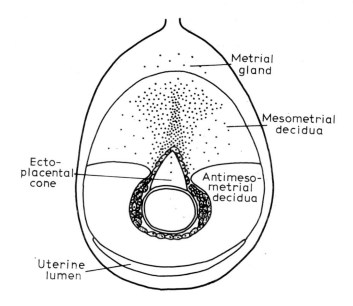

Fig. 6.33. Schematic drawing of a cross section through a pregnant rat uterus on the eleventh day of pregnancy. The black dots in metrial gland, mesometrial decidua and ectoplacental cone indicate the location and the frequency of granular cells at this stage. Similar granules in the cytoplasm of the trophoblastic cells surrounding the blastocyst are only seen in those cells facing the antimesometrial decidua. (Dallenbach-Hellweg *et al.*, *Amer. J. Anat.*)

decidua and metrial glands by treatment with progesterone, an effect that was inhibited by oestrogen. Using an antibody-fluorescence technique, Dallenbach-Hellweg *et al.* (1965) were able to localize the secreting cells in the metrial gland of the rat, the whole gland literally glowing with fluorescence if the animal had been treated on Day 15 with anti-relaxin serum and the tissue subsequently treated with a fluorescent anti-rabbit gamma-globulin serum. The ovary had no fluorescence.

Dallenbach-Hellweg *et al.* considered that the cells forming the wall of the maternal blood vessels of the metrial gland were transformed into "decidual cells" containing relaxin; their consequent close relations with the blood vessel would permit their endocrine function, the granules being emptied directly into the lumen of the vessel. In the electron microscope Wislocki *et al.* (1957) showed that the acidophilic granules of the metrial cell were characteristically different from those of the eosinophilic leucocyte, and suggested that they contained relaxin.

Corpora Lutea in the Sow

In the sow the corpora lutea seem to be an important source of relaxin, the concentration in them being some sixty times greater during pregnancy than in the luteal phase of the ovarian cycle. Belt *et al.* (1971) showed that the electron microscopical appearance of the luteal cells paralleled the secretion of the hormone; thus the endoplasmic reticulum increased in association with clusters of granules to reach a peak of activity at about 105–110 days. Between 44 and 26 hr before parturition the levels of relaxin in the luteal tissue declined to less than half their peak values, and this decline was accompanied by morphological regression, as revealed by a decrease in endoplasmic reticulum and granules.

The Cervix

This is the name given to the caudal narrowing of the uterus as it continues into the vagina (Fig. 6.34); the reduction in the amount of smooth muscle and increase in amount of collagenous tissue during this transition from uterus to vagina bring about a great decrease in extensibility so that the cervix must be regarded as a retaining device that ensures that the foetus is not born prematurely. Certainly, in some cases of patients with repeated late abortions and premature labour,

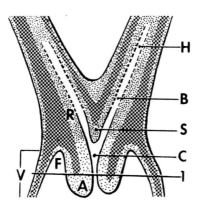

Fig. 6.34. Diagram of a midcoronal section of gestational tract of adult mouse showing lumen of each uterine horn (H) continuous with that of uterine corpus (B). Lumens are divided by midline septum (S) whose caudal edge marks the cranial limit of the uterine cervix (C) which contains a single lumen and projects into the upper vagina (V) forming deep vaginal fornices (F). Circularly arranged smooth muscle is cross-hatched; stroma of uterine horns and body indicated by (R); cervical and vaginal stroma by (A) and simple columnar epithelium by interrupted lines; squamous epithelium by continuous lines. (Leppi, *Anat. Rec.*)

biopsied pieces of cervix indicated an unusually high proportion of smooth muscle and only a little connective tissue (Roddick *et al.*, 1961).

Changed Distensibility

During birth, then, the foetus must expand this canal, and De Vaal (1947) observed that the cervices of rats became progressively more distensible during pregnancy; and similar changes have been observed in other species. Histologically there is an abrupt transition from the looser and more abundant cellular and vascular endometrial stroma of the uterine horns to the thick, interlacing bundles of compactly arranged collagenous fibres of the endocervical stroma. In non-pregnant animals the cervix was small with a canal that was practically undilatable; during pregnancy softening and dilatability developed and was most marked some 4–5 days before term; and these changes were reflected histologically in an altered distribution and arrangement of the collagenous fibres. At 10 days the network consisted of compactly arranged thick bundles of fibres, and at later stages these apparently split longitudinally to form a more reticular network that permitted the expansion of the interfibrillar ground substance, which is presumably a mucopolysaccharide. This less compact arrangement of fibres and cells was more clearly distinguished in the electron microscope (Leppi and Kinnison, 1971), whilst intense synthetic activity taking place in the fibrocytes was indicated by the prominent Golgi apparatus, and may have been indicative of synthesis of ground-substance.

Collagen and Glycosaminoglycans

Danforth *et al.* (1974) have made chemical estimations of the human cervix before and during pregnancy. There was a significant increase in the water-content of the tissue in the post-partum cervix from 74·4 per cent to 78·4 per cent, and this was associated with a decrease in the amount of collagen that could be extracted by a sodium chloride solution from 7·9 per cent in non-pregnant to 3·4 per cent in post-partum cervices. Thus, the changed structure of the tissue was not simply a result of loosening of the collagen fibres but one of dissolution, presumably by the action of collagenase. In addition there was a large increase in a typical constituent of the ground-substance, namely glycosoaminoglycans; moreover, a new material, not present in non-pregnant cervices, appeared, different from the glycosoaminoglycans extracted from these by the presence of large amounts of mannose and xylose. Hence, as Danforth *et al.* have emphasized, the pre-partum relaxation of the cervix is an active process involving considerable biochemical changes in this tissue.

Rapidity of Changes in Cervix

Fitzpatrick (1977) has pointed out that the cervix of the ewe changes within a period of five hours from an unyielding compact structure to a relaxed and yielding one. During this time the collagen content falls from 50 per cent of the dry matter to 30–40 per cent at parturition and the dry matter from 20 per cent to 11 per cent.

Role of Prostaglandins

Studies on humans with unfavourable cervices have shown that some softening may be achieved by infusions of PGF_{2a} but experimental studies on the ewe and goat have not given unequivocal results probably because of the antagonistic action of progesterone (Fitzpatrick, 1977); thus, in some sheep, intra-aortic infusion of the PG produced obvious softening but in others there was no obvious effect; when the prostaglandin was infused into the cervix, rather than intravascularly, more definite softening was achieved. In the goat, which depends on its corpora lutea throughout pregnancy, removal of these leads to abortion; however, if indomethacin is given parturition is prevented and cervical dilatation is prevented. If progesterone is given to the lutectomized goat pregnancy is once again maintained and also cervical softening and prostaglandin secretion are blocked. We have already seen that there is a linear relationship between the levels of prostaglandin in amniotic fluid of women in late pregnancy and the dilatation of the cervical canal, but of course this need not be causal.

Role of Prostaglandin in Relaxin Release. Sherwood et al. (1976) have suggested that the luteolytic action of PGF_{2a} in the pig might also be accompanied by a release of relaxin into the blood. Thus, the concentration of relaxin in the pig's blood becomes high only on the day preceding parturition (Sherwood et al., 1975), and if the release of PGF_{2a} and luteolysis are the cause of parturition it could be that the release of relaxin is likewise due to the prostaglandin. Sherwood et al. (1976) found that a dose of PGF_{2a} into the pregnant sow sufficient to induce parturition caused an eight-fold increase in blood relaxin. However, immediately preceding parturition, the blood concentration was low, possibly due to impaired synthesis.

Relaxin and the Guinea-Pig. There is evidence that progesterone is not the "hormone of pregnancy" of the guinea-pig; at any rate exogenous progesterone does not prolong the luteal phase of the ovarian cycle, nor yet does it delay parturition (Zarrow et al., 1963). Porter (1972) has shown by a cross-circulation technique that there is a substance in the pregnant guinea-pig's blood that modulates uterine contractions in the non-pregnant animal; the effects were unlikely to be due to progesterone and could be mimicked by relaxin, which might therefore be concerned in inducing the uterine contractions of labour as well as relaxation of the cervix and changes in the symphysis pubis.

Oestrogen and Relaxin

As with the changes in the symphysis pubis, so the effects of pregnancy could be largely mimicked in the ovariectomized animal by treatment with oestrogen and relaxin, although the degree of extens-

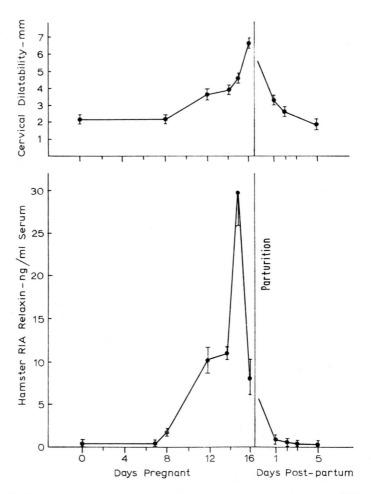

Fig. 6.35. Concentration of immunoreactive relaxin in serum and dilatability of the uterine cervix of the pregnant hamster. (O'Byrne *et al., Endocrinology.*)

ibility was not equal to that taking place in normal pregnancy (Leppi, 1964; Zarrow and Yochim, 1961). As Leppi pointed out, the changes in connective tissue morphology under the influence of hormones is analogous with those taking place in the capon's comb and the sexual skin of the primate under hormonal influence; thus, the dermal layer of the capon's comb becomes oedematous and swollen under the influence of androgens, and the perineal skin undergoes a similar change under the influence of oestrogens.

Secretion of Relaxin. As with pubic relaxation, so the softening of the cervix may be related to secretion of relaxin; Figure 6.35 shows the correlation between cervical distensibility and blood-relaxin in the guinea-pig. In the sow, the rise in relaxin level occurred only two to three days before parturition, rising to a peak at minus 14 hr and falling rapidly after parturition (Sherwood *et al.*, 1975).

Mechanical Studies

Harkness and his collaborators have studied the progressive changes in dimensions and distensibility of the cervix taking place during

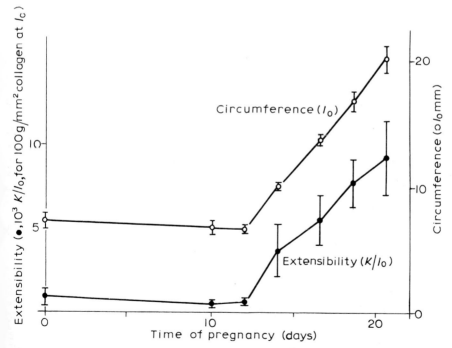

Fig. 6.36. Time-course of change in circumference (○) and extensibility (●) of the rat's cervix. (Harkness and Nightingale, *J. Physiol.*)

pregnancy or under the influence of hormones. They emphasize that the strength of a tissue is governed largely by its collagen content, so that an increase in its dimensions without a corresponding increase in proportion of collagen leads to diminished strength, i.e. greater distensibility. The parallelism between the two parameters in the rat's cervix is illustrated in Figure 6.36.

The Uterus. Cullen and Harkness (1964) pointed out that the changes in the cervix during pregnancy are essentially those that take place in the body of the uterus, namely an increase in absolute size and a decrease in proportion of collagen, effects that are likewise under the influence of the steroid hormones. The greatest increase in volume of the uterine horn was obtained under the influence of relaxin. Thus, the uterus, like the cervix, increases its size in preparation for the growth of the foetus on the one hand and the ultimate passage of the foetus through the birth canal on the other. Both events are manifest as an expansion of the connective-tissue framework and consequently in a decrease in tensile strength or an increase in distensibility.

REFERENCES
(see pp. 545 and 546 for additional references)

Abel, M. H., Krane, E. J., Thomas, A. L. and Nathanielsz, P. W. (1977). Plasma prolactin concentrations prior to delivery and the effect of cortisol induced delivery. *J. Endocrinol.*, **72**, 36 P.

Adams, W. M. and Wagner, W. C. (1970). The role of corticoids in parturition. *Biol. Reprod.*, **3**, 233–228.

Aiken, J. W. (1972). Aspirin and indomethacin prolong parturition in rats—evidence that prostaglandins contribute to expulsion of the foetus. *Nature*, **240**, 21–25.

Ainsworth, L. and Ryan, K. J. (1966). Steroid hormone transformations by endocrine organs from pregnant mammals. I. *Endocrinology*, **79**, 875–883.

Albano, J. D. M. *et al.* (1976). The development of ACTH-sensitive adenylate-cyclase activity in the foetal rabbit adrenal: a correlated biochemical and morphological study. *J. Endocrinol.*, **71**, 333–341.

Anderson, A. B. M., Flint, A. P. F. and Turnbull, A. C. (1974). Mechanism of action of glucocorticoids in induction of ovine parturition: effect on placental steroid metabolism. *J. Endocrinol.*, **61**, xxxvi.

Anderson, A. B. M., Flint, A. P. F. and Turnbull, A. C. (1975). Mechanism of action of glucocorticoids in induction of ovine parturition: effect on placental steroid metabolism. *J. Endocrinol*, **66**, 61–70.

Anderson, A. B. M., Laurence, K. M. and Turnbull, A. C. (1969). The relationship in anencephaly between the size of the adrenal cortex and the length of gestation. *J. Obstet. Gynaec. Brit. Commonw.*, **76**, 196–199.

Anderson, M. L., Long, J. A. and Hayashida, T. (1975). Immunofluorescence studies on the localization of relaxin in the corpus luteum of the pregnant rat. *Biol. Reprod.*, **13**, 499–504.

Andersson, B. (1951). Some observations on the neuro-hormonal regulation of milk-ejection. *Acta physiol. scand.*, **23**, 1–7.

Atkinson, L. E. *et al.* (1975). Circulating levels of steroids and chorionic gonadotropin during pregnancy in the rhesus monkey, with special attention to the rescue of the corpus luteum in early pregnancy. *Biol. Reprod.*, **12**, 335–345.

Barnafi, L. and Croxatto, H. (1963). The *in vitro* effect of progesterone and estrogens on the oxytocin response of rat uterus. *Acta physiol. latinam.*, **13**, 26–29.

Bashore, R. A., Smith, F. and Gold, E. M. (1970). Placental transfer and metabolism of 4-^{14}C-cortisol in the pregnant monkey. *Nature, Lond.*, **228**, 774–776.

Bassett, J. M. and Thorburn, G. D. (1969). Foetal plasma corticosteroids and the initiation of parturition in sheep. *J. Endocrinol.*, **44**, 285–286.

Batra, S. and Bengtsson, L. P. (1976). A highly efficient procedure for the extraction of progesterone from uterus and its compatability with subsequent radioimmuno-assay. *J. Ster. Biochem.*, **7**, 599–603.

Batra, S. and Bengtsson, B. (1978). Effects of diethylstilboestrol and ovarian steroids on the contractile responses and calcium movements in rat uterine smooth muscle. *J. Physiol.*, **276**, 329–342.

Bedford, C. A., Challis, J. R. G., Harrison, F. A. and Heap, R. B. (1972). The role of oestrogens and progesterone in the onset of parturition in various species. *J. Reprod. Fert.*, Suppl. 16, 1–23.

Belt, W. D., Anderson, L. L., Cavazos, L. F. and Melampy, R. M. (1971). Cyto-plasmic granules and relaxin levels in porcine corpora lutea. *J. Endocrinol*, **89**, 1–10.

Bengtsson, L. P. and Schofield, B. M. (1963). Progesterone and the accomplishment of parturition in the sheep. *J. Reprod. Fert.* **5**, 423–431.

Biggers, J. D., Curnow, R. N., Finn, C. A. and McLaren, A. (1963). Regulation of the gestation period in mice. *J. Reprod. Fert.*, **6**, 125–138.

Bland, K. P. and Donovan, B. T. (1969). Control of luteal function during early pregnancy in the guinea-pig. *J. Reprod. Fert.*, **20**, 491–501.

Caldwell, B. V., Tillson, S. A., Brock, W. A. and Speroff, L. (1972). The effects of exogenous progesterone and estradiol on prostaglandin F level in ovariectomized ewes. *Prostaglandins*, **1**, 217–228.

Carminati, P., Luzzani, F., Soffientini, A. and Lerner, L. J. (1975). Influence of day of pregnancy on rat placental, uterine, and ovarian prostaglandin synthesis and metabolism. *Endocrinology*, **97**, 1071–1079.

Cawood, M. L., Heys, R. F. and Oakey, R. E. (1976). Corticosteroid production of the human foetus: evidence from analysis of urine from women pregnant with a normal or an anencephalic foetus. *J. Endocrinol*, **70**, 117–126.

Challis, J. R. G. (1971). Sharp increase in free circulating oestrogens immediately before parturition in sheep. *Nature, Lond.*, **229**, 208.

Challis, J. R. G., Davies, I. J. and Ryan, K. J. (1975). The effects of dexamethasone and indomethacin on the outcome of pregnancy in the rabbit. *J. Endocrinol.*, **64**, 363–370.

Challis, J. G. *et al.* (1974). The concentrations of progesterone, estrone and estradiol-17β in the peripheral plasma of the rhesus monkey during the final third of gesta-tion, and after the induction of abortion. *Endocrinology*, **95**, 547–553.

Challis, J. R. G., Robinson, J. S. and Thorburn, G. D. (1977). Fetal and maternal endocrine changes during pregnancy and parturition in the rhesus monkey. *Ciba Symp.*, **No. 47**, pp. 211–227.

Chard, T. (1972). The posterior pituitary in human and animal parturition. *J. Reprod. Fert.*, Suppl. **16**, 121–138.

Chard, T. (1973). The posterior pituitary and the induction of labour. In *Endocrine Factors in Labour*. Memoirs of the Society of Endocrinology. Eds. A. Klopper and J. Gardner. C.U.P., 61–76.

Chez, R. A., Hutchinson, D. L., Salazar, H. and Mintz, D. H. (1970). Some effects of fetal and maternal hypophysectomy in pregnancy. *Am. J. Obstet. Gynec.*, **108**, 643–650.

Cohen, H. (1963). Relaxin: studies dealing with isolation, purification, and characterization. *Trans. N.Y. Acad. Sci.*, **25**, 313–336.

Cox, D. F. (1964). Relation of litter size and other factors to duration of gestation in
the pig. *J. Reprod. Fert.*, **7**, 405–407.

Crelin, E. S. (1969). The development of the bony pelvis and its changes during
pregnancy and parturition. *Trans. N.Y. Acad. Sci.*, **31**, 1049–1058.

Crelin, E. S. and Newton, E. V. (1969). The pelvis of the free-tailed bat: sexual
dimorphism and pregnancy changes. *Anat. Rec.*, **164**, 349–357.

Csapo, A. (1956). Progesterone "block". *Amer. J. Anat.*, **98**, 273–291.

Csapo, A. (1969). The four direct regulatory factors of myometrial function. *Ciba
Study Group No. 34*, 13–42.

Csapo, A. (1969). Introduction. *Ciba Study Group No. 34*, 4–42.

Csapo, A. I. (1976). Prostaglandins and the initiation of labor. *Prostaglandins*, **12**,
149–164.

Csapo, A. I. (1977). The "see-saw" theory of parturition. *Ciba Symp. No. 47*,
159–195.

Csapo, A., Erdos, T., de Maltos, C. R., Grames, E. and Moscowitz, C. (1965).
Stretch-induced uterine growth, protein synthesis and function. *Nature*, **207**,
1378–1379.

Csapo, A. I. and Wiest, W. G. (1969). An examination of the quantitative relationship between progesterone and the maintenance of pregnancy. *Endocrinology*, **85**,
735–746.

Cullen, B. M. and Harkness, R. D. (1964). Effects of ovariectomy and of hormones on
collagenous framework of the uterus. *Amer. J. Physiol.*, **206**, 621–627.

Currie, W. B. (1975). Secretion rate of prostaglandin F during induced labor in
goats. *Prostaglandins*, **9**, 867–879.

Currie, W. B. and Thorburn, G. D. (1973). Induction of premature parturition in
goats by prostaglandin $F_2\alpha$ administered into the uterine vein. *Prostaglandins*, **4**,
201–212.

Currie, W. B. and Thorburn, G. D. (1977). Parturition in goats. *J. Endocr.*, **73**,
263–278.

Dalle, M. and De Lost, P. (1976). Plasma and adrenal cortisol concentrations in
foetal, newborn and mother guinea-pigs during the perinatal period. *J. Endocrinol.*, **70**, 207–214.

Dallenbach, Hellweg-G., Battista, J. V. and Dallenbach, F. D. (1965). Immuno-
histological and histochemical localization of relaxin in the metrial gland of the
pregnant rat. *Amer. J. Anat.*, **117**, 433–450.

Danforth, D. N. *et al.* (1974). The effect of pregnancy and labor on the human
cervix: changes in collagen, glycoproteins, and glycosaminoglycans. *Amer. J.
Obstet. Gynec.*, **120**, 641–649.

Davies, I. J., Challis, J. R. G. and Ryan, K. J. (1974). Progesterone receptors in the
myometrium of pregnant rabbits. *Endocrinol.*, **95**, 165–173.

Davies, L. J. and Ryan, K. J. (1972). Comparative endocrinology of gestation.
Vit.Horm., **30**, 223–279.

De Vaal, O. M. (1946). Extensibility of ostium uteri in the rabbit. *Acta physiol.
pharmacol., Neerl.*, **14**, 79–81.

Drost, M. and Holm, L. W. (1968). Prolonged gestation in ewes after foetal adrenalectomy. *J. Endocrinol.*, **40**, 293–296.

Dupouy, J. P., Coffigny, H. and Magre, S. (1975). Maternal and foetal corticosterone
levels during late pregnancy in rats. *J. Endocrinol.*, **65**, 347–352.

Fitzpatrick, R. J. (1977). Dilatation of the uterine cervix. *Ciba Symp. No. 47*, 31–39.
Flint, A. P. F. *et al.* (1976). Bilateral adrenalectomy of lambs *in utero*: effects on maternal hormone levels at induced parturition. *J. Endocrinol.*, **69**, 433–444.
Gardner, W. U. (1936). Sexual dimorphism of the pelvis of a mouse, the effect of estrogenic hormones upon the pelvis and upon the development of scrotal hernias. *Amer. J. Anat.*, **59**, 459–483.
Gardner, W. U. and Pfeiffer, C. A. (1938). Skeletal changes in mice receiving estrogen. *Proc. Soc. exp. Biol. Med.*, **37**, 678–679.
Goy, R. W., Hoar, R. M. and Young, W. C. (1957). Length of gestation in the guinea-pig with data on the frequency and time of abortion and stillbirth. *Anat. Rec.*, **128**, 747–757.
Gustavii, B. (1972). Labour, a delayed menstruation? *Lancet*, 2, 1149–1150.
Gustavii, B. (1974). Sweeping of the fetal membranes by a physiologic saline solution: effect on decidual cells. *Am. J. Obstet. Gynec.*, **120**, 531–536.
Gustavii, B. (1975). Release of lysosomal acid phosphatase into the cytoplasm of decidual cells before the onset of labour in humans. *Br. J. Obstet. Gynecol.*, **82**, 177–181.
Gustavii, B. and Brunk, U. (1972). A histological study of the effect on the placenta of intra-amniotically and extra-amniotically injected hypertonic saline in therapeutic abortion. *Acta. obstet. gynec. scand.*, **51**, 121–125.
Hall, K. (1947). The effects of pregnancy and relaxin on the histology of the pubic symphysis in the mouse. *J. Endocrinol.*, **5**, 174–182.
Hall, K. (1949). The role of progesterone in the mechanism of pelvic relaxation in the mouse. *Quart. J. exp. Physiol.*, **35**, 65–75.
Hall, K. and Newton, W. H. (1947). The effect of oestrone and relaxin on the X-ray appearance of the pelvis of the mouse. *J. Physiol.*, **106**, 18–27.
Harkness, R. D. and Nightingale, M. A. (1962). The extensibility of the cervix uteri of the rat at different times of pregnancy. *J. Physiol.*, **160**, 214–220.
Harney, P. J., Sneddon, J. M. and Williams, K. I. (1974). The influence of ovarian hormones upon the motility and prostaglandin production of the pregnant rat uterus. *J. Endocrinol*, **60**, 343–351.
Heap, R. B. *et al.* (1977). Progesterone and oestrogen in pregnancy and parturition: comparative aspects and hierarchical control. *Ciba Symp. No. 47*, 127–156.
Hendricks, C. H. (1964). Patterns of fetal and placental growth: the second half of normal pregnancy. *Obstet. Gynecol.*, **24**, 357–365.
Hillier, K., Calder, A. A. and Embrey, M. P. (1974). Concentrations of prostaglandin $F_2\alpha$ in amniotic fluid and plasma in spontaneous and induced labours. *J. Obstet. Gynaec. Brit. Commonw.*, **81**, 257–263.
Hindson, J. C., Schofield, B. M. and Ward, W. R. (1969). The effect of progesterone on recorded parturition and on oxytocin sensitivity in the sheep. *J. Endocrinol.*, **43**, 207–213.
Hisaw, F. L. (1926). Experimental relaxation of the pubic ligament of the guinea-pig. *Proc. Soc. exp. Biol. Med.*, N.Y., **23**, 661–663.
Hisaw, F. L. *et al.* (1944). Importance of the female reproductive tract in the formation of relaxin. *Endocrinology*, **34**, 122–134.
Hisaw, F. L. and Zarrow, M. X. (1950). The physiology of relaxin. *Vit. Horm.*, **8**, 151–178.
Holm, L. W., Parker, H. R. and Galligan, S. J. (1961). Adrenal insufficiency in postmature Holstein calves. *Amer. J. Obst. Gynecol.*, **81**, 1000–1008.
Illingworth, D. V., Heap, R. B. and Perry, J. S. (1970). Changes in the metabolic

clearance rate of progesterone in the guinea-pig. *J. Endocrinol.*, **48**, 409–417.

Jones, C. T., Ritchie, J. W. K. and Flint, A. P. F. (1977). Some experiments on the role of the foetal pituitary in the maturation of the foetal adrenal and the induction of parturition in sheep. *J. Endocrinol*, **72**, 251–257.

Karim, S. M. M. (1968). Appearance of prostaglandin F2a in human blood during labour. *Brit. med. J.* (**iv**), 618–621.

Karim, S. (1971). Action of prostaglandin in the pregnant woman. *Am. N.Y. Acad. Sci.*, **180**, 483–498.

Karim, S. M. M. and Devlin, J. (1967). Prostaglandin content of amniotic fluid during pregnancy and labour. *J. Obstet. Gynaec. Brit. Commonw.*, **64**, 230–234.

Karim, S. M. M. and Filshie, G. M. (1970). Therapeutic abortion using prostaglandin F_2a. *Lancet* (**1**), 157–159.

Karim, S. M. M. and Filshie, G. M. (1970). Use of prostaglandin E_2 for therapeutic abortion. *Brit. med. J.*, (**3**), 198–200.

Karim, S. M., Trussell, R. R., Patel, R. C. and Hillier, K. (1968). Response of pregnant human uterus to prostaglandin-F_2a-induction of labour. *Brit. Med. J.*, (**iv**), 621–623.

Keirse, M. J. N. C., Flint, A. P. F. and Turnbull, A. C. (1974). F Prostaglandins in amniotic fluid during pregnancy and labour. *J. Obstet. Gynaec. Brit. Commonw.*, **81**, 131–135.

Kirton, K. T., Phariss, B. B. and Forbes, A. D. (1970). Some effects of prostaglandins E_2 and F_2a on the pregnant rhesus monkey. *Biol. Reprod.*, **3**, 163–168.

Klopper, A. (1973). The role of oestrogens in the onset of labour. In *Endocrine Factors in Labour*. Memoirs of the Society of Endocrinology. Eds. A. Klopper and J. Gardner, C.U.P., 47–59.

Koering, M. J. and Kirton, K. T. (1973). The effects of prostaglandin F_2a on the structures and function of the rabbit ovary. *Biol. Reprod.*, **9**, 226–245.

Kuriyama, H. and Suzuki, H. (1976). Effects of prostaglandin E_2 and oxytocin on the electrical activity of hormone-treated and pregnant rat myometria. *J. Physiol.*, **260**, 335–359.

Landesman, R. *et al.* (1971). The relaxant action of vitodrine, a sympathomimetic amine, on the uterus during term labour. *Am. J. Obstet. Gynec.*, **110**, 111–114.

Lanman, J. T. *et al.* (1975). Ovarian and placental origins of plasma progesterone following fetectomy in monkeys (*Macaca mulatta*) *Endocrinology*, **96**, 591–597.

Larkin, L. H., Fields, P. A. and Oliver, R. M. (1977). Production of antisera against electrophoretically separated relaxin and immunofluorescent localization of relaxin in the porcine corpus luteum. *Endocrinology*, **101**, 679–685.

Lee, C. and Jacobsen, H. I. (1971). Uterine estrogen receptor in rats during pubescence and the estrous cycle. *Endocrinology*, **88**, 596–601.

Leppi, T. J. (1964). A study of the uterine cervix of the mouse. *Anat. Rec.*, **150**, 51–66.

Leppi, T. J. and Kinnison, P. A. (1971). The connective tissue ground substance in the mouse uterine cervix: an electron microscopic histochemical study. *Amer. J. Anat.*, **170**, 97–117.

Lewis, R. B. and Schulman, J. D. (1973). Influence of acetylsalicylic acid, an inhibitor of prostaglandin synthesis, on the duration of human gestation and labour. *Lancet*, (**2**), 1159–1160.

Liggins, G. C. (1968). Premature parturition after infusion of corticotrophin or cortisol into foetal lambs. *J. Endocrinol.*, **42**, 323–329.

Liggins, G. C. *et al.* (1977). Parturition in the sheep. *The Fetus and Birth. Ciba Symp.*, No. 47, pp. 5–25.

Liggins, G. C., Fairclough, R. J., Grieves, S. A., Kendall, J. Z. and Knox, B. S. (1973). The mechanism of initiation of parturition in the ewe. *Rec. Progr. Horm. Res.*, **29,** 111–150.

Liggins, G. C., Forster, C. S., Grieves, S. A. and Schwartz, A. L. (1977). Control of parturition in man. *Biol. Reprod.*, **16,** 39–56.

Liggins, G. C., Kennedy, P. C. and Holm, L. W. (1967). Failure of initiation of parturition after electrocoagulation of the pituitary of the fetal lamb. *Amer. J. Obst. Gynecol.*, **98,** 1080–1086.

Liggins, G. C. and Vaughan, G. S. (1973). Intravenous infusion of salbutamol in the management of premature labour. *J. Obstet. Gynaec. Brit. Commonw.*, **80,** 29–32.

Liskowski, L., Wolf, R. C., Chandler, S. and Meyer, R. K. (1970). Urinary estrogen excretion in pregnant rhesus monkeys. *Biol. Reprod.*, **3,** 55–60.

Madill, D. and Bassett, J. M. (1973). Corticosteroid release by adrenal tissue from foetal and newborn lambs in response to corticotrophin stimulation in a perfusion system *in vitro. J. Endocrinol.*, **58,** 75–87.

Manaugh, L. C. and Novy, M. J. (1976). Effects of indomethacin on corpus luteum function and pregnancy in rhesus monkeys. *Fert. Steril.*, **27,** 588–598.

Maule Walker, F. M. and Poyser, N. L. (1974). Production of prostaglandins by the early pregnant guinea-pig uterus *in vitro. J. Endocrinol.*, **61,** 265–271.

Milic, A. B. and Adamsons, K. (1969). The relationship between anencephaly and prolonged pregnancy. *J. Obstet. Gynaec. Brit. Commnw.*, **76,** 102–111.

Mitchell, M. D., Flint, A. P. F. and Turnbull, A. C. (1975). Stimulation by oxytocin of prostaglandin F levels in uterine venous effluent in pregnant and puerperal sheep. *Prostaglandins*, **9,** 47–55.

Morton, D. G. (1942). Observations of the development of pelvic conformation. *Amer. J. Obstet. Gyn.*, **44,** 799–816.

Mueller-Heubach, E., Myers, R. E. and Adamsons, K. (1972). Effect of adrenalectomy on pregnancy length in the rhesus monkey. *Am. J. Obstet. Gynec.*, **112,** 221–226.

Nathanielsz, P. W., Abel, M. and Smith, G. W. (1972). Initiation of parturition in the rabbit by intra-aortic infusion of prostaglandin F_2a. *J. Endocrinol.*, **55,** 617–618.

Nathanielsz, P. W., Jack, P. M. B., Krane, E. J., Thomas, A. L., Ralter, S. and Rees, L. H. (1977). The role and regulation of corticotropin in the fetal sheep. *Ciba Sym.*, No. 47., pp. 73–91.

Newton, W. H. (1935). "Pseudo-parturition" in the mouse, and the relation of the placenta to post-partum oestrus. *J. Physiol.*, **84,** 196–207.

Newton, W. H. and Lits, F. J. (1938). Criteria of placental endocrine activity in the mouse. *Anat. Rec.*, **72,** 333–345.

Novy, M. J. (1977). Endocrine and pharmacological factors which influence the onset of labour in rhesus monkeys. *Ciba Symp.*, **No. 47.,** 259–288.

Novy, M. J., Cook, M. J. and Manaugh, L. (1974). Indomethacin block of normal onset of parturition in primates. *Am. J. Obstet. Gynec.*, **118,** 412–416.

O'Byrne, E. M., Sawyer, W. K., Butler, M. C. and Steinetz, B. G. (1976). Serum immunoreactive relaxin and softening of the uterine cervix in pregnant hamsters. *Endocrinology*, **99,** 1333–1335.

Pickles, V. R., Hall, W. J., Best, F. A. and Smith, G. N. (1965). Prostaglandin in endometrium of menstrual fluid from normal and dysmenorrhoeic subjects. *J. Obst. Gynaec. Brit. Commw.*, **72,** 185–192.

Pinnell, S. R. and Crelin, E. S. (1963). Fate of pelvic bone autotransplanted to the tibia in estrogen-treated adult female mice. *Anat. Rec.*, **145**, 345 (Abstr.).

Porter, D. G. (1972). Myometrium of the pregnant guinea-pig: the probable importance of relaxin. *Biol. Reprod.*, **7**, 458–464.

Poyser, N. L., Horton, E. W., Thompson, C. J. and Los, M. (1970). Identification of prostaglandin $F_2\alpha$ released by distension of guinea-pig uterus *in vitro*. *J. Endocrinol.*, **48**, xliii.

Roberts, J. S., McCracken, J. A., Gavagan, J. E. and Soloff, M. S. (1976). Oxytocin-stimulated release of prostaglandin $F_2\alpha$ from ovine endometrium *in vitro*: correlation with estrous cycle and oxytocin-receptor binding. *Endocrinology*, **99**, 1107–1114.

Roberts, J. S. and Share, L. (1968). Oxytocin in plasma of pregnant, lactating and cycling ewes during vaginal stimulation. *Endocrinology*, **83**, 272–278.

Roberts, J. S. and Share, L. (1969). Effects of progesterone and estrogen on blood levels of oxytocin during vaginal distention. *Endocrinology*, **84**, 1076–1081.

Roberts, J. S. and Share, L. (1970). Inhibition by progesterone of oxytocin secretion during vaginal stimulation. *Endocrinology*, **87**, 812–815.

Roddick, J. W., Buckingham, J. C. and Danforth, D. N. (1961). The muscular cervix—a cause of incompetency in pregnancy. *Obst. Gynecol.*, **17**, 562–565.

Saldivar, J. T. and Melton, C. E. (1966). Effects *in vivo* and *in vitro* of sex steroids on rat myometrium. *Amer. J. Physiol.*, **211**, 835–842.

Schofield, B. M. (1964). Myometrial activity in the pregnant guinea-pig. *J. Endocrinol.*, **30**, 347–354.

Selye, H., Borduas, A. and Masson, G. (1942). Studies concerning the hormonal control of deciduomata and metrial glands. *Anat. Rec.*, **82**, 199–209.

Selye, H. and McKeown, T. (1935). Studies on the physiology of the maternal placenta in the rat. *Proc. Roy. Soc. B.*, **119**, 1–29.

Sharma, S. C., Hibbard, B. M., Hamlett, J. D. and Fitzpatrick, R. J. (1973). Prostaglandin $F_2\alpha$ concentrations in peripheral blood during the first stage of normal labour. *Brit. Med. J.* (i), 709–711.

Sherwood, O. D., Chang, C. C., Bevier, G. W. and Dzuik, P. J. (1975). Radioimmunoassay of plasma relaxin throughout pregnancy and at parturition in the pig. *Endocrinology*, **97**, 834–837.

Sherwood, O. D. *et al.* (1976). Relaxin concentrations in pig plasma following the administration of prostaglandin $F_2\alpha$ during late pregnancy. *Endocrinology*, **98**, 875–879.

Sommerville, I. F. (1969). Discussion of Csapo's paper: *Ciba Study Group*, No. 34, pp. 13–42.

Steinetz, B. G. *et al.* (1960). Bioassay of relaxin using a reference standard. *Endocrinology*, **67**, 102–115.

Strauss, J. F. *et al.* (1975). On the role of prostaglandin in parturition in the rat. *Endocrinology*, **96**, 1040–1043.

Szego, C. M. *et al.* (1971). The lysosomal membrane complex. *Biochem. J.*, **123**, 523–538.

Talmage, R. V. (1947). Changes produced in the symphysis pubis of the guinea-pig by the sex steroids and relaxin. *Anat. Rec.*, **99**, 91–113.

Thorburn, G. D. *et al.* (1972). Parturition in the goat and sheep: changes in corticosteroids, progesterone, oestrogens and prostaglandin. F. *J. Reprod. Fert.*, Suppl. **16**, 61–84.

Turnbull, A. C. and Andersen, A. B. M. (1969). The influence of the foetus on myometrial contractility. *Ciba Study Group*, **34**, 106–112.

Turnbull, A. C. *et al.* (1974). Significant fall in progesterone and rise in oestradiol levels in human peripheral plasma upon onset of labour. *Lancet*, (**1**), 101–104.

Umo, I., Fitzpatrick, R. J. and Ward, W. R. (1976). Parturition in the goat: plasma concentrations of prostaglandin F and steroid hormones and uterine activity during late pregnancy and parturition. *J. Endocrinol*, **68**, 383–389.

Van Rensburg, S. J. (1963). Endocrinological aspects of habitually aborting Angora-goat ewes. *S. A. Med. J.*, **37**, 1114–1115.

Wahl, L. M., Blandau, R. J. and Page, R. C. (1977). Effect of hormone on collagen metabolism and collagenase activity in the pubic symphysis ligament of the guinea-pig. *Endocrinology*, **100**, 571–579.

Waltman, R., Tricomi, V. and Palav, A. (1972). Midtrimester hypertonic saline induced abortion effect of indomethacin on induction/abortion time. *Amer. J. Obstet. Gynecol.*, **114**, 829–831.

Weiner, R. and Kaley, G. (1972). Lysosomal fragility induced by prostaglandin F_2a. *Nature, London*, **236**, 46–47.

Wiest, W. G. (1970). Progesterone and 20-a-hydroxypregn-4-en-3-one in plasma, ovaries and uterus during pregnancy in the rat. *Endocrinology*, **87**, 43–48.

Willman, E. A. and Collins, W. P. (1976). Distribution of prostaglandins E_2 and F_2a within the foetoplacental unit throughout human pregnancy. *J. Endocrinol*, **69**, 413–419.

Winters, A. J., Colston, C., MacDonald, P. C. and Porter, J. (1975). Fetal plasma prolactin levels. *J. clin. Endocrinol.*, **41**, 626–629.

Wislocki, G. B., Weiss, L. P., Burgos, M. H. and Ellis, R. A. (1957). The cytology, histochemistry and electron microscopy of the granular cells of the metrial gland of the gravid rat. *J. Anat.*, **91**, 130–140.

Yoshinaga, K., Hawkins, R. A. and Stocker, J. F. (1969). Estrogen secretion by the rat ovary *in vivo* during the estrous cycle and pregnancy. *Endocrinology*, **85**, 103–112.

Zarrow, M. X. (1948). The role of the steroid hormones in the relaxation of the symphysis pubis of the guinea-pig. *Endocrinology*, **42**, 129–140.

Zarrow, M. X., Anderson, N. C. and Callantine, M. R. (1963). Failure of pro-gestogèns to prolong pregnancy in the guinea-pig. *Nature, London*, **198**, 690–692.

Zarrow, M. X. *et al.* (1960). Local action of placental progestogen on uterine muscu-lature of the rabbit. *Fert. Steril.*, **11**, 370–372.

Zarrow, M. X. and Yochim, J. (1961). Dilatation of the uterine cervix of the rat accompanying changes during the estrous cycle, pregnancy and following treat-ment with estradiol, progesterone and relaxin. *Endocrinology*, **69**, 292–304.

CHAPTER 7

Lactation

Although lactation, i.e. the secretion of milk in quantities to feed the newborn, obviously begins after parturition, the development of the mammary glands to permit this secretion takes place during pregnancy, so that logically this consideration of control mechanisms should perhaps be included in the previous section. Furthermore, the lactation of pregnancy is imposed on a preliminary development of the mammary glands during the prepuberal stage, the feature of this growth being proliferation of ducts, as opposed to the *lobulo-alveolar growth* that is associated with pseudopregnancy and pregnancy. From the aspect of control mechanisms, mammogenesis has been divided into the earlier ductal growth and the later lobulo-alveolar growth, leading finally into a developmental stage culminating in the secretion of milk. This *lactogenesis* has several aspects, namely the trigger that provokes secretion in quantity, the maintenance of secretion, or *galactopoiesis*, and the milk-ejection reflex that permits the release of the secretion from the gland.

Prolactin

Pigeon Crop-Sac Assay

It was early recognized that the pituitary contained a principle that would stimulate mammary growth and promote secretion of milk, a convenient mode of assay being to treat the female pigeon with pituitary extracts and observe the changes in the crop-sac, i.e. the organ that produces a form of milk in this bird. Thus, Riddle *et al.* (1933), when they demonstrated the existence of a pituitary principle, different from growth hormone and free from gonadotrophin hormones, injected their preparations into the tissue adjacent to the gland and measured the weight after some days. On average, the unstimulated gland had a weight of 130–200 mg, whilst after optimal treatment with the extract, the weight could be as high as 1000 mg. When the logarithm of the dose was plotted against the crop-gland weight, a straight line was obtained. They showed that preparations that stimulated the crop-gland were the only ones that stimulated lactation in rats and guinea-pigs, and that the hormone was effective after castration and hypo-physectomy. They called the hormone *prolactin*.*

Cells of Origin

The anterior pituitary cells responsible for secreting prolactin are recognizable by their morphological changes taking place during lactation; they belong to the group of acidophils, and are characterized by their affinity for erythrosin and in the electron microscope, by the large size of their granules (700 mμ diameter compared with 335–450 mμ for cells secreting growth hormone). When a lactating rat is separated from its litter for some hours the uptake of dye is much more intense, indicating accumulation of secretion; within half an hour of placing the litter in the cage there is almost complete disappearance of the granules. In the electron microscope there is clear evidence in the Golgi apparatus of accumulation of material indicative of active synthesis of new granular matter (Pasteels, 1963).

Subsequent preparative studies have led to the isolation of the pure protein, that from sheep having a molecular weight of 23,300 containing 198 amino-acid residues whose sequence has been derived (Fig. 7.2, p. 365).

Prolactin, Growth Hormone and Human Chorionic Lactogen

The main question at issue has been whether there is, in primates including the human, a specific pituitary hormone, *prolactin*, or whether the human growth hormone, HGH, and human chorionic somato-trophic hormone (HCS) or human placental lactogen (HPL), isolated from the placenta, are not the hormones also responsible for stimulation of mammary gland growth and lactation; these hormones having very definite lactogenic and mammotrophic effects and being, moreover, very similar chemically to the prolactin isolated from sub-primate pituitaries.

---

* Prolactin has also been called luteotrophic hormone; this is because, in the rat and mouse, it promotes formation of the corpora lutea and secretion of progestins. It may thus produce a condition of pseudopregnancy in rats when injected. The name is unfortunate since this action is limited to only a few species; however, its luteotrophic action in the rat or mouse provides a useful method of assay, based on the increase in luteal cell size.

Separation of Prolactin

Recently, however, extracts of anterior pituitary have been subjected to gel-filtration; this gives a single peak, which may be distinguished into material that is readily precipitated by an antiserum to sheep prolactin rather than by one to growth hormone (GH), or human placental lactogen (HPL). This suggests that the human pituitary does, indeed, contain a prolactin different from GH, the two being distinguished immunologically, the pituitary prolactin being similar to that of other species, and therefore precipitable by antibodies to them. The human pituitary prolactin was separated from growth hormone by passing through a Sephadex column containing attached to it an anti-HPL; this attracted the GH strongly so that practically all of it was kept back while the pituitary prolactin passed through.

Blood Levels. Using this separative technique Friesen *et al.* (1972) obtained the values for human blood at different ages and in different physiological conditions illustrated in Table I.

Cultured Pituitary. *In vitro* studies of the pituitary also permit a ready distinction between GH and prolactin synthesis because of the differing modes of release of the hormones; thus prolactin release is held in check by a hypothalamic *inhibitory* factor, PIF, by contrast with the *releasing* factor, GRF, for growth hormone; consequently the *in vitro* cultured gland will tend to release its synthesized prolactin into the medium since it is no longer under the influence of a hypothalamic inhibitory factor, whilst GH will tend to accumulate in the gland (Fig. 7.1).

Human Placental Lactogen

Some 66 per cent of the protein synthesized by the human placenta *in vitro* represents human placental lactogen (HPL) so called by Josimovitch and MacLaren (1962) but also called by Li (1972) human chorionic somatomammotrophin (HCS).* The lactogenic activities of HPL and HGH are identical (Chadwick *et al.*, 1961) whereas the growth promoting activities differ by as much as 100 to 1000-fold (Friesen, 1965) so that the two hormones are different, and thus different from pituitary prolactin. The amino-acid sequences of all three, HPL, HGH, prolactin, are remarkably similar (Figs 7.2–7.4), as we should expect in view of the difficulty in separating them; so far as HPL and

---

* The name *human chorionic somatomammotrophin* (HCS) indicates the origin of the hormone and both its growth-promoting (somatotrophin) and lactogenic, activities (mammotrophin); a preference is expressed for the simpler term, *human placental lactogen* (HPL) by those who have to use it often (see Beck, 1972).

Table I

Serum prolactin values in monkeys and humans measured by radioimmunoassay
(Friesen *et al.*, 1972)

	Number of subjects	Serum prolactin ng/ml	
		Mean	Range
Children	36	11	0–17
Adults (males)	42	10	0–28
Menstrual cycle	9		
Follicular		10	4·2–21
Luteal		11	4·9–42
Acromegaly	13	16	3–26
Galactorrhoea	21	100	12–1800
Idiopathic hypopituitary	14	10	0–32
Pregnancy			
First trimester	24	25	7–47
Second trimester	96	50	6–350
Third trimester	102	134	36–600
Term	19	207	11–600
Postpartum (6 weeks)	25	10	0–14
Newborn			
Cord blood	19	258	120–500
One week	7	192	40–400
Nursing mother			
1–3 weeks			
Before suckling	6	14	8–20
30 min after		259	175–400
60 min after		133	75–200
3–5 days			
Before suckling	2	130	120–140
30 min after		230	200–260
60 min after		185	180–190
Insulin (10 patients)	No response		
(2 patients)	Maximum increment of 20 ng/ml		
Arginine (6 patients)	No response		
Anaesthesia and surgery			
Human subjects	4		6–150
Pregnant monkeys	6		30–1000
Non-pregnant monkeys	4		2–60

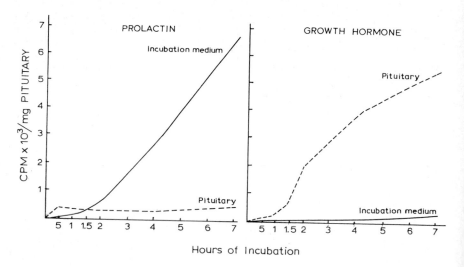

Fig. 7.1. Differing behaviours of the incubated pituitary gland in relation to synthesis and release of prolactin (left) and growth hormone (right). Note that large amounts of prolactin are released into the medium, presumably due to absence of prolactin inhibiting factor (PIF). (MacLeod and Leymeyer, *Ciba Symp.*)

HGH are concerned these have 190 amino-acid residues* and the similarity extends over a very large range, the differences between them being manifest at the amino-terminal end, so that we may assume that GH activity is probably associated with this end, whilst lactogenic activity is determined by another region. It is very likely indeed that the two molecules are derived from a common ancestor, as predicted by Sherwood (1967); they are also closely related to ovine prolactin but their relation to human pituitary prolactin has not yet been established. If growth hormone activity depends on only a small portion of the molecule, there is a good possibility of finding a fragment with high activity that could be easily synthesized (Sherwood *et al.*, 1972).

Species Differences. Lyons was the first to imply that the rat placenta contained mammotrophic as well as luteotrophic activity; Kelly *et al.* (1976) have summarized the placental lactogenic activity found in a number of species during the course of pregnancy, namely man, monkey, baboon, chinchilla, hamster, rat, mouse, sheep and goat. In the hamster, goat, sheep, monkey and human the level remained

* 160 of these amino-acid residues are in identical positions, whilst differences in amino acids, in the remaining positions, could have been brought about by a single base-change in the code determining them (Li, 1972).

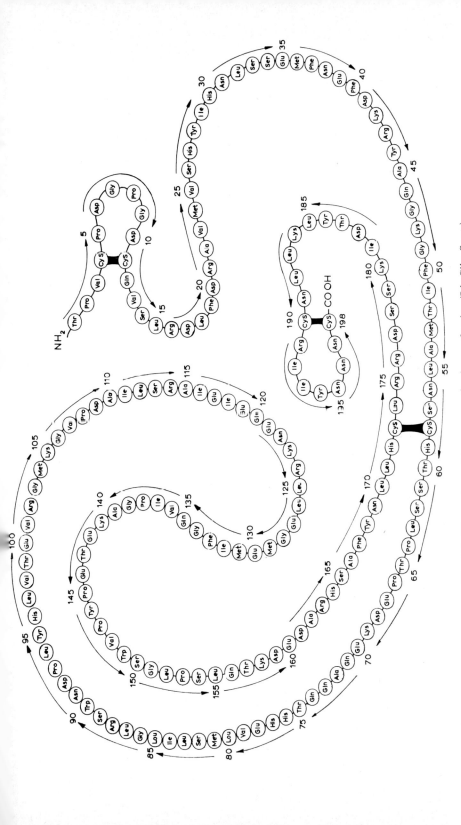

Fig. 7.2. The amino-acid sequence of ovine prolactin. (Li, *Ciba Symp.*)

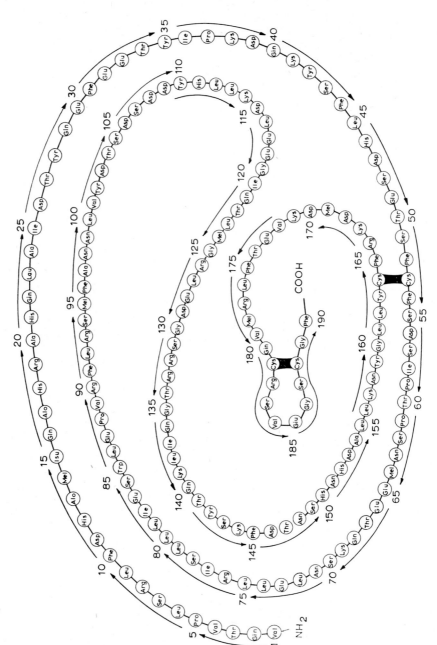

Fig. 7.3. The amino-acid sequence of human placental lactogen (HPL) or human chorionic somatomammotrophin (HCS). (Li, *Ciba Symp.*)

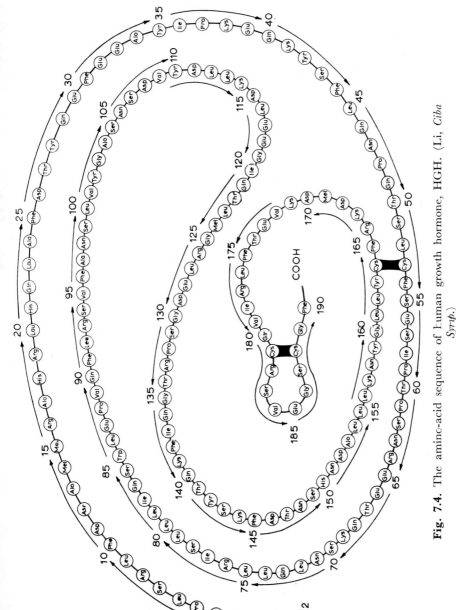

Fig. 7.4. The amino-acid sequence of human growth hormone, HGH. (Li, *Ciba Symp.*)

elevated until term, whereas in the guinea-pig the level declined after reaching a peak at mid-pregnancy; in the mouse and rat there were two peaks of activity, whilst in the cow the level was low at all times.

Prolactin Receptors

Turkington (1970) showed that RNA synthesis of isolated mammary cells could be accelerated by prolactin when it was covalently linked to sepharose particles, and concluded that the action was on the cell membrane; and subsequent studies on cell membrane fractions derived from mammary gland homogenates (Shiu and Friesen, 1974) indicate that the "receptor" for prolactin is attached to cell membranes, being found in the material separated out by 105,000 g centrifugation rather than in the supernatant as with other types of receptor material. Binding to this membrane material is apparently obligatory for prolactin function, since an anti-prolactin not only inhibited binding to membrane receptor material but also inhibited the incorporation of ^{3}H-leucine into casein, and the amino-acid transport into explants of pseudopregnant rabbits' mammary glands (Shiu and Friesen, 1976).

Rabbit Mammary Gland. Figure 7.5 shows the prolactin receptor

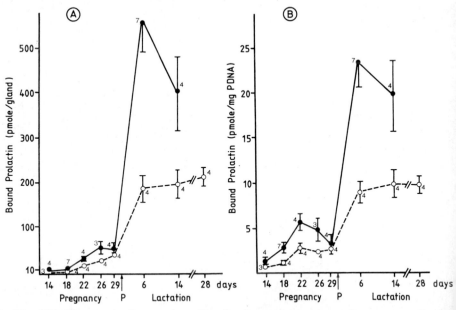

Fig. 7.5. Changes in the number of prolactin receptors per total mammary gland (panel A) or per mg DNA phosphorus, PDNA, (panel B) during the course of pregnancy and lactation in the rabbit. Dashed lines, control animals; full lines, animals treated with bromcryptine. (Djiane *et al.*, *Endocrinology*.)

activity, measured in terms of number of binding sites per milligramme of protein, of mammary glands of the rabbit during pregnancy and postnatally. There is a sharp increase over the period immediately following parturition. In order to show that the changes in binding of the labelled prolactin were due to changes in the receptor material rather than to changes in occupancy of the sites already present, the depressor of prolactin secretion, bromocriptine (CB 154) was given into a separate set of animals; when this is present, the increased binding activity becomes much more apparent.

Rat Mammary Gland. In a study on the rat, Holcomb *et al.* (1976), observed a very low degree of binding of ovine ^{125}I-prolactin to slices of mammary gland tissue; as illustrated by Figure 7.6 at parturition there was a striking increase in binding which was maintained throughout the 22 days of lactation. The low binding during pregnancy was surprising since, during the second half of this period, mammary development is rapid, and it was surmised that the high levels of placental lactogen prevailing at this period (Kelly *et al.*, 1976) were masking the binding sites, in the sense that they competed for the

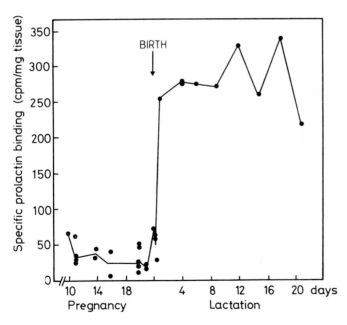

Fig. 7.6. Specific binding of ^{125}I-labelled prolactin to rat mammary gland during pregnancy and lactation. Specific binding is the difference between radioactivity bound in the presence of excess unlabelled prolactin and that bound in its absence. (Holcomb *et al.*, *Biochim. biophys. Acta.*)

radioactive material in the medium. By removing the placenta at different stages of pregnancy, the concentration of placental lactogen could be caused to fall, and as Figure 7.7. shows, removal resulted in striking increases in binding. Following birth, of course, lactation is accompanied by high levels of circulating prolactin, but these do not appear to affect the binding capacity as manifest in Figure 7.6.

Hormonal Requirements for Mammogenesis

Tubular Development

In immature animals, injections or local implantations of oestrogen and progesterone will induce tubular proliferation, but this effect of the ovarian hormones relies on the presence of an intact pituitary, so that it depends either on a synergistic action with pituitary hormones or perhaps on a feedback effect on the hypothalamus. Experimentally, then, we may remove the ovaries and pituitary of immature rats (Lyons *et al.*, 1958) and try to find a substitution therapy that will permit the normal prepuberal proliferation of ducts. Oestrogen was found to be useless alone, whereas growth hormone (GH) was highly

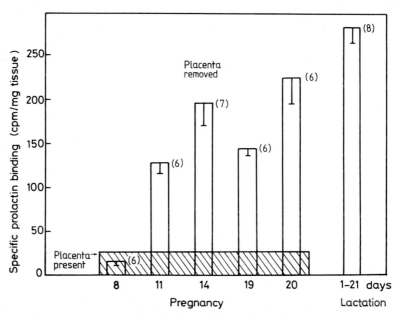

Fig. 7.7. Specific binding of ^{125}I-labelled prolactin to pregnant rat mammary gland after ovariectomy and hysterectomy. Each bar represents the mean $\pm$ S.E. The shaded area indicates the mean level of prolactin binding in mammary tissue of intact animals. Parentheses indicate numbers of animals tested. (Holcomb *et al.*, *Biochim. biophys. Acta.*)

effective, whilst optimum action was obtained by combined oestrogen and growth hormone treatment. When rats were adrenalectomized as well—*triply operated*—then the best growth was obtained with GH, oestrogen and a glucocorticoid, DCA.

Lobulo-Alveolar Growth

In the pregnant animal, ovariectomy prevents this stage in mammary development, which can be brought about by replacement therapy with oestrogen plus progesterone; furthermore, in virgin females, dosage with these ovarian hormones can bring about some lobulo-alveolar growth, but once again only in the presence of an intact pituitary. When the animal is hypophysectomized as well as ovariectomized—doubly operated—then prolactin together with oestrogen and progesterone restore full growth. In view of the role of growth hormone in the development of the tubular system, which continues during pregnancy, it is reasonable to expect this hormone to be involved in this stage of development. In fact, in their study of doubly operated *male* rats in which implants of hormone mixtures were made locally in the mammary glands, Lyons *et al.*, showed that full lobulo-alveolar growth required all four hormones studied, namely prolactin (MH), growth hormone (STH), oestrogen (OE), and progesterone (P) (Table II).

TABLE II

Mammary reactions in hypophysectomized male rats to pellets containing mammotrophin (MH), growth hormone (GH), oestrogen (OE) and progesterone (P) in cholesterol or in the listed combinations* implanted in right mammary gland. (Lyons *et al.*, 1958)

	No. of	MH	GH	OE	P	Lobulo-alveolar growth Right ++	+	—	Left +	—
Group	animals	(mg.)	(mg.)	(μg.)	(mg.)					
1	5	28	2·3	28	42	5				5
2	4	28	—	28	42	1	2	1		4
3	6	28	2·3	—	42			6		·6
4	4	28	2·3	28	—			4		4
5	5	—	2·3	28	42			4		4
6	4	28	—	—	42			4		4
7	4	—	—	—	42			4		4

* OE; P; MH; GH; OE + P; OE + MH; OE + GH; MH + GH; all negative for lobulo-alveolar growth.

Lactational Mammary Growth

Lyons *et al.* pursued their study of the hormonal requirements for mammogenesis to the stage leading to the actual production of milk as evidenced histologically by the presence of lobules filled with milk. From a histological point of view at any rate this process of development from a prolactational to a lactating mammary gland is gradual during the last part of pregnancy when the two processes seem to compete for dominance. As we shall see, however, when lactation is assessed biochemically by the synthesis of lactose, the onset is more sudden. Lyons *et al.*'s experiments on replacement therapy for operated animals in which the earlier stages of growth had been achieved indicated the need for prolactin and corticosteroids to bring the glands to the stage where milk secretion could be seen. Figure 7.8 summarizes the general concept of mammogenic control put forward by Lyons *et al.* on the basis of their studies. Thus oestrogen secretion resulting from FSH secretion synergizes with growth hormone (STH) and corticoids (C), secreted under the influence of ACTH, in inducing mammary duct growth. Prolactin (MH) in the rat provokes the secretion of progesterone which, with oestrogen, prolactin and growth hormone bring about the lobulo-

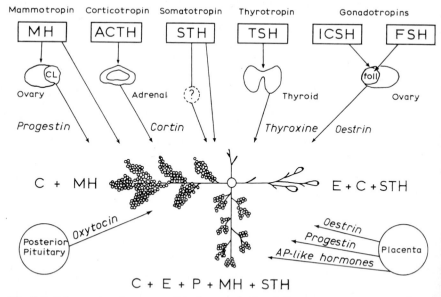

Fig. 7.8. Scheme showing some of the hormones that influence mammary growth and lactation. In the mammary diagram: upper = rudimentary gland; right = prepubertal to pubertal gland; lower = gland of pregnancy (prolactational); left = lactating gland. (Lyons *et al., Rec. Progr. Horm. Res.*)

alveolar growth. For the final stages leading to lactogenesis the dominance of prolactin and C are important; growth hormone and thyroid secretions are unnecessary although they undoubtedly contribute to the normal metabolic condition of the rat required to sustain production of milk.

The Ovarian Hormones

During pregnancy, the concentrations of progesterone and oestrogen are high in the blood, and mammogenesis, as we have seen, depends on appropriate levels of one or both of these. In general, both oestrogen and progesterone are necessary to stimulate extensive lobulo-alveolar development, but the situation is different in different species; thus in the rat, mouse, cat and rabbit oestrogen evokes growth of the ducts alone, alveolar development being only achieved by very high doses or prolonged treatment. In the guinea-pig, goat and cow lobulo-alveolar growth is achieved by physiological doses of oestrogen. As with other aspects of reproduction, it seems that there is an optimal balance between oestrogen and progesterone concentrations (Benson *et al.*, 1957).

Practically, the possibility of inducing mammary gland development in sterile cows has been examined in some detail. Thus oestrogen alone, injected into sterile heifers and cows, can produce considerable udder development and secretion of milk, whilst a combination of this with progesterone was usually more effective (Folley, 1956).*

Oestrogen Receptors. When the steroid hormones have a direct action on a target organ such as the uterus, homogenates of the target tissue are found to contain specific protein receptors; and this is true of the mammary gland. Figure 7.9 shows the changes in the estimated number of receptors in homogenates of mammary gland taken from the lactating rat; there is no increase in concentration of the complex between oestrogen and the receptor, indicating a low concentration of oestrogen during lactation. Autoradiographically, the labelled oestradiol was seen mostly over the epithelial linings of the acini and ducts (Sander and Attramodal, 1968). According to Shyamala and Nandi (1972) the equilibrium constant of the complex between the receptor isolated from the cytosol of mouse lactating mammary tissue, $K = 10^{-10}M$, is the same as that for the receptor isolated from the uterus.

---

* As Cowie and Tindal (1971), p. 126, have pointed out, much of the mammary development obtained by ovarian hormone treatment of sterile animals took place after cessation of the treatment, i.e. during the period when the animal was being milked, so that the major role in inducing mammogenesis and lactation in non-pregnant animals is probably exerted by the pituitary hormones released in response to the milking stimulus (p. 398).

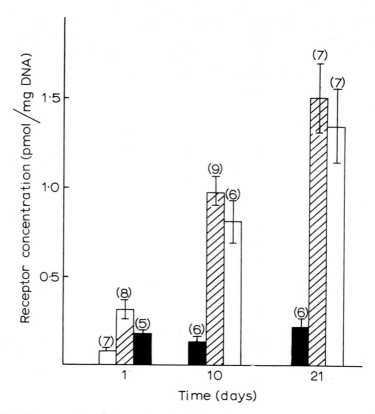

Fig. 7.9. Concentrations of cytoplasmic and nuclear receptor-oestrogen complex in rat mammary gland during lactation. Open bars represent saline-injected rats and cross-hatched bars those injected with 25 μg of oestradiol. Solid bars represent nuclear receptor-complex in saline-injected animals. (Hsueh *et al.*, *J. Endocrinol.*)

In vivo, the labelled oestradiol was bound to the nucleus but *in vitro* only to the cytoplasmic material.

Prolactin and Growth Hormone

In general, the concentration of prolactin in the blood is low during pregnancy (Table I, and Figure 7.10) but increases sharply at parturition; Sinha *et al.* (1974) have shown that in the second half of pregnancy in the mouse there is extensive cell division and protein synthesis which continue well into lactation; these activities are correlated with DNA and RNA synthesis, as illustrated by Figure 7.11 where plasma levels of prolactin and growth hormone are correlated with DNA and RNA contents of the mammary glands. The point of main interest is the fall

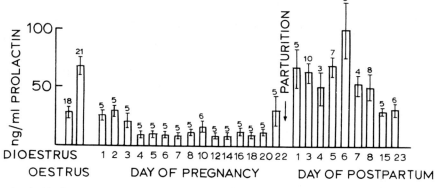

Fig. 7.10. Serum prolactin concentration of rats during pregnancy and after parturition. Mean and standard error of mean are indicated at top of each bar. Number of rats in each group is indicated at bottom of each bar. (Amenomori *et al.*, *Endocrinology*.)

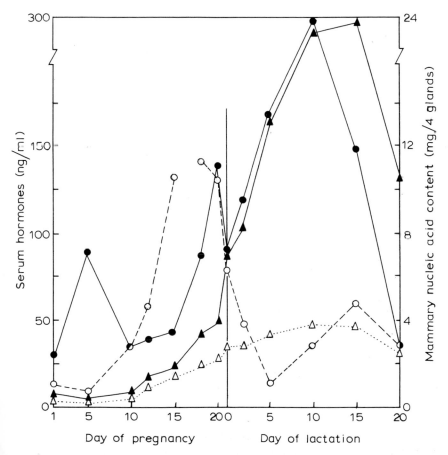

Fig. 7.11. Changes in mammary DNA (△) and RNA (▲) content of mice during pregnancy and lactation in relation to the serum concentrations of prolactin (●) and growth hormone (○). (Sinha *et al.*, *J. Endocrinol*.)

in growth hormone concentration at parturition and the large rise in prolactin concentration suggesting that mammary development in the mouse largely depends on growth hormone; certainly, when growth hormone was neutralized by antibodies, mammary development was inhibited.

Effects of Antisera. Thus, anti-growth hormone was more effective in suppressing milk production in mice, as measured by the gain in weight of suckling pups, than anti-prolactin, whilst a combination of both antisera was most effective. The anti-growth hormone decreased the DNA content of the mammary glands more than the anti-prolactin whereas the anti-prolactin caused a fall in the RNA/DNA ratio suggesting that its action was on protein synthesis whilst that of growth hormone was on cell proliferation.

Direct Mammatrophic Effect of Prolactin

Lyons (1942) injected prolactin directly into the galactophore of the rabbit's nipple; such injections into virgin rabbits ovariectomized and pretreated with oestrogen and progesterone for four weeks, had striking effects on the histological appearance of the mammary tissue indicative of growth and of secretion of milk. Without prolactin treatment, the mammary glands showed prolactational growth with progesterone and oestrogen, equivalent to that of a 3-week pregnant rabbit, so that prolactin becomes essential at about this time.

THE LACTOGENIC TRIGGER

The concept of a change in the biochemistry or physiology of the mammary gland taking place over the course of several hours and leading from a condition of no secretion to one of full secretion of milk is embodied in the expression "lactogenic trigger". In certain species, such as the rat and mouse, this concept may well be justified, since the secretion of milk, recognized biochemically by the presence of lactose in the mammary gland in detectable amounts, occurs only 12 hours prepartum (Chadwick, 1962); in the cow and goat, however, lactose appears in the blood and urine days before parturition, and lactose synthetase activity and the ability to synthesize the constituents of milk appear between 30 and 7 days prepartum, and in the sheep 3–4 weeks prepartum (Mellenberger and Bauman, 1974).

Plasma Progesterone

According to Kuhn (1969), suckling itself is not normally a factor in *initiation* of lactogenesis although, as we shall see, it can influence maintenance, or galactopoiesis. Kuhn investigated the possibility that the lactogenic trigger consisted in the fall in plasma progesterone that is such a prominent feature associated with parturition (p. 295).

Ovariectomy

He carried out ovariectomy on rats 2–3 days prepartum and, as Figure 7.12 shows, there was a steady rise in lactose concentration in the mammary glands; hysterectomy had the same effect, after a longer delay. These effects could be due to the fall in progesterone level that certainly follows ovariectomy; the effects of hysterectomy could operate through the lowered blood-progesterone. Prolactin injections had no effect on ovariectomized rats, presumably because there was adequate secretion, but it did inhibit lactose secretion in the hysterectomized animals, presumably because it is luteotrophic in the rat and was therefore promoting progesterone secretion.

Parallel Plasma Changes

In Figure 7.13, the changes in a number of parameters during the period leading up to and succeeding parturition are shown; the steady decline in progesterone is associated with a parallel rise in the metabolic derivative, 20α-OH-progesterone, so that there is a profound change in the ratio of these two progestogens at parturition. It will be recalled that Wiest et al. (1968) had postulated that control over progestational activity was exerted through the conversion of the highly active

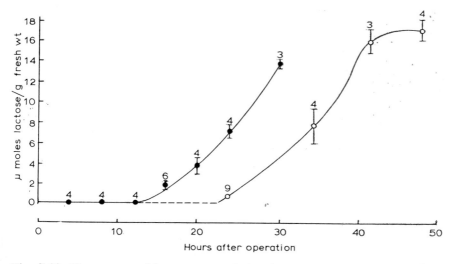

Fig. 7.12. Time-course of lactose accumulation in pregnant rat mammary tissue after ovariectomy (●) and hysterectomy (○) 2–3 days pre-partum. Bars and numbers indicate S.E. and numbers of rats. (Kuhn, *J. Endocrinol.*)

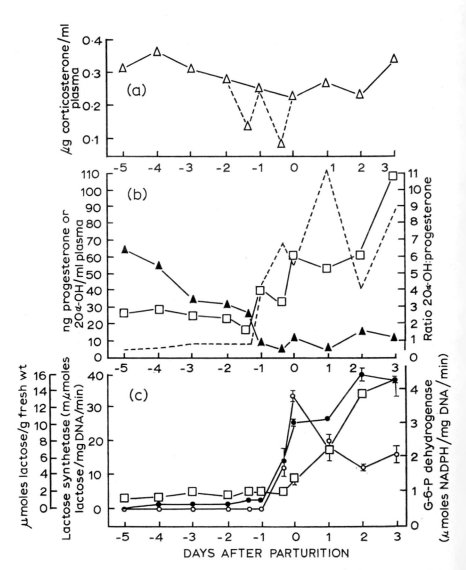

Fig. 7.13. Changes in concentrations of plasma hormones and of mammary tissue substances in normal parturient rat. (*a*) Concentrations of plasma corticosterone at 8.00–9.30 hr (– – △ – –) and 16.00–17.30 hr (—△—). (*b*) Concentrations of plasma progesterone (▲), 20α-OH (□) and the ratio 20α-OH: progesterone (- - - - -). (*c*) Concentrations of lactose (○) and activities of lactose synthetase (●) and glucose-6-phosphate (G-6-P) dehydrogenase (■) in mammary tissue. Each value is a mean of five rats and all parameters were determined in the same rats. (Kuhn, *J. Endocrinol.*)

progesterone to its much less active derivative, 20α-OHP, and it could well be that this represented the trigger for induction of lactation, the control by progesterone being exerted through an inhibition of lactose synthetase activity.

Oestrogen

We may note that Kuhn found that lactogenesis occurred in both ovariectomized and hysterectomized rats so that it seems unlikely that oestrogen is involved in initiation of milk secretion.

Prolactin as Trigger?

In opposition to this view of progesterone as the trigger for lactogenesis, we have that of the secretion of prolactin; however, this is unlikely in the rat and other species where prolactin has luteotrophic activity, since this activity would tend to raise progesterone levels at parturition, whereas in fact, the level falls. Furthermore, injection of prolactin to rats in late pregnancy does not induce lactogenesis (Talwalker et al., 1961), unless death or abortion of the foetuses is first induced (Meites and Turner, 1947).

Maintenance of Lactation. The evidence indicating a role for prolactin in initiating lactogenesis is, according to Kuhn, evidence for its role in *maintenance* of an already established lactation. Thus, Bintarningsih et al. (1958) injected hypophysectomized pregnant rats with LH and cortisol and caused secretion of milk, but as Kuhn pointed out this was, in effect, an action to maintain lactation that had been induced by the reduction in ovarian hormone concentrations resulting from hypophysectomy; they were in fact measuring the *galactopoietic* action of the hormones. Finally, we must note that hypophysectomy itself during the second half of pregnancy does not prevent lactogenesis occurring at the normal time (see, for example, Newton and Richardson, 1940*), although, as we shall see, the operation does prevent the maintenance of secretion, so that the hypophysectomized mother is not able to maintain her young.

Possible Steroid Feedback

Oestrogen-Induced Prolactin Secretion

Kuhn considered that progesterone had a direct inhibitory action on the mammary gland; an alternative hypothesis, namely that the changed balance in oestrogen-progesterone levels in the blood at term

* It must be emphasized that the placenta of the mouse secretes prolactin, so that mammogenesis, and possibly initiation of milk secretion, might be brought about by this source of hormone.

reacts on the hypothalamus, has been considered by Chen and Meites (1970). Thus, as Figure 7.14 shows, oestrogen stimulates prolactin secretion in rats,* an effect that was slightly inhibited by progesterone; the rise in plasma prolactin was accompanied by a rise in pituitary content, indicating enhanced synthesis of the hormone. A more direct demonstration of release of prolactin from the pituitary as a result of oestrogen injections is illustrated by Figure 7.15 which shows the serum prolactin of rats grafted with anterior pituitaries under the

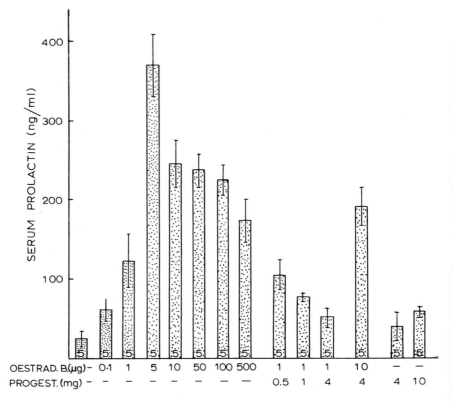

| OESTRAD. B.(μg) | – | 0·1 | 1 | 5 | 10 | 50 | 100 | 500 | 1 | 1 | 1 | 10 | – | – |
| PROGEST. (mg) | – | – | – | – | – | – | – | – | 0.5 | 1 | 4 | 4 | 4 | 10 |

Fig. 7.14. Showing the increases in serum prolactin produced by doses of oestradiol benzoate; progesterone has a slight inhibitory action. (Meites *et al., Rec. Progr. Horm. Res.*)

* *In vitro*, small doses of oestrogen stimulate synthesis of prolactin by the isolated rat pituitary (Nicoll and Meites, 1964). Pasteels' (1963) study of the changed histology of the anterior pituitary as a result of doses of about 2·5 mg of oestrogen revealed a transition to a condition equivalent to that of the end of pregnancy, manifest as a multiplication of the erythrosinophil cells, hypertrophy of the Golgi apparatus and the endoplasmic reticulum, and degranulation of most of the cells. Progesterone had similar but less intense effects.

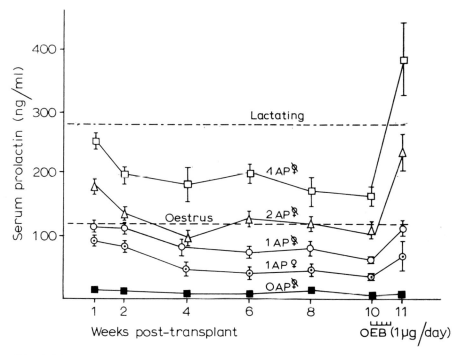

Fig. 7.15. Serum prolactin levels in rats bearing 1, 2 or 4 anterior pituitary (AP) grafts under the kidney capsule. Note large increases when oestradiol benzoate (OEB) was injected from the 10th day onwards. All rats were ovariectomized, except one group with one AP transplant (Chen *et al.*, *Neuroendocrinology*).

kidney capsule and thus not under the influence of the hypothalamus; compared with animals with no transplant (black squares) the levels of prolactin are high, and injection of oestradiol benzoate caused a sharp rise in serum prolactin, this time, presumably, by a direct action on the transplanted pituitary. Thus, the low level of oestrogen in the blood during pregnancy, which in the rat does not rise till the day of parturition (Yoshinaga *et al.*, 1969), might well be the cause of the low levels of prolactin, whilst the rise in oestrogen at parturition would cause prolactin release and act as a trigger for lactogenesis.

Oestrogen and PIF. The action of oestrogen on the secretion of prolactin is well established; thus Ratner and Meites (1964) took pituitaries from rats and incubated these with hypothalamic extracts of animals that had been injected with oestradiol. The hypothalamic extract from animals treated with oestrogen had no capacity to inhibit release of prolactin from the incubated pituitaries, so that the oestradiol must have completely exhausted the hypothalamus of its PIF, presumably by suppressing its synthesis.

Isolated Pituitary. In this way, then, oestrogen promotes prolactin release. Studies on the effects of steroids on secretion of prolactin by the isolated pituitary *in vitro* showed that only oestrogen increased production of the hormone, so that any effect of corticosteroids was probably through the hypothalamus.

Hypothalamic Oestrogen Implant. Implantation of oestrogen into the basal tuberal region of the hypothalamus, i.e. near the median eminence, provoked secretion of prolactin in female rats, as revealed by its luteotrophic effect and by its effects on the crop-gland of the pigeon (Ramirez and McCann, 1964). Similar effects were obtained by implantation into the pituitary, so that an inhibition of PIF synthesis or release seems to be the action of oestrogen, at any rate in large doses. Even in male rats, Deis (1967) was able to induce mammary development leading to large cystic formations full of secretion, by implants of oestrogen in the hypothalamus. However, as Kuhn has argued, the effects of increased prolactin secretion, in whatever manner evoked, might well be on maintenance rather than initiation of secretion.

Ovariectomy. If ovarian secretion of oestrogen is a significant factor in promoting prolactin release, then we must expect a changed pattern of release in the ovariectomized pregnant animal; in fact Simpson *et al.* (1973) have been unable to find any alteration, the plasma concentration in normal and ovariectomized rats showing a gentle rise during late pregnancy.

Direct Inhibitory Action of Oestrogen

It may well be that a direct inhibitory action of oestrogen on the mammary tissue tends to obscure the issue; according to Bruce and Ramirez (1970) implantation of oestradiol into the mammary glands of lactating rats inhibited milk secretion, whereas implantation in the pituitary accentuated it, this latter effect being, presumably, due to facilitation of prolactin release.* A synthetic oestrogen, stilboestrol, is used to inhibit lactation in humans, and according to Fields (1945) this is a genuine effect rather than the consequence of the cessation of suckling which, of course, is associated with the treatment. Thus stilboestrol given to nursing mothers suckling their babies caused a definite reduction of milk-yield by comparison with untreated controls

---

* Injections of oestradiol into lactating rats decrease the specific binding of mammary gland tissue to prolactin (Bonet *et al.*, 1977).

Induction of Lactation by Suckling

Unmarried Nursemaids

There have been numerous reports of the induction of lactation in non-pregnant women by repeated suckling. Thus, Foss and Short (1951), in a review of abnormal lactation, quote the case of three unmarried nursemaids, who had never been pregnant, who, after repeated suckling their mistresses' babies, eventually secreted milk.

Virgin Goats

Cowie *et al.* (1968) subjected virgin ovariectomized goats to regular twice-daily "milking"; this led to mammary growth and, after several months, to a milk-yield of 0·8 to 2·0 litres (Fig. 7.16). Transection of the pituitary stalk prevented this development, and replacement therapy with growth hormone was ineffectual. Since prolactin secretion presumably is not prevented by section of the pituitary stalk, this effect of section may have been due to the absence of ACTH.

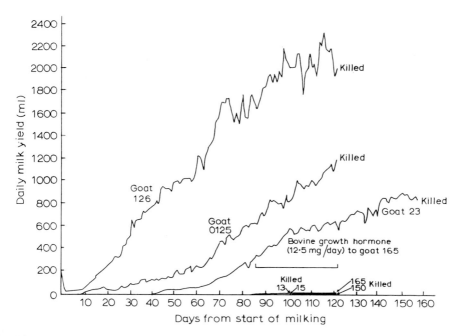

Fig. 7.16. Effects of twice-daily "milking" in milk yield of virgin ovariectomized goats. The yields of the four pituitary-stalk sectioned animals (13, 15, 150 and 165) are so low that they cannot be accurately plotted. (Cowie *et al.*, *J. Endocrinol.*)

Self-Licking

It is possible that the self-licking of nipples, described by Roth and Rosenblatt (1966) in rats, acts as a stimulus for mammogenesis and lactogenesis.

Studies on Organ Culture

Minimal Hormonal Requirements

The *in vitro* study of mouse mammary gland permits an excellent assessment of the contributions of the various hormones to mammogenesis and lactation. Pioneering studies of Juergens *et al.* (1965) and of Rivera (1964) showed that the minimal requirements for survival of explants of pregnant mammary tissue were insulin and an adrenal cortical steroid, aldosterone being more effective than cortisol. The minimum requirements for initiation of mammary secretion, as shown by alveolar secretory appearance, were aldosterone plus insulin plus a pituitary hormone, either prolactin or growth hormone; on a molar basis the two hormones were equally effective.

Protein Synthesis

Turkington (1968) showed that, if only insulin and hydrocortisone were in the medium, the epithelial cells proliferated by mitotic division but did not produce any protein, i.e. no α-lactalbumin, or casein. When human placental lactogen, HPL, or ovine prolactin, was added, synthesis took place in enlarged alveolar ducts. It appeared that this synthesis occurred only in those epithelial cells that had been formed by division, i.e. in relatively non-differentiated cells. This was shown by adding the mitotic inhibitor, colchicine, before incubation and no synthesis was obtained; if it was added after preliminary incubation, and at the time of adding prolactin, then synthesis was normal. Actinomycin D prevented any effect of prolactin, even 12 hours after its addition to the medium, so that the action of the hormone is to induce the synthesis of the necessary proteins. These are caseinogen, and the lactose synthetase complex which, as we have seen (Volume 2) consists of two separate proteins, namely galactosyl transferase (Protein A) and the constituent of milk, α-lactalbumin (Protein B). During the latter half of pregnancy the activity of galactosyl transferase rises nearly to maximum values, whilst that of α-lactalbumin remains low, but rises to a maximum after parturition. If the appearance of galactose synthetase in the gland were rate-limiting for lactogenesis at parturition, then the onset of lactogenesis could be regulated by the induction of the synthesis of α-lactalbumin.

Progesterone Inhibition. Turkington and Hill (1969) showed that, *in vitro* at any rate, progesterone exerted a strong inhibitory action on the development of α-lactalbumin in organ-cultured mammary tissue; this is shown by Figure 7.17, where it is seen that the appearance of galactosyl transferase is unaffected by the presence of progesterone, whereas that of α-lactalbumin is completely inhibited; the concentration of progesterone employed was of the same order as that found in the pregnant rat.

Critical Enzyme

The general concept of the initiation of secretion of milk being the result of the production of a critical enzyme for synthesis of one or more constituents is an attractive one, and has been investigated from several aspects, the most fundamental being the attempt to correlate the appearance of specific enzymes in the mammary gland with the

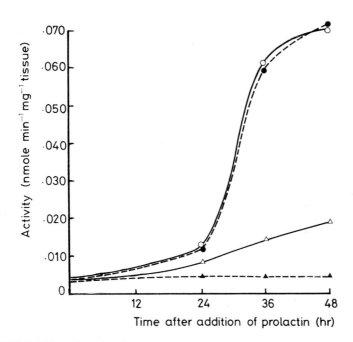

Fig. 7.17. Inhibition of production of α-lactalbumin by progesterone, leaving galactosyltransferase activity unaffected. Mammary explants from mice midway through pregnancy were incubated in a medium with insulin and hydrocortisone for 72 hr; then the medium was changed to include prolactin with progesterone (closed symbols) or without progesterone (open symbols). Circles, galactosyltransferase activity; triangles, α-lactalbumin activity. (Turkington and Hill, *Science.*)

onset of lactation. In general, such correlations have been successful, although the *abrupt* appearance of a critical enzyme acting as a trigger may be too simple a view. Thus, Figure 7.18 shows the development of activity of the A and B components of the lactose synthetase complex during pregnancy and subsequent lactation. In agreement with Turkington, Palmiter (1969) found that development of B comes later and is thus rate-limiting for synthesis of lactose, but it will be seen that, although activity in late pregnancy is small, it increases before parturition and therefore probably before the large fall in plasma progesterone.

Golgi Vesicles. It could have been argued that the trigger was not so much the induction of the enzyme, or a component such as α-lactalbumin, but the passage of this from the endoplasmic reticulum to the Golgi apparatus (Brew, 1969), in which case the amount actually present in a homogenate would be irrelevant. However, Jones (1972) prepared Golgi vesicles from mouse mammary glands and found no significant change in the concentration of α-lactalbumin isolated between the 16th day of pregnancy and the 6th day of lactation. He estimated that the mouse mammary gland could, in this preparturi-

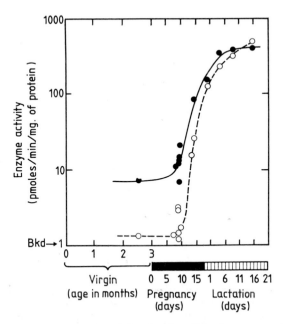

Fig. 7.18. Development of the A and B subunits of the lactose synthetase complex during various stages of pregnancy in the mouse mammary gland. (●) A-subunit activity; (○) B-subunit activity. (Palmiter, *Biochem. J.*)

tional state, produce 4·4 mg lactose per mg DNA* whereas in fact it only contained 0·057 mg/mg DNA, so that there is a real problem as to what stops this synthesis in the foetal mouse.

Rabbit Lactose Synthesis. In the rabbit the synthesis of lactose is measurable by Day 22 of pregnancy (gestation period 31 days), so here again an abrupt change from incapacity to synthesize to capacity to synthesize does not take place. Mellenberger and Bauman (1974a) examined the changes in synthetic ability quantitatively, measuring the ability of the excised mammary tissue to synthesize lactose. Some results are shown in Table III, and it will be seen that there was no significant synthesis until between Day 15 and Day 24. The RNA/

TABLE III

Illustrating the biphasic increase in lactose synthesis by excised mammary tissue of the rabbit taking place between the 15th day of pregnancy and the 22nd day of lactation (Mellenberger and Bauman, 1974a)

	Days Pregnant			Days of Lactation				
	15	24	29	2	5	8	15	22
Lactose								
Synthesis	5	125	125	435	725	960	1500	665
	5	100	170	385	795	1350	1515	635
(n mol of ^{14}C-) glucose (incorporated into lactose per 3 hr per 100 mg wet wt.)								

DNA ratio in the tissue ran parallel with synthetic ability, and when the activities of different enzymes were estimated it was found that those of lactose synthetase and UDP-glucose-4-epimerase correlated with lactose synthesis.

Milk-Fat Syntheses

Studies on synthesis of milk-fat revealed an essentially similar story (Mellenberger and Bauman, 1974b), an early preparturitional phase of secretion being succeeded by a postparturitional phase, as with lactose synthesis. The authors suggested that the second phase could be associated with suckling (p. 409).

* The amount of DNA in a tissue is a measure of the number of cells, and provides a better standard of reference than the weight of the tissue. The RNA/DNA ratio is an index to protein synthetic capacity when this is changing rapidly.

Mid-Pregnant and Virgin Mice

The studies on the development of synthetic capactiy in the explanted mammary tissue have been carried further and have revealed some more interesting features. Thus, when mammary tissue from mid-pregnant mice is compared with that of mature virgin animals, there are not only obvious morphological and biochemical differences, but also differences in the requirements for developing synthetase activity. Thus in the mid-pregnant mammary tissue there are a number of epithelial cells that have already acquired the morphology of secreting cells, with well developed rough surfaced endoplasmic reticulum, Golgi apparatus, etc. Biochemically, moreover, there is definitely some lactose synthetase activity. When the tissue is incubated with the hormone triad, prolactin, insulin and hydrocortisone, there is rapid multiplication of cells with development of synthetic activity, whilst the morphology of the epithelial cells develops to that of actively secreting cells (Fig. 7.19). With the virgin tissue there is neglible lactose synthetase with very little B-protein activity, and there are very few cells with the morphology characteristic of secretory activity. When these cells are incubated with the hormonal triad, cell division accompanied by DNA synthesis occurs, but this takes some 24 hours longer, so that peak protein synthesis—which apparently depends first on cell division as we have seen—is delayed correspondingly (Stockdale and Topper, 1966). It appears, then, that in both virgin and pregnant mammary tissue, DNA synthesis and cell division are coupled to the development of protein synthesis, and thus to lactose synthesis. This is somewhat surprising, in view of the active cell proliferation that must have been going on in the pregnant mammary tissue before being taken for study, compared with the quiescent condition of the virgin tissue.

Critical Mitosis. Owens *et al.* (1973) considered that the early work on pregnant tissue, employing colchicine to prevent cell division, was open to other interpretations, and they repeated the experiments using specific inhibitors of DNA synthesis, namely 5-fluorodeoxyuridine or cytosine arabinoside. These inhibitors were able to block development of synthetase activity in virgin tissue but not in mid-pregnant tissue, so that it was concluded that the "critical mitosis", necessary for the epithelial cell to develop synthetase activity under the influence of hormones, had already taken place in pregnant tissue but not in virgin tissue. Thus mammary development in virgin tissue might be controlled by inhibition of this "critical mitosis".

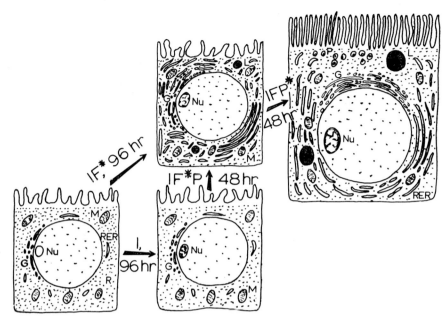

Fig. 7.19. Schematic representation of some of the ultrastructural changes taking place in the alveolar epithelial cells of explants from mid-pregnant mouse mammary glands with added hormones. Asterisks indicate the hormone thought to be primarily involved in bringing about certain ultrastructural changes. I, insulin; F, hydrocortisone; P, progesterone; Nu, nucleolus; R, ribosomes; G, Golgi apparatus; M, mitochondria; P, protein granules; RER, rough endoplasmic reticulum. (Mills and Topper, *J. Cell Biol.*)

Insulin Sensitivity. Another feature of the virgin tissue is the time required for it to develop its insulin sensitivity; thus we have seen that insulin is required for the medium if it is to support cell division, whilst the remaining hormones of the triad are necessary for the development of secretion. It turns out that virgin tissue must be incubated for one to several days before insulin will permit the mitosis.

Asynchrony in Protein Synthesis. Finally, Vonderhaar (1973), showed that, when the time-course of development of the three essential proteins was studied, namely casein, A-protein and B-protein, in virgin mouse tissue, an asynchrony in development became manifest, as shown in Figure 7.20(a) where it will be seen that B-protein (α-lactalbumin) reaches its maximum some 24 hr after A-protein and later still after casein. When the mice were "primed" with prolactin for three days before their tissue was taken, then, as Figure 7.20(b) shows, synthesis of all three proteins was synchronous, and similar to that taking place in tissue from pregnant animals.

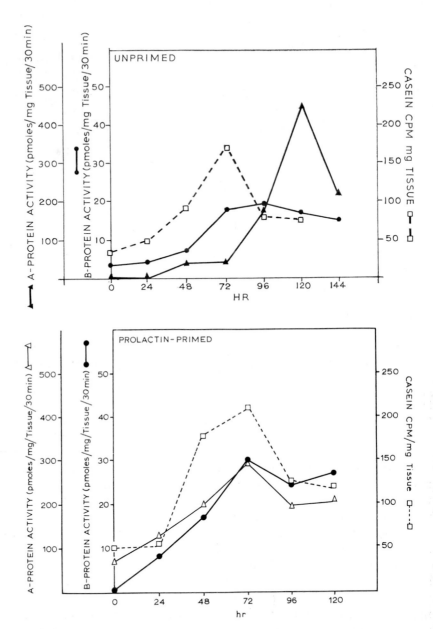

Fig. 7.20. (a) Time-course of casein synthesis and the development of lactose synthetase A- and B-protein activities in explants of mammary glands from unprimed mature virgin mice. (□), casein synthesis; (△), A-protein activity; (●), B-protein activity. (b) Time-course of casein synthesis and development of A- and B-protein activities in explants of mammary glands from prolactin-primed mature virgin mice. (□), casein synthesis; (△), A-protein activity; (●), B-protein activity. (Vonderhaar, *J. biol. Chem.*)

Fat Synthesis

Lipoprotein Lipase

We have seen how the synthesis of fat in the milk requires the development within the gland of the hydrolytic enzyme lipoprotein lipase (Volume 2), which permits the tissue to extract the lipid from the circulating blood. In the non-lactating animal the activity of this liproprotein lipase is negligible, but it increases 2–3 days before parturition and remains high during lactation, and decreases rapidly when suckling stops. In adipose tissue the reverse changes take place, so that the mammary gland, during lactation, has a preferential access to the circulating fat. In the hypophysectomized rat this reciprocal activity is well shown; Figure 7.21 shows that the lipoprotein lipase activity of the mammary gland falls rapidly after hypophysectomy, due presumably to the loss of prolactin since early treatment with this hormone causes a return to full activity within 48 hours. So far as adipose tissue is concerned, the effect of hypophysectomy is to cause a striking *increase* in lipoprotein lipase activity, indicating that the pituitary, presumably acting through prolactin, inhibits the development of this activity while it promotes it in the mammary gland.

Summary

To summarize, then, the biochemical studies, whilst indicating a parallelism between the appearance of synthetic machinery and intensity of lactation, do not confirm the notion of an abrupt induction of synthesis of a critical enzyme. The experimental finding that progesterone inhibits onset of lactation, and with it the synthesis of α-lactalbumin, remains as an important clue to the control mechanism, however.

HORMONAL REQUIREMENTS FOR MAINTENANCE OF LACTATION

The pioneering studies of Meites, Nelson, Young and others have indicated that prolactin is not the only pituitary hormone necessary for maintenance of an established lactation; in addition, growth hormone and the adrenal corticotrophic hormone, ACTH, are necessary, so that adequate lactation in the hypophysectomized animal requires a mixture of all three hormones, although the requirements vary with the species.

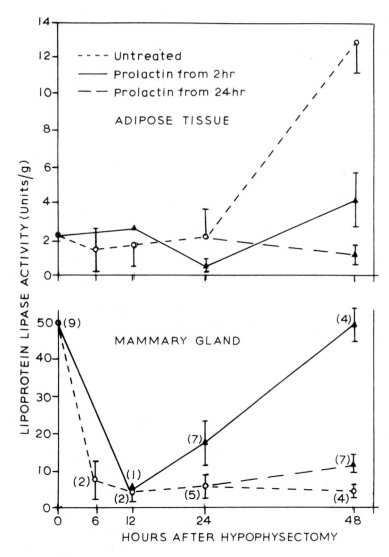

Fig. 7.21. Illustrating reciprocal relations between adipose tissue and mammary glands with respect to lipoprotein lipase activity. The activity of the mammary gland falls rapidly after hypophysectomy; treatment with prolactin causes a return to full activity within 48 hr. Hypophysectomy causes a striking *increase* in lipoprotein lipase activity in adipose tissue, which is prevented by prolactin treatment (Zinder *et al.*, *Amer. J. Physiol.*)

Hypophysectomy

The effects of hypophysectomy on milk secretion in goats are striking, and illustrated by Figure 7.22, which also shows how, in these animals, lactation can be restored to normal by an appropriate regimen of hormones; included in the replacement therapy were prolactin, growth hormone, dexamethasone (a glucocorticoid), and tri-iodo-thyronine.

Replacement Therapy

Systematic studies on variations in the administered hormone indicated that prolactin, growth hormone and adrenocorticosterone were the essential triad; thyroid hormone and insulin appeared to improve the yields in the presence of this triad, but they were inactive on their own. More recently, Cowie *et al.* (1964), cut the pituitary stalk in goats; if care is taken to prevent repair of the section, the anterior pituitary loses its connections with the hypothalamus, and thus probably

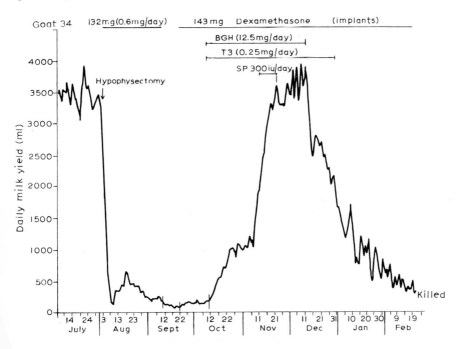

Fig. 7.22. Effects of hypophysectomy on milk yields in the goat, and the effects of hormonal replacement therapy. BGH = bovine growth hormone; T3, tri-iodo-thyronine; SP = sheep prolactin. Horizontal lines indicate periods over which the hormones were administered. (Cowie and Tindal, *Physiology of Lactation.*)

only secretes prolactin, due to the release from the action of PIF. Lactation occurred in the goats, but at about 50 per cent of normal, and a combination of bovine growth hormone and dexamethasone brought milk yield back to normal. Prolactin had no effect, presumably because there was adequate secretion. Thus, in the goat it would seem that the minimal requirements in the hypophysectomized animal are prolactin, growth hormone and corticosteroids.

Rabbit. In the rabbit the replacement after hypophysectomy is less complex; thus, as Figure 7.23 shows, injections of prolactin alone were adequate to restore yields; with time, the yields tended to decline, but they could be restored by giving growth hormone. The changes in composition of the milk that took place after hypophysectomy consisted in a rise in protein and fat content and a fall in galactose; in essence these are the changes taking place during the gradual decline in lactation as the young are weaned, so that hypophysectomy consisted in a telescoping of the normal events into a few days.

Rat. In rats, Cowie et al. (1960) grafted pituitaries under the kidney capsule of lactating rats, the pituitary tissue being taken from 8-day-old sucklings. The animals were then mated and hypophysectomized on the fourth day of their second lactation. Animals with grafts, and presumably secreting prolactin, showed a slight maintenance of milk secretion. Thus, the milk secretion of the hypophysectomized animals without grafts was $2 \cdot 0 \pm 0 \cdot 4$ g; that of animals with three grafts was $11 \cdot 5 \pm 2 \cdot 4$ g and that of animals with three grafts and ACTH treatment was $33 \cdot 8 \pm 5 \cdot 7$ g. Normal secretion would have been in the region of 90–100 g.

Oxytocin

In connection with the effects of hypophysectomy, it must be appreciated that removal of the posterior lobe hormones means removal of oxytocin, which, as we shall see, is involved in the response to suckling, so that unless the animal is treated with oxytocin the yields of milk in response to suckling will be very small. It may be that the failure to obtain complete replacement therapy by treatment with ACTH, growth hormone and prolactin in rats is due to the inadequacy of the oxytocin injections as substitutes for the release of this hormone during suckling (Cowie, 1957).

Role of Adrenal Steroids

Talwalker et al. (1961) were able to induce a copious secretion of milk in pregnant rats and rabbits by injections of hydrocortisone; this type of experiment forms the basis of another theory for the triggering

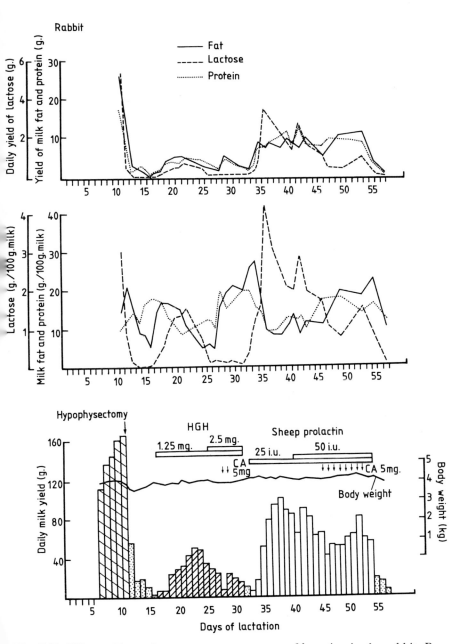

Fig. 7.23. Effects of hypophysectomy on parameters of lactation in the rabbit. Replacement therapy, indicated on bottom graph, consisted of purified anterior pituitary hormones and cortisol acetate. (Cowie *et al.*, *J. Endocrinol.*)

of lactogenesis, namely that the low level of corticosteroids in the blood during *pregnancy* acted as a brake on lactogenesis; an increase in the level of corticosteroids, resulting from increased ACTH secretion, would be the trigger. Whereas there is no doubt that glucocorticoids are necessary to maintain full secretion in adrenalectomized animals, and that secretion can be induced in pregnant and pseudopregnant animals by appropriate injections (Chadwick and Folley, 1962), the notion that lactogenesis is provoked by a rise in blood corticosteroids is not completely substantiated.

Plasma Levels

The study of Kuhn (Fig. 7.13, p. 378) failed to indicate any change in plasma levels at parturition; on the other hand, Figure 7.24 from Voogt *et al.* (1969) shows a pronounced rise in the plasma level of corticosterone, the principal adrenal corticoid in the rat, associated with parturition and subsequent lactation.

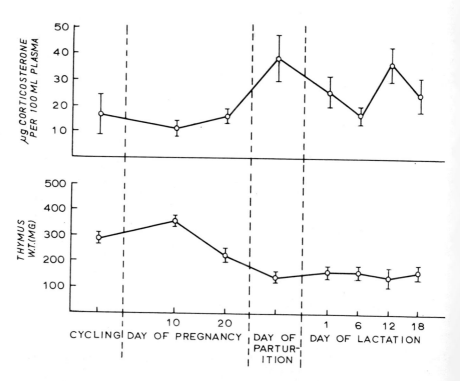

Fig. 7.24. Influence of oestrous cycle, pregnancy, parturition and lactation on plasma corticosterone and thymus weight in the rat. (Voogt *et al.*, *Amer. J. Physiol.*)

Triply Operated Animals

Again, so far as the mouse is concerned, Nandi and Bern (1961) have made out a strong case for the role of a rise in adrenal corticosteroid as the final stimulus to lactogenesis. Thus, in triply operated animals (ovaries, pituitary and adrenals removed), lobulo-alveolar development typical of pregnancy could be induced by the aid of a mammogenic combination of oestrogen, progesterone, growth hormone and prolactin; if a dose of glucocorticoid was then given, milk secretion was initiated. Again, in normal pregnant mice mammary secretion was induced prematurely by injections of corticosteroids. Although Nandi and Bern based their conclusions as to the effectiveness of hormone substitutions on the light-microscopical appearance of the mammary glands of their mice, there is little doubt that they were assessing the secretory powers of the tissue; thus Wellings (1969) made electron microscopical estimates of the area of granular endoplasmic reticulum, Golgi apparatus, and mitochondria in sections in triply operated animals that had been given replacement therapy, and the nearest approximation to normal appearance was obtained by growth hormone, prolactin and corticosteroids.

Effect of Non-Specific Stress

The rise in corticosterone observed at parturition may well be the result of stress. At any rate, Nicoll et al. (1960) stressed rats by severe cold, intense light, and so on; they were primed with oestrogen, and the stress imposed on them brought about some degree of lactation.

Role of the Placenta

HPL Secretion

The secretion of a placental lactogen represents the taking over, at any rate partially, of the mammogenic function of the pituitary by the conceptus, analogous with the secretion of the gonadotrophic hormone. As we have indicated, the human hormone is well identified as a different entity from the pituitary prolactin. Its concentration in the blood may be determined by radioimmunoassay, and the changes during pregnancy are illustrated in Figure 7.25 from Saxena et al. (1968).

Rat. In the rat Ray et al. (1955) found that placental extracts had pigeon crop-sac activity, and in rats it was lactogenic; thus hypophysectomized-ovariectomized animals were treated with oestrogen, progesterone and placental extract to induce mammary development; then they were given a regimen of placental extract with or without

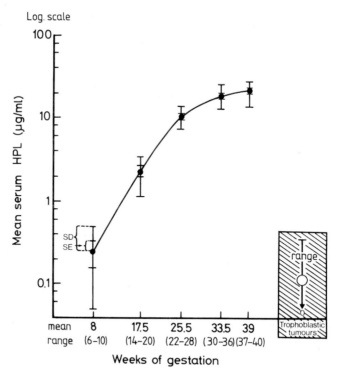

Fig. 7.25. Serum placental lactogen concentrations (HPL) in normal human pregnancy expressed as a function of the gestation period plotted on a semilogarithmic scale. The mean and range in patients with various forms of trophoblastic disease are presented for comparison in the hatched rectangle. (Saxena *et al., Amer. J. Obst. Gynaec.*)

hydrocortisone, and lactation was induced. When Desjardins *et al.* (1968) removed foetuses and placentas from pregnant rats on the 12th to 16th days of pregnancy, they reduced the mammary gland weights to those of non-pregnant controls; thus, the maternal hormone secretion is clearly inadequate to prevent mammary involution during the second half of pregnancy in the rat.*

Role of Suckling in Lactation

Although suckling is probably not the trigger for initiating lactogenesis, there is no doubt that galactopoiesis, i.e. the maintenance of secretion, is increased by the suckling stimulus in some, but not all,

* Kohmoto and Bern (1970) have shown that mouse placental extracts have some mammotrophic activity in organ culture.

species. Thus lactation, i.e. the production of milk, quite rapidly terminates after weaning—the removal of the young—and is associated with involution of the mammary glands.

Severed Mammary Ducts

That galactopoiesis was dependent on suckling was first suggested by the experiments of Selye (1934), who cut the main ducts of nipples of rats three days after parturition; as a result, the young suckled, but were unable to obtain milk and they were kept alive by being suckled by an alternative mother with normal nipples. The mammary glands of these nipple-less mothers were all turgid with milk when examined, three to four days after initiation of suckling, whereas the glands of control mothers, not allowed to suckle their young, contained no milk, and there was involution of the alveoli. Thus the mere act of suckling, presumably stimulating the nerves of the nipple, without milk withdrawal, was sufficient to maintain lactation.

Innervation of Teat

The teat and its immediate surroundings are profusely innervated by unmyelinated sensory fibres, being most concentrated at the teat; here the fibres end in the dermis, by contrast with the more superficial epidermal endings in other regions of the body, so that it is probable that the more vigorous deformation of the skin involved in suckling is required to initiate the reflex effects.

Prolonged Lactation

Perhaps the most striking demonstration of the power of the suckling stimulus to maintain lactation was provided by Bruce (1958) who continually replaced a lactating mother rat's litters by 3-day-old rats taken from other mothers, and in this way, as Figure 7.26 shows, lactation with normal yield of milk was maintained for as long as 9–12 months. The oestrous cycle of this rat was prolonged to an average of 17·8 days giving rise to what Selye had called a "suckling pseudo-pregnancy". If the rat was mated after 50 to 100 days of continuous lactation there was no delay in implantation, such as occurs when mating at postpartum oestrus in the presence of a suckling litter (p. 236) However, lactation could not be maintained by the suckling stimulus, until the birth of the new litter, so that production of milk fell during the week before parturition. After a mouse had been weaned, Bruce showed that lactation could be started up within a few days if suckling was renewed within ten days; if 109 days had been allowed to elapse, three pups had to suckle for 8 weeks before adequate milk production

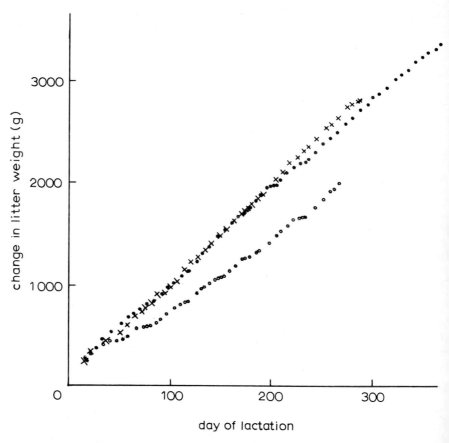

Fig. 7.26. Prolonged lactation in the normal rat. Different symbols indicate different rats. Each point represents one foster litter. Milk output was measured by increase in litter weight. (Bruce, *Proc. Roy. Soc.*)

could be established. When she tried to induce lactation in virgin rats, Bruce was unsuccessful, although Selye and McKeown (1934), working on mice, had been able to induce some mammary gland development by suckling, although not in ovariectomized animals.

Afferent Pathway

Eayrs and Baddeley (1956) sectioned tracts of the spinal cord of rats in an attempt to define the pathway of the afferent stimulus; section of the dorsal columns produced some impairment, but there was partial recovery so that these are not essential. When the dorsal roots of the spinal segments appropriate to the position of the nipples were cut, lactation was abolished; hemisection of the cord also abolished lactation if the lesions were on the side of the nipples, but not on the other. The most serious of the localized lesions were in the lateral funiculi so that it is likely that the sensory messages ascend in these, on the same side. We shall see that milk ejection, as opposed to milk secretion, depends on a very definite reflex, involving sensory stimulation of the nipples and brought about by the release of oxytocin by the posterior pituitary. In any experiment involving blocking the sensory route, therefore, it must be ensured that the failure of the pups to obtain milk is not due to the absence of the milk-ejection reflex. Eayrs and Baddeley showed that their lesions were effective when their animals were treated with oxytocin, so that they were, indeed, measuring alterations in lactogenesis.

Number of Sucklings

Studies on the effects of the number of suckling pups on milk secretion indicated that this was a factor; thus Tucker (1964) used as

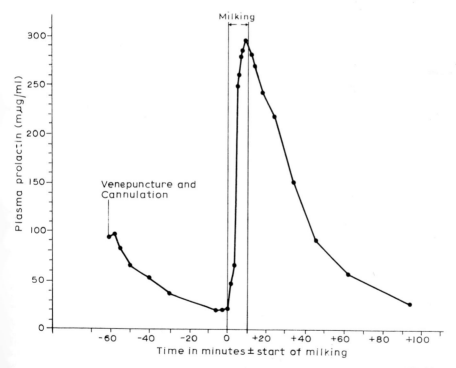

Fig. 7.27. Effect of milking on plasma prolactin concentration in a cow. (Johke, *Endocrinol. Japon.*)

his criterion the amounts of RNA and DNA in the mammary glands of lactating rats; when they suckled four pups the DNA content per gland was 18 mg and that of RNA 31 mg; when 10 pups suckled the amounts were 24 and 61 respectively. The increased RNA/DNA ratio from 1·7 to 2·8 suggested a greater protein synthesis with the larger number of sucklings.

Prolactin Release. The essential feature of the suckling stimulus is the release of prolactin into the blood; thus, Figure 7.27 shows the steep rise in plasma prolactin in the cow at the onset of milking. It will be recalled that Ratner and Meites (1964) found that hypothalamic extracts from rats injected with oestradiol had no PIF-activity, and the same is true if the hypothalamus is taken from suckling animals. Again Minaguchi and Meites (1967) examined the effects of suckling on the ability of the hypothalamus to release prolactin and LH from incubated pituitaries and obtained the figures shown in Table IV.

TABLE IV

Effects of lactation on release of prolactin and luteinizing hormone from incubated pituitaries in response to extracts of hypothalamus. Hypothalamic extracts were taken from cycling and lactating rats; the pituitaries were from 18-day post partum lactating rats (Minaguchi and Meites, 1967)

	Cycling Controls	Suckling Rats	
Prolactin Release (IU/100 mg)	0·67 ± 0·11	1·15 ± 0·12	+ 72%
Prolactin Release (IU/100 mg)	0·52 ± 0·09	1·24 ± 0·17	+ 138%
Luteinizing Hormone Release (μg/mg)	0·221	0·067	
	0·132	0·081	

Thus, more prolactin was released by the hypothalami of suckled animals; by contrast the release of LH was inhibited.*

Figure 7.28 shows the serum prolactin levels of female rats with and without their litters at the fourth day post-partum, whilst Figure 7.29 shows the changes in plasma levels in humans before, during and after suckling; in all cases the rise did not begin until suckling had actually begun so that pyschic factors seem not to be important; moreover, a

* More recently, Smith (1978) has studied the response of the rat's pituitary to LHRH during di-oestrus and subsequent pregnancy, and found no difference in sensitivity of the pituitary during suckling, so that the depression of blood-LH during suckling is due mostly to a decrease in hypothalamic secretion of LHRH. Changes in the intensity of the suckling stimulus, from two to eight pups, had little effect on sensitivity.

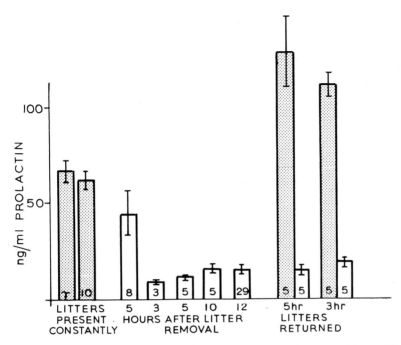

Fig. 7.28. Serum prolactin levels of 4th day post-partum mother rats with or without litters. Bars at right (litters returned) show serum prolactin in rats with return and no return of litters. (Amenomori *et al., Endocrinology.*)

breast pump was as effective as suckling in inducing the rise in prolactin (Frantz *et al.*, 1972). According to these authors, suckling is a necessary stimulus for maintenance of lactation in woman as in the rat, so that the repeated rises in prolactin secretion act as the galactopoietic factor.

Other Pituitary Hormones

Sar and Meites (1969) have shown that, in addition to lowering the hypothalamic content of PIF, a period of suckling induced a decrease of some 33–70 per cent in the pituitary prolactin and a 51–54 per cent decrease in growth hormone content. It is possible that this release of growth hormone is at least partly responsible for the increased food-intake that occurs when a mother suckles her young. So far as ACTH is concerned, Voogt *et al.* (1969) described a four-fold increase in plasma concentration of corticosterone, the major glucocorticoid of the rat's adrenal gland, associated with a 75 per cent fall in pituitary content of ACTH in consequence of a period of 30 minutes of suckling after a period of 12 hours of non-suckling.

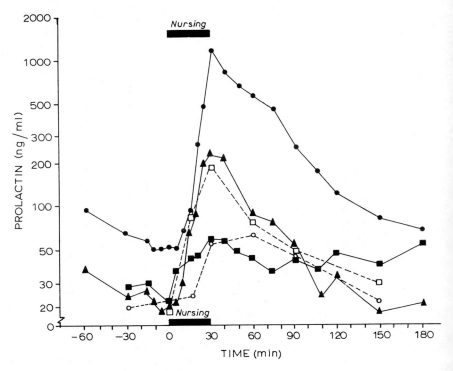

Fig. 7.29. Plasma prolactin values before, during, and after nursing in five women 23–150 days post-partum. In all cases prolactin did not begin to rise until after the onset of suckling, and in all except one the highest value was obtained at the end of the 30-minute nursing period. (Frantz *et al.*, *Rec. Progr. Horm. Res.*)

Effect of Number of Sucklings

When the animal suckles there is a decrease in the pituitary content of prolactin which builds up in the inter-suckling periods; when Mena and Grosvenor (1968) increased the number of suckling rats from two to ten, the rise in prolactin content of the pituitary, due to stopping suckling for $8\frac{1}{2}$ hr, increased from a three-fold one to a ten-fold one, whilst the maximum fall occurred when there were six pups as opposed to ten.

In general, then, these experiments suggest that the suckling stimulus acts as the means of adapting the prolactin (and oxytocin, Fuchs and Wagner, 1963) requirements of the pituitary to the milk requirements during lactation.

Effects of Denervation

In some species, however, notably the goat and cow, the suckling stimulus is not necessary to maintain lactogenesis. Thus in the goat Denamur and Martinet (1959) were able to obtain normal lactation after cutting off the mammary glands from their nervous connections with the hypothalamus, for example by completely removing the gland except for its connections with the inguinal artery and vein, denervating these vessels to remove any sympathetic connections.

After replacement, the milk yield was 59,425 ml compared with 95,221 ml. Again, when they cut the spinal cord at T11 and carried out a lumbar sympathectomy, the milk production was actually greater than normal.

Anaesthesia

A study by Tindal *et al.* (1963) suggests that rabbits can produce normal supplies of milk in the absence of this reflex. They anaesthetized lactating rabbits deeply with barbiturates at the time of suckling; if no oxytocin was administered the young failed to gain weight after each suckling, by contrast with the waking animals, indicating the absence of milk ejection. Thus, the anaesthetic had abolished the milk-ejection reflex (p. 409). When oxytocin was administered, the pups obtained normal amounts of milk at each suckling. Tindal *et al.* argue that the reflex release of prolactin or other lactogenic hormone must have been blocked by the anaesthesia so that this experiment suggests a non-neural mechanism for initiation of secretion of lactogenic hormones in this species, as with the goat.

Spinal Transection

In support of this, Mena and Beyer (1963) showed that spinal transection of rabbits at T2 and cutting the sympathetic chain, whilst it abolished the milk-ejection reflex, did not greatly reduce the milk yield if oxytocin was given. Injections of ACTH improved milk yield under these conditions, but this was clearly related to the improved appetite and general health of the animal associated with this treatment.

Exteroceptive Stimuli

Grosvenor and Turner (1958) noticed that, although suckling between Days 2–14 post-partum caused a reduction in prolactin in the pituitary gland, suggestive of secretion into the blood, suckling was ineffective on Day 21 post-partum. The reason for this turned out to be that the animals studied 21 days post-partum had been exposed to the neighbourhood of other female rats and their litters, an exposure that had depleted their pituitaries of prolactin, so that when tested for the effects of suckling they had already lost too much for an effect to be measurable (Mena and Grosvenor, 1972). The sensitiveness of the pregnant rat to the presence of others takes time to develop; thus, at 7 days, post-partum exposure to her own or other pups does not influence prolactin release; at 14 days post-partum, however, the mother's own pups or similarly aged pups, placed under her, will induce prolactin release, but she does not respond to much younger or older pups. Again, on Day 14 the mother does not respond at a distance to her pups or other mothers with their litters placed in the same room, whereas she does on Day 21. Thus the mechanisms of reflex release of prolactin are highly complex; they are related to the age of the pups, so that if a mother suckles pups that are older than her own, the time required for her to become sensitive to exteroceptive stimuli is shortened. We may note that exteroceptive stimuli from pups can also affect release of oxytocin and corticotrophin (ACTH); thus Zarrow *et al.* (1972) noted that the plasma corticosterone levels of lactating rats increased after exposure (without suckling) to pups, an effect that was blocked by olfactory bulbectomy.

Mammogenesis

It is possible that stimulation of the nipples during pregnancy contributes to development of mammary tissue. In the rat Roth and Rosenblatt (1966) noticed that licking of the nipple region was greatly increased during pregnancy, at the expense of other parts of the body. When they prevented licking by placing rubber collars round the necks of the pregnant rats the mammary tissue at term contained about half the secretory tissue as that of controls with collars that were notched and did not prevent access to the nipples.

Dispensability of Prolactin

The secretion of prolactin may be inhibited by the drug 2-bromo-α-ergocryptine, but this does not diminish the yield of milk in lactating goats (Hart, 1973); since milking is associated with a release of growth hormone into the blood (Hart and Flux, 1973) and since growth hormone, tri-iodothyronine and dexamethasone can maintain an established lactation in the hypophysectomized goat (Cowie, 1971),* there is little doubt that in this species prolactin is not essential for *maintaining* lactation, as opposed to initiating it.

* In *Lactation*, pp. 132–136. Ed. Falconer. London: Butterworth, quoted by Hart and Flux (1973).

HYPOTHALAMIC CONTROL
Effects of Lesions

It has been indicated on several occasions that the synthesis and release of prolactin from the anterior pituitary rely on a *disinhibition* of these processes, which are normally held in check by an inhibitory factor, PIF. Thus, lesions in the hypothalamus that prevent the normal release of, say, luteinizing hormone, LH, should block the occurrence of events that rely on release of this, such as ovulation. A similar lesion should actually stimulate events that depend on release of prolactin, since we are releasing the pituitary from the inhibitory action of PIF. Gale and Larsson (1963) carried out "radiation hypophysectomy" in goats, causing lesions that blocked release of ACTH, etc. As Figure 7.30 shows, the effect was to reduce milk yield and to block

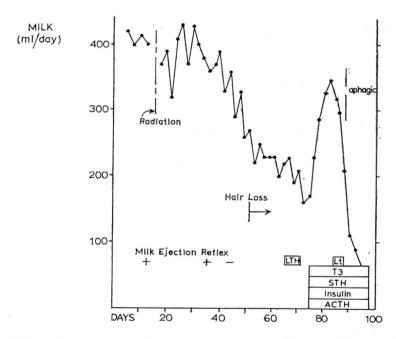

Fig. 7.30. Radiation of the median eminence caused a decline in milk production which was associated with the loss of the milk ejection reflex and the development of diabetes insipidus. Very marked hair loss from the neck and back also occurred in this animal. Administration of LTH for seven days failed to improve the depressed lactation, but when ACTH, T3, STH, and insulin were given, milk production was rapidly restored. Addition of prolactin (LTH) to this regimen failed to improve lactation further. The onset of aphagia shortly thereafter caused a quick deterioration of the health of the animal. (Gale and Larsson, *Acta physiol. scand.*)

the milk-ejection reflex, presumably through interruption of the release of oxytocin; the yield was restored by a combination of ACTH, insulin, growth hormone and tri-iodothyronine (T3), but prolactin (LTH in the diagram) was ineffective presumably because it was already being secreted maximally through loss of the hypothalamic inhibition. Towards the end of the period of study, aphagia intervened and accounts for the late fall in milk yield.

Prolactin Release

Direct proof of prolactin release in response to hypothalamic lesions was provided by Chen et al. (1970), their lesions being in the region of the median eminence, whilst the induction of lactation in ovariecto-mized oestrogen-primed rabbits was achieved by Haun and Sawyer (1960) the lesions once again being in the region of the arcuate nucleus and posterior part of the median eminence. Again, in *male* rats de Voe et al. (1966) obtained a whitish colour of the mammae and a milky secretion that could be readily expressed from the cut gland, associated with alveolar development, by lesions in the median eminence. The effects were transient, however.

Prolactin and LH. In several physiological states there is a recipro-cal relation between release of prolactin, on the one hand, and luteinizing hormone, on the other; Clemens et al. (1971) found that when they stimulated the preoptic anterior hypothalamic region (PO/AH) they increased LH secretion but decreased that of prolactin, so it may be that this centre is concerned with the reciprocal relation between the two hormones. They considered that there were regions in the hypothalamus that excited release of PIF and others that inhibited this; thus the medial basal hypothalamus, when stimulated, caused release of LH and prolactin, so that this region presumably contained many neurones that inhibit PIF release; by contrast the PO/AH region contains many neurones that stimulate PIF release and thus inhibit secretion of prolactin.

A similar duality in hypothalamic control has been described by Wolinska et al. (1977) in the ewe, a stimulatory region being located in the anterior part of the medial basal hypothalamus (MBH) whereas an inhibitory part was in the caudal part of the same hypothalamic region. The evidence was derived from the effects of lesions on the mammo-trophic and lactotrophic processes, the structural changes in prolactin cells, and the blood prolactin concentrations in pregnant and lactating ewes.

Destruction of the Afferent Pathway

We have seen that spinal section prevents lactation by cutting off the afferent pathway in the suckling reflexes; Beyer *et al.* (1962) placed lesions in the brain-stem interrupting the dorsal longitudinal bundle thereby probably cutting off the sensory pathway from the mammary glands to the hypothalamus; these were effective in blocking lactation in cats.

Suckling and the Milk-Ejection Reflex

"Let Down"

Although the act of suckling must produce a considerable negative pressure in the teat favouring withdrawal of milk, the actual removal of the secretion from the mammary gland relies on the development of a positive pressure within the ducts due to a contraction of the myoepithelial cells in the epithelium lining the alveoli and small ducts. It is the sudden development of this pressure during the operation of milking that the farmer describes as the "let down".

Myoepithelial Cells. Thus, the act of suckling does not provide the driving force for emptying the alveoli and ductules of the gland, but instead it evokes a reflex contraction of muscular elements in the gland. At first it was considered that this was related to vascular engorgement or contraction of the large ducts, but Richardson's histological studies on the myoepithelial cells of the epithelial lining to the alveoli and small ducts showed that these were most probably responsible. Thus, smooth muscle, as classically described, is confined to the teat and large ducts, and contractions or relaxations could serve to expel the milk here, but in most species where the volume of milk in this reserve space is relatively small the emptying of these would be inadequate. The myoepithelial cells* cover the stromal surface of the epithelium, ducts and cisterns of the entire gland, and when the gland is collapsed after milking the myoepithelium is tightly contracted on the collapsed alveoli, whilst in the distended gland it is stretched out into elongated cells; it never loses contact at any point with the surface of the alveolar epithelium when contraction occurs, whilst folds in the epithelium follow the direction of the main processes of the myoepithelial cells. All these morphological changes confirm that the myoepithelium plays an active role in contracting the alveoli and are responsible for the "let down". According to Linzell, contraction of the

---

* Myoepithelial cells of the mammary gland have their analogue in the salivary gland, the so-called basket cells lying similarly beneath the epithelial cells and the basement membrane; they presumably assist in expelling saliva from acini into the ducts.

myoepithelium lining the large ducts causes a widening of their lumen and thus favours the flow of the milk ejected from the alveoli and smaller ducts.

Suckling in the Rabbit

The importance of the reflex is more obvious in animals such as the rabbit or rat than in animals like the cow and goat, where there is a considerable storage capacity for milk that can be removed without operation of the reflex. Cross and Harris (1952) described suckling in the rabbit; when the doe is placed with her litter it goes at once to the nest and nuzzles and pulls the young towards her teats; the pups lie supine and show vigorous suckling movements, wriggling their bodies about actively. After 30–90 sec the pups become still, and distinct sounds of gulping and smacking of lips can be heard reaching a peak in 1–2½ minutes. The doe moves away after 2–5½ minutes. If the litter is removed after the first half minute of activity it is found, by weighing the pups, that they have obtained only a gramme or two of milk compared with the 280 grammes or more that are finally taken. Thus, the vigorous initial suckling was not accompanied by milk removal but was acting as the stimulus that only later culminated in the "let down" that permitted milk removal. If the mother was anaesthetized, the pups were unable to obtain any milk, but this block was removed by injecting the mother with a pituitary extract, the active principle in this respect being oxytocin.

Oxytocin

This peptide hormone is liberated by the neurosecretory cells of the hypothalamus and there is no doubt that the liberation of oxytocin is the final common pathway for the act of ejection of milk from the mammary gland.

Figure 7.31 shows the response to a pituitary extract (A) compared with the effects of electrical stimulation of the hypothalamus (B). Electrolytic lesions in the supraopticohypophyseal tract caused a failure of milk ejection. The long latent period in the response to electrical stimulation, and the fact that the response outlasts the stimulus, are strong arguments in favour of a humoral mechanism for the effector pathway in the reflex, whilst the blockage of the reflex by cord transection (p. 405) indicates the afferent pathway in the cord.

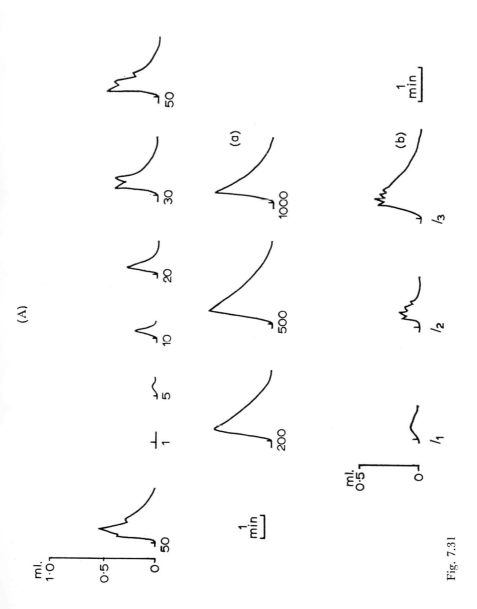

(A)

(a)

(b)

Fig. 7.31

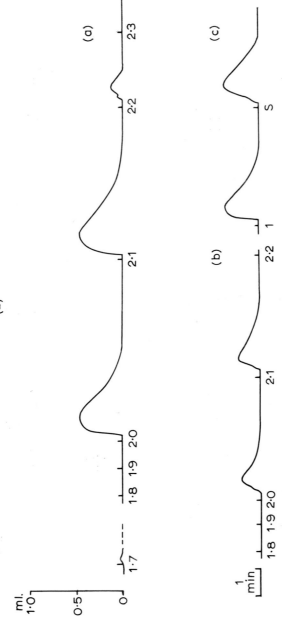

Fig. 7.31. A. Kymograph records of milk-ejection response to intravenous injections of posterior pituitary extracts. (a) Responses to different numbers of mU of whole pituitary extract. (b) Responses to 50 mU of: I_1, the purified vasopressor fraction; I_2 the purified oxytocic fraction; and I_3, the whole pituitary extract. B. Milk-ejection responses to electrical stimulation of midline structures in hypothalamus and pituitary gland. (a) Stimuli applied to vertical plane of infundibular stem. Numbers indicate depth of electrode tip below surface of skull in cm. (b) Stimuli applied in vertical plane of median eminence. (c) I, intravenous injection of 50 mU of whole pituitary extract; S, re-stimulation of infundibular stem. (Cross and Harris, *J. Endocrinol.*)

Hypothalamic Stimulation
Release of Oxytocin into the Blood

Direct evidence for the release of oxytocin into the blood, either during natural milking or as a result of electrical stimulation of the hypothalamus, has been provided by several studies, the oxytocin in the blood being assayed by its ability to cause a rise in pressure in a cannulated teat, or within the uterus. It will be recalled that the hypothalamic regions containing large numbers of large-celled neurosecretory neurones are the supraoptic nucleus (SON) and the paraventricular nucleus (PVN), and it is generally believed, but by no means unequivocally proved, that the SON is responsible for secretion of vasopressin (ADH) whilst the PVN secretes oxytocin. What is certain is that separate neurones synthesize and release the neurohumours, and also that vasopressin-releasing neurones are present in high concentration in the SON, but the rigid compartmentation of the oxytocin-secreting neurones in the paraventricular nucleus, on the one hand, and the vasopressin-secreting neurones in the supraoptic nucleus on the other, is not so certain.

Vasopressin and Oxytocin Neurones. Bisset *et al.* (1967) stimulated the hypothalamus in three transverse planes, as indicated in Figure 7.32; in each plane the electrode was moved to different loci

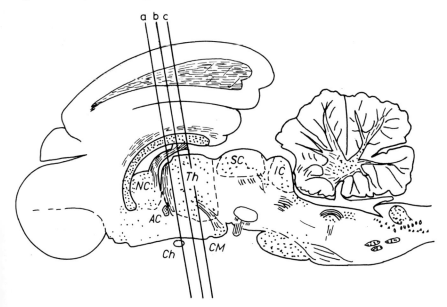

Fig. 7.32. Illustrating planes in which stimulating electrode moved. AC, anterior commissure; Ch, chiasma; CM, mammillary body; IC, inferior colliculus; NC, caudate nucleus; SC, superior colliculus; Th, thalamus. (Bisset *et al.*, *Proc. Roy. Soc.*)

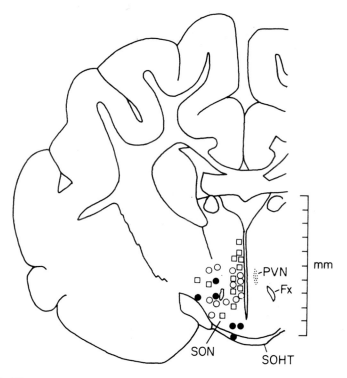

Fig. 7.33. Diagrammatic coronal section at level (b) showing positions of electrode tip at which stimuli were applied. □ No detectable release of vasopressin or oxytocin. ○ Moderate amounts of vasopressin or oxytocin. ● Large amount of vasopressin with little or no oxytocin. SON = Supra-optic nucleus. PVN = Paraventricular nucleus. SOHT = Supra-opticohypophyseal tract. Fx = Fornix. Divisions on scale are millimetres. (Bisset *et al.*, *Proc. Roy. Soc.*)

and some results of stimulation in one plane are illustrated in Figure 7.33; it will be seen that, although points were found where large amounts of vasopressin without detectable oxytocin were released, there was no point where only oxytocin to the exclusion of vasopressin was released when stimulated. Analysis of the different regions for their content of oxytocin or vasopressin gave the results shown in Table V; the Vasopressin/Oxytocin ratio was high in the SON compared with the PVN suggesting local concentrations of the neurone-type in the two nuclei, but not exclusive representation in one or the other.

TABLE V

Total amounts (m–u) of vasopressin and oxytocin in different parts of the
hypothalamico-pituitary axis (Bisset *et al.*, 1967)

Site	Vasopressin	Oxytocin	V/O
Paraventricular Nucleus	2·3	1·3	1·8
	6·5	3·3	2·0
Supraoptic Nucleus	25	5·7	4·4
	192	21	9·1
Rest of Hypothalamus	4·4	1·6	2·8
	8·9	6·5	1·1
Median Eminence	7·0	17·3	0·4
	2·36	8·9	2·7
Pituitary	4000	1000	4·0
	15,290	8157	1·9

Paraventricular Nucleus. In a later study, Bisset *et al.* (1971)
concluded that oxytocin was only released by stimulating the paraventricular nucleus, so that the supraoptic nucleus was concerned
exclusively with release of vasopressin. However, stimulation of the
PVN also released some vasopressin, and this is consistent with an
antidiuretic response obtained by stimulating this nucleus in the cat
(Koella, 1949)*.

Unit Recordings

Response to Nipple Stimulation

By placing a recording electrode in the paraventricular nucleus
Brooks *et al.* (1966) were able to obtain discharges in single units in
response to stimulation of the nipples, and we may assume that these
were the hypothalamic neurosecretory neurones involved in the milk-
ejection reflex. Figure 7.34 shows the increase in discharge-frequency
in a unit in response to stimulation of the nipples by intermittent suction
over the period indicated by the black bar; in the upper record the
pressure developed in the cannulated mammary duct is also shown.
There was, of course, no direct evidence that the neurones giving rise to
these discharges were neurosecretory, liberating oxytocin at their
endings in the neurohypophysis.

* It must be appreciated that vasopressin has some oxytocic action, as measured in the
milk-ejection reflex. It is known that the response to haemorrhage consists in release of only
vasopressin (Ginsburg and Smith, 1959), yet removal of blood from the cat gives its blood
milk-ejecting activity, so that an allowance must always be made for the contribution of
vasopressin to any oxytocic action, which is about 14 per cent of its antidiuretic activity.

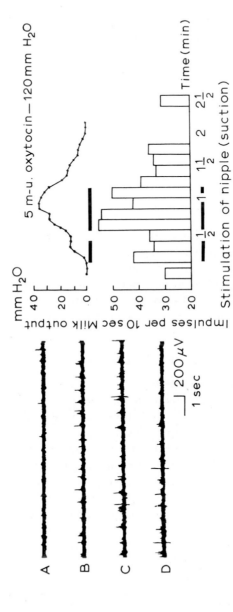

Fig. 7.34. Effects of suction stimulations of nipple (——) on paraventricular nucleus neurone activity in 10 sec intervals (bar graph) and on mammary duct pressure (upper right). Left: cellular discharge; A, control; B and C, during; and D, 3 min after suction stimulation of nipple. (Brooks *et al.*, *J. Physiol.*)

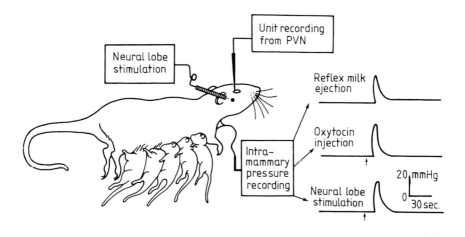

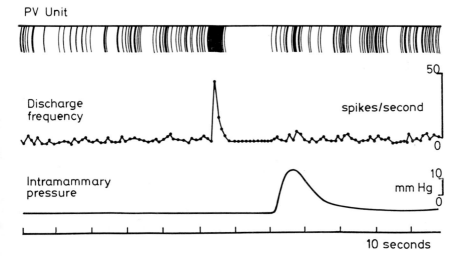

Fig. 7.35. Above: Experimental arrangement with a suckled rat with a stimulating electrode in the neural lobe, a recording microelectrode in the paraventricular nucleus, and a cannulated mammary gland. To the right are shown milk-ejection responses to suckling, intravenous injection of oxytocin, and stimulation of the neural lobe. Below: Record of an experiment in a suckled rat showing burst discharge of a neurosecretory cell occurring 18 sec before a reflex milk-ejection response. (Cross, *Frontiers in Neuroendocrinology.*)

Identification of Neurones

A considerable step forward was made in the analysis of unit activity in the PVN by using the technique of antidromic stimulation to identify neurones that had their endings in the posterior pituitary. The technique is illustrated by Figure 7.35; an electrode in the PVN records activity in individual neurones; the stimulating electrode in the neurohypophysis activates the axons and terminals of hypothalamic neurones, and the generated action potentials pass antidromically to the hypothalamus. If an action potential is obtained in the recording electrode this suggests that we are stimulating its axon or ending in the pituitary, thus identifying the hypothalamic unit as a neurosecretory cell. If the antidromically recorded action potentials conform to several criteria, this makes the identification more certain; thus there is the possibility that the neurone has been activated synaptically rather than directly, but if it responds faithfully to high frequency of stimuli with a similar frequency of action potentials a synaptic transmission is unlikely; again, the latency of synaptic transmission varies very considerably with the mode of stimulation, whereas the latency of an antidromically evoked response is reasonably constant. Finally, if the neurone from which recordings are being made is spontaneously active, then it should be possible to make the spontaneous action potential cancel out the antidromic action potential (Fig. 7.36).

Paraventricular Units

Using these criteria Sundsten *et al.* (1970) were able to identify units in the paraventricular nucleus of the lactating rabbit. They found that a quarter of the spontaneously firing units in the paraventricular nucleus did not respond antidromically, and so were not neurosecretory neurones; again, 52 per cent of the units responding antidromically did not fire spontaneously. Thus by merely studying units with spontaneous activity, a great many neurosecretory cells would be missed, whilst examining responses in any unit in the PVN would not be certain to give responses of oxytocin-releasing neurones. The mean latency of the antidromic stimulus was 19 msec corresponding to a conduction velocity of 0·7 m/sec characteristic of mammalian non-myelinated C-fibres. As Figure 7.37 shows, stimulation of one of these units caused a rise in pressure in the teat comparable with that obtained by an intravenous injection of oxytocin.

Recurrent Inhibition. Cross and his colleagues remarked on the phasic bursts of activity that often occurred in paraventricular neurones, suggesting that the discharge brought itself to an end by recurrent inhibition, analogous with the inhibition of motor neurones mediated by Renshaw cells. If the axons running from the hypothalamic nuclei

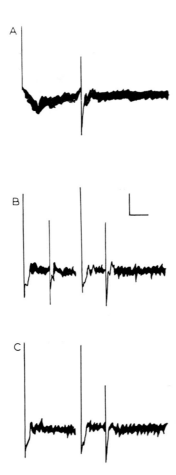

Fig. 7.36. Illustrating cancellation of antidromic spike by spontaneous action
potential. A, photograph of three oscilloscope sweeps superimposed to show the
constant latency of the antidromic action potential; the initial deflection is a large
stimulus artifact and the second is the antidromic potential. B, a train of two pulses
each followed by an antidromic action potential. C, the same train of two impulses
triggered by a spontaneous action potential. In this case the first antidromic action
potential collides with the spontaneous spike and disappears but the second appears
as expected (scale mark: A, 0·2 mV and 5 msec; B and C, 0·2 mV and 10 msec).
(Dyball, *J. Physiol.*)

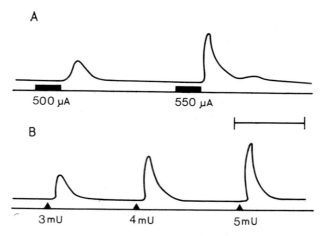

Fig. 7.37. Comparing responses to neural lobe stimulation and oxytocin injections. A: Milk-ejection responses to neural lobe stimulation; current in μamp. B: Milk-ejection responses to intravenous administration of oxytocin. Doses in mU. Calibration: 1 min. (Sundsten *et al.*, *Exp. Neurol.*)

to the hypophysis possess recurrent collaterals, it should be possible to inhibit spontaneous activity in identified neurosecretory cells by stimulation of the pituitary. In fact Hayward and Jennings, working on the unanaesthetized monkey with implanted electrodes, obtained just such an inhibition (Fig. 7.38) in neurones of the supraoptic nucleus.*

Unit Activity during Reproductive Cycle. Figure 7.39 from Negoro *et al.* (1973a) shows the mean firing rates of antidromically identified units in the rat's paraventricular nucleus during the course of an oestrous cycle, during pregnancy, and during lactation. Rates are high at pro-oestrus, at parturition, and during lactation, and it may be that these reflect the influence of oestrogens in the blood, since oestradiol, injected into ovariectomized animals, increased the rate of firing. Distension of the vagina causes a reflex release of oxytocin (p. 336); the reflex increase in unit activity brought about by this stimulus is influenced by the phases of the reproductive cycle, and the results are once again consistent with an increase in reflex sensitivity

---

* Some characteristic discharge features of the magnocellular neurosecretory cells of the hypothalamus, as studied in the unanaesthetized monkey, have been discussed in some detail by Hayward and Jennings (1973). They are distributed not only in the SON and PVN but also along the tract of Grieving that connects the PVN with the neurohypophysis, and they include neurones that secrete coherin in addition to those that secrete vasopressin and oxytocin, with their appropriate carriers, or neurophysins (Goodman and Hiatt, 1972). Whether the characteristics of discharge correlate with the character of the neurosecretion remains to be seen, however.

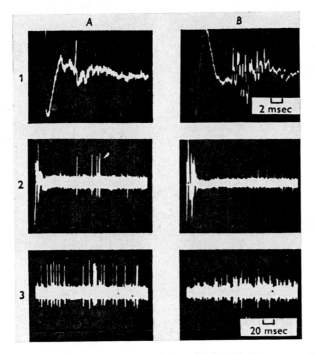

Fig. 7.38. Inhibition of "spontaneous" activity of "identified" magnocellular neuro-endocrine cells by electrical stimulation of the pituitary gland. Single pulses delivered to the pituitary evoke in (1A) an antidromic potential with a latency of 7·2 msec and in (1B) a group of multiple spikes with latencies varying from 7 to 10 msec. At a slower sweep speed and with ten superimposed traces the same stimulus to the pituitary inhibits "spontaneous" activity for 80 msec (2A) and 125 msec (2B) after the antidromic potentials. Ten superimposed sweeps without pituitary stimulation represent spontaneous activity during control periods of firing (3A and 3B). (Hayward and Jennings, *J. Physiol.*)

with increased secretion of oestrogen and a lowered sensitivity with progesterone (Negoro *et al.*, 1973b).

Reflex Discharges

Anaesthesia of the rabbit and goat blocks the milk-ejection reflex so that the reflex activation of single units of the hypothalamus is not practicable in these species. Wakerley and Lincoln (1973) found that, in the anaesthetized rat, which suckles her young for long periods at a time, discharges of hypothalamic neurones could be observed when suckling. Milk-ejection, as indicated by a rise in mammary pressure, usually began within 20 minutes of suckling and then occurred in phasic bursts rather than continuously. When activity in single, identified, paraventricular units was recorded, there was a striking burst of spikes beginning some 18 sec before the rise in intramammary

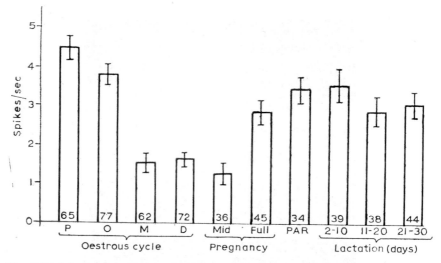

Fig. 7.39. Mean firing rate of antidromically identified units in the paraventricular nucleus recorded at various stages of the reproductive cycle. The vertical lines indicate the standard errors. P = Pro-oestrus, O = Oestrus, M = Metoestrus, D = Dioestrus, PAR = <24 h after parturition. The number of units recorded is shown at the base of each bar. (Negoro, *et al., J. Endocr.*)

pressure, as illustrated by Figure 7.40, where it is seen that spike-frequency increases by as much as 40-fold; the close coincidence in the curves of spike-frequency against time, on successive bursts, indicates the reproducibility in behaviour.

Delay in Discharge. It is interesting that background unit activity was quite unaffected by the initial period of suckling, and it was only when this had continued for some ten minutes or more that the rise in mammary pressure, and burst of unit activity, occurred The results indicate that the release of oxytocin is an all-or-none effect, not closely related in time to the suckling stimulus, and so is dependent to a large extent on an inherent rhythmicity of the hypothalamus. This agrees well with the observation that oxytocin is apparently not liberated into the blood continuously but in spurts. For example, Fox and Knaggs (1969) found that when they sampled the peripheral venous blood in the lactating woman during suckling, only occasional samples contained high concentrations of oxytocin, and the same was observed by McNeilly (1972) during suckling or hand-milking in the goat.*

* McNeilly points out that release of oxytocin into the blood may not be a necessary concomitant of milking in the goat, the manual stimulation of the myoepithelial cells being, in this species, adequate.

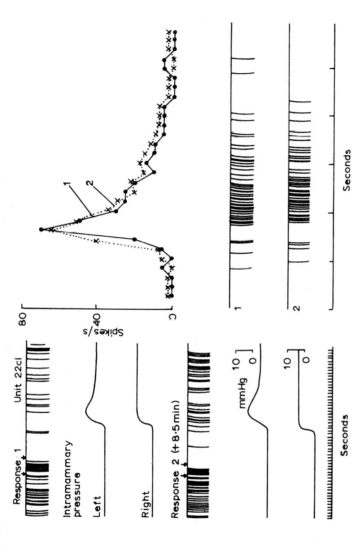

Fig. 7.40. Response of a single paraventricular unit to successive suckling stimuli (Responses 1 and 2). At the left-hand side, the electrical record is above the pressure records in left and right mammary ducts. On the right, the responses, between the arrows, have been expanded and shown in graphic form. Note the remarkable similarity in pattern of the two responses. (Wakerley and Lincoln, *J. Endocrinol.*)

Supraoptic Neurones

By applying the same technique of study to supraoptic neurones, many were found that responded in essentially the same manner to the suckling stimulus; and when multiple unit recordings were made simultaneously a synchronized burst in all responsive neurones was observed. Interestingly, the proportion of the identified neurones responding to the suckling stimulus, namely 48 per cent, was not greatly different from the 58 per cent found in the paraventricular nucleus, so that in the rat, at any rate, the conventional separation of function—paraventricular causing oxytocin secretion, supraoptic causing vasopressin secretion—cannot be sustained.

De-afferentation of the Hypothalamus

By separating the hypothalamus from its connexions other than that to the pituitary, Dyer *et al.* (1973) showed that spontaneous discharge in paraventricular units was decreased, and extraction of this region gave an increase in oxytocin-activity, perhaps due to a failure to release the hormone in consequence of the reduced nervous traffic. These authors found that the antidiuretic activity of the PVN was actually greater than the oxytocin activity, the ratio being about 3 compared with 4·45 for the rest of the hypothalamus.

Effects of Acetylcholine and Noradrenaline

Moss *et al.* (1972) applied drugs iontophoretically to paraventricular neurones, identified by antidromic stimulation as being neurosecretory. Acetylcholine excited 91 per cent of the neurosecretory cells and noradrenaline inhibited spontaneous activity in 83 per cent. When the effects on non-neurosecretory cells were studied, just the reverse was found so that if a preliminary identification had not been carried out it would have been said that half the cells were excited by acetylcholine and half inhibited. This sensitivity to neurotransmitters is consistent with the histological identification of noradrenergic and cholinergic neurones in the paraventricular nucleus. An interesting finding was that the neurones were highly sensitive to locally applied oxytocin (Moss *et al.*, 1972).

Positive Feedback. It might be argued that this sensitivity formed the basis for a positive feedback, so that release of oxytocin promoted further release; however, Dyball and Dyer (1971) were unable to find any effect of intravenous injections of oxytocin on electrical activity of paraventricular neurones. It is unlikely, therefore, that this sensitivity has physiological significance, but it emphasizes the necessity for retaining the hormone in granules within the neurones.

Negative Feedback through Oxytocin?

It has been suggested that the spurt-like release of oxytocin taking place during continuous suckling is due to a negative feedback operating through the level of oxytocin in the blood. Lincoln (1974) cannulated a nipple and observed the pulsatile ejection of milk that occurred in response to suckling of the remaining nipples. Injections of exogenous oxytocin failed to modulate the normal rhythm of ejection, so that it is unlikely that oxytocin, itself, governs its rhythmic release.

Emotional Inhibition

When a lactating doe was forcibly restrained and allowed to suckle, the milk-yield, as measured by the change in weight of the pups, was reduced by from 20 to 100 per cent. With a number of the animals the injection of 50 mU of oxytocin restored milk removal to normal, but in others this replacement therapy was only effective after anaesthesia of the doe (Cross, 1955).

Prostaglandins

Species Variability

The effects of prostaglandins on suckling vary with the species; as stimulators of smooth muscle activity they could obviously affect the milk-ejection process, and as substances that influence neurotransmission they could also influence the release of oxytocin.

Rats. In lactating rats Prilusky and Deis (1970) observed an inhibition of milk-ejection proportional to the dose of PGF_{2a}, an effect that was apparently mediated by a central block of oxytocin release. When release of oxytocin was inhibited by pentobarbitone, PGF_{2a} had a direct action on the mammary glands.

Women. On the other hand, in women, Cobo et al. (1974) found that PGF_{2a} caused an increase in pressure in the mammillary duct; the dependence on dose was not so clear-cut as that with oxytocin-induced pressure changes, and the long latency suggested that the stimulation of milk-ejection, reflected by this increased pressure, was secondary to an effect on oxytocin release.

Guinea-Pig. In the guinea-pig, McNeilly and Fox (1971) found that close intra-arterial injection of prostaglandins caused an increase in intra-mammillary pressure, the order of potency being:

$$E_1 > E_2 > F_{2a} > F_{1a}$$

The prostaglandins did not inhibit the effects of oxytocin; in fact their effects were unlikely to be indirect since the latent period for a response was no longer than that for oxytocin injected in the same way.

Rabbit. In the rabbit, Türker and Kiren (1969) found that PGE_1 inhibited the effect of oxytocin on the mammary gland, but it had no effect on its own, so that it was presumably interfering with the action of oxytocin on the myoepithelial cells of the gland.

REFERENCES

(See p. 546 for additional references)

Amenomori, Y., Chen, C. L. and Meites, J. (1970). Serum prolactin levels in rats during different reproductive states. *Endocrinology,* **86,** 506–510.

Beck, J. C. (1972). Conclusions. *Lactogenic Hormones.* Ciba Symp., pp. 403–404.

Benson, G. K., Cowie, A. T., Cox, C. P. and Goldzveig, S. A. (1957). Effects of oestrone and progesterone on mammary development in the guinea-pig. *J. Endocrinol.,* **15,** 126–144.

Beyer, C., Mena, F., Pacheco, D. and Alcaraz, M. (1962). Blockage of lactation by brain-stem lesions in the cat. *Amer. J. Physiol.,* **202,** 465–468.

Bintarningsih, Lyons, W. R., Johnson, R. E. and Li, C. H. (1958). Hormonally-induced lactation in hypophysectomized rats. *Endocrinology,* **63,** 540–548.

Bisset, G. W., Clark, B. J. and Errington, M. L. (1971). The hypothalamic neuro-secretory pathways for the release of oxytocin and vasopressin in the cat. *J. Physiol.,* **217,** 111–131.

Bisset, G. W., Hilton, S. M. and Poisner, A. M. (1967). Hypothalamic pathways for independent release of vasopressin and oxytocin. *Proc. Roy. Soc. B.,* **166,** 422–442.

Bodenheimer, T. S. and Brightman, M. W. (1968). A blood-brain barrier to peroxidase in capillaries surrounded by perivascular spaces. *Amer. J. Anat.,* **122,** 249–267.

Bonet, H. G., Gómez, F. and Friesen, H. G. (1977). Prolactin and estrogen binding sites in the mammary gland of the lactating and non-lactating rat. *Endocrinology,* **101,** 1111–1121.

Brew, K. (1969). Secretion of α-lactalbumin into milk and its relevance to the organization and control of lactose synthetase. *Nature,* **223,** 671–672.

Brooks, C. McC., Ishikawa, T., Koizumi, K. and Lu, H.-H. (1966). Activity of neurones in the paraventricular nucleus of the hypothalamus and its control. *J. Physiol.,* **182,** 217–231.

Bruce, H. M. (1958). Suckling stimulus and lactation. *Proc. Roy. Soc. B.,* **149,** 421–423.

Bruce, J. O. and Ramirez, V. D. (1970). Site of action of the inhibitory effect of oestrogen upon lactation. *Neuroendocrinol.,* **6,** 19–24.

Chadwick, A. (1962). The onset of lactose synthesis after injection of prolactin. *Biochem. J.,* **85,** 554–558.

Chadwick, A. and Folley, S. J. (1962). Lactogenesis in pseudopregnant rabbits treated with ACTH. *J. Endocrinol.,* **24,** xi-xii.

Chadwick, A., Folley, S. J. and Gemzell, C. A. (1961). Lactogenic activity of human pituitary growth hormone. *Lancet,* **(2),** 241–243.

Chen, C. L. *et al.* (1970). Serum prolactin levels in rats with pituitary transplants in hypothalamic lesions. *Neuroendocrinol.,* **6,** 220–227.

Chen, C. L. and Meites, J. (1970). Effects of estrogen and progesterone on serum and pituitary prolactin levels in ovariectomized rats. *Endocrinology,* **86,** 503–505.

Clemens, J. A., Shaar, C. J., Kleber, J. W. and Tandy, W. A. (1971). Recriprocal control by the preoptic area of LH and prolactin. *Exp. Brain Res.,* **12,** 250–253.

Cobo, E., Rodriguez, A. and Villamizar, M. de (1974). Milk ejecting activity induced by prostaglandin $F_2\alpha$. *Am. J. Obstet. Gynec.,* **118,** 831–836.

Cowie, A. T. (1957). The maintenance of lactation in the rat after hypophysectomy. *J. Endocrinol.,* **16,** 135–147.

Cowie, A. T. *et al.* (1964). Lactation in the goat after section of the pituitary stalk. *J. Endocrinol.*, **28**, 253–265.

Cowie, A. T., Hartmann, P. E. and Turvey, A. (1969). The maintenance of lactation in the rabbit after hypophysectomy. *J. Endocrinol.*, **43**, 651–662.

Cowie, A. T., Knaggs, G. S., Tindal, J. S. and Turvey, A. (1968). The milking stimulus and mammary growth in the goat. *J. Endocrinol.*, **40**, 243–252.

Cowie, A. T. and Tindal, J. S. (1971). *The Physiology of Lactation.* Arnold, London.

Cowie, A. T., Tindal, J. S. and Benson, G. K. (1960). Pituitary grafts and milk secretion in hypophysectomized rats. *J. Endocrinol.*, **21**, 115–123.

Cross, B. A. (1955). Neurohormonal mechanisms in emotional inhibition of milk ejection. *J. Endocrinol.*, **12**, 29–37.

Cross, B. A. (1973). Unit responses in the hypothalamus. *Frontiers in Neuroendocrinology.* Eds. W. F. Ganong and L. Martini, pp. 133–171.

Cross, B. A. and Harris, G. W. (1952). The role of the neurohypophysis in the milk-ejection reflex. *J. Endocrinol.*, **8**, 148–161.

Deis, R. P. (1967). Mammary gland development by hypothalamic and hypophyseal estrogen implants in male rats. *Acta physiol. Latinam.*, **17**, 115–117.

Denamur, R. and Martinet, J. (1959). Les stimulus nerveux mammaires. Sont ils nécessaires à l'entretien de la lactation chez la chèvre. *C.r. Acad. Sci. Paris*, **248**, 743–746.

Desjardins, C., Pape, M. J. and Tucker, H. A. (1968). Contribution of pregnancy, fetuses, fetal placentas and deciduomas to mammary gland and uterine development. *Endocrinology*, **83**, 907–910.

Djiane, J., Durand, P. and Kelly, P. A. (1977). Evolution of prolactin receptors in rabbit mammary gland during pregnancy and lactation. *Endocrinology*, **100**, 1348–1356.

Dyball, R. E. J. (1971). Oxytocin and ADH secretion in relation to electrical activity in antidromically identified supraoptic and paraventricular units. *J. Physiol.*, **214**, 245–256.

Dyball, R. E. J. and Dyer, R. G. (1971). Plasma oxytocin concentration and paraventricular neurone activity in rats with diencephalic islands and intact brains. *J. Physiol.*, **216**, 227–235.

Dyer, R. G., Dyball, R. E. J. and Morris, J. F. (1973). The effect of hypothalamic deafferentation upon the ultrastructure and hormone content of the paraventricular nucleus. *J. Endocrinol.*, **57**, 509–516.

Eayrs, J. T. and Baddeley, R. M. (1956). Neural pathways in lactation. *J. Anat.*, **90**, 161–171.

Fields, H. (1945). The influence of stilbestrol upon lactation. *Amer. J. Obst. Gyn.*, **49**, 385–390.

Folley, S. J. (1956). *The Physiology and Biochemistry of Lactation.* Oliver & Boyd, Edinburgh and London.

Foss, G. L. and Short, D. (1951). Abnormal lactation. *J. Obst. Gynecol. B.E.*, **58**, 35–46.

Fox, C. A. and Knaggs, G. S. (1969). Milk-ejection activity (oxytocin) in peripheral venous blood in man during lactation and in association with coitus. *J. Endocrinol.*, **45**, 145–146.

Frantz, A. G., Kleinberg, D. L. and Noel, G. L. (1972). Studies on prolactin in man. *Rec. Progr. Horm. Res.*, **28**, 527–590.

Friesen, H. (1965). Further purification and characterization of a placental protein with immunological similarity to human growth hormone. *Nature*, **208**, 1214–1215.

Friesen, H., Belanger, C., Guyda, H. and Huang, P. (1972). The synthesis and secretion of placental and pituitary prolactin. In *Lactogenic Hormones.* Ciba Symp., pp. 83–103. Churchill-Livingstone, London.

Fuchs, A.-R. and Wagner, G. (1963). Quantitative aspects of release of oxytocin by suckling in unanaesthetized rabbits. *Acta endocrinol.*, **44**, 581–592.

Gale, C. C. and Larsson, B. (1963). Radiation induced "hypophysectomy" and hypothalamic lesions in lactating goats. *Acta physiol. scand.*, **59**, 299–318.

Ginsburg, M. and Smith, M. W. (1959). The fate of oxytocin in male and female rats. *Brit. J. Pharmacol.*, **14**, 327–333.

Grosvenor, C. E. (1965). Evidence that exteroceptive stimuli can release prolactin from the pituitary gland of the lactating rat. *Endocrinology*, **76**, 340–344.

Grosvenor, C. E. and Turner, C. W. (1958). Assay of lactogenic hormone. *Endocrinology*, **63**, 530–534.

Grosvenor, C. E. and Turner, C. W. (1958). Pituitary lactogenic hormone concentrations and milk secretion in lactating rats. *Endocrinology*, **63**, 535–539.

Hart, I. C. (1973). Effect of 2-bromo-α-ergocryptine on milk yield and the level of prolactin and growth hormone in the blood of the goat at milking. *J. Endocrinol.*, **57**, 179–180.

Hart, I. C. and Flux, D. S. (1973). The release of growth hormone in response to milking in the goat during early and late lactation. *J. Endocrinol.*, **57**, 177–178.

Haun, C. K. and Sawyer, C. H. (1960). Initiation of lactation in rabbits following placement of hypothalamic lesions. *Endocrinology*, **67**, 270–272.

Hayward, J. N. and Jennings, D. P. (1973). Activity of magnocellular neuroendocrine cells in the hypothalamus of unanaesthetized monkeys. *J. Physiol.*, **232**, 515–543.

Holcomb, H. H., Costlow, M. E., Buschow, R. A. and McGuire, W. L. (1976). Prolactin binding in rat mammary gland during pregnancy and lactation. *Biochim. biophys. Acta*, **428**, 104–112.

Johke, T. (1970). Factors affecting the plasma prolactin level in the cow and the goat as determined by radioimmunoassay. *Endocrinol. Japon.*, **17**, 393–401.

Jones, E. A. (1972). Studies on the particulate lactose synthetase of mouse mammary gland and the role of α-lactalbumin in the initiation of lactose synthesis. *Biochem. J.*, **126**, 67–78.

Juergens, W. G., Stockdale, F. E., Topper, Y. J. and Elias, J. J. (1965). Hormone-dependent differentiation of mammary gland *in vitro*. *Proc. Nat. Acad. Sci. Wash.*, **54**, 629–634.

Kelly, P. A., Tsushima, T., Shiu, R. P. C. and Friesen, H. G. (1976). Lactogenic and growth hormone-like activities in pregnancy determined by radioreceptor assays. *Endocrinology*, **99**, 765–774.

Koella, W. (1949). Die Beeinflussung der Harnsekretion durch hypothalamische Reizung. *Helv. physiol. pharmac. Acta*, **7**, 498–514.

Kohmoto, K. and Bern, H. A. (1970). Demonstration of mammatrophic activity of the mouse placenta in organ culture and by transplantation. *J. Endocrinol.*, **48**, 99–107.

Kuhn, N. J. (1969). Progesterone withdrawal as the lactogenic trigger in the rat. *J. Endocrinol.*, **44**, 39–54.

Li, C. H. (1972). Recent knowledge of the chemistry of lactogenic hormones. In *Lactogenic Hormones*. Ciba Symp. Churchill-Livingston, pp. 7–22.

Lincoln, D. W. (1974). Does a mechanism of negative feedback determine the intermittent release of oxytocin during suckling? *J. Endocrinol.*, **60**, 193–194.

Lincoln, D. W. and Wakerley, J. B. (1974). Electrophysiological evidence for the activation of supraoptic neurones during the release of oxytocin. *J. Physiol.*, **242**, 533–554.

Lyons, W. R. (1942). The direct mammatrophic action of lactogenic hormone. *Proc. Soc. exp. Biol. Med., N.Y.*, **51**, 308–311.

Lyons, W. R., Li, C. H. and Johnson, R. E. (1958). The hormonal control of mammary growth and lactation. *Rec. Prog. Horm. Res.*, **14**, 219–248.

MacLeod, R. M. and Lehmeyer, J. E. (1972). Regulation of the synthesis and release of prolactin. In *Lactogenic Hormones*. Ciba Symp. Churchill-Livingston, London.

McNeilly, A. S. (1972). The blood levels of oxytocin during suckling and hand-milking in the goat with some observations on the pattern of hormone release. *J. Endocrinol.*, **52**, 177–188.

McNeilly, A. S. and Fox, C. A. (1971). The effects of prostaglandins on the guinea-pig mammary gland. *J. Endocrinol*, **51**, 603–604.

Meites, J. and Turner, C. W. (1947). The induction of lactation during pregnancy in rabbits and the specificity of the lactogenic hormone. *Amer. J. Physiol.*, **150**, 394–399.

Mellenberger, R. W. and Bauman, D. E. (1974a). Lactose synthesis in rabbit mammary tissue during pregnancy and lactation. *Biochem. J.*, **142**, 659–665.

Mellenberger, R. W. and Bauman, D. E. (1974b). Fatty acid synthesis in rabbit mammary tissue during pregnancy and lactation. *Biochem. J.*, **138**, 373–379.

Mena, F. and Beyer, C. (1963). Effect of high spinal section on established lactation in the rabbit. *Amer. J. Physiol.*, **205**, 313–316.

Mena, F. and Grosvenor, C. E. (1968). Effect of number of pups upon suckling-induced fall in pituitary prolactin concentration and milk ejection in the rat. *Endocrinology*, **82**, 623–626.

Mena, F. and Grosvenor, C. E. (1972). Effect of suckling and of extero receptive stimulation upon prolactin release in the rat during late lactation. *J. Endocrinol.*, **52**, 11–22.

Mills, E. S. and Topper, Y. J. (1970). Some ultrastructural effects of insulin, hydrocortisone, and prolactin on mammary gland explants. *J. Cell Biol.*, **44**, 210–328.

Minaguchi, H. and Meites, J. (1967). Effects of suckling on hypothalamic LH-releasing factor and prolactin inhibiting factor, and on pituitary LH and prolactin. *Endocrinology*, **80**, 603–607.

Moss, R. L., Urban, I. and Cross, B. A. (1972). Microelectrophoresis of cholinergic and aminergic drugs on paraventricular neurons. *Amer. J. Physiol.*, **223**, 310–318.

Nandi, S. and Bern, H. A. (1961). The hormones responsible for lactogenesis in BALB/cCrgl mice. *Gen. comp. Endocrinol.*, **1**, 195–210.

Negoro, H., Visessuwan, S. and Holland, R. C. (1973a). Unit activity in the paraventricular nucleus of female rats at different stages of the reproductive cycle and after ovariectomy, with or without oestrogen or progesterone treatment. *J. Endocrinol.*, **59**, 545–558.

Negoro, H., Visessuwan, S. and Holland, R. C. (1973b). Reflex activation of paraventricular nucleus units during the reproductive cycle and in ovariectomized rats treated with oestrogen or progesterone. *J. Endocrinol.*, **59**, 559–567.

Newton, W. H. and Richardson, K. C. (1940). The secretion of milk in hypophysectomized pregnant mice. *J. Endocrinol.*, **2**, 322–328.

Nicoll, C. S. and Meites, J. (1964). Prolactin secretion *in vitro*: effects of gonadal and adrenal cortical steroids. *Proc. Soc. exp. Biol. Med.*, *N.Y.* **117**, 579–583.

Nicoll, C. S., Talwalker, P. K. and Meites, J. (1960). Initiation of lactation in rats by nonspecific stress. *Amer. J. Physiol.*, **198**, 1103–1106.

Owens, I. S., Vonderhaar, B. K. and Topper, Y. J. (1973). Concerning the necessary coupling of development to proliferation of mouse mammary epithelial cells. *J. biol. Chem.*, **248**, 472–477.

Palmiter, R. D. (1969). Hormonal induction and regulation of lactose synthetase in mouse mammary gland. *Biochem. J.*, **113**, 409–417.

Pasteels, J. L. (1963). Recherches morphologiques et expérimentales sur la sécrétion de prolactine. *Arch. de Biol.*, **74**, 439–553.

Prilusky, J. and Deis, R. P. (1970). Inhibitory effect of prostaglandin F_2a on oxytocin release and on milk ejection in lactating rats. *J. Endocrinol*, **69**, 395–399.

Ramirez, V. D. and McCann, S. M. (1964). Induction of prolactin secretion by implants of estrogen into the hypothalamo-hypophyseal region of female rats. *Endocrinology*, **75**, 206–214.

Ratner, A. and Meites, J. (1964). Depletion of prolactin-inhibiting activity of rat hypothalamus by estradiol or suckling stimulus. *Endocrinology*, **75**, 377–382.

Ray, E. W., Averill, S. C., Lyons, W. R. and Johnson, R. E. (1955). Rat placental hormonal activities corresponding to those of pituitary mamatropin. *Endocrinology*, **56**, 359–373.

Riddle, O., Bates, R. W. and Dykshorn, S. W. (1933). The preparation, identification and assay of prolactin—a hormone of the anterior pituitary. *Amer. J. Physiol.*, **105**, 191–216.

Rivera, E. M. (1964). Differential responsiveness to hormones of C_3H and A mouse mammary tissue in organ culture. *Endocrinology*, **74**, 853–864.

Roth, L. J. and Rosenblatt, J. S. (1966). Mammary glands of pregnant rats: development stimulated by licking. *Science*, **151**, 1403–1404.

Sander, S. and Attramodal, A. (1968). An autoradiographic study of oestradiol incorporation in the breast tissue of female rats. *Acta Endocrinol.*, **58**, 235–242.

Sar, M. and Meites, J. (1969). Effects of suckling on pituitary release of prolactin, GH, and TSH in postpartum lactating rats. *Neuroendocrinol.*, **4**, 25–31.

Saxena, B. N., Refetoff, S., Emerson, K. and Selenkov, H. A. (1968). A rapid radioimmunoassay for human placental lactogen. *Amer. J. Obst. Gynecol.*, **101**, 874–885.

Selye, H. (1934). On the nervous control of lactation. *Amer. J. Physiol.*, **107**, 535–538.

Sherwood, L. M. (1967). Similarities in the chemical structure of human placental lactogen and pituitary growth hormone. *Proc. Nat. Acad. Sci. Wash.*, **58**, 2307–2314.

Sherwood, L. M., Handwerger, S. and McLaurin, W. D. (1972). The structure and function of human placental lactogen. In *Lactogenic Hormones*. Ciba Symp. Churchill-Livingston, London, pp. 27–44.

Shiu, R. P. C. and Friesen, H. G. (1974). Properties of a prolactin receptor from the mammary gland. *Biochem. J.* **140**, 301–311.

Shiu, R. P. C. and Friesen, H. G. (1976). Blockade of prolactin by an antiserum to the receptors. *Science N.Y.*, **192**, 259–261.

Shyamala, G. and Nandi, S. (1972). Interactions of 6,7-^{3}H-17β-estradiol with the mouse lactating mammary tissue *in vivo* and *in vitro*. *Endocrinology*, **91**, 861–867.

Simpson, A. A., Simpson, M. H. W. and Kulkarni, P. N. (1973). Prolactin production and lactogenesis in rats after ovariectomy in late pregnancy. *J. Endocrinol.*, **57**, 425–429.

Sinha, Y. N., Selby, F. W. and Vanderlaan, W. P. (1974). Relationship of prolactin and growth hormone to mammary function during pregnancy and lactation in the C3H/ST mouse. *J. Endocrinol.*, **61**, 219–229.

Smith, E. R. and Davidson, J. M. (1968). Role of estrogen in the cerebral control of puberty in female rats. *Endocrinology*, **82**, 100–108.

Smith, M. S. (1978). Hypothalamic-pituitary responsiveness during lactation in the rat: estrogen-induced luteinizing hormone surges. *Endocrinology*, **102**, 121–127.

Stockdale, F. E. and Topper, Y. J. (1966). The role of DNA synthesis and mitosis in hormone-dependent differentiation. *Proc. Nat. Acad. Sci. Wash.*, **56,** 1283–1289.

Sundsten, J. W., Novin, D. and Cross, B. A. (1970). Identification and distribution of paraventricular units excited by stimulation of the neural lobe of the hypophysis. *Exp. Neurol.*, **26,** 316–329.

Talwalker, P. K., Nicoll, C. S. and Meites, J. (1961). Induction of mammary secretion in pregnant rats and rabbits by hydrocortisone acetate. *Endocrinology*, **69,** 802–808.

Tindal, J. S., Beyer, C. and Sawyer, C. H. (1963). Milk ejection reflex and the maintenance of lactation in the rabbit. *Endocrinology*, **72,** 720–724.

Tucker, H. A. (1964). Influence of number of suckling young on nucleic acid content of lactating rat mammary gland. *Proc. Soc. exp. Biol. Med., N.Y.*, **116,** 218–220.

Türker, R. K. and Kiran, B. K. (1969). Interaction of prostaglandin E_1 with oxytocin on mammary gland of the lactating rabbit. *Eur. J. Pharmacol.*, **8,** 377–379.

Turkington, R. W. (1968). Induction of milk protein synthesis by placental lactogen and prolactin *in vitro*. *Endocrinology*, **82,** 575–583.

Turkington, R. W. (1970). Stimulation of RNA synthesis in isolated mammary cells by insulin and prolactin bound to Sepharose. *Biochem. biophys. Res. Commun.*, **41,** 1362–1367.

Turkington, R. W. and Hill, R. L. (1969). Lactose synthetase: progesterone inhibition of the induction of α-lactalbumin. *Science*, **163,** 1458–1460.

de Voe, W. F., Ramirez, V. D. and McCann, S. M. (1966). Induction of mammary secretion by hypothalamic lesions in male rats. *Endocrinology*, **78,** 158–164.

Vonderhaar, B. K., Owens, I. S. and Topper, Y. J. (1973). An early effect of prolactin on the formation of α-lactalbumin by mouse mammary epithelial cells. *J. biol. Chem.*, **248,** 467–471.

Voogt, J. L., Sar, M. and Meites, J. (1969). Influence of cycling, pregnancy, labor, and suckling on corticosterone-ACTH levels. *Amer. J. Physiol.*, **216,** 655–658.

Wakerley, J. B. and Lincoln, D. W. (1973). The milk-ejection reflex of the rat. *J. Endocrinol.*, **57,** 477–493.

Wellings, S. R. (1969). Ultrastructural basis of lactogenesis. In *Lactogenesis. The Initiation of Milk Secretion at Parturition*. Eds. M. Reynolds and S. J. Folley. University of Penna Press, Philadelphia, pp. 5–25.

Wiest, W. G., Kidwell, W. R. and Balogh, K. (1968). Progesterone catabolism in the rat ovary: a regulatory mechanism for progestational potency during pregnancy. *Endocrinology*, **82,** 844–859.

Wolińska, E., Polkowska, J. and Domański, E. (1977). The hypothalamic centres involved in the control of production and release of prolactin in sheep. *J. Endocrinol*, **73,** 21–29.

Yoshinaga, K., Hawkins, R. A. and Stocker, J. F. (1969). Estrogen secretion by the rat ovary *in vivo* during the estrous cycle and pregnancy. *Endocrinology*, **85,** 103–112.

Zarrow, M. X., Schlein, P. A. and Denenberg, V. H. (1972). Corticosterone release in the lactating rat following olfactory stimulation from rat pup. *Endoc. Soc. Program.*, 53rd Meeting, San Francisco, Cal., U.S.A. (Quoted by Mena and Grosvenor, 1972).

CHAPTER 8

Puberty and Old Age

Pubertal Period

The ability to reproduce, as evidenced by the power of the female ovary to produce follicles capable of maturing and producing ripe ova, and of the male to produce spermatozoa, does not develop in a continous fashion, so that after birth there is a period of quiescence in which development either does not occur at all, or is very slow. The maturation process that brings into action these processes of ovulation and spermatogenesis takes place at a definite age and over a definite period, according to the species. We may describe this *pubertal period* as the phase of bodily growth during which the gonads secrete hormones in sufficient amount to cause accelerated growth of the genital organs and the appearance of secondary sexual characteristics. In the human, this period is spread over some two to four years, and occurs between the ages of 11 and 15 in boys and 9 and 13 in girls. In a short-lived animal like the rat the transition from immaturity through puberty takes place in a day or two and occurs at the age of about 40 days.

Changes in the Gonads

In the adult male the bulk of the testis is made up of the seminiferous tubules, containing the spermatogenic cells, and the interstitial cells, which secrete the steroid hormones, mainly testosterone. With the onset of puberty the mitotic activity in spermatogonia becomes pronounced and spermatogenesis really begins. The interstitial cells of Leydig, secreting testosterone, make up the bulk of the testis in early embryonic life, but at birth there are very few, and puberty is characterized by the reappearance of these cells. So far as the ovary is concerned, there is little structural change during early childhood, the total weights steadily increasing (Fig. 8.1), and the event that characterizes the onset of puberty is essentially the onset of ovulation; as Figure 8.1 shows, uterine weight exhibits a rapid increase, in the human, at about 10 years.

Accessory Reproductive Organs

As we have seen, prenatally there is a differentiation of an essentially female form to one of maleness, in the genetic male, governed by secretions of the foetal testis. *In utero* the foetus is subject to the influence of maternal steroid hormones so that at birth some of the accessory organs, whose growth and development are governed by steroid hormones, may be precociously developed, e.g. the mammary glands, but after birth when the influence of the maternal hormones is withdrawn there is regression, e.g. of the infantile uterus or phallus. Puberty is characterized by the rapid development of the accessory organs, so that the most obvious feature of puberty in the female is undoubtedly mammary development.

Secondary Sex Characters

These are perhaps the most striking features of the onset of puberty, e.g. the growth of the mane of the lion, the antlers of deer, comb of the cock, and so on. In the human, axillary and pubic hair growth are characteristic features. Secretion of sweat and scent glands is another factor; so far as the human is concerned, the axillary sweat glands are

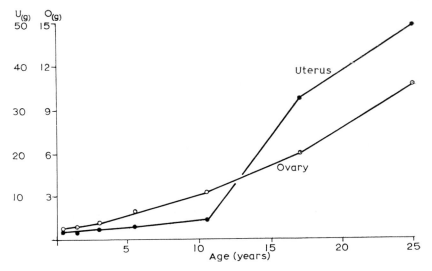

Fig. 8.1. The growth of the uterus (U) and ovary (O) in the human. Ordinates are weight in grammes. (Donovan and Van der Werff ten Bosch, *Physiology of Puberty.*)

inactive in children and become functional at puberty; acne is a typical feature of puberty, and results from the secretion of androgens in both male and female. In subhuman primates an interesting characteristic is the sexual skin; and changes in this take place just before puberty in many species.

Describing the Onset of Puberty

Stages in the Male

In the human, the changes associated with puberty do not take place simultaneously, and not always in the same order. In general, however, it has been possible to describe a series of stages that represent successive steps towards complete maturation. Thus, Lee *et al.* (1970) have described four stages in boys namely *prepuberty*, when there is lack of or only downy pubic hair and no enlargement of the penis and testes over juvenile sizes; *early puberty*, when early testicular and penile enlargement and pigmented pubic hair are present; *late puberty* with a moderate amount of pubic hair, the presence of axillary hair, further genital enlargement or the presence of hair on upper lip, breast hyperplasia or uneven voice pitch; *postpuberty*, if penis and testes are within the range of normal for adults and if a reasonable combination of the following are present: male hairline, hair on chin or chest, deepened voice, male escutcheon, absence of breast hyperplasia.

Stages in the Female

A corresponding set of stages for the female consisted of *prepuberty*, with no pubic hair or breast development; *early puberty*, marked by pigmented pubic hair and/or budding breasts with a marked general growth-spurt. *Late puberty* began at menarche; girls at this stage had moderate pubic hair, axillary hair and a secondary mound of developing breast. *Post-pubertal* girls menstruated regularly, had adult breasts without a secondary mound, and adult axillary and pubic hair. The value of assessing these pubertal stages is that, when changes in hormonal levels in the blood or urine are studied, it is found that they correlate much better with pubertal stage than with chronological age. The difference in age at which the three main characteristics of puberty develop is illustrated by Figure 8.2; thus menstruation (menarche) develops latest and is clearly the clue to maturation.

Objective Estimates

In the male the weights of the testes and prostate provide good indices to development, but of course cannot be measured in the intact animal; Figure 8.3 shows the changes in the human male with age;

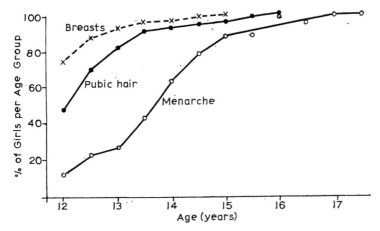

Fig. 8.2. Showing the variations in age of onset of the characteristic signs of puberty in 1,484 girls. (Donovan and Van der Werff ten Bosch, *Physiology of Puberty.*)

and the steep rise in weight of the testes between the ages of 11–12 is indicative of the onset of puberty; the rise in weight of the prostate takes place later. In the female, there are corresponding changes in the weight of ovaries and uterus (Fig. 8.1).

Experimental Animals

In the female rat, the most commonly studied experimental animal, the sign of puberty is taken as the opening of the vaginal canal, which precedes the first ovulatory cycle by a day or two and occurs at around Day 40. The time of onset can be altered by various influences, perhaps the most striking effect being that of the presence of a male, or its smell, seen in mice.

Social Factors Governing the Age of Puberty

Humans

It is well known that the average age at which girls begin to menstruate varies amongst different communities or in the same community with its social development. Thus, Figure 8.4 illustrates the secular trend in the age at menarche derived from historical records of different countries, and a progressive decrease in the age of onset is manifest. It would be tedious to describe the various facts and theories relating to this point, so far as the human is concerned, and the interested reader is referred to the monograph by Donovan and Van der Werff ten Bosch (1965).

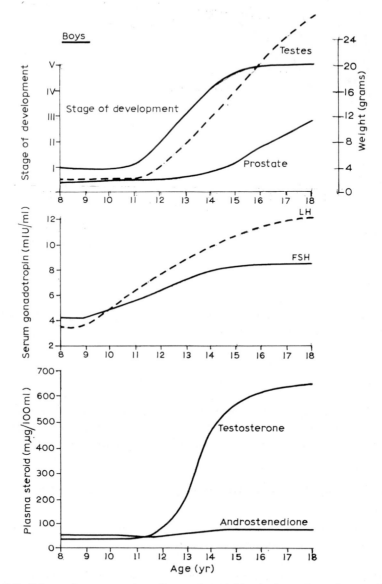

Fig. 8.3. Schematic presentation of several variables in boys throughout puberty. (Odell and Moyer, *Physiology of Reproduction*.)

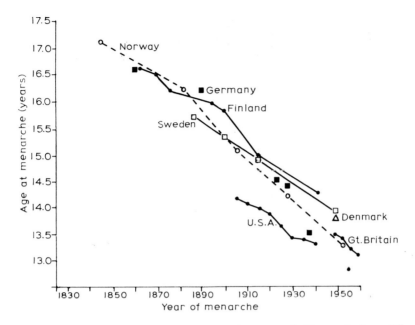

Fig. 8.4. The secular trend in the age at menarche as revealed by results from different countries. (Donovan and Van der Werff ten Bosch, *Physiology of Puberty.*)

Proximity to the Male

We have seen how pregnancy can be blocked by exposing a newly mated female mouse to a strange male, or the odour of one, a blockage due to the inhibition of implantation. Vandenbergh (1967) found that, by exposure of weanling females to an adult male, they accelerated the first oestrus and mating and vaginal opening, as shown by Table I.

TABLE I

Acceleration of puberty in female rats by exposure to an adult male (Vandenbergh, 1967)

Treatment	Vaginal Opening	1st Oestrus	1st Mating
Male Day 21	30·5	37·1	40·6
Male Day 30	30·4	41·9	44·4
Male Day 38	32·4	45·6	46·6
No Male	34·9	57·1	

When males were kept in separate rooms, in cages, the soiled bedding from these, sprinkled in the cages of the females, was sufficient to accelerate puberty (Table II) (Vandenbergh, 1969).

TABLE II

Effects of exposure of weanling female rats to males on puberty (Vandenbergh, 1969)

Treatment	Vaginal Opening	1st Oestrus
Intact Male	30·9	39·6
Castrated Male	39·3	54·6
Male behind mesh	31·2	40·6
Male Odour*	29·5	42·3
Male Odour†	32·7	45·4
No Male Odour	34·9	54·6

* The male was exposed to a female kept on heat with oestrogen.
† The male was kept alone.

Grouping of Females

Stiff et al. (1974) have pointed out that the clock-like regularity with which laboratory mice can pass through their ovulatory cycles depends very much on the conditions under which they are kept. Thus if females are grouped together in the absence of any male their cycles become irregular and in fact puberty may be considerably delayed; thus, the average time for vaginal opening in the isolated female was 23·3 days whereas when they were grouped it was 28·6 days. In all, then, there are three main "social" conditions, single females, grouped females, and females in the presence of a male. Table III illustrates some results of Stiff et al. The difference between grouped females without a male and a single female with male amounts to as much as 17 days, a remarkable effect of social conditions and equivalent to adding three months to a human pregnancy!

TABLE III

Effects of grouping of females on development of puberty (Stiff et al., 1974)

	Time to vaginal opening in days	
	Single Females	Grouped Females
Male Present	28·0	35·7
No Male Present	35·9	55·1

Pheromones

LH-Surge. The agent responsible for accelerating puberty belongs to the class of pheromones discussed earlier (p. 242). The acceleratory effect on the female is apparently exerted through increased hypophyseal activity since circulating levels of luteinizing hormone, LH, rise some fourfold after exposure to an adult male, a "surge" that is followed by a rise in plasma oestrogens. The mere LH-surge is, of itself, not sufficient to bring on sexual maturity since withdrawal of the stimulus immediately after the onset does not lead to maturity; the animal must be exposed for some 36 hr for this to occur. Apparently the action of the pheromone helps to mature the positive feedback necessary for ovulation by increasing the secretion of ovarian oestrogen. According to Vandenbergh et al. (1975), the active substance in the urine is non-volatile and so must influence the female through contact; it is apparently a polypeptide with a molecular weight of about 860 daltons.

Synchronization of Oestrus. Bronson (1971) suggested that the pheromone responsible for synchronization of oestrus in females (p. 244) was the same as that for accelerating puberty, but this cannot be true if this latter is non-volatile; furthermore the pheromone responsible for acceleration of puberty is not secreted by the preputial gland whereas that responsible for synchronization is (cf. Vandenbergh et al., 1975).

Diet

A high protein diet favoured maturation of mice, but this was relatively insignificant compared with the more powerful effects of male proximity (Vandenbergh et al., 1972).

Secretion of the Hormones

The pituitary gonadotrophic hormones promote growth and development of the gonads, namely the testes and ovaries; they also promote the secretion of their characteristic steroid hormones, oestrogens and androgens; and development of the accessory reproductive organs, penis prostate, uterus, etc., relies on these steroid secretions. Moreover, the adrenal gland synthesizes and secretes the sex steroids, e.g. oestrogen, androsterone, dehydroepiandrosterone, and so on. Thus, the levels of steroids and of gonadotrophins in the blood will doubtless be reflected in the changes taking place with development.

Steroid Hormones

So far as the steroids are concerned, the secretion of both "male" and "female" steroids by boys and girls is about the same up to the age of approaching puberty (Fig. 8.5); only after the age of 10–11 does the difference in sexes become manifest, with the male excretion of androgens (in this case 17-ketosteroids) exceeding that of the female, whilst the male excretion of oestrogens fails to show the pubertal rise characteristic of the female. The source of the prepubertal sex steroids is

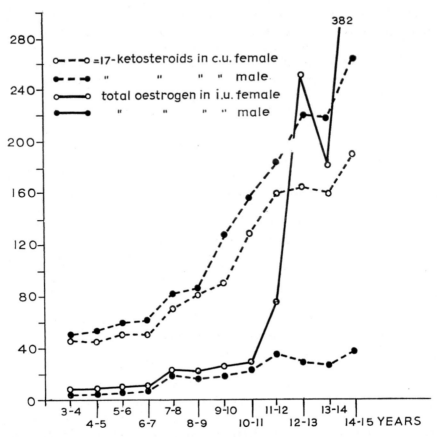

Fig. 8.5. Excretion of 17-ketosteroids and oestrogens in girls and boys of different ages. Note that it is the excretion of oestrogens by the female that shows the sharp rise at about 10–11 years. (Nathanson *et al.*, *Endocrinology*.)

undoubtedly the adrenal gland, and even post-puberty the secretion of 17-ketosteroids (androsterone and dehydroepiandrosterone) is predominantly due to this gland, rather than the gonads. After puberty in the male, the relatively larger secretion of these androgens may be due to the contribution of the testes. The steroid secretory patterns described by Figure 8.5 are derived from urinary excretion; the development of highly sensitive radio-immunoassays for steroid hormones has permitted the estimation of the plasma levels; Figure 8.6 shows the linear increase in concentration of oestradiol in the plasma of girls when plotted against pubertal stage.

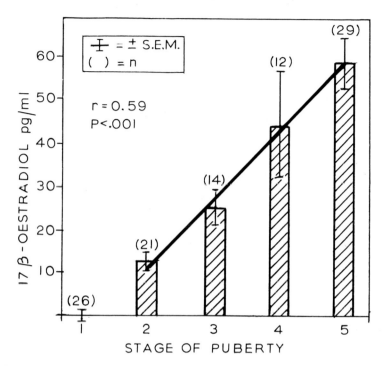

Fig. 8.6. Plasma levels of oestradiol-17β in women throughout puberty. (Jenner *et al.*, *J. clin. Endocrinol.*)

The Gonadotrophins

Corresponding with the absence of significant difference between boys and girls before puberty, so far as steroid hormone secretion is concerned, we find negligible differences in the patterns of secretion of FSH (ICSH) and LH between the human sexes up to the onset of puberty, which occurs earlier in the female (Fig. 8.7). The changes in serum FSH over the critical pubertal period are shown in Fig. 8.8, where the concentrations have been plotted against stage in sexual development, some five stages being defined, essentially similar to those described earlier. Changes in serum LH are shown in Figures 8.9 and 8.10. The excretion of 17-ketosteroids paralleled the serum LH-levels in boys, but whether it is the adrenal trophin that provokes secretion of these androgens is by no means certain.

Changes in the Rat. Figures 8.11 and 8.12 show the changes in blood LH and FSH taking place with age in rats; the pubertal rise in LH in the female is striking; there is also an early (Day 10) rise in FSH in the *female* which may be related to a delay in onset of negative

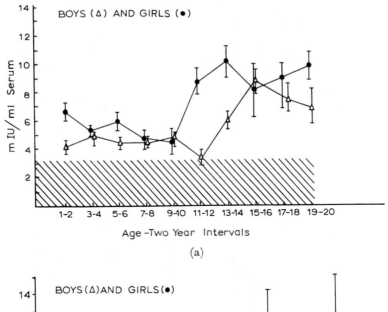

(a)

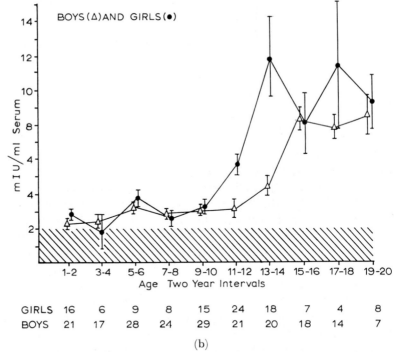

	1-2	3-4	5-6	7-8	9-10	11-12	13-14	15-16	17-18	19-20
GIRLS	16	6	9	8	15	24	18	7	4	8
BOYS	21	17	28	24	29	21	20	18	14	7

(b)

Fig. 8.7. (a) Serum concentrations of LH in boys and girls aged 1 through 20 analysed at 2-year intervals. (b) Serum concentrations of FSH in boys and girls aged 1 through 20 analysed at 2-year intervals.

The shaded areas are below the mean minimal detectable doses. Vertical lines indicate standard errors. (Lee *et al., J. clin. Endrocinol.*)

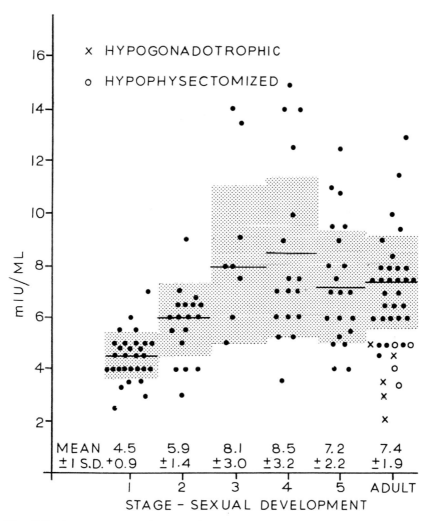

Fig. 8.8. Serum concentrations of FSH in males as a function of stage in sexual development. (Raiti *et al.*, *Metabolism.*)

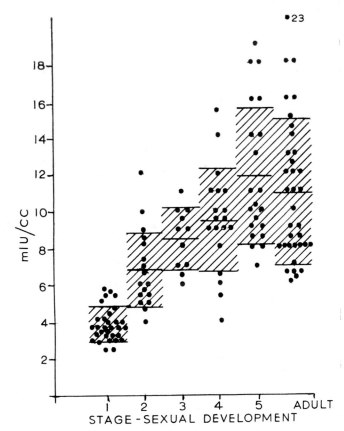

Fig. 8.9. Serum-LH in males as a function of sexual development. (Johanson *et al.*, *J. Pediatr.*)

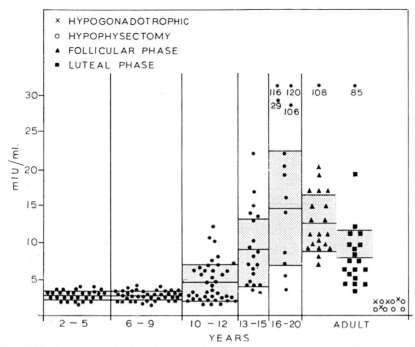

×	HYPOGONADOTROPHIC
o	HYPOPHYSECTOMY
▲	FOLLICULAR PHASE
■	LUTEAL PHASE

Fig. 8.10. Serum-LH in females as a function of age. Representative values for hypogonadotrophic subjects are included. (Johanson *et al., J. Pediatr.*)

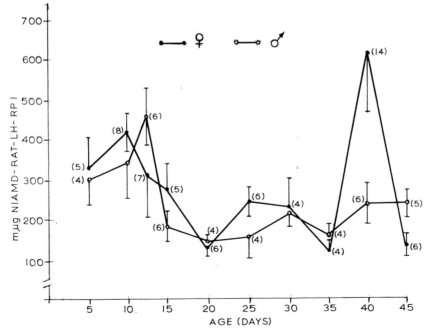

Fig. 8.11. Plasma levels of LH as a function of age in immature female and male rats. Numbers in parentheses indicate animals used; at 5 days each number represents the pool of 2–3 rats. The 40-day group includes 9 of 14 rats with opened vagina. (Ojeda and Ramirez, *Endocrinology.*)

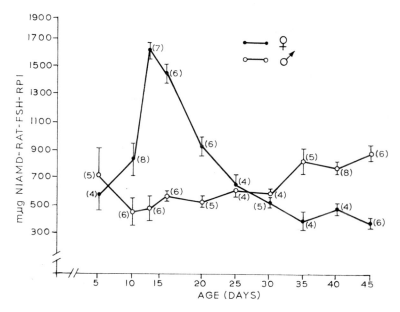

Fig. 8.12. Plasma levels of FSH in growing immature female and male rats. Numbers in parentheses indicate numbers of animals used. At five days, each number represents the pool of 2–3 rats. Vertical lines indicate standard error. (Ojeda and Ramirez, *Endocrinology.*)

feedback from plasma steroids; thus, Ojeda and Ramirez (1972) found that the response to hemicastration, consisting in a raised FSH-level due to release from negative feedback, occurred as soon as ten days of age in the male whereas the corresponding phenomenon in the female occurred by 15–20 days after birth. Figure 8.13 illustrates results of a more recent study in which gonadotrophin levels were examined from an earlier date; this reveals an initially very high level of FSH in the blood of the male rat immediately after birth; on the graph are indicated the stages in testicular development at different ages. Meijs-Roelofs (1975) concentrated on the days preceding and following first ovulation and, as Figure 8.14 shows, there was a peak of LH, FSH, prolactin, progesterone and oestradiol immediately preceding the first ovulation. There was little to differentiate the onset of first ovulation from later ones.

When portal blood was sampled in the anaesthetized rat, Sarkar *et al.* measured a surge in immunoreactive LHRH at the same time as the LH-surge, whilst pituitary responsiveness to the releasing hormone

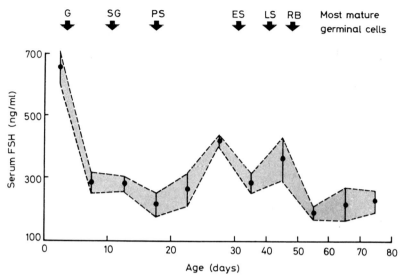

Fig. 8.13. Serum-FSH levels during development in the male rat. The times at which different germinal cells appear in the testis are represented: G = gonocytes; SG = spermatogonia; PS = primary spermatocytes; ES = early spermatids; LS = late spermatids; RB = residual bodies. (Lee *et al.*, *J. Reprod. Fert.*)

was higher on the day before vaginal opening than on the two succeeding days. The precocious ovulation induced by injection of pregnant mare's serum was likewise accompanied by a surge of LHRH in the hypophyseal portal blood (Fink and Sarkar, 1978). Thus, the first (puberal) surge of LH in the rat is similar in origin to that occurring in the mature animal, and could also be due to the development of a positive feedback of oestrogen (p. 136).*

Anti-LHRH Administration. The importance of a continuous secretion of gonadotrophins during the pre-pubertal period is demonstrated by the effects of administration of an anti-serum to LH–RH. According to Bercu *et al.* (1977), who treated male rats with the antiserum on Days 1, 5, 10, 15 and 34, injections in the early stages produced a permanent decrease in fertility (but not complete sterility) in the male, associated with decreased size of penis and seminal vesicles. It would seem, therefore, that the normal development of

---

* As in the adult, manipulation of the action of catecholamines, through injections of the catecholamine synthesis inhibitors, α-methyl-*p*-tyrosine or 6-hydroxydopamine, had no effect on vaginal opening; however, the LH-surge in response to PMS could be reduced by α-methyl-*p*-tyrosine, and restored by administering dihydroxyphenylserine at the same time (Fink and Sarkar, 1978).

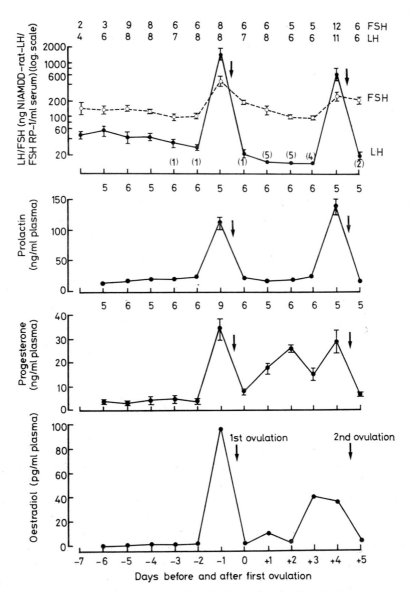

Fig. 8.14. Serum levels of LH and FSH and plasma levels of prolactin, progesterone and oestradiol in female rats from 7 days before until 5 days after the first day of ovulation (Day 0). Numbers on the top of each graph represent numbers of animals. Numbers in parentheses represent numbers of animals with undetectable LH-levels (<15 ng/ml). (Meijs-Roelofs *J. Endocrinol.*)

pituitary control over gonadal development requires the activity of the hypothalamic releasing hormone from as early as Day 1.

The Gonadostat Hypothesis

Evidence from a number of sources indicates that the final maturation of the reproductive system leading to complete puberty and the beginning of reproductive activity involves the secretion of hormones, activated through the hypothalamic-pituitary complex.

Hypothalamic Lesions

Thus, Donovan and Van der Werff ten Bosch (1956) were the first to show that a lesion in the hypothalamus could bring on precocious oestrus; normally the ferret does not mate in winter but, by making a lesion in the anterior hypothalamic region, just behind the optic chiasma, they were able to mate the lesioned females with testerosterone-treated males. Later (1959) they made systematic lesions in groups of rats at different ages, and as Table IV shows, they were able

TABLE IV

The effect of hypothalamic lesions placed at different ages on vaginal opening in the rat (Donovan and Van der Werff ten Bosch, 1959)

| Age at operation (days) | Age at vaginal opening of earliest rat (days) | | | Average age at vaginal opening of earliest ⅓ of all rats | | |
	Blank	Lesion	Advance-ment (days)	Blank	Lesion	Advance-ment (days)
14–15	34	30	4			
14*	32	27	5	32·9	28·3	4·6
10*	33	30	3	37·9	32·5	5·4
3–4*	35	28	7	39·3	33·7	5·6

* Rats of the same inbred albino strain.

to advance the date of vaginal opening by some 4–5 days, the age at which the lesion was made being a matter of indifference between 3 and 15 days. The important region for lesioning appeared to be the ventral part of the anterior hypothalamus. Again Gellert and Ganong (1960) placed lesions in the arcuate nucleus* in the posterior tuberal

* There is not complete agreement on the sites in the hypothalamus necessary for precocious sexual maturation; the studies of Schiavi (1964) indicate that both anterior lesions

region and obtained precocious maturation of females as indicated by
the time of vaginal opening and the first oestrus (Table V).

TABLE V

Mean ages of vaginal opening and first oestrus in rats with hypothalamic lesions
(Gellert and Ganong, 1960)

Lesion	Age (days) Vaginal Opening	1st Oestrus
Normal controls	41·0	46·4
Arcuate nucleus	32·5	34·8
Control	39·7	42·2
Anterior Hypothalamus	37·0	39·3

Parabiosis

As a working hypothesis, then, we may suppose that the puberal
secretions of gonadotrophins that allow sexual activity are inhibited by
the hypothalamus. A number of other experiments support this
hypothesis; for example Kallas, (1929) joined two infantile rats
(weighing 15–20 g) together in parabiotic union, i.e. made "Siamese
twins" of them so that there was intercommunication of their vascular
supplies. Ovariectomy of the one was associated with hypertrophy of the
ovaries in the parabiotic partner, and premature puberty. Frequently,
the normal partner of the union showed complete oestrus, with
cornification of the vaginal epithelium; the uterus was enlarged and
full of secretion. The appearance was "exactly like that of an animal
injected with anterior pituitary extract".

Negative Feedback. It will be recalled that secretion of gonadotrophins by the anterior pituitary are normally held in check by a
negative feedback from the gonads, so that secretion of steroid hormones,
such as oestrogen, tends to inhibit FSH secretion and thus to hold

(caudal part of optic chiasma, suprachiasmatic nuclei and anterior hypothalamus) and
posterior lesions of the medial hypothalamus (premammillary region, caudal part of ventromedial and dorsomedial nuclei and caudal and dorsal parts of arcuate nucleus) are effective.
He discussed results that are in conflict with this, e.g. of Gellert and Ganong who found no
effect of anterior lesions on precocity. Bauer (1954) found, in a study of precocious puberty
in 24 human subjects, evidence of lesions in the posterior hypothalamus in 17 cases, 9
affecting the mammillary bodies. Corbin and Schottelius (1960) obtained a *delayed* puberty
by lesions in the mammillary body and ventromedial nucleus; this was associated with
obesity, obviously due to lesions in the satiety centre (Volume 3). Anterior hypothalamic
lesions give precocious puberty associated with persistent oestrus, so that Corbin and
Schottelius consider that the anterior hypothalamus contains an inhibitory system, as
postulated by Donovan and Van der Werff ten Bosch, whilst the posterior hypothalamus
contains an excitatory system.

down the blood-level of oestrogen. This negative feedback is revealed, as we have seen, by the appearance of "castration cells" in the anterior pituitary indicative of enhanced secretion of hormones when the gonads are removed; or again, if one gonad is removed the compensatory hypertrophy of the other is brought about by enhanced secretion of gonadotrophins permitted by the reduction in negative feedback caused by the lowered steroid secretion. Thus we may interpret the experiments on the parabiotic rats by saying that the reduced level of steroid in the castrated member of the pair induced hypersecretion of its pituitary, which then caused development of the gonads and premature puberty in the normal member of the pair. Another type of experiment involves injections of antiserum to gonadotrophins (Kupperman *et al.*, 1942); if this treatment is brief, the gonadal atrophy that first appears is followed by gonadal hypertrophy and advancement of puberty. Thus if the animals were injected from Days 10–19 of life, and then killed 1, 3, 5, 7, 9, 12 and 15 days later, the ovarian and uterine weights were as shown in Table VI.

TABLE VI

Ovarian and uterine weights of rats treated from Days 10–19 with antigonadotrophic serum and killed at the stated ages (Kupperman *et al.*, 1942)

Day		Weight (g)	
		Ovaries	Uterus
20	Injected	5·3	12·4
24	Control	8·9	19·0
24	Injected	15·0	39·0
	Control	11·0	20·0
28	Injected	27·0	119·0
	Control	13·0	26·5
34	Injected	54·0	148·0
	Control	16·0	29·0

Development of Negative Feedback

There is little doubt that the negative feedback on the pituitary develops at an early age. Florsheim and Rudko (1968) concluded from their studies that the portal system of communication between hypothalamus and pituitary was established at birth or before, in the rat; and this is consistent with the high levels of gonadotrophins in the blood of prenatal rats. These high levels, and their subsequent fall during early life, suggest the early development of a negative feedback. Caligaris *et al.* (1972) found that the response to ovariectomy, i.e. the rise in LH-secretion, appeared very early; thus if they spayed a rat on Day 1, and examined it 10 days later, they found a rise in blood LH, suggesting that at this early stage the negative feedback was functional.*

Changed Feedback Sensitivity

Ramirez and McCann (1963) found that the sensitivity of immature rats to oestradiol injections, measured by the ability to prevent the rise in LH-secretion occurring at ovariectomy, was greater than that of mature rats; and they suggested that this decreased sensitivity to negative feedback permitted the increased gonadotrophin secretion that was necessary to bring about the final stages of puberty. The gonadostat was thus set at a different level.

More exhaustive studies by Odell and his collaborators (see, for

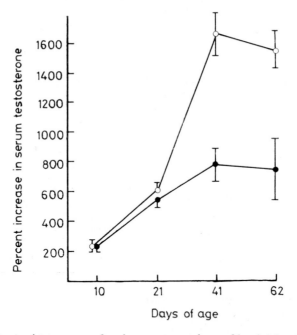

Fig. 8.15. Change in response of male rats to two doses of luteinizing hormone as a function of age. O——O, 30 μg/100 g body wt; ●——●, 10 μg/100 g body wt (NIH-LH-B7). (Odell and Swerdloff, *Rec. Progr. Horm. Res.*)

* Evidence on the time at which negative feedback appears seems to be conflicting; thus Baker and Kragt (1969) concluded from their study of the effects of unilateral castration that it did not appear before 25 days. They point out that the concentration of FSH in the plasma reaches a peak at 20–25 days, suggesting the onset of negative feedback at this point.

example, Odell and Swerdloff, 1976) have shown that the picture
is not so simple; the immature rat does, indeed, show a greater sensitivity
to steroid hormone feedback, but only in respect to the secretion of
FSH; thus the inhibitory effects of testosterone on LH-secretion by
male rats were the same in 10-day-old, 21-day-old and 75-day-old rats.
Odell *et al.* (1974) suggested that an important factor in maturation,
at any rate in the male, was an increased sensitivity of the testes to LH,
a sensitivity that was increased as a result of the high level of FSH in
the circulating blood. They showed that FSH could, indeed, increase
the rat's responsiveness to LH (Odell *et al.*, 1973). Moreover, when
the effects of LH on the rat's blood level of testosterone were measured
at different ages, the response increased strikingly, especially after
Day 21 (Fig. 8.15). It was suggested that the priming with FSH
increased the number of FSH-receptors in the testis.*

Oestrogen Injections. So as to be able to control the injection of
oestrogen into rats and to sample their blood at required moments,
Steele and Weisz (1974) implanted cannulae in a vein and in this way
they were able to demonstrate a very definite change in prepuberal
sensitivity to negative feedback by oestradiol. Thus, a much smaller
dose of oestradiol was required to suppress the level of LH in pre- than in
post-puberal rats. The duration of the suppression of blood-LH was
inversely proportional to the dose of oestradiol so that large doses
produced such a short suppression that it could have been missed by
Swerdloff.

Oestrogen Implantation. The negative feedback of oestrogen in
immature rats was more directly demonstrated by implanting small
amounts chronically into the median eminence region and this
inhibited uterine and ovarian development (Smith and Davidson,
1968). When oestradiol was implanted acutely into the anterior
hypothalamic preoptic area, however, there seemed to be a positive
feedback, so that puberty was brought on precociously. It will be
recalled, however, that lesions in this area carried out by Donovan and
Van der Werff ten Bosch were without effect on the timing of
puberty, but this finding may well be of significance in revealing the
onset of positive feedback, which, as we shall see, may be the ultimate
element in the first onset of ovulation indicative of puberty.

---

* The interested reader should consult an article by Odell and Swerdloff (1976) which
describes elaborate experiments on the sensitivity of the male pituitary to LH as a factor in
maturation. Smith *et al.* (1977) stabilized the plasma testosterone levels of rats by implants
of the hormone; using this technique, they found a greater suppression of LH-secretion, for a
given level of plasma testosterone, in prepuberal rats than in mature animals.

Positive Feedback. The Gonadotrophic Trigger
Oestrogen-Progesterone Balance

Against the view of a slowly developing insensitivity to negative feedback, it has been argued that the real element in acquisition of maturity is the development of a *positive* feedback of oestrogen on the hypothalamus-pituitary complex, a feedback that leads to a surge of LH-secretion and that only begins to develop at the 22nd day of age in the rat, after certain neural structures have developed. Thus, Caligaris *et al.* (1972) found that the secretion of gonadotrophins, e.g. LH, could be raised by progesterone after priming the rats with oestrogen, provided that they were greater than 22 days old. This is illustrated by Figure 8.16, which shows the effects of either an

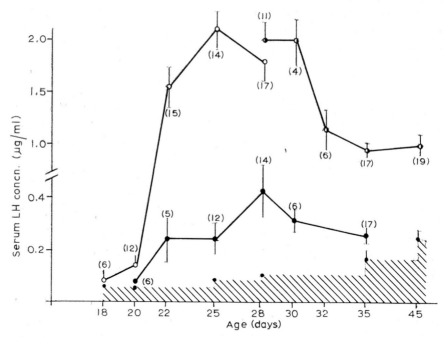

8.16. Release of luteinizing hormone (LH) induced by the injection of oestradiol benzoate or progesterone into spayed oestrogen-primed rats. Ovariectomized rats, primed with 10 μg oestradiol benzoate were injected with a second dose of oestrogen 2 days later (●) or with 1 mg progesterone 3 days later (○). Serum LH was determined on the 3rd day following the priming dose. Each point is the mean value and the vertical lines indicate ± S.E.M. The stippled area indicates the LH values of spayed oestrogen-primed animals. The numbers of animals are shown in parentheses. White-and-black points indicate animals spayed at the time of the priming dose (◑). (Caligaris *et al.*, *J. Endocrinol.*)

injection of progesterone (open circles) or of oestradiol (closed circles) on the serum-LH of ovariectomized rats primed with oestrogen A. maximal responsiveness to injection of steroid occurs around Days 22–28. Thus, Caligaris *et al.* considered that the pituitary-hypothalamus is ready to respond at ten days before puberty, but that it is only the attainment of a correct steroid hormone background that is required to trigger the first ovulation.

Releasing Hormone

The changed responsiveness to steroid hormones could be due to an increased sensitivity of the pituitary to releasing hormone. Ramirez and Sawyer showed that onset of puberty was accompanied by a sharp drop in the pituitary level of LH with a rise in blood concentration. When they extracted the median eminence-stalk region they found (1966), shortly prior to vaginal opening, an abrupt rise in LHRH-content followed by an equally sharp drop on the first day of vaginal opening. This pattern of releasing hormone content could be advanced by as much as a week by doses of oestradiol beginning on Day 26. This increased sensitivity of the pituitary to LH releasing hormone caused by oestradiol occurs very early after birth; thus Wilkinson *et al.* (1977) measured the *in vitro* release of LH from hemipituitaries taken from immature rats primed with oestradiol before removal. The priming potentiated the release as early as 5 days after birth.

Human Responses. Figure 8.17 shows the striking increase in

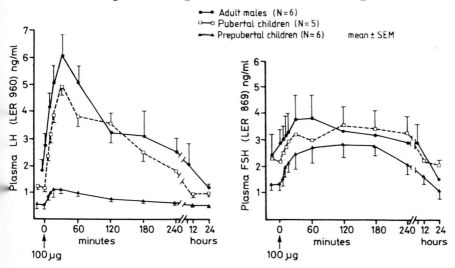

Fig. 8.17. Effects of 100 μg of LHRH on plasma LH and FSH in pre-pubertal (□) and pubertal (○) children. (Roth *et al., J. clin. Endocrinol.*)

release of LH in response to releasing hormone on transition from the pre-puberal (2–13 years) to the puberal (12–18 years) condition in male humans. Interestingly, this change in responsiveness was confined to the release of LH, that of FSH being unaffected by age (Roth *et al.*, 1972).

Monkey. Dierschke *et al.* (1974) have studied the sensitivity of the female monkey to oestradiol injections during maturation. In the mature animal an injection of oestradiol will cause a surge of LH provided it is given during the follicular stage of the menstrual cycle, but in the immature animal this positive feedback does not occur. They found that an oestrogen-induced LH-surge could not be induced until some 4–8 months after the menarche, and this is concordant with the fact that in monkeys and women the menarche precedes fertility by as much as a year, the absence of fertility being due to failure to ovulate. The pituitaries of six premenarche monkeys responded to synthetic LHRH by release of FSH similarly to adult pituitaries. As described by Dierschke *et al.*, the course of maturation in the monkey is as follows:

During premenarche the tonic secretions of LH and FSH are low by comparison with those of the adult; the cyclic system of secretion is inoperative even though the anterior pituitary seems competent to store and, in response to exogenous releasing hormone, to secrete adequate quantities of FSH and perhaps LH. Circulating levels of oestrogens are low, rarely greater than 10 pg/ml and those of pro-gestogen are undetectable. Menarche occurs at 2 years, with "break-through bleeding" in response to continuous oestrogen stimulation since no cyclic changes in serum oestrogen are observed. During the next several months, blood-oestrogen rises slowly, probably in response to the slowly rising FSH and LH; and spontaneous pulses of LH and FSH may occur, asynchronously with each other. At one year after menarche the oestrogen level in the blood rises above the threshold for triggering the surge of LH and FSH, and the system is now competent to respond to oestrogen by a positive feedback, and we now have regular ovulatory cycles.

Hormone-Induced Precocity

The positive action of steroid hormones, leading to the release of "ovulating hormone", e.g. LH or a mixture of FSH and LH, is well established in the sense that, by appropriate treatment of immature animals with either oestrogen or a combination of oestrogen-priming and progesterone injection, the onset of the first ovulation and vaginal opening may be brought forward by many days. For example Ying and Greep (1971) found that repeated doses of oestradiol benzoate induced premature ovulation in rats, and even a single dose given at 30 days caused over 60 per cent of rats mated with fertile males to become pregnant with normal lactation. This dose of oestrogen was followed some 54–56 hours later by a peak in LH and FSH secretion, so that the phenomenon can be described as one of positive feedback (Ying *et al.*, 1971). As we should expect, then, the gonadotrophic hormones, themselves, should be able to induce premature puberty; this is achieved by priming for a few days with pregnant mare's serum (PMS) which presumably accelerates maturation of the follicles; puberty is induced by a final single dose of PMS or of progesterone which McCormack and Meyer considered to result in endogenous release of LH, the ovulating trigger. According to these authors, ovulation can be induced as early as the age of 18 days by this treatment.*

Role of FSH

It is possible, then, that the role of the high level of FSH found pre- and neonatally in the rat (Fig. 8.13, p. 447) is to promote development of the ovary so that by Day 21 or so it is fully capable of ovulation if provided with the appropriate stimulus. An acute, or perhaps gradual, increase in release of LH acting on ovaries primed neonatally with FSH, might be the sufficient trigger to bring about the first pre-ovulatory changes in pituitary gonadotrophin secretion acting through secretion of oestrogen. In this connection it has been claimed that the sensitivity of the hypothalamic-pituitary axis to oestrogen in a positive feedback mechanism develops rapidly towards the onset of puberty, and it may be this sensitivity, together with the increased LH-secretion, that brings about the first ovulatory cycle (Bronson and Desjardins, 1974).

Male Exposure

Bronson and Desjardins (1974) profited by the studies of Vandenbergh (p. 437) which permitted acceleration of onset of puberty by exposure of the female to male odours. The initial response to the male's presence turned out to be an increase in secretion of LH, but not of FSH, in immature females, the initial rise in LH being *followed* by a large rise in circulating oestrogen. Following this, some two days later, the normal pre-ovulatory changes in serum gonadotrophins and plasma progesterone take place at the appropriate time of the day. The results of this study are illustrated by Figure 8.18 which shows the large rise in oestradiol following the initial rise in LH due to male

---

* Zarrow and Wilson (1961) have shown that the follicle of the mouse can respond to PMS plus HCG by ovulation as early as Day 14; in the rat the earliest is 17–18 Days. The point is that the response to a given dose of gonadotrophic hormone falls with age.

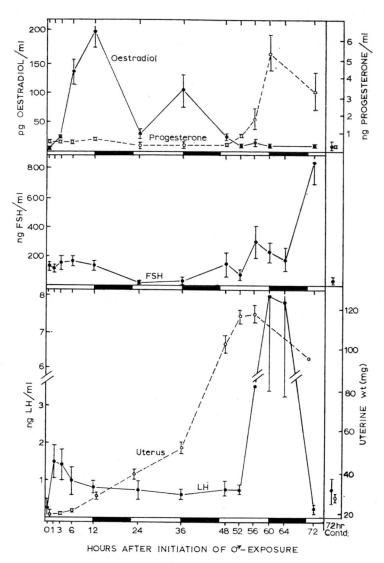

Fig. 8.18. Changes in levels of oestradiol, progesterone, FSH and LH in blood of mice, together with uterine weight, after exposure to the male. (Bronson and Desjardins, *Endocrinology.*)

exposure. The level of progesterone only rises in preparation for the first ovulation of the first oestrus, and in consequence of the large pre-ovulatory rise in LH.

Male Exposure and Oestradiol. Bronson and Desjardins considered that the peaks in blood oestrogen, occurring on the first and second days of exposure to the male, were the decisive factors in bringing on the first ovulation in the immature female rat; this response to LH-release, following male exposure, is made possible by the high levels of blood-FSH in the immature rat. They showed that the first two days of high blood-oestrogen, following male exposure, were not sufficient to induce ovulation on the third day; the female had to be exposed to the male for the third day, so that the regimen seemed to be a full three days of male exposure; during the first two the animal was being primed by the two surges of oestrogen; during the third day the system received a final change, perhaps altered sensitivity to progesterone. In a later study Bronson (1975) showed that oestrogen injections could substitute for male exposure during the 72-hour period during which male exposure produced this premature maturity. Thus, as Figure 8.19 shows, single injections of oestradiol benzoate on either Day 1 or Day 2 followed by exposure to the male on

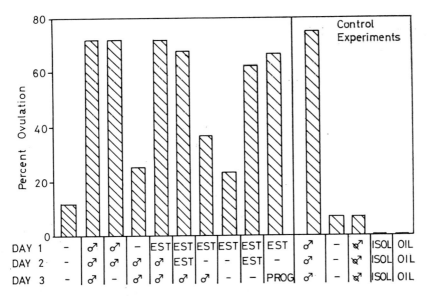

Fig. 8.19. Percentages of females that ovulated in response to three days of experimental procedures including exposure to males and/or injection of steroid (primary experiment on the left). The dash refers to no treatment on a particular day, i.e. the females were simply re-grouped in the male-free room. (Bronson, *Endocrinology*.)

Day 3, were just as effective in bringing on first ovulation as three days of male exposure. It will be seen, also, that male exposure for the first two days, only, is just as effective as male exposure on all three days. Again, two days of oestrogen are not as effective as two days of oestrogen plus a third of male exposure, or of progesterone treatment. Thus, on the third day, male exposure may be equivalent to progesterone secretion.

Equivalence of Oestradiol and Brain Lesions

The regions of the hypothalamus in which lesions induce premature puberty are rich in LH-releasing hormone, and Ruf *et al.* (1974, 1975) emphasized that the primary change is a raised blood-oestrogen, in consequence of LHRH release; and they were able to induce a rise in plasma LH to a value twenty times that in control animals either by treatment with oestradiol followed by progesterone or by a lesion followed by progesterone. Uterine growth, an index to oestrogen secretion, was affected in the same way, so that the lesion was equivalent to an injection of 10 μg of oestradiol benzoate. If this viewpoint is correct, then the suggestion that the lesions removed the neural substrate for negative steroid feedback becomes unnecessary, and the feature common to all the techniques for inducing premature puberty, including PMSG injections (Wilson *et al.*, 1974), consists in a raising of the level of oestrogen. Thus, according to Ruf *et al.* (1975), small lesions in the hypothalamus that spare the sites that specifically take up oestrogen and progesterone are efficient inducers of puberty.

FSH–LH Switch

Ruf *et al.* (1976) observed in their studies of the effects of hypothalamic lesions on maturation in the female rat that injections of oestradiol synergized with the lesion in advancing puberty, but that oestradiol alone did not influence maturation; however, it did change the LH/FSH ratio in the blood, suggesting a change from almost exclusive FSH secretion to one of LH secretion; thus FSH is required for maturation of the ovaries whilst LH is required for cyclicity.*

"Surge" Mode of Secretion

Foster *et al.* (1975) noted that patterns and levels of circulating LH in the sheep, resembling pre-ovulatory surges, occurred only shortly before the first ovulation, and they suggested that it was the development of the "surge" mode of secretion of gonadotrophins, or positive

---

* Hohlweg (1936) showed that oestrogen could induce luteinization in immature rats provided they reached a certain weight; below this only pituitary gonadotrophin would do it.

feedback to oestradiol, represented the acquisition of puberty. However, their recent study (Foster and Karsch, 1975) has shown that LH-surges of increasing magnitude may be induced in very young lambs in response to oestradiol implantation, so that by 27 weeks the induced surge was comparable with that in anoestrous adults without, however, inducing ovulation. The results thus emphasize that the "surge" mode of secretion of LH, in response to oestradiol, develops long before puberty, so that the first ovulation depends on the ability of the ovaries to produce the oestradiol-stimulus rather than on the pituitary-hypothalamus axis to respond to it. Foster and Karsch showed, too, that the negative feedback of progesterone, revealed by block of the oestradiol-induced discharge of LH, occurred long before puberty in the 12-week-old lamb.

Correlation of Structural Changes with Gonadotrophin Secretion

At this point it is worth considering a combined morphological and endocrinological study of maturation in the female rat since the results have led to a description of the events that favours the development of an increased sensitivity of the maturing ovary to gonadotrophins, associated with increased sensitivity of the hypothalamic-pituitary axis to steroid hormones, so that the emphasis is rather on the role of the ovary in bringing about the first ovulation. Knudsen et al. (1974) have measured a number of morphological features in the female rat from 22 days to the first preovulatory FSH–LH surge which, in their Holtzman strain of rats, occurred at Day 38.

Uterus

The weight increased significantly at Day 32 rising rapidly over the next two days (from about 80 g to more than 160 g). Associated with the gain in weight there were obvious histological developments in both endometrium and myometrium; thus, prior to Day 32 the appearance was of a quiescent tissue, with epithelium devoid of mitotic figures and a slit-like lumen of the uterus. From Day 32 there was hypertrophy of all tissues, the lumen becoming distended to the size seen in pro-oestrus; the widths of myometrium and endometrium stroma increased sharply whilst the height of the luminal epithelium followed the same pattern (Fig. 8.20). Between Days 30 and Days 34 there was a sharp rise and fall in the number of mitotic figures in the surface epithelium of the endometrium.

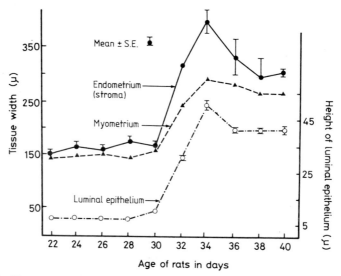

Fig. 8.20. Changes in histology of the uterus with age. (Knudsen *et al.*, *Anat. Rec.*)

Ovary

Changes in the ovary paralleled those in the uterus, the most prominent was the increase, from Day 32, in the percentage of large vesicular Type 6–7 follicles, i.e. follicles with thick theca interna and capable of secreting oestrogens.

Serum Gonadotrophins

In Figure 8.21 the serum LH and FSH levels are plotted and these must be compared with Figure 8.20 showing changes in the uterus. A significant rise in LH occurs between Days 32 and 34 to be followed by a large pre-ovulatory peak on Day 38 in which FSH shares; it will be noted that FSH was high throughout the whole period studied, whereas LH was barely detectable until the first rise on Day 32. Knudsen *et al.* concluded from these results that the rapid rise in uterine activity at Day 32 was due to a corresponding increase in ovarian activity, the large follicles producing sufficient oestrogen to cause these uterine changes. They emphasize that the uterus is the most sensitive index to ovarian function. A further factor that may account for the striking uterine and ovarian changes at Day 32 might well be an increased ovarian responsiveness to gonadotrophic hormone with age, an increase that seems to be well established (see, for example, Price and Ortiz, 1944). This would account for the increased follicular development without any change in FSH levels.

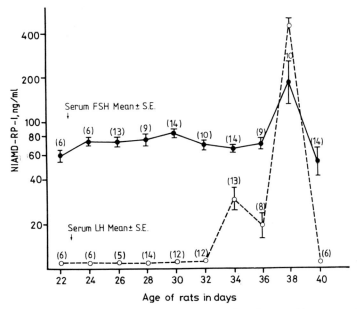

Fig. 8.21. Serum-LH and -FSH concentrations in female rats. Compare with Fig. 8.20.

Prolactin

Before Day 30 levels of prolactin were low (less than 10 ng/ml) but prolactin shared in the LH-peak, reaching a maximum of 120 ng/ml rather sooner at Day 35.

Importance of Ovarian Secretion

According to Kudsen *et al.*, the changes in gonadotrophin secretion revealed by Figure 8.21, preceding as they do the ovarian and uterine changes, reflect a response of the pituitary to the ovarian hormones that are apparently secreted in larger quantities before the changes in gonadotrophic hormone levels, the peak at Day 38 representing the culmination of this positive feedback. In general, then, the development of puberty follows from ovarian-follicular maturation with an apparent change in ovarian responsiveness to steady levels of gonadotrophs. Included in the changes is an increase in hypothalamic sensitivity to steroid feedback. Correlated with these changes are concomitant changes in uterine growth and activity, so that the ovary, on this view, acts as the "Zeitgeber" for the gonadotrophin surge and day of first ovulation. In this way follicular maturity and uterine receptivity are linked to the pre-ovulatory release of gonadotrophins.

Changing Sensitivity to Negative Feedback

Although there is not agreement as to the hypothesis of Ramirez and McCann, according to which puberty is achieved by a lowering of sensitivity of the hypothalamic-pituitary axis to negative feedback, the original observations seem to be valid and are substantiated by the work of Eldridge et al. (1974);* thus, Figure 8.22 compares the decreases in serum gonadotrophin in response to oestradiol after ovariectomy in mature and immature animals.

Oestradiol Receptors

The increased secretion of pituitary gonadotrophins taking place around puberty is caused through increased activity on the part of the hypothalamus; if, as seems likely, this itself is triggered off by the level of circulating steroid hormones in the blood, then we might expect the density of steroid-receptors in the hypothalamus to change during the critical period. The fact that feedback occurs soon after birth indicates that the neonatal hypothalamus is already sensitive, and the study of Kato et al. (1971, 1974) has shown that receptors are present at birth and that they increase in density to reach a maximum at 28 days. The in vivo uptake of ^{3}H-labelled oestradiol by the anterior hypothalamus and pituitary showed some signs of being greatest at 15–20 days (Presl et al., 1970). Such a finding, of course, gives no clue to the type of response to oestrogen mediated through the receptors; as we have seen, the feedback is negative for most of the pre-puberal period, whilst at the later stages a positive influence seems to predominate.†

Adrenal Gland

Adrenalectomy

Adrenalectomy delays the onset of puberty in the rat if carried out between Days 18 and 25, but not if delayed to Day 30. Thus Gorski and Lawton (1973) found a control period of $37 \cdot 9 \pm 0 \cdot 3$ days, and this was lengthened to $44 \cdot 4 \pm 0 \cdot 8$ days in adrenelactomized animals. Auto-transplantation of the adrenals under the skin, allowing adrenal cortical but not medullary secretions, partially reversed the time to $41 \pm 0 \cdot 6$ days. This suggests that the adrenal gland is a source of oestrogen

---

* They point out that the failure of Swerdloff et al. (1971) to find a change in sensitivity was probably due to their failure to start replacement therapy in their ovariectomized rats immediately.

† Parker and Mahesh (1976) emphasize that the receptors in the cytosol of the pituitary and hypothalamus may be expected to decrease at puberty if, at this time, there is an increased secretion of oestradiol; the increased uptake by the cytoplasmic receptors would be followed by migration to the nuclear acceptors (p. 178); this is what they did, in fact, observe.

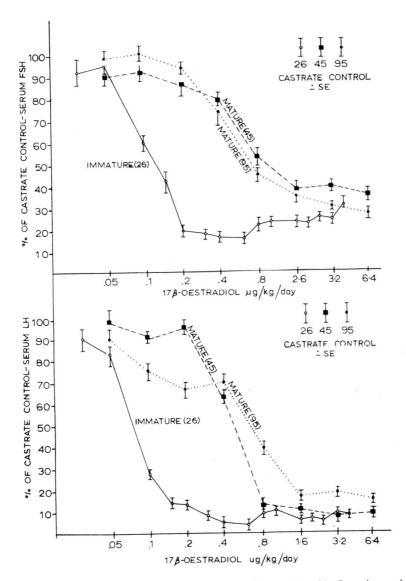

Fig. 8.22. Serum FSH (Top graph) and serum LH (Bottom graph) of ovariectomized mature and immature rats treated with oestradiol for 5 days post-operative. (Eldridge *et al.*, *Endocrinology*.)

required for the development of puberty, but according to Meijs-Roelofs and Kramer (1977) the effects of adrenalectomy are rather due to defective growth of the body as a whole; their work showed no influence on uterine growth so that the oestrogen-like products of the adrenal gland are not biologically active as oestrogens.

Rhythmic Secretion

Ramaley (1976) has emphasized the daily rhythm in the level of corticosterone and progesterone in the blood, due to adrenal secretion, in the immature rat. This rhythmicity, however, is not important for the achievement of puberty, which can be attained in adrenalectomized rats at the normal times by maintenance of a steady blood-level intermediate between the highest and lowest daily values.

Hypothalamic De-afferentation

Complete de-afferentation of the hypothalamus in 20-day-old rats, whereby the hypothalamus is left only with its connexions with the pituitary, caused reduced development of testes, seminal vesicles and prostates, the weights of these organs being lower than normal (Collu et al., 1974). Some evidence of an inhibitory action of the hypothalamus on development of the accessory sex organs was also obtained, since with only a frontal de-afferentation, cutting off the anterior region of the brain, the weights of these organs were greater than normal.

Role of Prolactin

Prolactin is a lactogenic hormone (p. 360), promoting the development of the mammary glands and the secretion of milk. In certain species, notably the rat, this hormone has a luteotrophic action, so that it was often called luteotrophic hormone (LtH), its luteotrophic action being typically manifest in the induction of pseudopregnancy in rats by injection.

Prepuberal Rise

The concentration of prolactin in the plasma of rats rises sharply before the onset of vaginal opening, although the content in the pituitary does not rise until later during the first oestrous cycle, suggesting release of preformed hormone rather than accelerated synthesis. As Voogt et al. (1970) emphasize, the rise in prolactin secretion does not necessarily implicate this hormone in the induction of puberty in the rat; they showed, however, that oestrogen is a powerful stimulant to prolactin secretion as well as an inducer of puberty, so that it is possible that the release of prolactin is a part of the complex of hormonal changes leading up to puberty and the first ovulatory cycle.

Median Eminence Implants

It is worth noting, in this context, that implantation of prolactin in the median eminence of rats hastens puberty by some five days (Clemens *et al.*, 1969) presumably by activating a short feedback loop in the hypothalamico-pituitary circuit that also leads to a rise in FSH and LH release, but since this depresses prolactin secretion, the effect is presumably exerted through the raised FSH and LH concentrations in the blood (Voogt and Meites, 1971). In general, where pituitary hormone content is concerned, prolactin, on the one hand, and LH and FSH on the other, show a reciprocal behaviour, a fall in the one being accompanied by a rise in the others.*

Prolactin and Testicular Sensitivity to LH

Bartke and Dalterio (1976) point out that, in rats, the level of testosterone in the blood increases sharply between Days 30 to 60, although during this period the concentration of LH in the plasma increases only slightly whilst that of FSH actually declines. During the same period, however, the concentration in serum and testis of prolactin-binding proteins increases. Since the plasma prolactin concentration increases with development, it is possible that prolactin contributes to the increased sensitivity of the testis to LH, which is known to occur during maturation. In their study, Bartke and Dalterio raised the plasma prolactin levels by giving a homograft of pituitary to rats, and they found that the secreted testosterone, in response to a given dose of LH, was increased. Dowd and Bartke (1972) showed that the serum prolactin levels of male rats increased from 2·1 ng/ml at 21 days, to 8·2 at 15 days to 73 at 56 days. Further evidence relating to the importance of prolactin in maturation is the fact that genetically prolactin-deficient mice are rarely fertile; whilst treatment of these mice with prolactin enables them to sire litters (Bartke and Dalterio, 1976).

---

* Meites *et al.* (1972) have summarized the actions of prolactin in the rat. It is *luteolytic* to the previous generation of corpora lutea in an ovarian cycle; in combination with LH it is *luteotrophic*, and so sustains pseudopregnancy and early pregnancy. In the absence of either LH or prolactin, the function of the corpus luteum is lost. The requirement for each hormone, however, changes with age and pregnancy (Morishige and Rothchild, 1974). Oestrogen is the principal stimulator of prolactin secretion, causing the rise in prolactin in the blood at puberty, the pro-oestrous surge of prolactin in the ovarian cycle, and the rise in prolactin at parturition. We may note that injections of prolactin can bring on early vaginal opening in rats (Wuttke *et al.*, 1976).

OLD-AGE AND REPRODUCTION
Women
Pituitary Secretions

The human female loses the power to reproduce between the ages of 40 and 50, and for a long time it was considered that this decline, along with many other changes taking place in both sexes with age, was due to a failure of the pituitary to secrete hormones in sufficient quantity. There is no doubt, however, that in menopausal women the pituitary secretions of gonadotrophins are actually larger, and this is apparently due to the failure of the normal negative feedback exerted by the level of oestrogens, comparable with the situation in unilateral gonadectomy (p. 131). The steep rise in concentration of LH in the blood of women in their fifth decade is illustrated by Figure 8.23. Thus, the primary event is not deficient pituitary secretion but deficient ovarian secretion.

Blood Oestrogen Concentrations

The average levels of oestradiol in the blood of cycling women is $65-137\mu g/ml$, depending on the stage of the cycle; for oestrone it is

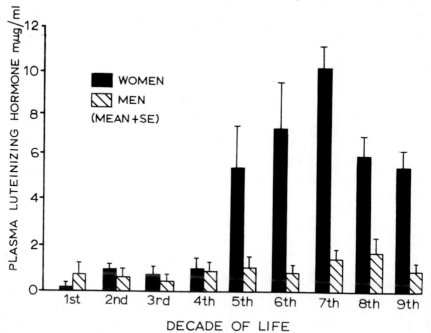

Fig. 8.23. Levels of luteinizing hormone in plasma of women and men as a function of age. Note post-menopausal rise in women. (Schalch *et al.*, *J. clin. Invest.*)

79–107 (Korenman *et al.*, 1969; Tulchinsky and Korenman, 1970). The mean value for a group of women aged 74–89 was 6·5 ± 0·7 for oestradiol and 24·8 ± 3·5 μg/ml for oestrone (Longcope, 1971). Thus there is about a 90 per cent decrease in oestrogen levels in the post-menopausal woman.

Oestrogen Secretion

The metabolic clearance and production of oestrogens in post-menopausal women have been described by Longcope (1971). So far as clearance is concerned, this decreases by about 25 per cent in post-menopausal women. The calculated mean rate of production of oestradiol in post-menopausal women was only 6 μg/day, compared with 80–500 μg/day in cycling women; moreover, the production probably arises from the peripheral conversion of precursors of adrenal origin, so that the post-menopausal ovary may secrete little or no oestradiol. The secretion of oestrone, on the other hand, is better maintained, being some 40 μg/day compared with some 90–300 μg/day in cycling women; this probably arises from androstenedione, which is secreted by the adrenal and the ovary.

Cycle Length

Treloar *et al.* (1967) showed that the human menstrual cycle was of reasonably constant length between the ages of 20 and 40; in the earlier and later periods of life, namely the 5–7 years post-menarche and the 6–8 years before menopause, cycle length was more variable. Figure 8.24 from this study presents an attempt to represent the change in inter-menstrual interval throughout the menstrual life of the "average individual". The upper curve gives the mean menstrual interval for each "person-year" of the study, and the appropriate standard deviations are shown in the lower curve. Thus, variability is a feature of the post-menarchical and premenopausal years. In Table VII the cycle-lengths of selected groups have been broken down into the follicular

TABLE VII

Characteristics of menstrual cycle-length in normal women throughout reproductive life and in disorders of follicular maturation (Sherman and Korenman, 1975)

Group	Follicular Phase	Luteal Phase	Total Cycle-Length
Age 18–30	16·9 ± 3·7	12·9 ± 1·8	30·0 ± 3·6
40–41	10·4 ± 2·9	15·0 ± 0·9	25·4 ± 2·3
46–51	8·2 ± 2·8	15·9 ± 1·3	23·2 ± 2·9
Long Follicular Phase	33·0 ± 2·2	14·2 ± 2·2	47·2 ± 4·3

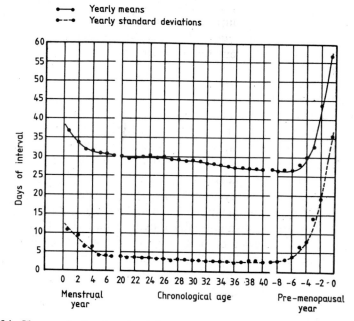

Fig. 8.24. Changes in the length of the menstrual cycle with age. The lower curve gives the standard deviations. (Treloar *et al.*, *Int. J. Fert.*)

and luteal phases; the progressive average shortening of the cycle as menopause is approached is associated with a shortened period of follicular maturation, as measured by the time from the first day of menstruation to the LH-peak. The subjects with a long follicular phase were young with irregular menses, and their lengthened follicular phase is striking. In general, the hormonal levels in the various groups followed similar courses except for the FSH and oestradiol levels in premenopausal women (Fig. 8.25); in these the FSH levels were uniformly high with normal LH levels, suggesting a diminished inhibitory feedback on the pituitary possibly acting through an "inhibin" rather than oestradiol (p. 147).

Men

Gonadotrophins and Testosterone

In the male, too, the secretion of gonadotrophic hormones increases with age (Fig. 8.26) and this is accompanied by, and presumably related to causally, a decline in the level of testosterone in the blood (Fig. 8.27). The binding capacity for sex steroid binding globulin (SSBG) increases with age after 60. These hormonal changes correlate

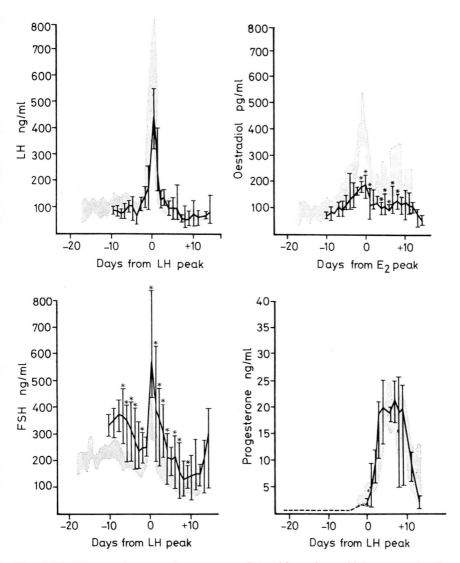

Fig. 8.25. Mean and range of serum gonadotrophic and steroid hormones in six pre-menopausal women with regular menses compared with the mean ± 2 S.E.M. in 10 cycles in women aged 18–30. An asterisk indicates a statistically significant difference between the values of the two groups. (Sherman and Korenman, *J. clin. Endocrinol.*)

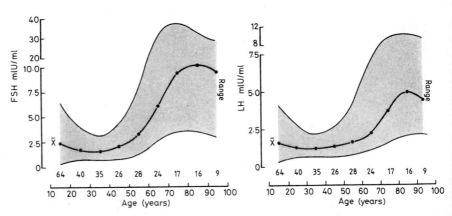

Fig. 8.26. Changes in the blood-levels of FSH and LH in men with age. (Baker *et al.*, *Rec. Progr. Horm. Res.*)

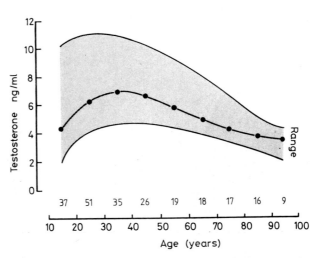

Fig. 8.27. Changes in the level of testosterone in the blood of men as a function of age. (Baker *et al.*, *Rec. Progr. Horm. Res.*)

with an increasing incidence of the clinical features of hypogonadism, namely testicular atrophy, loss of body hair and diminished sexual performance. Physiologically, then, the changes in senescence emphasize the role of the testes in controlling the pituitary secretions in a negative feedback through steroid secretion and probably also through inhibin (p. 147).

Prostate Hypertrophy

It may be that the hypertrophy of the prostate, characteristically seen in older men, is likewise due to diminished feedback from the gonadal steroid secretion; certainly, injections of oestrogens and androgens can decrease prostatic hypertrophy (Scott, 1953).

REFERENCES

(See p. 546 for additional references)

Bakcr, F. D. and Kragt, C. L. (1969). Maturation of the hypothalamic-pituitary-gonadal negative feedback system. *Endocrinology*, **85**, 522–527.

Bartke, A. and Dalterio, S. (1976). Effects of prolactin on the sensitivity of the testis to LH. *Biol. Reprod.*, **15**, 90–93.

Bauer, H. G. (1954). Endocrine and other clinical manifestations of hypothalamic disease. *J. clin. Endocrinol.*, **14**, 13–31.

Bercu, B. B. *et al.* (1977). Permanent impairment of testicular development after transient immunological blockade of endogenous luteinizing hormone releasing hormone in the neonatal rat. *Endocrinology*, **101**, 1871–1879.

Bronson, F. H. (1971). Rodent pheromones. *Biol. Reprod.*, **4**, 344–357.

Bronson, F. H. (1975). Male-induced precocial puberty in female mice: confirmation of the role of estrogen. *Endocrinology*, **96**, 511–514.

Bronson, F. H. and Desjardins, C. (1974). Circulating concentrations of FSH, LH, estradiol, and progesterone associated with acute, male-induced puberty in female mice. *Endocrinology*, **94**, 1658–1668.

Caligaris, L., Astrada, J. J. and Taleisnik, S. (1972). Influence of age on the release of luteinizing hormone induced by oestrogen and progesterone in immature rats. *J. Endocrinology*, **55**, 97–103.

Clemens, J. A., Sar, M. and Meites, J. (1969). Inhibition of lactation and luteal function in postpartum rats by hypothalamic implantation of prolactin. *Endocrinology*, **84**, 868–872.

Collu, R., Motta, M., Massa, R. and Martini, L. (1974). Effect of hypothalamic deafferentations on puberty in the male rat. *Endocrinology*, **94**, 1496–1501.

Corbin, A. and Schottelius, B. A. (1960). Effects of posterior hypothalamic lesions on sexual maturation of immature female albino rats. *Proc. Soc. exp. Biol. Med.*, **103**, 208–210.

Donovan, B. T. and Van der Werff Ten Bosch, J. J. (1956). Oestrus in winter following hypothalamic lesions in the ferret. *J. Physiol.*, **132**, 57–58P.

Donovan, B. T. and Van der Werff Ten Bosch, J. J. (1959). The hypothalamus and sexual maturation in the rat. *J. Physiol.*, **147**, 78–92.

Donovan, B. T. and Van der Werff Ten Bosch, J. J. (1965). *Physiology of Puberty*. Arnold, London.

Dowd, A. J. and Bartke, A. (1972). Serum levels of prolactin LH and FSH, and testis cholesterol content in rats from one to ten weeks of age. *Biol. Reprod.*, **7**, 115 (Abstr.).

Eldridge, J. C., McPherson, J. C. and Mahesh, V. B. (1974). Maturation of the negative feedback control of gonadotropin secretion in the female rat. *Endocrinology*, **94**, 1536–1540.

Fink, G. and Sarkar, D. K. (1978). Mechanism of first surge of luteinizing hormone and vaginal opening in the normal rat, and effect of neonatal androgen. *J. Physiol.*, **282**, 34–35P.

Florsheim, W. H. and Rudko, P. (1968). The development of portal system function in the rat. *Neuroendocrinol.*, **3**, 89–98.

Foster, D. L. and Karsch, F. J. (1975). Development of the mechanism regulating the preovulatory surge of luteinizing hormone in sheep. *Endocrinology*, **97**, 1205–1209.

Foster, D. L., Lemons, J. A., Jaffe, R. B. and Niswender, G. D. (1975). Sequential patterns of circulating luteinizing hormone and follicle-stimulating hormone in female sheep from early postnatal life through the first estrous cycle. *Endocrinology*, **97**, 985–994.

Gellert, R. J. and Ganong, W. F. (1960). Precocious puberty in rats with hypothalamic lesions. *Acta Endocrinol.*, **33**, 569–576.

Gorski, M. E. and Lawton, I. E. (1973). Adrenal involvement in determining the time of onset of puberty in the rat. *J. Endocrinol*, **93**, 1232–1234.

Hohlweg, W. (1934). Veranderungen des Hypophysenvorderlappens und der Ovariums nach Behandlung mit grossen Dosen von Follikelhormon. *Klin. Wchschr.*, **13**, 92–95.

Jenner, M. R., Kelch, R. P., Kaplan, S. L. and Grumbach, M. M. (1972). Hormonal changes in puberty. IV. *J. clin. Endocrinol.*, **34**, 521–530.

Johanson, A. T. *et al.* (1969). Serum luteinizing hormone by radioimmunoassay in normal children. *J. Ped.*, **74**, 416–424.

Kallas, H. (1929). Puberté précoce par parabiose. *C.r. Soc. Biol. Paris*, **100**, 979–980.

Kato, J., Atsumi, Y. and Inaba, M. (1971). Development of estrogen receptors in the rat hypothalamus. *J. Biochem.*, **70**, 1051–1053.

Kato, J., Atsumi, Y. and Inaba, M. (1974). Estradiol receptors in female rat hypothalamus in the developmental stages and during pubescence. *Endocrinology*, **94**, 309–317.

Knudsen, J. F., Costoff, A. and Mahesh, V. B. (1974). Correlation of serum gonadotrophins, ovarian and uterine histology in immature and prepubertal rats. *Anat. Rec.*, **180**, 497–507.

Korenman, S. G., Perrin, L. E. and McCallum, T. P. (1969). A radio-ligand binding assay for estradiol measurement in human plasma. *J. clin. Endocrinol.*, **29**, 879–883.

Kupperman, H. S., Meyer, R. K. and Finerty, J. C. (1942). Precocious gonadal development occurring in immature rats following a short-time treatment with antigonadotropic serum. *Amer. J. Physiol.*, **136**, 293–298.

Lee, V. W. K., Kretser, D. M. de, Hudson, B. and Wang, C. (1975). Variation in serum FSH, LH and testosterone levels in male rats from birth to sexual maturity. *J. Reprod. Fert.*, **42**, 121–126.

Lee, P. A., Midgley, A. R. and Jaffe, R. B. (1970). Regulation of human gonadotropins. VI. *J. clin. Endocrinol.*, **31**, 248–253.

Longcope, C. (1971). Metabolic clearance and blood production rates of estrogens in postmenopausal women. *Am. J. Obstet. Gynec.*, **111**, 778–781.

Meijs-Roelofs, H. M. A. (1975). Gonodotrophin and steroid levels around the time of first ovulation in the rat. *J. Endocrinol*, **67**, 275–282.

Meijs-Roelofs, H. M. A. and Kramer, P. (1977). Effect of adrenalectomy on the release of follicle-stimulating hormone and the onset of puberty in female rats. *J. Endocrinol*, **75**, 419–426.

Meites, J. *et al.* (1972). Recent studies on functions and control of prolactin secretion in rats. *Rec. Progr. Horm. Res.*, **28**, 471–516.

Morishige, W. K. and Rothchild, I. (1974). Temporal aspects of the regulation of corpus luteum function by luteinizing hormone, prolactin and placental luteotrophin during the first half of pregnancy in the rat. *Endocrinology*, **95**, 260–274.

Nathanson, I. T., Towne, L. E. and Aub, J. C. (1941). Normal excretion of sex hormones in childhood. *Endocrinology*, **28**, 851–865.

Odell, W. D. and Swerdloff, R. S. (1976). Etiologies of sexual maturation: a model system based on the sexually maturing rat. *Recent. Prog. Horm. Res.*, **32**, 245–277.

Odell, W. D., Swerdloff, R. S., Bain, J., Wollesen, F. and Grover, P. K. (1974). The effect of sexual maturation on testicular response to LH stimulation of testosterone secretion in the intact rat. *Endocrinology*, **95**, 1380–1384.

Odell, W. D., Swerdloff, R. S., Jacobs, H. S. and Hescox, M. A. (1973). FSH induction of sensitivity to LH: one cause of sexual maturation in the male rat. *Endocrinology*, **92**, 160–165.

Ojeda, S. R. and Ramirez, V. D. (1972). Plasma level of LH and FSH in maturing rats: response to hemigonadectomy. *Endocrinology*, **90**, 466–472.

Parker, C. R. and Mahesh, V. B. (1976). Hormonal events surrounding the natural onset of puberty in female rats. *Biol. Reprod.*, **14**, 347–353.

Presl, J. et al. (1970). Changes in uptake of ^{3}H-estradiol by the female rat brain and pituitary from birth to sexual maturity. *Endocrinology*, **86**, 899–902.

Ramaley, J. A. (1976). The role of corticosterone rhythmicity in puberty. *Biol. Reprod.*, **14**, 151–156.

Ramirez, V. D. (1972). Maturation of the gonadotrophin control system. *Acta. Endocrinol.*, Supp. **166**, 170–176.

Ramirez, V. D. and McCann, S. M. (1963). Comparison of the regulation of luteinizing hormone (LH) secretion in immature and adult rats. *Endocrinology*, **72**, 452–464.

Ramirez, V. D. and Sawyer, C. H. (1966). Changes in hypothalamic luteinizing hormone releasing factor (LHRF) in the female rat during puberty. *Endocrinology*, **78**, 958–964.

Roth, J. C., Kelch, R. P., Kaplan, S. L. and Grumbach, M. M. (1972). FSH and LH response to luteinizing hormone releasing factor in pre-pubertal and pubertal children, adult males. *J. clin. Endocr. Metab.*, **35**, 926–930.

Ruf, K. B., Kitchen, J. H. and Wilkinson, M. (1976). Synergistic effects of oestrogen and brain stimulation on precocious sexual maturation in the female rat. *Acta. Endocrinol.*, **82**, 225–237.

Ruf, K. B., Wilkinson, M. and de Ziegler, D. (1975). Brain lesions and precocious puberty in rats. *Nature*, **257**, 404–405.

Ruf, K. B., Younglai, E. V. and Holmes, M. J. (1974). Induction of precocious sexual development in female rats by electrochemical stimulation of the brain. *Brain Res.*, **78**, 437–446.

Schiavi, R. C. (1964). Effect of anterior and posterior hypothalamic lesions on precocious sexual maturation. *Amer. J. Physiol.*, **206**, 805–810.

Scott, W. W. (1953). What makes the prostate grow? *J. Urol.*, **70**, 477–488.

Sherman, B. M. and Korenman, S. G. (1975). Hormonal characteristics of the human menstrual cycle throughout reproductive life. *J. clin. Endocr. Metabl.*, **55**, 699–706.

Smith, E. R., Damassa, D. A. and Davidson, J. M. (1977). Feedback regulation and male puberty: testosterone-luteinizing hormone relationships in the developing rat. *Endocrinology*, **101**, 173–180.

Smith, E. R. and Davidson, J. M. (1968). Role of estrogen in the cerebral control of puberty in female rats. *Endocrinology*, **82**, 100–108.

Steele, R. E. and Weisz, J. (1974). Changes in sensitivity of the estradiol-LH feedback system with puberty in the female rat. *Endocrinology*, **95,** 513–520.

Stiff, M. E., Bronson, F. H. and Stetson, M. H. (1974). Plasma gonadotrophins in prenatal and prepubertal female mice; disorganization of pubertal cycles in the absence of a male. *Endocrinology*, **94,** 492–496.

Swerdloff, R. S., Jacobs, H. S. and Odell, W. D. (1972). In *Gonadotropins*. Eds. B. B. Saxena, C. G. Beiling and H. M. Gandy, p. 546. J. Wiley, N.Y. (Quoted by Eldridge *et al.*, 1974.)

Treloar, A. E. *et al.* (1967). Variation of the human menstrual cycle through reproductive life. *Internat. J. Fert.*, **12,** 77–126.

Tulchinsky, D. and Korenman, S. G. (1971). A radio-ligand assay for plasma estrone; normal values and variations during the menstrual cycle. *J. clin. Endocrinol.*, **31,** 76–80.

Vandenbergh, J. G. (1967). Effect of presence of a male in the sexual maturation of female mice. *Endocrinology*, **81,** 345–349.

Vandenbergh, J. G. (1969). Male odor accelerates female sexual maturation in mice. *Endocrinology*, **84,** 658–660.

Vandenbergh, J. G., Whitsett, J. M. and Lombardi, J. R. (1975). Partial isolation of a pheromone accelerating puberty in female mice. *J. Reprod. Fert.*, **43,** 515–523.

Voogt, J. L., Chen, G. L. and Meites, J. (1970). Serum and pituitary prolactin levels before, during, and after puberty in female rats. *Amer. J. Physiol.*, **218,** 396–399.

Voogt, J. L. and Meites, J. (1971). Effects of an implant of prolactin in median eminence of pseudopregnant rats on serum and pituitary, LH, FSH and prolactin. *Endocrinology*, **88,** 286–292.

Wilson, C. A., Horth, C. E., Endersby, C. A. and McDonald, P. G. (1974). Changes in plasma levels of oestradiol, progesterone and luteinizing hormone in immature rats treated with pregnant mare serum gonadotrophin. *J. Endocrinol.*, **60,** 293–304.

Wuttke, W., Döhler, K. D. and Gelato, M. (1976). Oestrogens and prolactin as possible regulators of puberty. *J. Endocrinol.*, **68,** 391–396.

Ying, S.-Y., Fang, V. S. and Greep, R. O. (1971). Estradiol benzoate (EB)-induced changes in serum luteinizing hormone (LH) and follicle-stimulating hormone (FSH) in immature female rats. *Fert. Ster.*, **22,** 799–801.

Ying, S.-Y. and Greep, R. O. (1971). Effect of age of rat and dose of a single injection of estradiol benzoate (EB) on ovulation and the facilitation of ovulation by progesterone (P). *Endocrinology*, **89,** 785–790.

Ying, S.-Y. and Greep, R. O. (1971). Effect of a single low dose of estrogen on ovulation, pregnancy and lactation in immature rats. *Fert. Ster.*, **22,** 165–169.

Zarrow, M. X. and Wilson, E. D. (1961). The influence of age on superovulation in the immature rat and mouse. *Endocrinology*, **69,** 851–855.

CHAPTER 9

Neurotransmitters, Nervous Action and Cyclic AMP

Hypothalamic Concentrations

A number of experiments have suggested that a catecholaminergic system is involved in gonadotrophic hormone release, and thus in the control over ovulation and lactation. For example, Stefano and Donoso (1967) observed changes in the concentration of noradrenaline in the hypothalamus during the oestrous cycle, and this was confirmed by Selmanoff *et al.* (1976) who used more refined techniques; the supra-ventricular nucleus showed a rise in concentration during the morning and afternoon of pro-oestrus, suggesting that noradrenergic neurones were concerned in the surge of LH on the afternoon of pro-oestrus.

Ovulation

Again Everett (1964) showed that dibenamine, an inhibitor of adrenergic transmission, administered to rabbits within one minute after copulation prevented ovulation, but if this was delayed for three minutes it was ineffective. Again, Rubinstein and Sawyer (1969) treated rats for a prolonged period with reserpine, which depletes the brain of mono-amines, and obtained a complete blockage of ovulation induced by electrical stimulation of the hypothalamus. When the animals were treated with a mono-amine-oxidase inhibitor, which permitted accumulation of mono-amines in the brain, ovulation in response to electrical stimulation was restored, although not spon-taneous ovulation. If reserpine treatment was confined to the morning of pro-oestrus, spontaneous ovulation was blocked, and the threshold for electrical stimulation in the PO/AH region was raised, but not the threshold for stimulation of the median eminence region, suggesting that the aminergic neurone(s) concerned in the stimulus to ovulation was in the rostral hypothalamus. We may presume that the catecholaminergic neurones of the arcuate nucleus and anterior periventricular hypothalamic nucleus exert their action on the hypothalamic neurones that liberate releasing hormones at their terminals, by means of axo-axonic synapses (Fuxe and Hökfelt, 1970).

Dopamine and Noradrenaline
Tubero-infundibular Neurones

Histological studies have identified a dopaminergic system of neurones in the tubero-infundibular region (Fuxe and Hökfelt, 1970; Björklund and Nobin, 1973) which terminates in the external layer of the median eminence, in addition to networks of noradrenaline- and 5-HT-containing nerve terminals in the hypothalamus and preoptic area. The effects of dopamine on gonadotrophin release, as manifest in several studies, strongly suggest that it is this dopaminergic system that is of fundamental importance, not only in relation to ovulation but also in relation to the release of prolactin, the hormone concerned primarily with lactation.

Isolated Pituitary

In vitro studies by Schneider and McCann (1970) in which they incubated isolated pituitary together with median eminence tissue showed that addition of dopamine caused release of LH into the medium, and they suggested that dopaminergic neurones might activate the small-celled hypothalamic neurones that secreted releasing hormone into the portal circulation.

Ventricular Injections

In extension of this work they injected amines into the IIIrd ventricles of 4-day-cycle rats and measured changes in LH-concentration in the plasma. 5-HT and noradrenaline were without effect, but dopamine, injected on Di-oestrus Day 2 or in Pro-oestrus, caused a five-to-tenfold rise in LH concentration (Fig. 9.1).

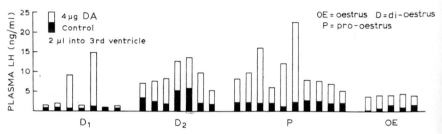

Fig. 9.1. Effect of intraventricular dopamine (DA) on plasma LH of cycling rats (Schneider and McCann, *Endocrinology.*)

LHRH Release

Direct evidence for release of releasing hormone by dopamine was provided by Rotsztejn et al., (1977) who measured the release from isolated fragments of rat hypothalamus in response to the amine; noradrenaline was ineffective. The region of the hypothalamus that gave ample release of relasing hormone was the medial and lateral palisade zones of the median eminence and this is where most of the tubero-infundibular dopaminergic terminals are concentrated. The release was blocked by pimozide.

Hypothalamus Sensitivity

When Kamberi et al. (1970) perfused dopamine into the cannulated portal vessel the level of LH in the blood was not affected, indicating that the sensitive region to the dopamine was the hypothalamus, rather than the pituitary. Pimozide, a dopamine receptor blocker, given subcutaneously, caused a gradual fall in the level of LH in the plasma, and when implanted in the median eminence-arcuate area it caused a small decline in the level of LH (Ojeda et al., 1974).

PO/AH Stimulation

Lichtensteiger and Keller (1974) stimulated various regions of the brain and observed changes in fluorescence of the dopaminergic neurones of the tubero-infundibular tract, an increased fluorescence being attributed to increased activity. The PO/AH hypothalamic region caused increased fluorescence, an effect that could be blocked by atropine, suggesting the presence of a cholinergic neurone in the pathway; at any rate atropine did not affect the fluorescence caused by direct stimulation of the arcuate nucleus. The release of LH in response to electrical stimulation of the PO/AH, could be partially blocked by drugs that block catecholamine synthesis, namely α-methyl-p-tyrosine, an effect that could be reversed by L-Dopa or dihydroxyphenylserine (Kalra and McCann, 1973).

Reserpine

If reserpine exerts its anti-ovulatory action by blocking transmission in the hypothalamus, we should expect infusion of dopamine or other transmitters, such as noradrenaline, to release the block and restore ovulation; Craven and McLeod (1973) however, were unable to restore ovulation in this way, the transmitter being infused directly into the arcuate-median eminence area through a stereotactically placed cannula. This suggests that the effects of reserpine might well be more peripheral, acting directly on the ovaries and blocking ovulation, either through an effect on the follicular circulation or perhaps on smooth muscle involved in release of the ovum. If such a peripheral action of reserpine occurred we should expect the effects of exogenous LH to be prejudiced; this was certainly true if rats were treated with reserpine in di-oestrus, an injection of 10 μg of LH at pro-oestrus failing to cause ovulation in any of the six rats studied, whereas the same dose given to controls caused ovulation in all. If reserpine treatment was postponed to pro-oestrus, when it blocked normal ovulation, it failed to influence the ovulation induced by the injection of LH (Table I).

TABLE I

The effect of intravenous administration of luteinizing hormone (LH) on ovulation in reserpine-treated female rats (Craven and McLeod, 1973)

No. of rats	Time of reserpine treatment	LH treatment at pro-oestrus 14.00 h (μg)	No. of ovulating rats/no. treated	No. of eggs/ovulation (means ± s.e.m.)
4	—	5	4/4	11·0 ± 0·5
5	—	10	5/5	10·4 ± 0·6
6	15.00 h dioestrus	10	0/6	—
6	11.00 h pro-oestrus	5	6/6	6·6 ± 0·8
6	11.00 h pro-oestrus	10	6/6	8·3 ± 0·5

Reserpine (5 mg/kg) was given subcutaneously

Ovarian Action

It could well be, then, that reserpine and the adrenergic system have two loci of action, a central hypothalamic one and a peripheral one in the ovaries. (Craven and McDonald, 1973.)

Thus Condon and Black (1976) showed that adrenaline, nora-drenaline and isoproterenol stimulated production of progesterone by bovine luteal tissue; the effects were abolished by propanolol, indicating the predominantly β-action. Propanolol also blocked the LH-stimulated production of progesterone, indicating that the synthesis of progesterone by the corpus luteum of the cow is activated by a β-adrenergic mechanism. These authors consider it unlikely that the adrenergic neurones would actually penetrate into the luteal cells.

In the first trimester of human pregnancy the intravenous infusion of vasopressin or theophyllamine causes a rise in plasma progesterone; the vasopressin effect is blocked by propanolol, suggesting a β-adrenergic control of luteal secretion in the human (Fylling, 1973).

Dopaminergic Inhibition

The experiments described so far suggest that the dopaminergic system of neurones in the hypothalamus is concerned with excitation of the arcuate system of neurones, inducing release of LHRH and thus causing ovulation. Against, or supplementary to, this view are many studies of Fuxe and his collaborators on the appearance of dopaminergic neurones during different physiological states. Thus in pregnancy, pseudopregnancy, and lactation, secretion of the gonadotrophins LH and FSH is suppressed, whilst the secretion of prolactin is increased. It will be recalled that secretion of prolactin is controlled by the hypothalamus in a negative fashion, the release being inhibited by a prolactin inhibiting factor PIF. Thus inhibition of the neurones releasing this hormone will promote prolactin secretion. Consequently, a generalized dopaminergic activation would result in decreased release of the gonadotrophins—FSH and LH—and an increased secretion of prolactin. Fuxe and Hökfelt suggest that impulses passing down these dopaminergic axons would exert a presynaptic-type of inhibition at axo-axonic synapses.

Dopamine Block. In support for this role of DA-neurones, Fuxe and Hökfelt quote experiments in which reserpine or DA-blocking agents actually caused an increase in LH secretion by the pituitary; again when secretion of FSH and LH is high, as in the castrated animal when the negative steroid feedback is abolished (p. 131), the activity of the dopaminergic neuronal system is low (Fuxe and Hökfelt, 1970); furthermore, the system is rapidly reactivated by injections of steroid hormones such as testosterone or oestradiol. Thus it may be that these dopaminergic neurones are involved in the negative feedback of the steroid hormones on the hypothalamus.*

Pro-Oestrus. Using a fluorescence technique to measure turnover of dopamine in hypothalamic neurones, Ahrén et al. (1971) found that the fluorescence in dopaminergic neurones decreased in pro-oestrus at a time when LH plasma levels were falling, so that they concluded that dopaminergic neurones were actually concerned in an *inhibition* of LH release.

---

* Drouva and Gallo (1976, 1977) have provided evidence for an inhibitory action of dopamine on the episodic release of LH in ovariectomized rats; thus selective stimulation of dopamine-receptors by apomorphine or ET 495 blocked the pulsatile release; blocking the dopamine receptors with d-butaclamol prevented these effects of apomorphine and ET 495. Again, selective depletion of noradrenaline by U-14,624 or FLA-63 blocked the pulsatile secretion of LH.

Adrenergic System

Rubinstein and Sawyer (1970) found that intraventricularly administered epinephrine was more effective in triggering ovulation in the pentobarbital-blocked rat at pro-oestrus than was dopamine.

Again Krieg and Sawyer (1976) implanted cannulae in the IIIrd ventricles of rats and introduced noradrenaline or dopamine when the animals were freely moving. Noradrenaline raised the level of LH in the blood rapidly due, doubtless, to a release of LHRH; dopamine, on the other hand, had no effect on plasma-LH. When the animals were anaesthetized, and unit activity in the arcuate nucleus measured, it was found that noradrenaline decreased spike discharges at doses that raised the blood-LH. These authors concluded that the noradrenergic system was stimulatory to neurones secreting LHRH, the decrease in neuronal activity of arcuate neurones when noradrenaline was infused into the IIIrd ventricle implying that certain elements of this nucleus are inhibited during release of LH, perhaps the tubero-infundibular dopaminergic neurones discussed above.*

In general, then, the situation regarding the dopaminergic and noradrenergic systems in the hypothalamus is confused so far as their involvement in the pre-ovulatory rise in LH-secretion and subsequent ovulation is concerned.

PROLACTIN SECRETION
Dopaminergic System

Effects on PIF

The effects of dopaminergic neurones on the secretion of prolactin seem to be better documented than effects on secretion of the gonadotrophic hormones, LH and FSH. In general, the experiments indicate that the dopaminergic system is excitatory on the hypothalamus, causing

---

* The reader may be referred to the paper by Krieg and Sawyer (1976) for a full discussion of the situation regarding the involvement of noradrenaline in release of LH. Adrenergic blocking agents prevent the copulation-induced ovulation in the rabbit and the spontaneous ovulation in rodents, acting by way of LHRH release. Inhibitors of noradrenaline synthesis, such as DDC, block the advancement of LH release by progesterone in pro-oestrous rats, and the LH-response to electrical stimulation of the PO/AH (Kalra and McCann, 1973); both of these responses are restored by dihydroxyphenyl serine (DOPS) which is converted directly to noradrenaline, bypassing the DDC inhibition. According to Krieg and Sawyer, dopamine has no effect on LH release, its effects being confined to the prolactin release system (Ojeda et al., 1974). They quote the work of Weiner et al. (1971) showing that noradrenaline and adrenaline give a biphasic response in the units in the median eminence, excitation followed by inhibition; the excitation is associated with LH release due to activation of LHRH-neurones close to the IIIrd ventricle. Krieg and Sawyer suggest that noradrenaline activates LHRH-neurones and depresses the inhibitory influences exerted on them by tubero-infundibular dopaminergic neurones in the arcuate nucleus.

an increased secretion of the inhibiting hormone PIF thereby *decreasing* secretion of prolactin. Thus, the dopamine-blocker 3,4-di-O-methyl dopamine, a metabolic product of L–DOPA, competes with dopamine at its receptor sites; when administered to rats the plasma levels of prolactin were raised (Smythe and Lazarus, 1973). Again L–DOPA, the metabolic precursor of dopamine, when administered intravenously, increases the levels of dopamine in dopaminergic neurones.

In rats Lu and Meites (1971) observed a reduced prolactin secretion after L–DOPA treatment, suggesting increased secretion of PIF; similarly, a monoamine oxidase inhibitor, such as pargyline, reduced prolactin secretion whereas methyldopa, which inhibits catecholamine synthesis, increased the secretion of prolactin manyfold. Essentially similar results were found by Donoso et al. (1971); thus drugs that inhibited noradrenaline or 5-HT synthesis had no effect on prolactin secretion, whereas L-methyl-dopa increased the secretion of prolactin; another blocker of catecholamine synthesis, DL-α-methyltyrosine caused a rise in plasma and pituitary prolactin levels.

Intraventricular Injections

Injection of dopamine into the IIIrd ventricle of normal male or lactating female rats resulted in a decrease in plasma prolactin, presumably due to release of the inhibiting hormone, since the portal blood from dopamine-treated animals contains the inhibiting hormone (Donoso et al., 1973). In a similar type of study Kamberi et al. (1971) found a reduced level of prolactin in the blood after intraventricular injection of dopamine; if a portal vessel was cannulated and the dopamine infused through this, there was no effect, suggesting that the point of attack is the hypothalamus rather than the pituitary, and this agrees with the finding of Kanematsu and Sawyer (1963) that implantation of a tube, containing at its tip reserpine, into the median eminence and posterior tuberal region of the oestrogen-primed ovariectomized rabbit, caused a release of prolactin; whereas implantation into the anterior pituitary had little effect.

Direct Action on Pituitary

However, there is evidence favouring a direct action of dopamine on the pituitary, inhibiting release of prolactin; in fact, it has been argued with some cogency that dopamine is the prolactin inhibiting factor, PIF. Macleod and Lehmeyer incubated pituitary glands *in vitro* and the incorporation of ^{3}H-leucine into the gland and its subsequent release as protein were used as a measure of prolactin secretion; 5.10^{-7}M dopamine caused 85 per cent inhibition.

Agents known to cause stimulation of alpha-adrenergic or dopamin-ergic receptors inhibited prolactin release, such as apomorphine or ergocryptine. The authors suggested, therefore, that dopamine might exert a direct control over the secretion of the anterior pituitary, so far as prolactin was concerned. In a later study Donoso *et al.* (1973) caused lesions in the median eminence in order to interrupt hypo-thalamic control. On administering L–DOPA they found a dramatic fall in plasma prolactin, suggesting that the gland itself synthesized high concentrations of dopamine from the L–DOPA, which then acted on the secretory cells. Thus dopamine can act by causing release of prolactin-inhibiting factor (PIF) or by directly inhibiting prolactin release, if these results can be substantiated.

Portal Infusion

A direct action on the pituitary was shown by cannulating a hypo-physeal portal vein and introducing hypothalamic extracts; in this way Takahara *et al.* (1974) obtained a suppression of prolactin release; similar suppression could be obtained by infusing dopamine and noradrenaline into the portal vein, and they showed that the degree of suppression produced by pituitary extracts was related to the amounts of these catecholamines in them.*

Portal Blood in Pregnancy

When Ben-Jonathan *et al.* (1977) collected portal blood they found a large rise in dopamine concentration throughout pregnancy, which may correspond with the low values of prolactin in the blood pertaining throughout pregnancy in the rat. They noted, however, that in di-oestrous females, when the prolactin concentration in the blood is also low, the dopamine in the blood was not raised.

Dual Action

Wiggins and Fernstrom (1977) made use of the fact that the plasma level of prolactin in female rats rises rapidly on the afternoon of pro-oestrus, so that inhibition of release can be measured with some ease. They found that L–DOPA, the precursor of dopamine, did, in fact, reduce the rise at pro-oestrus and, by making use of decarboxylase

---

* If the catecholamines were dissolved in saline instead of 5 per cent glucose they were ineffective, and it was concluded that this was due to oxidation in the saline medium.

inhibitors, i.e. substances that inhibited the formation of dopamine from DOPA,* they found that there were probably two mechanisms for dopamine action, a neurogenically activated release of PIF through hypothalamic dopaminergic neurones, and a release by a direct hormonal action of dopamine passing along the portal circulation to the prolactin-secreting cells.

PIF Equals Dopamine?

As mentioned above, it has been suggested that dopamine is, in fact, the prolactin inhibiting factor, which thus acts both as a hormone and neurotransmitter; as we have seen, the gonadotrophin releasing hormone, LHRH, may also be regarded as a neurotransmitter, not only in relation to the pituitary but in relation to other central neurones. The same is true for other peptides, such as TRH.

Dopamine Concentrations. Gibbs and Neill (1978) pointed out that adequate concentrations of dopamine in portal blood must pertain if it is to act as a release-inhibiting factor. They found a value of 9 ng/ml in di-oestrous rats; moreover, when they infused dopamine into the general circulation the portal circulation was some 70 per cent of that in a main artery, so that to demonstrate the effectiveness of arterial infusions they established an arterial concentration of 9–10 ng/ml. This did not affect the prolactin secretion of di-oestrous rats. When they were pre-treated with α-methyl-p-tyrosine, which blocks dopamine synthesis,† the blood-prolactin levels were raised, and now, under these conditions, the infusion of dopamine caused 70 per cent suppression of prolactin secretion. Similarly, when median eminence lesions had blocked hypothalamic input to the pituitary and the prolactin levels were raised, the dopamine infusions suppressed the increased prolactin secretion by 42 per cent.

Adrenergic System

Inhibition of Prolactin Release

So far as noradrenaline and adrenaline are concerned, the study of Birge et al. (1970) on isolated pituitary glands treated with the amines suggests a direct inhibitory effect on prolactin release, independent of any action of the hypothalamic neurones; thus treatment of the gland

---

* Wiggins and Fernstrom made use of two decarboxylase-inhibitors; MK–486 inhibits the conversion of DOPA to dopamine in tissues outside the central nervous system, but not within; it actually potentiated the effects of DOPA, probably by leaving more DOPA available to the hypothalamus; RO6–4602, which inhibited decarboxylase both peripherally and centrally, blocked the effect of DOPA.

† Given to parturient rats it increased the suckling-induced rise in blood prolactin (Voogt and Carr, 1975).

with $2\cdot4.10^{-6}$ to $2\cdot4.10^{-4}$M noradrenaline and $5\cdot4.10^{-3}$ to $5\cdot4.10^{-4}$M adrenaline inhibited release of prolactin from the gland whilst the content rose. There was no significant effect on synthesis or release of growth hormone which, under these conditions, was also synthesized in the isolated gland. Since both phentolamine and propanolol blocked the effects of the amines, the receptors sensitive to them were of both a- and β-type.

In a later study, Schally *et al.* (1976) described extracts from the porcine hypothalamus with very high prolactin-inhibiting activity; these contained 15 per cent of noradrenaline and 2 per cent of dopamine and the magnitude of the inhibition was proportional to the amount of noradrenaline in the extract. Synthetic noradrenaline and dopamine inhibited the *in vitro* release of prolactin from the pituitary.

Ergot Alkaloids

Ergot alkaloids inhibit early pregnancy and pseudopregnancy in the rat; these actions are due to luteotrophin inhibition so that the corpora lutea degenerate and progesterone secretion is inhibited; hence decidualization, etc are abolished (Zeilmaker and Carlsen, 1962). The effects may be reversed by treatment with progesterone or prolactin, which is luteotrophic in the rat (Shelesnyak, 1958). The anti-prolactin activity is seen, further, in the inhibition of lactation (Zeilmaker and Carlsen, 1962) and of the growth of experimental mammary tumours (Nagasawa and Meites, 1970).*

Wuttke *et al.* (1971) found that, when administered subcutaneously, ergocornine had no effect on the oestrous cycle although the prolactin surge was blocked. When implanted in the median eminence, ergocornine lowered the serum prolactin, increasing the hypothalamic content of prolactin inhibitory factor (**PIF**) suggesting a direct action on the hypothalamus. A direct action on the anterior pituitary may also be demonstrated, e.g. as an inhibition of release by the isolated pituitary (Lu *et al.*, 1971) or that liberated from a pituitary graft.

SEROTONIN OR 5-HYDROXYTRYPTAMINE (5-HT)

Hypothalamic Neurones

Neurones containing this biological amine in their terminals are present in the hypothalamus mainly in the nucleus suprachiasmaticus,

---

* Iproniazid was also effective, but this, unlike ergocornine, did not lower the prolactin levels in the blood which is presumably the mechanism by which the ergot alkaloid exerts its effects. Dopamine was without effect on the tumours.

in the middle third of the retrochiasmatic area and the corresponding area at the level of the median eminence (Fuxe and Hökfelt, 1970); their cell bodies are derived primarily from the midbrain.

Pineal. Melatonin, the hormone secreted by the pineal gland (p. 519) causes a rise in concentration of 5-HT in the brain (Anton-Tay *et al.*, 1968) so that the influence of the pineal on oestrous behaviour and the ovarian cycle may well be exerted through serotoninergic neurones.

Pitfalls

When studying the effects of 5-HT (and, for that matter, of other transmitters) on aspects of reproductive behaviour, care must be taken to distinguish a direct action, through mimicking the discharge of serotoninergic neurones, from an indirect one following the powerful vasoconstrictive action of the drug, which might well prevent access of hormones to their targets. Thus Schneider and McCann (1970) found no effect of 5-HT on LH-secretion when they injected it into the IIIrd ventricle in a site where the transmitter should be able to exert its effect by diffusion into the hypothalamus. Wilson and McDonald (1974) found that 5-HT when administered subcutaneously in pro-oestrus would inhibit ovulation; it would also antagonize the ovulatory action of LH in the Nembutal-blocked rat. The inhibitory effect of 5-HT could be antagonized by injections of oestradiol, and also of a peripheral vasodilator, dipyridamole, 10 or 60 minutes after 5-HT administration, so that Wilson and McDonald inclined to the opinion that the main action of systemically injected 5-HT is exerted through its vaso-constrictive properties.

Blood-Brain Barrier. This is reasonable since 5-HT, like other transmitters, does not cross the blood-brain barrier easily, so that to exert an effect on the hypothlamus it may well be necessary for it to be administered into the cerebrospinal fluid (Vol. 3) or directly into the nervous tissue. Alternatively, in order to raise the level of 5-HT in the brain by intravenous injections, a precursor, namely 5-hydroxytrypto-phan (5-HTP) is used; in this case Lu and Meites (1973) found that the injection raised the level of prolactin in the blood by a factor of nine-fold, whereas injection of 5-HT had no effect. Melatonin behaved similarly to 5-hydroxytryptophan, increasing plasma prolactin.

If, as is suspected (p. 362), the release of prolactin is controlled by both a releasing factor (PRF) and an inhibiting factor (PIF), then it could be, as argued by Boyd *et al.* (1976), that the serotoninergic fibres are causing release of the releasing factor, PRL, by the hypo-thalamic neurones.*

Serotonin Block

In general, then, we must interpret evidence based on subcutaneous or intravenous administration of 5-HT with some caution, and place more reliance on experiments in which the supposititous serotoninergic system is activated indirectly, or inactivated by drugs that inhibit 5-HT synthesis or block its action at its effector sites. Methysergide (MES), an ergot derivative of lysergic acid, is a competitive blocker of 5-HT, and it has been shown to prevent the inhibition of ovulation described above (Labhsetwar, 1971); it induces lordosis (Zemlan et al., 1973) and prevents oestrogen-induced release of prolactin (Caligaris and Taleisnik, 1974). However, according to Gallo et al. (1975) the stimulatory effects of MES on release of prolactin in ovariectomized and lactating rats is unlikely to be due to a blockage of serotonin receptors; and it may well be that MES also acts on the dopaminergic pathway, a possibility suggested by the finding that a β-blocker (MJ 1999) blocks both dopamine and serotonin responses (York, 1970).

According to Lawson and Gala (1976) it is necessary to block the intense inhibitory effects of dopamine on prolactin release in order to demonstrate any stimulatory effect of 5-HT.

Monamine-oxidase Inhibitors

These raise the levels of mono-amines, including 5-HT, in the tissues and potentiate the action of the mono-amine as a transmitter. Meyerson (1964) found that they inhibited ongoing lordosis in oestrogen-progesterone treated rats; and this may have been due to 5-HT accumulation.

A more direct proof of 5-HT involvement was provided by Kordon et al. (1968) who studied the effects of MAO-inhibitors on the super-ovulation induced in the immature rat by injections of pregnant mare's serum, PMS, followed by chorionic gonadotrophin, hCG. The effect could be due to accumulation within the brain of dopamine, nora-drenaline or 5-HT; by using different blocking agents it was shown that

---

* Evidence advanced in favour of a serotoninergic stimulation of prolactin releasing factor is that the rise in blood prolactin caused by reserpine treatment, which blocks PIF release, can be increased by ether-stress, presumably due to release of the releasing factor. The ether-induced rise in prolactin can be blocked by methysergide (Boyd et al., 1976). Clemens et al. (1977) have summarized evidence favouring the role of serotonin in control of prolactin secretion; they injected the serotonin precursor 5-hydroxytryptophan and increased the effectiveness of the serotonin so formed by injections of the potentiator fluoxetin; as a result, there was a rise in prolactin in the blood, not obtainable by injections of 5-hydroxytryptophan or of fluoxetin separately. The serotonin agonist, quipazine, raised prolactin blood levels.

the suppression of superovulation only occurred when 5-HT accumulated in the brain. By injecting an MOA-inhibitor (Nialamide) into the brain, superovulation was suppressed, but only when the site was the inediobasal tuberal part of the hypothalamus, injection into the pituitary being ineffective. Blockade of 5-HT synthesis with PCPA restored the superovulation (Kordon, 1969).

P-Chlorophenylalanine (PCPA)

This inhibits synthesis of 5-HT, and it also inhibits lordosis in female rats, once again emphasizing the probable inhibitory role of 5-HT in sexual behaviour.* When injected directly into the hypothalamus it had the same effects (Zemlan et al., 1973). When administered to rats that were deprived of their suckling young, a procedure that normally causes a rise in blood-prolactin (p. 402), this inhibitor of 5-HT synthesis prevented the rise, supporting the concept of a serotoninergic anti-gonadotrophin release pathway. Thus ovulation, which in the rat depends on prolactin and LH, was completely inhibited by administration of PCPA on Day 2; this was restored by administration of the 5-HT precursor, 5-HTP (Kordon et al., 1971/2).

Dual Aminergic Control over Ovulation

According to Labhsetwar (1971), the neuronal control over ovulation is achieved by a balance between a dopaminergic activating and a serotoninergic inhibiting system of neurones acting on the neurones responsible for liberating releasing hormone at their terminals. In his work, he found that subcutaneous 5-HT, given before pro-oestrus, interfered with ovulation but not with vaginal cornification, uterine ballooning or mating, an effect that was blocked by methysergide. The effect was central, since the LH-stores in the pituitary were high, the LH-surge having been blocked through hypothalamic inhibition. Since during a cycle, the LH in the rat's plasma remains constant and low, except during the ovulating surge, it would seem, on this basis, that the serotonin system is dominant during most of the cycle, and only at the time of ovulation does control pass to the dopaminergic system. The special factor that intervened to permit the dopaminergic system to take over could be the act of mating, in the induced type of ovulation, or the secretion of oestrogen, favouring increased pituitary sensitivity to releasing hormone, in those animals that ovulate cyclically.

---

* Héry et al. (1976) observed a rhythmic surge of LH in ovariectomized rats with oestrogen implants. When 5-HT synthesis was blocked by PCPA the afternoon rise in LH was prevented, whilst administration of 5-HTP caused the reappearance of the surge on the next afternoon. Methiopin, which blocks 5-HT receptors, prevented the afternoon surge whereas fluphenazine, a dopamine receptor blocker, had no effect.

CHOLINERGIC SYSTEM

The involvement of a cholinergic system in some aspects of reproduction has been indicated by several scattered studies.

Thus, Jacobson *et al.* (1950) found that acetylcholine and adrenaline blocked the pseudopregnancy of rats induced by electrical stimulation of the cervix, effects, however, that could be obtained in the presence of the inhibitors atropine and dibenamine. Again, Grosvenor and Turner (1958) showed that the suckling-induced surge of prolactin could be blocked by atropine and dibenamine, suggesting both adrenergic and cholinergic neurones in the reflex release of the lactogenic hormone.

Finally Zarrow and Quinn (1963) showed that dibenamine, atropine and Nembutal all blocked the superovulation in the immature rat caused by pregnant mare's serum (PMS).

Atropine

More recent studies have generally confirmed the involvement of a cholinergic system; thus Libertun and McCann (1973) were able to block the gonadotrophin surges by subcutaneous or IIIrd ventricle injections of atropine at 11·45–1·15 p.m. of pro-oestrus. They also prevented the post-castration rise in LH and FSH in males. Since it did not affect the pituitary response to LHRH, the effects were on the central nervous system.

Pilocarpine and Eserine

In a later study these authors (1974) observed a biphasic response to pilocarpine and eserine, an initial fall in blood LH and prolactin being followed by a larger rise (Fig. 9.2); atropine sulphate reduced the effects but not atropine methyl nitrate, a muscarinic blocker that does not cross the blood-brain barrier. Thus some central muscarinic synapse must have been involved.

Double Prolactin Surge

In the pseudopregnant or pregnant rat, there are two surges of prolactin in the blood, the one between 14.30 and 20.30 hr and the other between 23.30 and 05.30 hr. Atropine only blocked the nocturnal surge, so that the two surges seem to be differently controlled (McLean and Nikitovich-Winer, 1975).

Dual Mechanism

The inhibition of prolactin release by cholinergic drugs such as pilocarpine, described by Grandison *et al.* (1974), may apparently be mediated through an adrenergic system; thus reserpine, as we have

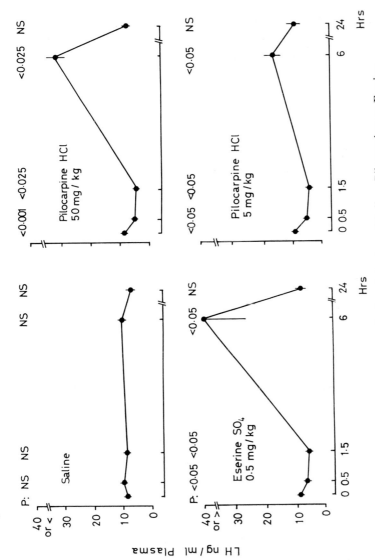

Fig. 9.2. Plasma LH prior to, and after, injection of Saline, Pilocarpine or Eserine into ovariectomized oestrogen-primed rats. P values indicate significant changes from the starting values at a particular time. NS = not significant. (Libertun and McCann, *Proc. Soc. exp. Biol. Med.*)

seen, raises the blood prolactin through its anti-adrenergic action. Pilocarpine, acting on its own, decreases blood prolactin, but it does not antagonize the action of reserpine, and so may well act through an adrenergic system. Grandison and Meites (1976) therefore suggest that adrenergic neurones that control prolactin release have acetylcholine receptors and are activated by cholinergic neurones. If these were excitatory on the adrenergic neurones, pilocarpine and acetylcholine would have the same inhibitory effect on prolactin release.

Pituitary Release of FSH

The release of FSH from isolated pituitary bodies is increased by addition of hypothalamic fragments; it is not increased by acetylcholine alone, but the hypothalamus-induced increase in release is markedly stimulated by the transmitter. Thus the releases were as follows:

Anterior Pituitary alone	$4\cdot64 \pm 0\cdot57$
A.P. + Acetylcholine	$4\cdot36 \pm 0\cdot73$
A.P. + Hypo. Fragments	$7\cdot16 \pm 0\cdot84$
A.P. + Hypo. Fragments + ACh	$16\cdot68 \pm 1\cdot88$

(Simonovic *et al.*, 1974.)

Release of LH

Fiorindo and Martini (1975) incubated rat pituitaries with hypothalamic fragments and failed to find a greater release of LH than that observed without the hypothalamic fragments (Fig. 9.3). However, when acetylcholine was added to the medium together with the hypothalamic fragments release of LH was increased (Fig. 9.3). Presumably acetylcholine was stimulating the release of LHRH from the hypothalamic fragments. The effects of acetylcholine could be completely blocked by atropine, whilst prostigmine enhanced the release in a medium containing anterior pituitary plus hypothalamic fragments. An alpha adrenergic blocking agent, phentolamine, failed to block the effects of acetylcholine. Fiorindo and Martini concluded that the acetylcholine action was probably muscarinic in view of the effects of atropine; moreover it seems that nicotinic drugs inhibit secretion of gonadotrophins (Blake *et al.*, 1972; Kanematsu and Sawyer, 1973).

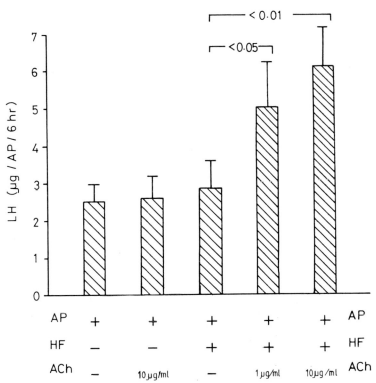

Fig. 9.3. Effect of acetylcholine (ACh) on the release of LH by anterior pituitaries (AP) incubated *in vitro* alone or with hypothalamic fragments (HF). (Fiorindo and Martini, *Neuroendocrinol.*)

GABA

The inhibitory transmitter of the central nervous system, γ-amino-butyric acid (GABA) has been suspected of being concerned in the release of several pituitary releasing hormones, including LHRH (Ondo, 1974). Schally *et al.* (1976, 1977) have examined the inhibitory action of hypothalamic extracts on prolactin release, and have shown that these "PIF-extracts" contain, in addition to dopamine and noradrenaline, GABA. They demonstrated that synthetic GABA would inhibit prolactin release from the isolated pituitary or cultured pituitary cells; there was no effect on LH release. Perphenazine, which inhibits the PIF-activity of catecholamines, has no effect on the influence of GABA. *In vivo* studies also revealed an inhibition of GABA release, but the authors raised the question as to whether physiological doses of the transmitter would be effective.[*]

[*] Bicuculline, the inhibitor of GABA-action, is a potent stimulator of prolactin release (Rivier and Vale, 1977).

Effects of Anaesthesia

Pass and Ondo (1977) have emphasized the effects of pentobarbitone anaesthesia on GABA responses; thus in their earlier study on anaesthetized rats, introduction of GABA into the ventricles caused increased secretion of LH and prolactin but not of FSH; when working on unanaesthetized animals, however, *only* prolactin was released, so that the release of luteinizing hormone, LH, seems to have been artefactual.

HISTAMINE

Hypothalamic Concentration

There are high concentrations of this amine, which is probably a neurotransmitter, within the hypothalamus especially the median eminence (see, for example, Brownstein *et al.* 1974), and there is some evidence of its being involved in the control of secretion by the pars distalis.

Effects on Prolactin Secretion

Libertun and McCann (1976) introduced histamine into the IIIrd ventricles of ovariectomized oestrogen-primed rats, producing a rise in prolactin secretion. The stress-induced rise in prolactin secretion was blocked by the classical anti-histamine blocker, diphenhydramine. These studies suggested that prolactin secretion was promoted by histamine, acting through the H1 receptor, which is blocked by the classical anti-histamines. Arakelian and Libertun (1977) studied the effects of histamine and its antagonists on the suckling-induced release of prolactin by lactating rats.

Histamine Antagonists. Figure 9.4 illustrates the typical sharp rise in serum prolactin caused by suckling, whilst Figure 9.5 shows the much smaller rise when the classical antagonist diphenhydramine was given intrajugularly (black line) and intraventricularly (dashed line). By contrast the H2 blocker, metiamide, had no effect on prolactin release in response to suckling, but it did cause a large increase in prolactin release in non-suckling rats, although the classical H1 antagonist had no effect on non-suckling rats. Therefore H1 and H2 antagonists had opposite actions, the former blocking prolactin release in suckling mothers and the latter causing release in non-suckling mothers.

Agonists. The use of agonists confirmed these findings; thus histamine given intraventricularly in a relatively high dose released a small amount of prolactin in non-suckling mothers, probably as the net result of a predominant H1 stimulation; the H2 agonist, 4-methyl-

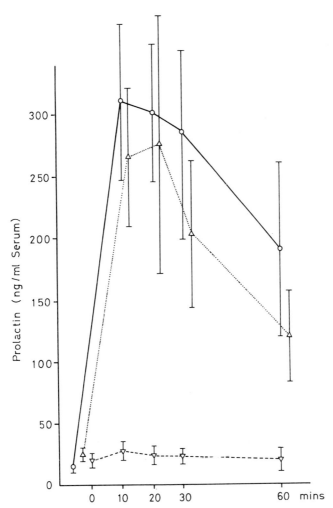

Fig. 9.4. Serum prolactin in suckling mothers injected intrajugularly with saline (solid line) or micro-injected with saline into the third ventricle (dotted line). In non-suckling mothers, intraventricularly injected with saline, prolactin remained low (broken line). Bars = S.E.M.; numbers in parentheses, numbers of animals. (Arakelian and Libertun, *Endocrinology.*)

histamine, blocked suckling-induced prolactin release. Thus, H2 receptors are related to an inhibition of prolactin release, so that an H2-antagonist, such as metiamide, causes an increased release in the non-suckling mother. The H1 receptor is concerned with facilitation of prolactin release, so that the classical anti-histamines block prolactin release, whether induced by stress or by suckling.

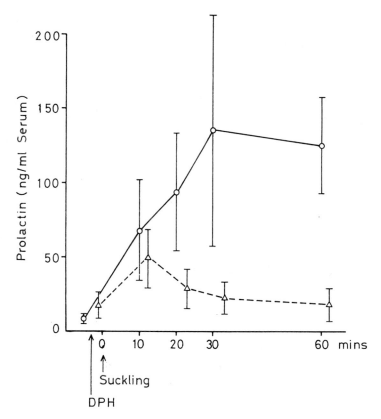

Fig. 9.5. Partial suppressive effect of diphenhydramine (DPH) by intrajugular route (solid line) or micro-injected into the ventricle (broken line) on prolactin release in suckling mothers. (Arakelian and McCann, *Endocrinology*.)

ROLE OF INNERVATION

The control mechanisms in the events of the ovulatory cycle have been described in terms of the release of hormones; the question has naturally arisen as to the role, if any, of the innervation of the ovary.

Innervation of Ovary

This innervation is mainly through the ovarian plexus, arising from the aortic and renal plexuses, constituting a meshwork of fibre bundles which invest both the ovarian artery and vein, forming the *extrinsic nerves* of the ovary. The *intrinsic nerves*, entering from the ovarian plexus, accompany the ovarian vessels; it was concluded by Kuntz (1945) that the entire efferent nerve supply was to the blood vessels, so that a nervous control over the follicles during ovulation, or over the secretory cells, was considered unlikely. However, if the innervation of the chicken ovary is at all comparable with that of the mammal, then Dahl's (1970) study of the thecal gland in this species certainly indicates an innervation of the steroid secreting cells, and he suggested that this might be concerned in the regulation of the seasonal activity of the reproductive behaviour that takes place among birds.

Smooth Muscle. The mammalian ovary contains smooth muscle; the perifollicular muscle has been examined in the electron microscope by Burden (1972); this is situated in the theca externa of large vesicular follicles which consists of collagen fibres, fibroblasts and smooth muscle cells, some 50 per cent of the cells in the tissue being smooth muscle. In the guinea-pig the nerve fibres run almost exclusively to the smooth muscle cells,* the preterminal axons having dense-cored vesicles. According to Jacobowitz and Wallach (1967) the amount of smooth muscle in the tissue varies with the species, being densest in the cat, less in man and still less in the monkey; histochemical assays of noradrenaline correlated with the relative densities of muscular tissue.†

Effects of Denervation

As reviewed by Bahr *et al.* (1974), the effects of denervation on ovulation in the mammal are ambiguous, perhaps because of the difficulty in ensuring adequate denervation; moreover, when this is achieved by tying off the arterial supply it is questionable whether the disruption of the blood supply is without an effect of its own. Again, transection of the cord may well not be at the correct level to achieve denervation. LePere *et al.* (1966), working on the baboon, denervated the ovary by vascular transection, i.e. by dissecting clear the vascular pedicle and transecting the aorta and vena cava above and below the entry of the ovarian vessels; these were then reanastomosed with interposition of a teflon graft. Alternatively, the blood vessels were stripped, removing all surrounding tissues down to and including the adventitia. All nerve fibres in the ovary disappeared when these treatments were combined with vagotomy; however, there was no change in the cycle length, suggesting an absence of nervous involvement in ovarian secretion or ovulation.

---

* In the rabbit, Owman and Sjöberg (1966) described only a few noradrenergic fibres running to non-vascular muscle of the ovary.

† The presence of a sensory innervation of the ovary is probably revealed by the "Mittelschmertz", namely the pain experienced by many women at the time of ovulation.

Autotransplanted Ovary

The fact that ovaries autotransplanted to ectopic sites, such as abdominal muscle or the eye, are capable of sustaining normal oestrous function (see, for example, Harris and Campbell, 1962) strongly suggests that cyclic ovulation and corpora lutea formation occur independently of innervation.* However, examination of the graft shows that, during the time required for it to become functional, not only is a new vascular circulation established but also a new set of adrenergic nerves. Thus, Jacobowitz and Laties (1970) employed the Hillarp fluorescence technique for identifying catecholaminergic neurones in ovaries transplanted into the anterior chamber of the eye. The ovary became attached to the iris by a connective-tissue bridge, and fluorescent varicose fibres could be seen coursing through the bridge. When full re-innervation had occurred this was similar to that in the ovary in its normal position, with some fibres running in close relation to the follicles. About 10–15 days are required for complete revascularization and re-innervation, and the same time is required for resumption of function (Jascczak and Hafez, 1973). However, the vascularization is obviously necessary for hormonal control over the transplanted ovary, so that it may be only coincidental that activity returns with innervation. The crucial test would be to transplant the ovary to the eye and remove the superior cervical ganglion, preventing re-innervation. More decisive experiments suggesting a role of adrenergic drugs on ovulation are those in which small quantities of a drug were injected directly into a follicle of the chick ovary, this being chosen because of its dense innervation.

Ovarian Injections

The a-blocker, dibenzyline (phenoxybenzamine), injected into the wall of the largest follicle of the ovary, blocked ovulation, whereas an inactive analogue—dibenzylaminoethanol—had no effect. The injection had to be at a critical time—some 14–15 hr before the expected ovulation—but not at 5–13 hr before. A dose of FSH and LH, in correct proportions, was able to overcome the blocking action (Ferrando and Nalbandov, 1969). Rather similar results were found in rabbits (Virutmassen et al., 1971), and Bahr et al. (1974) suggest that the gonadotrophins might well act through catecholamines, in accordance

---

* Jascczak and Hafez (1973) found that the transplanted ovary, in for example, the abdominal wall, would ovulate and the ovum might even find its way into the oviduct if the transplant was near the fimbria.

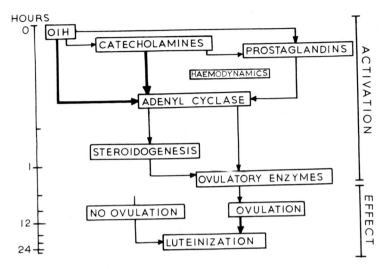

Fig. 9.6. Scheme relating the various changes and interactions initiated by release of ovulation-inducing hormone (OIH). OIH may act by different mechanisms that converge on a common path involving adenylate cyclase. (Bahr *et al.*, *Biol. Reprod.*)

with the scheme illustrated by Figure 9.6, where OIH indicates the particular gonadotroph or combination of hormones, that initiates ovulation.

Spontaneous Contractions

As to the mode of action of the nerves, we may assume that they influence the smooth muscle of the stroma of the ovary, causing contractions that might well be involved in follicular rupture. Certainly Rocerto *et al.* (1969) observed spontaneous contractions in the cat's ovary; these were increased by noradrenaline and adrenaline, an effect that was blocked by phentolamine; isoprenaline, the β-agonist, inhibited the spontaneous contractions, an effect that was blocked by propanolol.

Again, Walles *et al.* (1975) using strips of bovine follicle, showed that contraction was favoured by α-adrenergic agonists, whilst the contractions induced by the cholinergic drug, carbamyl choline, could be inhibited by the β-agonist isoprenaline.

In human subjects, too, the isolated ovary gives rise to spontaneous *in vitro* contractions (Diaz-Infante *et al.*, 1974); when this was oestrogen-dominated, i.e. pre-ovulatory, sensitivity to catecholamines was low, whereas in the progesterone-dominated tissue it was high. A cholinergic mechanism was suggested, also, since the potentiators prostigmine and bethanechol stimulated contraction, an effect inhibited by atropine, whilst Walles *et al.* (1975, 1976) identified cholinesterase histochemically in the bovine theca externa close to cells resembling those of smooth muscle. Strips of the follicle contracted in response to acetylcholine, an effect blocked by atropine.

CYCLIC AMP

In Volume 2 the role of cAMP as a second messenger has been discussed in considerable detail, so that here we need only indicate the stages in reproductive control that seem to involve this nucleoside.

Release of Pituitary Hormones

The release of pituitary hormones, by the action of the releasing hormones, is definitely a stage involving accumulation of cAMP in the pituitary cells (Robison *et al.* 1971), so that it is likely that ACTH, GH, TSH, LH and FSH are released from separate cells of the anterior pituitary, each of which possesses an adenylcyclase, sensitive to a different releasing hormone.

Steroidogenic Cells

Corpus Luteum

At the next stage, namely the action of the trophic hormone on the steroidogenic cells of the gonads, cAMP is also involved. For example, a study of Marsh *et al.* on steroidogenesis in the bovine corpus luteum demonstrated a striking increase in concentration of cAMP when physiological concentrations of LH were added (Table II) to slices of the tissue, an effect that was due to a stimulation of adenylcyclase activity rather than an inhibition of phosphodiesterase (Marsh, 1970).

Rat Ovary

When LH, FSH, and PGE were compared on the *in vitro* rat's ovary, Mason *et al.* (1973) found that all three were effective in increasing the production of cAMP; LH was very much more effective than FSH. The stage of the oestrous cycle at which the ovaries were removed had little influence on the susceptibility to the gonadotrophs and prostaglandin, although the luteal tissue of pregnancy gave only a very small response.

Ovarian Cycle. Goff and Major (1975) have shown in rabbit ovarian tissue that, *in vivo*, there are no great changes in concentration of cAMP during the course of the ovarian cycle; after administration of gonadotrophic hormone (hCG plus LH) there was a biphasic peak in cAMP concentration but this had returned to normal before ovulation. Thus, steroid synthesis seems not to be directly related to the cAMP in the ovary, although acute responses to gonadotrophins may well be

TABLE II

Effects of various substances on accumulation of cAMP in corpus luteum (Marsh et al., 1966)

Experiment and addition	Cyclic AMP (nanomoles/gm)
Experiment 1	
None	0·44
Luteinizing hormone (2·0 μg/ml)	14·0
Inactivated luteinizing hormone (2·0 μg/ml)	0·35
Experiment 2	
None	0·36
Luteinizing hormone (2·0 μg/ml)	6·76
Prolactin (2·0 μg/ml)	0·14
Experiment 3	
None	0·33
Luteinizing hormone (2·0 μg/ml)	2·77
Prolactin (2·0 μg/ml)	0·38
ACTH (0·3 unit/ml)	0·45
Epinephrine (0·01 μmol/ml)	0·43
Glucagon (10 μg/ml)	0·41

mediated through the second messenger. A similar failure to find cyclical variations in concentration of the rat ovary was reported by Mason et al. (1973), although, in vitro, LH increased the formation of cAMP in ovaries taken at all stages of the cycle.

Testis

In the testis, it will be recalled that the anterior pituitary hormones are necessary for maintaining the structural and functional integrity of this organ, so that injections of LH (ICSH) and FSH will partially restore spermatogenesis to the regressed hypophysectomized rat. It is generally agreed that LH stimulates the interstitial cells of Leydig to produce androgens whilst the action of FSH is more directly on the germ cells of the seminiferous tubules. Since the androgens, secreted in response to LH, have a trophic action on the seminiferous tubules, this hormone clearly has both a primary and a secondary action. Dorrington et al. (1972) teased the seminiferous tubules free from the rat's testis and showed that FSH increased the level of cAMP in the tissue, whilst LH had no effect. When the intact isolated testis, or homogenates of this, were examined, both LH and FSH stimulated production of cAMP (Murad et al., 1969), but when maximal effect was achieved with the one hormone, addition of the other did not increase cAMP production further. According to Sandler and Hall (1966) addition of cAMP to slices of rat testis increased the conversion of cholesterol to testosterone, thereby mimicking the effect of LH. In general, according to Braun and Sepsenwol (1974) there is an FSH-responsive adenylcyclase located in the cells of the seminiferous tubules and an LH-responsive cyclase located in the interstitial cells of Leydig.

Steroid Hormone Action

So far as the final stage in control is concerned, namely the action of the steroid hormone on its target tissue, there is not complete agreement, and the situation may well vary with the target cell and the particular hormone.

Uterus

Szego and Davis (1967, 1969) found a rise in cAMP concentration in the rat uterus within 30 sec of injecting oestradiol-17β into the ovariectomized animal, an effect that was blocked by propanolol, suggesting an action through a β-adrenergic mechanism.

Implantation. Mohla and Prasad (1970) argued that if oestrogen works through cAMP in so far as its uterine action is concerned, then the activation by oestrogen of synthesis of RNA by the blastocyst in delayed implantation might be mimicked by cAMP. They infused cAMP into the uterine lumen of rats in delayed implantation and collected the blastocysts later and studied their incorporation of ^{3}H-uridine into their nuclei as a measure of active synthesis. Incorporation was stimulated, suggesting a role of cAMP in activating the blastocyst prior to implantation. Using dibutyryl cAMP, Webb (1975) was able to induce implantation in ovariectomized rats provided they were pre-treated with progesterone.

Incubated Uteri. However, Zor *et al.* (1973) measured the accumulation of cAMP in the incubated uteri of immature rats. As Figure 9.7 shows, oestradiol had no effect on this accumulation although adrenaline and prostaglandins $PGF_2\alpha$ and PGE_2 had large effects. The same effects were obtained whether the substances were administered *in vivo* or added to the incubation medium. If we may accept this work, then, oestradiol does not work through cAMP; moreover it also does not work through synthesis of catecholamines or PG's since, in this system, these were active in causing accumulation of cAMP. It will be noted that the catecholamine action was blocked by propranolol whereas that of the PG's was not, indicating that the PG's do not work through the catecholamines.

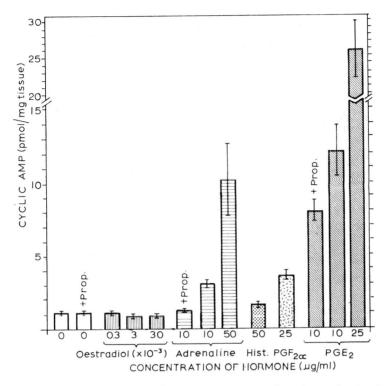

Fig. 9.7. Concentration of cyclic AMP in uteri from pre-pubertal rats after incubation with oestradiol, histamine (Hist.) or various stimulants of adenyl cyclase. PG = prostaglandin; Prop. = propranolol (10 μg/ml). Incubation was for 20 min at 37 °C in the presence of 10^{-2}M theophylline. Vertical bars indicate $\pm$ S.E.M. for 4–5 rats (experimental groups) or 10 rats (control group). (Zor *et al.*, *J. Endocrinol.*)

Rat Prostate

Mangan *et al.* (1973) have examined the ability of cAMP to mimic the actions of testosterone or 5α-dihydrotestosterone on the rat's prostate gland; they found that it failed to stimulate the gland nor yet did it influence its morphology; it was unable to simulate the effects of the androgens on stimulating RNA polymerase activity, nor yet to stimulate RNA synthesis. One point of analogy between cAMP and the steroids was in the activation of glucose-6-phosphate dehydrogenase, although the significance of this is reduced by the observation that the anti-androgen, cyproterone acetate, does not affect the androgen-stimulation of this synthesis. The authors suggested that the steroid promotes the synthesis of the RNA components necessary for enzyme synthesis, whereas cAMP controls the rate of this synthesis; thus cAMP would be implicated only at the translational level, whereas the steroid would be implicated at the transcriptional level, namely the production of RNA within the nucleus.

Prostaglandins

To revert to the involvement of prostaglandins, according to Kuehl *et al.* (1970) in the mouse ovary PGE_1, PGE_2 and LH all stimulate the production of cAMP; the PG-antagonist—7-oxa-13-prostanoic acid—blocked all three actions suggesting that LH acts through a prostaglandin receptor site. This contrasts with the action of PG's in *inhibiting* the hormonally induced formation of cAMP in adipose tissue.

REFERENCES

Ahrén, K., Fuxe, K. Hamberger, L., and Hökfelt, T. (1971). Turnover changes in the tubero-infundibular dopamine neurons during the ovarian cycle of the rat. *Endocrinology*, **88**, 1415–1424.

Anton-Tay, F., Chou, C., Anton, S. and Wurtman, R. J. (1968). Brain serotonin concentrations: elevation following intraperitoneal administration of melatonin. *Science*, **162**, 277–278.

Arakelian, M. C. and Libertun, C. (1977). H1 and H2 histamine receptor participation in the brain control of prolactin secretion in lactating rats. *Endocrinology*, **100**, 890–895.

Bahr, J., Kao, L. and Nalbandov, A. V. (1974). The role of catecholamines and nerves in ovulation. *Biol. Reprod.*, **10**, 273–290.

Ben-Jonathan, N. *et al.* (1977). Dopamine in hypophyseal portal plasma of the rat during the estrous cycle and throughout pregnancy. *Endocrinology*, **100**, 452–458.

Birge, C. A., Jacobs, L. S., Hammer, C. T. and Daughaday, W. H. (1970). Catecholamine inhibition of prolactin secretion by isolated rat adenohypophysis. *Endocrinology*, **86**, 120–130.

Björklund, A., and Nobin, A. (1973). Fluorescence histochemical and microfluorometric mapping of dopamine and noradrenaline cell groups in the rat diencephalon. *Brain Res.*, **51**, 193–205.

Blake, C. A. *et al.* (1972). Nicotine delays the ovulatory surge of luteinizing hormone in the rat. *Proc. Soc. exp. Biol. Med.*, **141**, 1014–1016.

Boyd, A. E., Spencer, E., Jackson, I. M. D. and Reichlin, S. (1976). Prolactin-releasing factor (PRF) in porcine hypothalamic extract distinct from TRH. *Endocrinology*, **99**, 861–871.

Braun, T. and Sepsenwol, S. (1974). Stimulation of ^{14}C-cyclic AMP accumulation by FSH and LH in testis from mature and immature rats. *Endocrinology*, **94**, 1028–1033.

Brownstein, M. J., Saavedra, J. M., Palkovits, M. and Axelrod, J. (1974). Histamine content of hypothalamic nuclei of the rat. *Brain Res.*, **77**, 151–156.

Burden, H. W. (1972). Ultrastructural observations on ovarian perifollicular smooth muscle in the cat, guinea-pig, and rabbit. *Amer. J. Anat.*, **133**, 125–129.

Caligaris, L. and Talesnik, S. (1974). Involvement of neurones containing 5-hydroxytryptamine in the mechanism of prolactin release induced by oestrogen. *J. Endocrinol.*, **62**, 25–33.

Clemens, J. A., Sawyer, B. D. and Cerimele, B. (1977). Further evidence that serotonin is a neurotransmitter involved in the control of prolactin secretion. *Endocrinology*, **100**, 692–698.

Condon, W. A. and Black, D. L. (1976). Catecholamine-induced stimulation of progesterone by the bovine corpus luteum *in vitro*. *Biol. Reprod.*, **15**, 573–578.

Craven, R. P. and McLeod, P. G. (1973). The effect of intrahypothalamic infusions of dopamine and noradrenaline on ovulation in reserpine-treated rats. *J. Endocrinol.*, **58**, 319–326.

Dahl, E. (1970). Innervation of the thecal gland of the domestic fowl. *Z. Zellforsch.*, **109**, 212–226.

Diaz-Infante, A. *et al.* (1974). *In vitro* studies of human ovarian contractility. *Obst. Gyn.*, **44**, 830–838.

Donoso, A. O., Bishop, W., Fawcett, C. P., Krulich, L. and McCann, S. M. (1971). Effects of drugs that modify brain monoamine concentrations on plasma gonadotropin and prolactin levels in the rat. *Endocrinology*, **89**, 774–784.

Donoso, A. O., Bishop, W. and McCann, S. M. (1973). The effects of drugs which modify catecholamine synthesis on serum prolactin in rats with median eminence lesions. *Proc. Soc. exp. Biol. Med. N.Y.*, **143**, 360–363.

Dorrington, J. H., Vernon, R. G. and Fritz, I. B. (1972). The effect of gonadotrophins on the 3',5'-AMP levels of seminiferous tubules. *Biochem. Biophys. Res. Comm.*, **46**, 1523–1528.

Drouva, S. V. and Gallo, R. V. (1976). Catecholamine involvement in episodic luteinizing hormone release in adult ovariectomized rats. *Endocrinology*, **99**, 651–658.

Drouva, S. V. and Gallo, R. V. (1977). Further evidence for inhibition of episodic luteinizing hormone release in ovariectomized rats by stimulation of dopamine receptors. *Endocrinology*, **100**, 792–798.

Everett, J. W. (1964). Central neural control of reproductive functions of the adenohypophysis. *Physiol. Rev.*, **44**, 373–431.

Ferrando, G. and Nalbandov, A. V. (1969). Direct effect on the ovary of the adrenergic blocking drug dibenzyline. *Endocrinology*, **85**, 38–42.

Fiorindo, R. P. and Martini, L. (1975). Evidence for cholinergic component in the neuroendocrine content of luteinizing hormone (LH) secretion. *Neuroendocrinol.*, **18**, 322–332.

Fuxe, K. and Hökfelt, T. (1970). Central monoaminergic systems and hypothalamic function. In *The Hypothalamus*. Eds. L. Martini, M. Motta and F. Fraschini, Academic Press, N.Y., pp. 123–138.

Fylling, P. (1973). Dexamethasone or propanolol blockade of induced increase in plasma progesterone in early human pregnancy. *Acta. Endocrinol.*, **72**, 569–572.

Gallo, R. V., Rabii, J. and Moberg, G. P. (1975). Effect of methysergide, a blocker of serotonin receptors, on plasma prolactin levels in lactating and ovariectomized rats. *Endocrinology*, **97**, 1096–1105.

Gibbs, D. M. and Neill, J. D. (1978). Dopamine levels in hypophysial stalk blood in the rat are sufficient to inhibit prolactin secretion *in vivo*. *Endocrinology*, **102**, 1895–1900.

Goff, A. K. and Major, P. W. (1975). Concentrations of cyclic AMP in rabbit ovarian tissue. *J. Endocrinol.*, **65**, 73–82.

Grandison, L., Gelato, M. and Meites, J. (1974). Inhibition of prolactin secretion by cholinergic drugs. *Proc. Soc. exp. Biol. Med.*, **145**, 1236–1239.

Grandison, L. and Meites, J. (1976). Evidence for adrenergic mediation of cholinergic inhibition of prolactin release. *Endocrinology*, **99**, 775–779.

Grosvenor, C. E. and Turner, C. W. (1958). Effects of oxytocin and blocking agents upon pituitary lactogen discharge in lactating rats. *Proc. Soc. exp. Biol. Med.*, **97**, 463–465.

Harris, G. W. and Campbell, H. J. (1966). The regulation of the secretion of luteinizing hormone and ovulation. In *The Pituitary Gland*. Vol. 2. Ed. G. W. Harris and B. T. Donovan, pp. 99–165. Butterworth, London.

Héry, M., Laplante, E. and Kordon, C. (1976). Participation of serotonin in the phasic release of LH. I. *Endocrinology*, **99**, 496–503.

Hökfelt, T. and Fuxe, K. (1972). Effects of prolactin and ergot alkaloids on the tubero-infundibular dopamine (DA) neurons. *Neuroendocrinol.*, **9**, 100–122.

Jacobowitz, D. and Laites, A. M. (1970). Adrenergic reinnervation of the cat ovary transplanted to the anterior chamber of the eye. *Endocrinology*, **86**, 921–924.

Jacobowitz, D. and Wallach, E. E. (1967). Histochemical and chemical studies of the autonomic innervation of the ovary. *Endocrinology*, **81**, 1132–1139.

Jacobson, A., Salhanick, H. A. and Zarrow, M. X. (1950). Induction of pseudopregnancy and its inhibition by various drugs. *Amer. J. Physiol.*, **161**, 522–527.

Jascczak, S. and Hafez, E. S. E. (1973). Hormonal and generative functions of transplanted ovaries in the rabbit and monkey. *Amer. J. Obstet. Gyn.*, **115**, 112–122.

Kalra, S. P. and McCann, S. M. (1973). Effect of drugs modifying catecholamine synthesis on LH release induced by preoptic stimulation in the rat. *Endocrinology*, **93**, 356–362.

Kamberi, I. A., Mical, R. S. and Porter, J. C. (1970). Effect of anterior pituitary perfusion and intraventricular injection of catecholamines and indoleamines on LH release. *Endocrinology*, **86**, 1–12.

Kamberi, I. A., Mical, R. S. and Porter, J. C. (1971). Effect of anterior pituitary perfusion and intraventricular injection of catecholamines on prolactin release. *Endocrinology*, **88**, 1012–1020.

Kanematsu, S. and Sawyer, C. H. (1963). Effects of intrahypothalamic implants of reserpine on lactation and pituitary prolactin content in the rabbit. *Proc. Soc. exp. Biol. Med. N.Y.*, **113**, 967–973.

Kanematsu, S. and Sawyer, C. H. (1973). Inhibition of progesterone-advanced LH surge at proestrus by nicotine. *Proc. Soc. exp. Biol. Med.*, **143**, 1183–1186.

Kordon, C. (1969). Effect of selective experimental changes in regional hypothalamic monoamine levels on superovulation in the immature rat. *Neuoroendocrinol.*, **4**, 129–138.

Kordon, C., Gogan, F., Hery, M. and Rotsztejn (1971/2). Interference of serotonin-containing neurons with pituitary gonadotropins release-regulation. *Gyn. Invest.*, **2**, 116–121.

Kordon, C., Javoy, F., Vassent, G. and Glowinski, J. (1968). Blockade of superovulation in the immature rat by increased brain serotonin. *Euro. J. Pharmacol.*, **4**, 169–174.

Krieg, R. J. and Sawyer, C. H. (1976). Effects of intraventricular catecholamines on luteinizing hormone release in ovariectomized-steroid-primed rats. *Endocrinology*, **99**, 411–419.

Kuehl, F. A. *et al.* (1970). Prostaglandin receptor sites: evidence for an essential role in the action of luteinizing hormone. *Science*, **169**, 883–886.

Kuntz, A. (1945). *The Autonomic Nervous System.* 3rd ed. Philadelphia, Lea and Febiger.

Labhsetwar, A. P. (1971). Effects of serotonin on spontaneous ovulation: a theory for the dual hypothalamic control of ovulation. *Acta. Endocrinol.*, **68**, 334–344.

Lawson, D. M. and Gala, R. R. (1976). The interaction of dopaminergic and serotonergic drugs on plasma prolactin in ovariectomized, estrogen-treated rats. *Endocrinology*, **98**, 42–47.

LePere, R. H., Benoit, P. E., Hardy, R. C. and Goldzieher, J. W. (1966). The origin and function of the ovarian nerve supply in the baboon. *Fert. Ster.*, **17**, 68–75.

Libertum, C. and McCann, S. M. (1973). Blockade of the release of gonadotropins and prolactin by subcutaneous or intraventricular injection of atropine in male and female rats. *Endocrinology*, **92**, 1714–1724.

Libertum, C. and McCann, S. M. (1974). Further evidence for cholinergic control of gonadotropin and prolactin secretion. *Proc. Soc. exp. Biol. Med.*, **147**, 498–504.

Libertum, C. and McCann, S. M. (1976). The possible role of histamine in the control of prolactin and gonadotropin release. *Neuroendocrinol.*, **20**, 110–120.

Lichtensteiger, W. and Keller, P. J. (1974). Tubero-infundibular dopamine neurons and the secretion of luteinizing hormone and prolactin. *Brain Res.*, **74**, 279–303.

Lu, K.-H. and Meites, J. (1971). Inhibition by L-Dopa and monoamine oxidase inhibitors of pituitary prolactin release: stimulation by methyldopa and d-amphetamine. *Proc. Soc. exp. Biol. Med. N.Y.*, **137**, 480–483.

Lu, K.-H. and Meites, J. (1973). Effects of serotonin precursors and melanotonin on serum prolactin release in rats. *Endocrinology*, **93**, 152–155.

Lu, K.-H., Koch, Y. and Meites, J. (1971). Direct inhibition by ergocomine of pituitary prolactin secretion. *Fed. Proc.*, **30**, 474.

Macleod, R. M. and Lehmeyer, J. E. (1974). Studies on the mechanism of the dopamine-mediated inhibition of prolactin secretion. *Endocrinology*, **94**, 1077–1085.

Mangan, F. R., Pegg, A. E. and Mainwaring, W. I. P. (1973). A reappraisal of the effects of adenosine, 3′:5′-cyclic monophosphate on the function and morphology of the rat prostate gland. *Biochem. J.*, **134**, 129–142.

Marsh, J. M. (1970). The stimulatory effect of luteinizing hormone on adenylcyclase in the bovine corpus luteum. *J. biol. Chem.*, **245**, 1596–1603.

Marsh, J. M., Butcher, R. W., Saward, K. and Sutherland, E. W. (1966). The stimulatory effect of luteinizing hormone on adenosine 3′,5′-monophosphate accumulation in corpus luteum slices. *J. biol. Chem.*, **241**, 5436–5440.

Mason, N. R., Schaffer, R. J. and Toomey, R. E. (1973). Stimulation of cyclic AMP accumulation in rat ovaries *in vitro*. *Endocrinology*, **93**, 34–41.

McLean, B. K. and Nikitovich-Winer, M. B. (1975). Cholinergic control over the nocturnal prolactin surge in the pseudopregnant rat. *Endocrinology*, **97**, 763–770.

Meyerson, B. J. (1964). Central nervous monoamines and hormone induced estrus behaviour in the spayed rat. *Acta physiol. Scand.*, **63**, Suppl. 241, 3–32.

Mohla, S. and Prasad, M. R. N. (1970). Stimulation of RNA synthesis in the blastocyst and uterus of the rat by adenosine 3′,5′-monophosphate (cyclic ApMP). *J. Reprod. Fert.*, **23**, 327–329.

Murad, F., Strauch, B. S. and Vaughan, M. (1969). The effect of gonadotrophins on testicular adenylcyclase. *Biochim. biophys. Acta.*, **177**, 591–598.

Nagasawa, H. and Meites, J. (1970). Suppression by ergocomine and iproniazid of carcinogen-induced mammary tumors in rats; effects on serum and pituitary prolactin levels. *Proc. Soc. exp. Biol. Med.*, **135**, 469–472.

Ojeda, S. R., Harms, P. G. and McCann, S. M. (1974). Effect of blockade of dopaminergic receptors on prolactin and LH release: median eminence and pituitary sites of action. *Endocrinology*, **94**, 1650–1657.

Ondo, J. G. (1974). Gamma-aminobutyric acid effects on pituitary gonadotropin secretion. *Science*, **186**, 738–739.

Owman, C. and Sjöberg, N.-O. (1966). Adrenergic nerves in the female genita tract of the rabbit with remarks on cholinesterase containing structures. *Z. f. Zellforsch.*, **74**, 182–197.

Pass, K. A. and Ondo, J. G. (1977). The effects of γ-aminobutyric acid on prolactin and gonadotropin secretion in the unanæsthetized rat. *Endocrinology*, **100**, 1437–1442.

Rivier, C. and Vale, W. (1977). Effects of γ-aminobutyric acid and histamine on prolactin secretion in the rat. *Endocrinology*, **101**, 506–511.

Robison, G. A., Butcher, R. W. and Sutherland, E. W. (1971). *Cyclic AMP*. Academic Press, N.Y.

Rocerto, T., Jacobowitz, D. and Wallach, E. E. (1969). Observations of spontaneous contractions of the cat ovary *in vitro*. *Endocrinology*, **84**, 1336–1341.

Rotsztejn, W. H., Charli, J. L., Pattou, E. and Kordon, C. (1977). Stimulation by dopamine of luteinizing hormone releasing hormone (LHRH) release from the mediobasal hypothalamus in male rats. *Endocrinology*, **101**, 1475–1483.

Rubinstein, L. and Sawyer, C. H. (1969). Role of catecholamines in stimulating the release of pituitary ovulating hormone(s) in rats. *Endocrinology*, **86**, 988–995.

Rubinstein, L. and Sawyer, C. H. (1970). Role of catecholamines in stimulating the release of pituitary ovulating hormone(s) in rats. *Endocrinology*, **86**, 988–995.

Sandler, R. and Hall, P. F. (1966). Stimulation *in vitro* by adenosine-3′,5′-cyclic monophosphate of steroidogenesis in rat testis. *Endocrinology*, **79**, 647–649.

Schally, A. V. *et al.* (1976). Purification of a catecholamine-rich fraction with prolactin release-inhibiting factor (PIF) activity from porcine hypothalami. *Acta. Endocrinol.*, **82**, 1–14.

Schally, A. V. *et al.* (1976). Re-examination of porcine and bovine hypothalamic fractions for additional luteinizing hormone and follicle stimulating hormone-releasing activities. *Endocrinology*, **98**, 380–391.

Schally, A. V., Redding, T. W., Arimura, A., Dupont, A. and Linthieum, G. L. (1977). Isolation of gamma-amino butyric acid from pig hypothalami and demonstrations of its prolactin release-inhibiting (PIF) activity *in vivo* and *in vitro*. *Endocrinology*, **100**, 681–691.

Schneider, H. P. G. and McCann, S. M. (1970). Mono- and indolamines and control of LH secretion. *Endocrinology*, **86**, 1127–1133.

Selmanoff, M. K., Pramik-Holdaway, M. J. and Weiner, R. L. (1976). Concentrations of dopamine and norepinephrine in discrete hypothalamic nuclei during the rat estrous cycle. *Endocrinology*, **99**, 326–329.

Shelesnyak, M. C. (1958). Maintenance of gestation in ergotoxin-treated pregnant rats by exogenous prolactin. *Acta Endocrinol.*, **27**, 99–109.

Simonovic, I., Motta, M. and Martini, L. (1974). Acetylcholine and the release of the follicle-stimulating hormone-releasing factor. *Endocrinology*, **95**, 1373–1379.

Smythe, G. A. and Lazarus, L. (1973). Blockade of the dopamine-inhibitory control of prolactin secretion in rats by 3,4-dimethoxyphenylethylamine (3,4,-di-O-methyl dopamine). *Endocrinology*, **93**, 147–151.

Stefano, F. J. E. and Donoso, A. O. (1967). Norepinephrine levels in the rat hypothalamus during the estrous cycle. *Endocrinology*, **81**, 1405–1406.

Szego, C. M. and Davis, J. S. (1967). Adenosine 3′,5′-monophosphate in rat uterus: acute elevation by estrogen. *Proc. nat. Acad. Sci. U.S.A.*, **58**, 1711–1718.

Szego, C. M. and Davis, J. S. (1969). Inhibition of estrogen-induced elevation of cyclic 3′,5′-adenosine monophosphate in rat uterus. I. *Mol. Pharmacol.*, **5,** 470–480.

Takahara, J., Arimura, A. and Schally, A. V. (1974). Suppression of prolactin release by a purified porcine PIF preparation and catecholamines infused into a rat hypophyseal portal vessel. *Endocrinology*, **95,** 462–465.

Virutmassen, P., Hickock, R. L. and Wallach, E. E. (1971). Local ovarian effects of catecholamines on human chorionic gonadotropin-induced ovulation in the rabbit. *Fert. Ster.*, **22,** 235–243.

Voogt, J. L. and Carr, L. A. (1975). Potentiation of suckling-induced release of prolactin by inhibition of brain catecholamine synthesis. *Endocrinology*, **97,** 891–897.

Walles, B., Edvinsson, L., Owman, C., Sjöberg, N.-O. and Spomong, B. (1976). Cholinergic nerves and receptors mediating contraction of the graafian follicle. *Biol. Reprod.*, **15,** 565–572.

Walles, B. *et al.* (1975). Evidence for a neuromuscular mechanism involved in the contractility of the ovarian follicular wall. *Biol. Reprod.*, **12,** 239–248.

Walles, B. *et al.* (1975). Mechanical response in the wall of ovarian follicles mediated by adrenergic receptors. *J. Pharmacol.*, **193,** 460–473.

Webb, F. T. G. (1975). Implantation in ovariectomized mice treated with dibutyryl adenosine 3′,5′-monophosphate (dibutyryl cyclic AMP). *J. Reprod. Fert.*, **42,** 511–577.

Weiner, R. I., Blake, C. A., Rubinstein, L. and Sawyer, C. H. (1971). Electrical activity of the hypothalamus: effects of intraventricular catecholamines. *Science*, **171,** 411–412.

Wiggins, J. F. and Fernstrom, J. D. (1977). L-Dopa inhibits prolactin secretion in proestrous rats. *Endocrinology*, **101,** 469–474.

Wilson, C. A. and McDonald, P. G. (1974). Inhibitory effect of serotonin on ovulation in adult rats. *J. Endocrinol.*, **60,** 253–260.

Wuttke, W., Cassell, E. and Meites, J. (1971). Effects of ergocomine on serum prolactin and LH, and on hypothalamic content of PIF and LRF. *Endocrinology*, **88,** 737–741.

York, D. H. (1970). Possible dopaminergic pathway from substantia nigra to putamen. *Brain Res.*, **20,** 233–249.

Zarrow, M. X. and Quinn, D. L. (1963). Superovulation in the immature rat following treatment with PMS alone and inhibition of PMS-induced ovulation. *J. Endocrinol.*, **26,** 181–188.

Zeilmaker, G. H. and Carlsen, R. A. (1962). Experimental studies on the effect of ergocomine methanesulphonate on the luteotrophic functions of the rat pituitary gland. *Acta Endocrinol.*, **41,** 321–335.

Zemlan, F. P., Ward, I. L., Crowley, W. R. and Margules, D. L. (1973). Activation of lordotic responding in female rats by suppression of serotonergic activity. *Science N.Y.*, **179,** 1010–1011.

Zor, U. *et al.* (1973). Mechanism of oestradiol action on the rat uterus: independence of cyclic AMP, prostaglandin E_2 and β-adrenergic action. *J. Endocrinol.*, **58,** 525–533.

CHAPTER 10

The Pineal Organ

Biological Clocks

A number of animal functions show rhythmic changes that are related to the diurnal, or annual, cycles in the proportions of lightness and darkness. Typical examples are exhibited by the breeding cycles of birds and many mammals, cycles that can be modified experimentally by alterations in the lighting periods. Diurnal fluctuations in behaviour are typically manifest in the feeding and motility rhythms of rats and mice (Volume 3).

Oestrous Cycle

Within the oestrous cycle of many species, such as the rat, there are diurnal fluctuations in hormonal output that lead to ovulation taking place at a quite specific time of the day in pro-oestrus. Thus we have seen that in the rat this most commonly takes place $3\frac{1}{2}$ to 5 hours before the onset of light (Everett, 1948). Figure 10.1 illustrates the

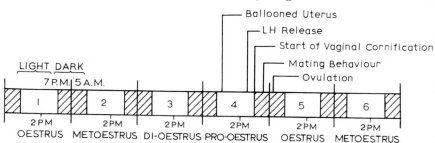

Fig. 10.1. The timing of the manifest events in the rat's oestrous cycle. (Schwartz, *Rec. Progr. Horm. Res.*)

timing of the events during the ovarian cycle of the 4-day rat, and it will be seen that the characteristic features—ballooned uterus due to filling with secretion, LH release, etc., take place at specific times in relation to the light–dark cycle.

Male Gonadotrophs

Again, when the technique for radioimmunoassay was applied to the study of plasma LH and FSH concentrations in men, characteristic variations were observed as in Figure 10.2.

Light–Dark Cycle

That many of these cyclic changes in function are dependent on an inherent "biological clock" rather than being the results of cyclical changes in the lighting stimulus, has been shown by experimentally interfering with the lighting cycle. Thus we may change from a normal alternation of light and dark to continuous darkness, in which event ovulation takes place approximately when expected, but if constant conditions of illumination are maintained for sufficiently long, the events in the cycle become more and more asynchronous, so that the cyclical activity depends not only on an inherent rhythmic biological clock but also on a rhythmic external stimulus.

Circadian Rhythm. The term circadian rhythm has come to be applied to this type of diurnal rhythm, so that other rhythms, dependent entirely on rhythmic external stimuli, fall outside this classification. An

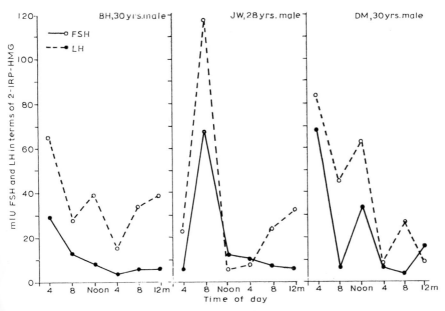

Fig. 10.2. FSH and LH levels in plasma obtained every 4 hr for 24 hr from 3 men. (Saxena *et al.*, *J. clin. Endocrinol.*)

experimental shifting of the timing of a rhythmically occurring event, such as ovulation, can be brought about by simply changing the phase of the cycle of lighting, but it of course does not take place immediately. Thus Carlyle and Carter (1961) found after some 20 days of retarding "dawn" in rats exposed to artificial light and dark, that there was a shift of the time of ovulation by the same amount (8 hr).

Persistent Oestrus. The most striking effect of changed lighting conditions is the condition of persistent oestrus that develops when rats are · exposed to constant illumination; this develops gradually and becomes manifest in a failure of ovulation, with persistent vaginal cornification, and is probably the result of an unremitting but slight hypersecretion of FSH and LH.* Carlyle and Carter found that, if rats were exposed to continuous illumination for 1, 2, 3 and 4 days respectively, the proportions of rats ovulating at the expected time were 14/14, 11/14, 9/13 and 1/14. Continuous dim green illumination for four days had little or no influence on ovulation.

Seasonal Reproductive Phenomena

The circadian rhythms described so far may well be accidental manifestations of cyclical hormonal changes, in which the pineal is involved, rather than important physiological mechanisms. Thus the rat is essentially a nocturnal animal, so that, as Reiter (1973a) has emphasized, the study of rhythmic changes in behaviour and gonado-trophin secretion under the artificial lighting conditions of a laboratory may well be the study of artefacts of cyclical changes in illumination to which the animal is not usually exposed. The seasonal changes in reproductive behaviour constitute rhythms involving much longer periods, and their physiological function in contributing to the birth of the young at the most propitious period—usually the Spring—is obvious. These are affected by the lighting period which, except at the equator, varies annually.

Junco Finch

In 1925 the Canadian zoologist Rowan drew attention to the influence of environmental lighting on the changes in gonadal size that take place in a junco finch; the testes of this migratory species enlarge and regress with a yearly cycle, growing rapidly in the Spring and atrophying in the autumn. Rowan showed that he could obtain enlargement of the atrophied testes in the winter by exposing the birds

---

* According to Daane and Parlow (1971) the levels of these gonadotrophins in the blood are not much higher than in oestrus. The first observable effect of the continuous light is the abolition of the "LH-surge" in pro-oestrus.

to an artificially lengthening day from October onwards. Such changes in weight strongly suggest a cyclical secretion of the gonadotrophins, LH and FSH.

The Hamster

This small animal is a hibernator, spending its winters in a torpid condition in a burrow where illumination is minimal. During the winter months there is a striking regression in its gonadal weight, and this is accompanied by an inhibition of spermatogenesis. Thus in November Czyba et al. (1964) found that the seminiferous tubules were active, most containing spermatozoa; by December spermatogenesis was incomplete, so that no cells beyond the spermatid stage were present, whilst in January all signs of spermatogenesis were absent. With the onset of Spring this reduction in sexual competence, which has an obvious survival value in preventing conception at an unsuitable time of year, is reversed. The involvement of the illumination, i.e. the shortening of the days in winter, is demonstrated experimentally by exposing animals in the laboratory to summer light–dark periods (Deanesley, 1938) whilst in the summer regression of the gonads could be induced by exposing the animals to winter lighting (Hoffman et al., 1965). Again, removal of the eyes from male or female hamsters resulted in degeneration of gonads and accessory sex organs, such as the uterus (Reiter, 1969).

Rabbit

Farrell et al. (1968) observed that the post-coital ovulation in the rabbit showed a seasonal variation, being maximal in Spring and minimal in October to December.

PINEAL GLAND

The pineal gland* or *epiphysis cerebri*, has long been suspected of being concerned with the animal's ability to regulate its behaviour in accordance with seasonal and diurnal variations in the light–dark periods but, of course, the anatomical location of this organ within the skull, deep in the brain of many birds and mammals, poses the problem of how the organ could respond to light or dark.

---

* The pineal or *epiphysis cerebri* is one of a family of invaginations or evaginations of the ventricular roof that become vascularized to form gland-like structures. Collectively they may be described as *circumventricular organs*, and they include the pituitary components and choroid plexuses. It must be emphasized that the *paraphysis* is not a pineal component but a structure formed in relation to the *telencephalon*, as opposed to the diencephalon, and in higher vertebrates remains an embryonic rudiment; in those submammalian species where it develops, it appears as an epithelium related to exchanges between blood and cerebrospinal fluid.

Lower Vertebrates

Phylogenetically the pineal organ has developed along two main pathways, so that in lower vertebrates, such as lizards and frogs, it is a saccular organ which is often divided into a more superficially placed extracranial "parietal eye" with a well-developed light-collecting apparatus and containing photoreceptors comparable with retinal rods and cones. This is connected with the deeper epiphysis, to which light-induced messages are transmitted and, from thence, by tracts to other parts of the brain. Here the function of the pineal as a light-transducer is obvious (Fig. 10.3).

Higher Vertebrates

In higher vertebrates, such as the mammals, but also in snakes and birds, the pineal develops, as before, from an outpouching of the ventricular roof, but becomes separated from the rest of the brain, in the sense that no tracts linking it through its stalk remain; instead it only receives an afferent sympathetic supply through the nervi conarii, which are postganglionic fibres from the superior cervical ganglion. In addition to this change in nervous connections, there has been a change in structure, the saccular form, containing light-sensitive elements, being replaced by a solid parenchymatous form, the characteristic and most numerous cell being the *pinealocyte*, a cell giving out long processes that come into relation with blood vessels or other cells. Sympathetic nerve terminals, recognized by catecholamine-containing vesicles, make a synaptic relation with the pinealocytes. The other main type of cell is the astrocyte. If light is to influence the pineal gland in these species with no parietal eye, it must do so through the retina and the visual pathway.

Pathway for Visual Excitation. The nervous pathway along which the visual input reaches the sympathetic outflow from the cord was determined by placing lesions at various points and studying the degree of inhibition of melatonin synthesis by light. The main visual pathways consist in the primary optic tracts leading to the lateral geniculate bodies, the pretectal nuclei and the superior colliculi. In the rat there are at least two other bundles leaving the optic chiasm, the *accessory optic tracts*, and it seems that the visual pathway concerned with pineal excitation is through the inferior accessory optic tracts, passing through the median forebrain bundle, which mediate the central control of autonomic function. The fibres relay in the terminal nuclei of Bochenek and finally descend through the brain-stem and upper spinal cord, emerging as preganglionic fibres to the superior cervical ganglia. The

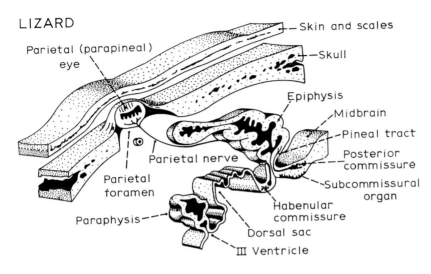

LIZARD

Skin and scales

Skull

Parietal (parapineal) eye

Epiphysis

Midbrain

Pineal tract

Posterior commissure

Parietal nerve

Subcommissural organ

Parietal foramen

Paraphysis

Habenular commissure

Dorsal sac

III Ventricle

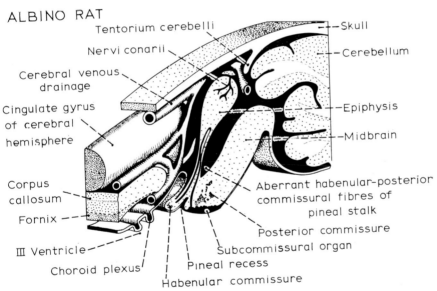

ALBINO RAT

Tentorium cerebelli

Skull

Nervi conarii

Cerebellum

Cerebral venous drainage

Cingulate gyrus of cerebral hemisphere

Epiphysis

Midbrain

Corpus callosum

Aberrant habenular-posterior commissural fibres of pineal stalk

Fornix

Posterior commissure

III Ventricle

Subcommissural organ

Choroid plexus

Pineal recess

Habenular commissure

Fig. 10.3. The pineal in lower vertebrates (*Top*) and mammal (*Bottom*). (Wurtman et al., *The Pineal*.)

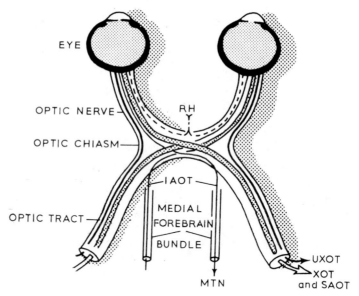

Fig. 10.4. Central visual projections in rat. XOT, crossed component of primary optic tract; UXOT, uncrossed (about 10% of total) component of primary optic tract; SAOT, superior accessory optic tract; RH, retino-hypothalamic fibres (dotted lines indicate that existence of such a primary visual projection is in doubt); IAOT, inferior accessory optic tract—completely crossed group of fibres originating in eye and leaving primary optic projections just behind optic chiasm to run through medial forebrain bundle and terminate in medial terminal nucleus, MTN, of accessory optic system. (Moore *et al.*, *Arch. Neurol.*)

postganglionic fibres terminate on or near the pineal parenchymal cells (Figs. 10.4 and 10.5). Thus interruption of this pathway, for example by removing the superior cervical ganglia, completely blocks the responses of the pineal to light or darkness (Wurtman *et al.*, 1964).

Gonadotrophin Secretion

It was shown by Fiske *et al.* (1960) that continuous exposure of rats to light caused a diminution in the weight of their pineal glands, suggesting an inhibition, by light, of their normal secretory activity. Continuous light increases the weights of the ovaries or, in the male, of the gonads and accessory organs such as seminal vesicles and prostate, suggesting an increased pituitary secretion of gonadotrophic hormones. Thus the effects of continuous light could be explained on the assumption that the pineal gland in the dark exerted an inhibitory action on the pituitary, depressing secretion of the gonadotrophins. The effect of light would be to block this inhibition.

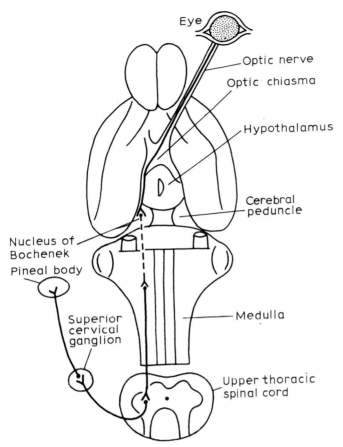

Fig. 10.5. Schematic diagram of pathway taken by visual impulses from the retina to the pineal gland of the rat. (Wurtman *et al.*, *The Pineal*.)

Pinealectomy

The inhibitory role of the pineal gland has been amply demonstrated, both by experiments involving pinealectomy and by injections of pineal extracts.*

Males

Thus to quote just a few experimental studies, Table I illustrates the changes in weight of gonads and accessory sex organs of rats; if, as

* Moszkowska *et al.* (1971) emphasize that the effects of pinealectomy in polycyclic species such as the rat, are often obscured by compensatory feed-back processes, so that it is easier to show the effects when these hormonostatic mechanisms are upset, as for example, when rats are treated postnatally with testosterone.

TABLE I

The effects of pinealectomy on the weight (± s.e.) of endocrine structures in male rats (Fraschini and Martini, 1970)

Groups	Final Body Wt, g	Pituitary Wt, mg	Testes Wt, g	Seminal Vesicles Wt, mg	Prostates Wt, mg
Sham-operated (20)	263·7±7·2	7·7±0·34	2·82±0·21	290·5±11·3	184·8±15·2
Pinealectomized (15)	258·3±5·4	8·1±0·49	3·24±·019	530·7±18·1	274·1±17·0

has been often claimed, changes in testicular weight are an index of plasma levels of FSH, whilst changes in weights of seminal vesicles and prostate are indications of the levels of LH (ICSH) in the blood, then the results of Table I indicate an increased secretion of both gonadotrophins as a result of pinealectomy. The hamster's seasonal reproductive rhythm is an annual one, and during the course of the year there are striking changes in weight of the gonads, reaching a maximum in about May. As Table II shows, when animals were exposed to a natural photoperiod throughout the year, the testicular weight varied

TABLE II

Effects of pinealectomy on gonadal weight in hamsters (From Reiter, 1973)

Month		Weight Normal	Weight Pinealectomized
January	Testes	605 mg	2542 mg
	Body	126 g	116 g
March	Testes	1452 mg	2668 mg
	Body	138 g	134 g
May	Testes	2481 mg	2564 mg
	Body	132 g	128 g

from some 605 mg in winter to 2481 in May; if another group of animals was pinealectomized, the testicular weight was high in winter and remained about the same throughout the year.

Females

In female rats Dunaway (1966) found that pinealectomy increased the proportion of animals that ovulated in response to pregnant mare's serum (PMS) and their ovaries contained more than the usual number of corpora lutea.

Persistent Oestrus Rats

In rats, de-afferentation of the basal hypothalamus induces a persistent oestrus due to separation from the PO/AH region that apparently governs cyclicity in the rodent. Ovulation may be restored in these animals by pinealectomy (Mess et al., 1973). The same effects can be induced by superior cervical ganglionectomy. According to Trentini et al. (1976) pinealectomy or superior cervical ganglionectomy increased the secretion of LH on the day of pro-oestrus in some 50 per cent of their rats; in the remaining animals no change was observed, possibly through missing the LH-peak values at time of sampling. The authors suggested that the pineal normally suppressed LH secretion acting through the serotoninergic system described by Kamberi et al. (1970); thus injections of serotonin or melatonin will suppress the ovulation-inducing effects of pinealectomy in the frontal hypothalamic de-afferented rat (Tima et al., 1973).*

Pineal Extracts

In general, pineal extracts were found to reverse the effects of pinealectomy, whilst given to normal animals they tended to reduce gonadal weight (Wurtman et al., 1959). They also antagonized the effects of continuous light in so far as they induced di-oestrus in animals with prolonged oestrus (Ift, 1962).

Transplants

Gittes and Chu (1965) were able to transplant pineal glands from one rat to another if they used a closely inbred strain (isogenic transplant); pinealectomy increased the duration of oestrus but transplants always reversed this effect; thus if the oestrus-index of sham-operated intact pineal controls was unity, that of pinealectomized sham-operated animals was 1·24 whilst that of pinealectomized animals with multiple intramuscular transplant of pineal gland was 0·99.

Melatonin

Amphibian Pigmentation

In 1917 McCord and Allen discovered that extracts of the pineal, when fed to amphibians, caused them to become pale, due to aggrega-

* Anovulatory rats due to neonatal androgenization could not be rendered ovulatory by pinealectomy or superior cervical ganglionectomy although the operations caused some thecal development and limited corpora lutea formation. The failure is probably due to a reduced sensitivity to LHRH in these androgenized rats (Ruzsas et al., 1977).

tion of the pigment granules in their dermal melanocytes;* the pineal extract was having an opposite action to the melanocyte stimulating hormone, MSH, of the intermediate lobe of the pituitary, and the presumptive hormone was called *melatonin*. The discovery was valuable for the elucidation of the chemical factor in the pineal since it provided an accurate assay of extracts during purification.

Biosynthesis

The active principle, melatonin, turned out to be a tryptophan derivative, (Lerner *et al.*, 1959) closely related to serotonin or 5-hydroxytryptamine, being formed from this hormone or neurohumour by two successive reactions, namely acetylation to acetylserotonin and subsequent O-methylation to methoxyacetyl serotonin, or 5-methoxy-N-acetyltryptamine (Fig. 10.6).

HIOMT. This final step is catalysed by the enzyma hydroxyindole-O-methyltransferase (HIOMT) and in mammals the pineal is the only organ (except the eye; Nagle *et al.*, 1974) containing this enzyme, which also catalyses the methylation of 5-hydroxytryptophol (Fig. 10.7).

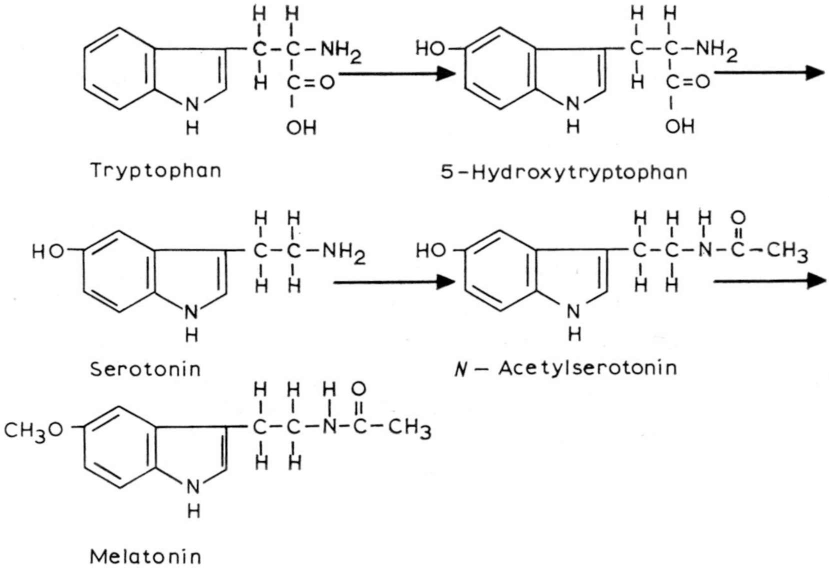

Fig. 10.6. Biosynthesis of melatonin from tryptophan in the pineal gland. The steps are catalysed by (1) tryptophan hydroxylase; (2) L-aromatic amino acid decarboxylase; (3), N-acetylating enzyme; (4) HIOMT. (Wurtman *et al.*, *The Pineal*.)

* McGuire and Moller (1966) noticed that it was only the dermal melanocytes of the frog that responded to melatonin, the epidermal melanocytes being unresponsive and thus similar to mammalian melanocytes.

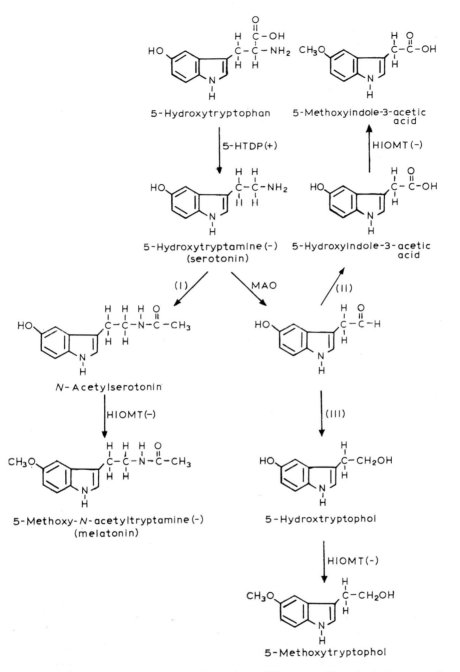

Fig. 10.7. Reactions of tryptophan derivatives. (Wurtman, *Neurochemical Aspects of Hypothalamic Function.*)

Cyclical Synthesis of Enzymes

The enzyme catalysing the acetylation of serotonin, N-acetyltransferase, is also present in the pineal, and it would seem from the studies of Wurtman, on the one hand, and Klein and Weller, on the other, that cyclical secretory activity of melatonin depends on cyclical synthesis or activation of these two enzymes. Thus Wurtman *et al.* (1965) found the following amounts of HIOMT activity* per milligramme of gland, during the oestrous cycle of rats:

Pro-oestrus	10·9
Oestrus	11·0
Metoestrus	22·4
Di-oestrus	25·4

Again, in a study of the macaque monkey, Quay (1966) found that when these were exposed to a standard photoperiod from 6 a.m. to 5 p.m., the concentration of serotonin in the pineal increased to a peak in the light, just before the onset of darkness, and then fell to about a quarter of this value, whilst the HIOMT activity was greatest at night. If the rise in HIOMT activity leads to accelerated conversion of serotonin to melatonin, this reciprocal relation between the two is understandable. Further studies on the cyclical changes in enzymatic activity in the pineal gland will be discussed later.

Melatonin as the Pineal Hormone

There is little doubt that melatonin can substitute, at least to some extent, for pineal secretion, or in normal animals mimic the effects that are attributed to pineal secretion. Thus Motta *et. al* (1967) showed that the rise in weight of the prostate and seminal vesicles following pinealectomy in rats was prevented by systemic administration of melatonin; however the increase in testicular weight was not suppressed, so that they concluded that there were probably at least two hormones secreted by the pineal, the one inhibiting pituitary secretion of FSH and the other the secretion of LH (ICSH); since melatonin did not affect testicular weight it was presumably not the hormone controlling FSH secretion.

Ovulation Block. If melatonin is responsible for inhibiting LH secretion by the pituitary, then we may expect it to block ovulation; when Collu *et al.* (1971) injected this into the lateral ventricles of rats around the critical period of ovulation, during pro-oestrus, they blocked ovulation completely in seven out of nineteen animals, whilst

---

* Expressed as picomoles of melatonin formed per hr from N-acetylserotonin and S-adenosylmethionine, the latter providing the methyl group.

in the animals that ovulated, the number of ova was 5 compared with a normal average of 11. When administered subcutaneously the hormone was without effect. Since other studies have shown that melatonin is effective when administered systemically, as opposed to intraventricularly, the failure to observe an effect was probably due to the short life of the hormone in the blood-stream (30 min) rather than a failure to cross the blood-brain barrier.

Role of the Pituitary and Hypothalamus

Before continuing with the nature of the pineal hormone(s) we may consider the mechanism of pineal action in more general terms. From what we have seen, the pineal seems to inhibit the secretion of pituitary gonadotrophins, the effect of light being to inhibit secretion of the pineal hormones and thus to release the pituitary from this inhibitory control.

Pinealectomy vs. Gonadectomy

Table III shows the effects of pinealectomy on the pituitary LH and FSH contents compared with the effects of castration. We have seen

TABLE III

Effects of pinealectomy and of castration on pituitary LH and FSH levels in male rats[1]
(Fraschini and Martini, 1970)

Groups [2]		Pituitary LH [3] μg/Pit	Pituitary FSH [3] μg/Pit
Controls	(20)	9·05 ± 0·75 (4)	64·35 ± 2·09 (3)
Pinealectomized	(15)	21·47 ± 2·10 (4) [4]	216·38 ± 16·70 (3) [4]
Castrated	(18)	25·63 ± 3·10 (6) [4]	372·98 ± 15·30 (3) [4]

[1] Values are means ±SE.

[2] No. of rats in parentheses.

[3] Microgram equivalents of NIH-LH-B-1 bovine or NIH-FSH-S-3 ovine per pituitary. No. of assays in parentheses.

[4] $P < 0.005$ vs. controls.

(p. 131) that castration, by removal of a negative feedback on the pituitary, promotes secretion of gonadotrophins, and this is reflected in increased pituitary content through accelerated synthesis. It will be seen that pinealectomy increases both LH and FSH contents, the stimulus being comparable to that of castration. If increased mitotic activity is a sign of increased secretion, then the study of Bindoni and Raffaele (1968) confirms that pinealectomy increases secretion; thus

the number of mitoses per 10,000 nuclei in intact animals was 13·0 ±
2·08; in sham-operated animals it was 11·1 ± 1·18 and in pinealectomized animals it was 28·6 ± 1·30.

Prolactin Secretion

The secretion of prolactin by the rat's pituitary is also under pineal
control. Thus Ronnekleiv *et al.* (1973) have shown that there is an early-
morning surge of prolactin secretion as manifest by the plasma level at
3.30 to 5.00 a.m. This is abolished by pinealectomy. Electron microscopical examination of the lactotrophic cells of the pituitary showed
that when animals were maintained in constant darkness there were
very few granules in these cells compared with the pituitaries of animals
kept in constant light. They suggested therefore that the pineal
inhibited the hypothalamic release of PIF, so that during darkness,
when the pineal is active, secretion of prolactin is promoted.

Blinding and Anosmia. Again Shiino *et al.* (1974) showed that
blinding and destruction of the olfactory bulbs raised the serum
prolactin, presumably by removal of pineal inhibition of PIF release;
when the blinding and anosmia were combined with pinealectomy,
the serum prolactin fell below that of sham-operated controls. The
prolactin-secreting cells of animals made blind and anosmic had lost
mature secretory granules and contained, instead, numerous immature
granules. In the hamster, blinding alone is sufficient to depress pineal
activity; in the rat, however, the maximum antigonadal potential of
the pineal can only be demonstrated by combining blinding with
anosmia (Reiter and Sorrentino, 1971). Blask and Reiter (1975)
compared normal with blind, blind plus anosmic and blind plus
anosmic plus pinealectomized rats; so far as LH was concerned,
combined blinding and anosmia reduced plasma LH levels by a
factor of about twenty, an effect that was largely reversed by pinealectomy. So far as sexual maturation is concerned, blinding or anosmia
alone have no significant effect; when both senses were destroyed there
was a very pronounced decrease in ovarian weight from about 35 mg/
100 g body-wt to about 20 mg; when sense-deprivation was combined
with pinealectomy the ovarian weight remained normal (Reiter and
Ellison, 1970).

Indoleamines and Pituitary Function

Kamberi *et al.* (1970) injected either catecholamines or indoleamines
into (a) a stalk portal vessel, (b) the third ventricle or (c) into a basilar
artery, and they measured the concentration of LH in systemic blood
by radioimmunoassay. Dopamine in the third ventricle caused a

twenty-fold rise in LH in the blood, noradrenaline and adrenaline being also effective, but in higher concentrations. Injections of serotonin and melatonin had the opposite effects, decreasing the level of LH in the blood, but the effects were by no means so striking. The amines were quite ineffective when administered by way of a stalk portal vessel or through the basilar artery, so that it is likely that the hormones were acting directly on the hypothalamus, rather than on the pituitary.

Dopamine-Indoleamine Antagonism. In a later study, Kamberi *et al.* (1971) showed that intraventricular melatonin or serotonin stimulated the release of prolactin by as much as 300 per cent of the pre-injection levels; secretion of FSH was suppressed to 75 per cent–85 per cent of pre-injection levels. Thus the dopamine and serotonin-melatonin systems are antagonistic in so far as their influence on the hypothalamicophypophyseal axis is concerned, both substances presumably acting by release from dopaminergic or serotoninergic nerve endings, and this raises the question as to whether melatonin exerts a direct action on the hypothalamus or works by modifying the concentration of serotonin in nerve terminals.

Indoleamine Inter-relationships. Thus, the indoleamines are closely inter-related metabolically, possibly through actions on the enzyme, *pyridoxal kinase*, which catalyses the formation of pyridoxal phosphate, which is a prosthetic group of the enzyme, *aromatic L-amino acid decarboxylase*, where it is a requirement for the formation of both GABA and serotonin. The activity of the kinase increases after injection of melatonin into rats, an effect that is blocked by actinomycin (Anton-Tay *et al.*, 1970).

Blood-Brain Barrier

The catecholamines do not cross the blood-brain barrier easily, so that failure to act through the basilar artery is understandable, and the same is true for serotonin, which probably exerts its central actions by being released locally from nerve terminals. Melatonin, with its hydroxy-group methylated crosses the blood-brain barrier with less difficulty (Wurtman *et al.*, 1964), and the finding of definite effects when exhibited by way of the general circulation tends to confirm this, so that the failure of Kamberi *et al.* to find an effect of melotonin when administered by way of the basilar artery is perhaps surprising, but, as Wurtman (1971) has emphasized, the rapid metabolic breakdown of melatonin by the liver requires that melatonin, to have a measurable action, must be perfused into the vascular system for some time, or else administered by repeated subcutaneous or intraperitoneal injections. Thus, Wurtman *et al.* (1963) found that daily subcutaneous or intraperitoneal injections of melatonin delayed onset of puberty in female rats.

Cerebrospinal Fluid. In this connection we must consider the possibility that the secreted pineal hormones exert their action through the cerebrospinal fluid; from here they would obtain ready access to the hypothalamus and midbrain, and would be immune from the metabolic inactivation by the liver.*

Brain Implants

Implantation of pineal tissue, or of melatonin in the median eminence of rats causes a reduction in the pituitary LH-content; thus castration induced the typical rise in LH-content, and this (Fig. 10.8) was reduced from $25 \cdot 6 \pm 3 \cdot 0$ μg/pituitary to $13 \cdot 9 \pm 1 \cdot 2$ by an implantation of melatonin (Fraschini *et al.*, 1968). Implants of 5-hydroxy-tryptophol were more effective than melatonin (Fraschini *et al.*, 1971). The implants had no effect on pituitary FSH-content whereas serotonin and 5-methoxytryptophol were effective. Thus it seems likely from this work of Fraschini *et al.* that other indoleamines, in addition to melatonin, are acting as pineal hormones.

Midbrain. Of some interest was the finding that implants of melatonin in the midbrain were also effective in reducing pituitary levels of LH, and this is consistent with the finding that intraperitoneal or intraventricular injections of melatonin are taken up selectively by the midbrain (Anton-Tay and Wurtman, 1969). They also caused an increase in the level of serotonin in the midbrain (Anton-Tay *et al.*, 1968) and since serotoninergic axons, arising from cell bodies in the midbrain, constitute a major input to the hypothalamus by way of the median forebrain bundle, it is possible that melatonin exerts at least part of its action through this serotoninergic pathway.

Other Active Pineal Principles

Besides melatonin, the pineal gland contains several other indoleamines derived from the progenitor, tryptophan (Fig. 10.7, p. 521). Of these, serotonin is a well-established neurohumour, and some of the pineal serotonin is concentrated within nerve terminals where it

---

* Whilst there is little doubt that in early aquatic vertebrates the pineal secretions were, indeed, emptied into the ventricular system, from whence they were carried, by characteristic brain-cells, called *tanycyctes*, to the responsive neurones, it is very unlikely that in the higher vertebrates this form of transport has any great significance. With the development of a vascularized brain, the role of distribution of hormones to the neurones has been taken over by the blood vessels. As we have seen (Vol. 3) the vessels in the brain and spinal cord are invested by processes from astrocytes, which may well be the phylogenetic descendants of the tanycytes; in which case, as Quay (1970) has suggested, the transport of hormones from the capillaries to the neurones might well take place through the astrocytes.

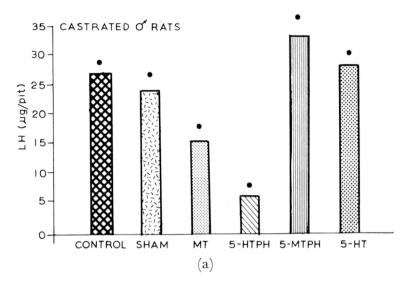

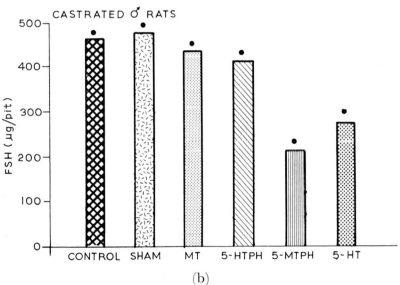

Fig. 10.8. Effects of implants of indole and methoxyindole derivatives in the median eminence on pituitary LH (a) and FSH (b) contents of castrated male rats. MT, melatonin; 5-HTPH, 5-hydroxytryptophol; 5-MTPH, 5-methoxytryptophol; 5-HT, 5-hydroxytryptamine (serotonin). (Fraschini *et al.*, *Ciba Symp.*)

presumably acts as a transmitter. The close involvement of serotonin in melatonin synthesis, however, demands that it be also present in the pinealocyte.

Catecholamines

The pineal possesses the enzymes necessary for the conversion of tyrosine to dopamine and noradrenaline (Fig. 10.9) these being found in relatively large amounts in the gland; the noradrenaline is located

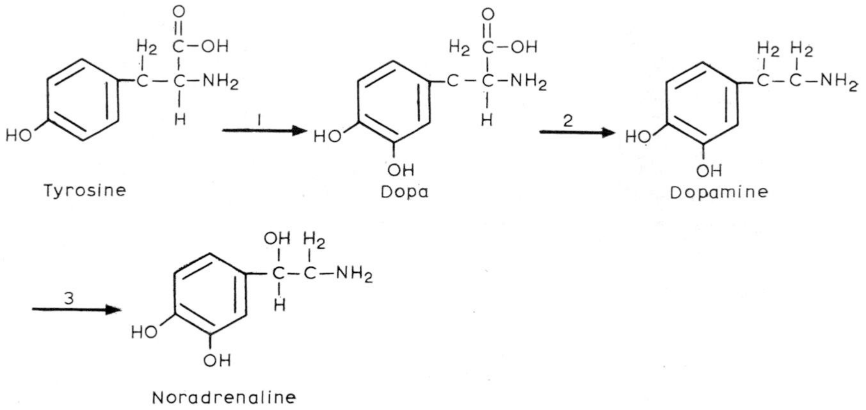

Fig. 10.9. Pathways in the formation of noradrenaline in the pineal. Step 1 is cata⁻lysed by tyrosine hydroxylase; Step 2 by aromatic amino acid decarboxylase; Step 3 by dopamine-β-oxidase. (Wurtman *et al.*, *The Pineal*.)

within the sympathetic nerve endings (Wolfe *et al.*, 1962), as is serotonin, the relative concentrations of these two transmitters varying with the species; thus, the sympathetic nerves of the rabbit, cat and hamster, for example, store mostly noradrenaline whilst those of the rat and guinea-pig contain mostly serotonin. Dopamine seems to be concentrated in pineal parenchymal cells.*

Prolactin Release. It was early noted that pineal extracts caused release of prolactin, and, according to Blask and Reiter (1975), this is by virtue of a control of the release of inhibiting and releasing factors (PIF and PRF). This could be through secretion of melatonin since, according to Kamberi *et al.* (1971), melatonin injected into the third ventricle causes a rise in the prolactin-titre. Blask *et al.* (1976) obtained

* Of some interest is the recent discovery of gonadotrophin releasing hormone (LHRH) in concentrations some four times or more greater than in the hypothalamus (White *et al.*, 1974); the gland also sequesters the releasing hormone if given intravenously. Since the main action of the pineal is anti-gonadal, this finding presents a paradox. Thus, White *et al.* were able to induce ovulation in the oestrous rabbit with extracts of pineal gland.

both PIF and PRF activities in extracts of bovine, rat and human pineal glands, as revealed by both *in vitro* and *in vivo* studies.

The quantities of melatonin, TRH and oestrogen in the extracts were much too small to have had any effect on prolactin secretion, so that they concluded that the pineal could influence prolactin concentrations through stimulating PIF and/or PRF release, or directly by virtue of its own PIF and/or PRF activities. Thus the pineal may serve not only as supplemental sources of gonadotrophic releasing hormone but also of prolactin inhibitory and/or releasing hormone. If, as seems probable (p. 485) PIF is indeed dopamine, then the presence of this catecholamine in extracts of pineal gland could account for its action on prolactin release.

Arginine Vasotocin

Benson *et al.* (1972) have questioned the importance of melatonin in pineal secretion, pointing out that the quantities found in the gland are very small, of the order of 3.10^{-4} μg/mg tissue compared with the milligramme quantities required to produce physiological effects; by purifying extracts of pineal they were able to obtain a preparation that was 60–70 times as effective, weight-for-weight, as melatonin in antagonizing the compensatory hypertrophy of the ovary after unilateral ovariectomy (p. 131). In a subsequent study Orts and Benson (1973) took steps to remove any melatonin from their extracts, and found these just as effective. They considered that the active principle was a polypeptide of molecular weight 500 to 1000 but rejected the possibility that this was 8-arginine vasotocin (AVT) since they considered that the latter had no anti-gonadotrophic activity. Cheesman (1970) purified anti-gonadotrophic extracts* of pineal gland and isolated arginine vasotocin from them:

$$\text{Cys-Tyr-Ile-Glu(NH}_2)\text{-Asp(NH}_2)\text{-Cys-Pro-Arg-Gly-NH}_2$$

Orts (1977) has also described non-melatonin containing fractions of pineal extracts; these lower blood-LH; one of the extracts, F_3, seemed to be arginine vasotocin; F_4 and F_5 were smaller peptides. Vaughan *et al.* (1974) have also described several pineal fractions with similar action to that of arginine vasotocin; they point out that their A_3 is different from Moszkowska *et al.*'s F_3 in being strongly soluble in acetone.

* Moszkowska *et al.* (1971) have also fractionated pineal extracts, obtaining one, which they call F3, with a molecular weight less than 700 and which inhibits gonadotrophin secretion by the pituitary *in vitro* in a similar manner to the inhibition produced by crude pineal extracts. Another fraction, F2, had opposite effects to those of F3.

Effects of Exogenous AVT. Several other studies have demonstrated that this substance, which is present in the brain in many mammals including that of the human foetus (Pavel *et al.*, 1973/4) and cerebrospinal fluid (Pavel, 1970), can inhibit the development of gonads and accessory organs in mice and hamster. According to Pavel *et al.* (1973) it is some million times more potent than melatonin in inhibiting compensatory ovarian hypertrophy when administered intraventricularly.

As Figure 10.10 shows, an intravenous injection of as little as 0·1 μg per hour to rats at pro-oestrus will block the pre-ovulatory surge of LH. In 94 per cent of the rats examined the LH-surge was blocked although ovulation was blocked in a smaller percentage, namely 78 per cent, probably because in some animals the LH-surge was only delayed. When administered into the ventricles the hormone was effective in very low doses—0·2 to 0·0001 μg/hr (Osland *et al.*, 1977).

Site of Action. So far as the mechanism of action is concerned, it would seem that arginine vasotocin acts at a high neuronal level, preventing the preovulatory release of LH-releasing hormone rather

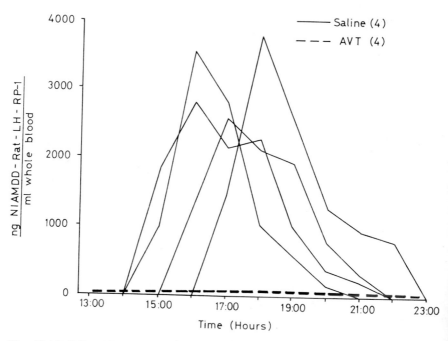

Fig. 10.10. Effect of arginine vasotocin on the pre-ovulatory surge of LH in pro-oestrous rats. Arginine vasotocin was given intravenously hourly from 1300 hr through 1600 hr. (Cheesman *et al.*, *Endocrinology*.)

than exerting an action on the pituitary; thus it had no effect on tonic secretion of LH in di-oestrous or ovariectomized rats, nor yet did it suppress the LHRH-induced rise in LH-secretion in ovariectomized animals. Arginine vasotocin had no effect on the release of LH and ovulation induced by PGE_2 administration, and this agrees with the absence of effect on the LHRH-induced release of LH, since it is considered that the prostaglandin acts directly on the LHRH-neurone, causing it to release LHRH. Hence the action of arginine vasotocin is at a higher level than that of the LHRH-releasing neurone of the hypothalamus.

When the medial pre-optic area was stimulated electrochemically at pro-oestrus, arginine vasotocin did not affect the surge of LH that follows this procedure (p. 115), so that it is unlikely that this hypothalamic area is the target for vasotocin action.

Pinealectomy. Arginine vasotocin can reverse the effects of pinealectomy; thus Pavel *et al.* (1975) injected minute amounts into the third ventricles of mice and thereby prevented the rise in pituitary prolactin content that occurs after pinealectomy; they calculated that only 120,000 molecules of the hormone were required per mouse. It was assumed that arginine vasotocin reversed the effects of pinealectomy by inhibiting synthesis and/or release of prolactin inhibiting factor (PIF); this would allow prolactin to accumulate in the gland and spill into the circulation. Thus treatment of the pituitary *in vitro* with arginine vasotocin causes release of prolactin (Vaughan *et al.*, 1975).

Melanocyte-Stimulating Hormone (MSH)

The pituitary secretion of this hormone is controlled through an inhibitory factor, MIF, as well as through a releasing factor, MRF. Melatonin decreases the pituitary MSH content whilst pinealectomy increases it; and it has been suggested that melatonin is involved in the light- and dark-induced changes in MSH secretion.

Some experiments of Pavel have led to the suggestion that arginine vasotocin exerts its action by first causing the liberation of melatonin from the pineal gland. Thus, he found (Pavel *et al.*, 1975) that arginine vasotocin, injected into the IIIrd ventricles of mice, prevented the rise in pituitary MSH that occurs after pinealectomy or 24 hours after exposure to constant light. The same dose of arginine vasotocin into normal mice decreased the MSH content of the pituitary. Pavel (1973) showed that arginine vasotocin, injected into the IIIrd ventricle, caused release of melatonin into the cerebrospinal fluid. Thus, it could be that melatonin is a "second messenger", its liberation being controlled by arginine vasotocin.

Changes in Synthetic Activity in the Pineal

Pineal HIOMT

We have indicated that the cyclical changes in pineal activity that are presumably at the basis of the circadian and other rhythms in gonadotrophin secretion, are reflected in changes in activity of the methylating enzyme, HIOMT, which completes the synthesis of melatonin from serotonin. Figure 10.11 illustrates the rise in activity in the evening, reaching its peak at about midnight. When animals are kept in constant darkness the variation is suppressed, the level being high in early evening and at midnight (Fig. 10.12). In constant light, the rhythm is also suppressed, this time with low activity at both times. Removal of the superior cervical ganglia or blinding the animals

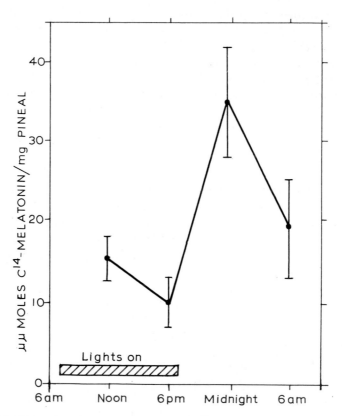

Fig. 10.11. Diurnal changes in the rat's pineal hydroxyindole-0-methyltransferase (HIOMT) activity. Vertical lines are standard errors of mean. (Axelrod *et al.*, *J. biol. Chem.*)

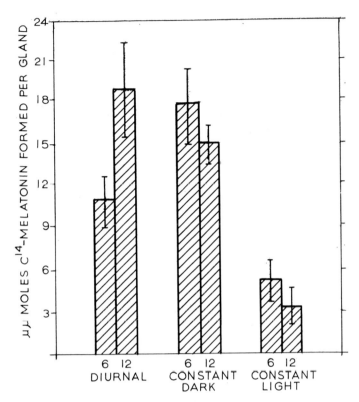

Fig. 10.12. Effect of constant light or darkness on diurnal rise in rat's pineal hydroxyindole-0-methyltransferase activity. The numbers 6 and 12 denote 6 p.m. and midnight respectively. (Axelrod *et al.*, *J. biol. Chem.*)

likewise abolished the rhythm, indicating that at this stage the cyclical event is closely related to the environmental change. According to the study by Quay (1963) during the period between 6 p.m. and midnight when HIOMT activity increases, the melatonin content of the gland increases three-fold whilst the serotonin content decreases by a factor of seven. Thus, the inverse effects on melatonin and serotonin suggest that synthesis of the one is at the expense of the other. A similar circadian rhythm in the content of 5-hydroxyindole-3-acetic acid, the methylation of which may also be catalysed by HIOMT (Fig. 10.7) doubtless reflects changes in HIOMT activity (Quay, 1964).

Pineal Noradrenaline

If the 24-hour rhythm is governed by activity in the sympathetic nerves we may expect to find a rhythm in noradrenaline content of the pineal, since this neurohumour is definitely localized to the nerve endings in the gland (Wolfe *et al.*, 1962). This rhythm, was, indeed, found (Fig. 10.13) the content rising during the darkness to reach a peak with the onset of light at 6 a.m.

Acetyltransferase Activity

The studies of Klein and his collaborators have emphasized the prominent role of the penultimate step in melatonin synthesis, namely the acetylation of serotonin, suggesting that this may well be the rate-determining step (Klein and Weller, 1973). As Figure 10.14 shows, there is a large increase in activity of this enzyme during darkness; it is some 180° out of phase with the rhythm of pineal serotonin content, but in phase with the rhythm of melatonin and noradrenaline contents.

The rhythmic variation in enzymatic activity persisted in the darkness after 6 days but it was much more variable from one animal to another indicating a disturbance in the mechanism. Earlier Klein *et al.*

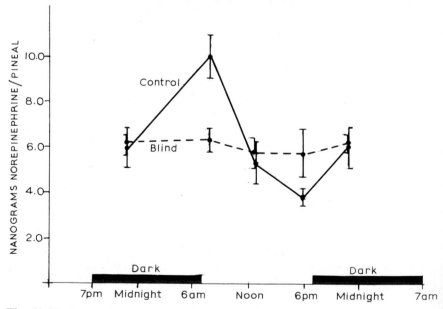

Fig. 10.13. Twenty-four-hour variation in rat's pineal norepinephrine levels, and the effect of blinding. Rats were blinded and maintained with sham-operated controls under diurnal lighting for 5 days. (Wurtman *et al.*, *J. Pharmacol.*)

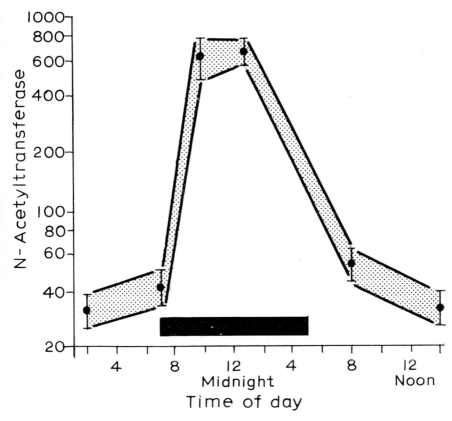

Fig. 10.14. The circadian rhythm in rat pineal N-acetyltransferase activity. The lighting-cycle (light 14 hr: dark 10 hr) was automatically regulated. The black bar represents the dark period. Enzyme activity is given as picomoles of ^{14}C-serotonin N-acetylated per gland homogenate per hr. (Klein and Weller, *Science.*)

(1970) had shown that noradrenaline or dibutyryl cAMP, added to cultured glands *in vitro*, caused a 6–10-fold increase in acetyltransferase activity, whilst there was only a small increase in HIOMT, unaffected by cAMP. The stimulated transferase activity was partially inhibited by cycloheximide, suggesting that the enzyme was being synthesized.

Noradrenaline as Zeitgeber. Thus, the *Zeitgeber* for the rhythm might well be a dark-induced release of noradrenaline; this would stimulate adenylcyclase in the pinealocyte, increasing the level of cAMP, and this would stimulate production of acetyltransferase leading to melatonin synthesis. The onset of light would inhibit any of these preliminary steps. In the absence of a signal from a lighting shift we must presume that some other mechanism regulates the cyclic synthesis of melatonin, since the rhythm is not destroyed in blinded animals. The rhythm is, however, almost abolished in animals deprived of the sympathetic innervation of the pineal (Klein *et al.*, 1971), so that the basic process seems to be the conversion in the dark of serotonin to N-acetylserotonin at a high rate through the sympathetic activation of the transferase. This process tends to be inhibited by light, so that in continuous light there is a high level of serotonin and a low level of transferase activity, whilst both rhythms persist in blinded animals and in normals in the dark. The problem remains, then, to locate the source of rhythmic sympathetic discharge.*

Pharmacology of Melatonin

Apart from their action in relation to reproductive behaviour, melatonin or pineal extracts tend to decrease spontaneous motor activity, measured, for example, by treadmill behaviour in rats, whilst pinealectomy increases this activity (Reiss *et al.*, 1963). When Marczynski *et al.* (1964) implanted crystals of melatonin in the pre-optic region of the hypothalamus of cats, they induced a characteristic sleep-pattern in the EEG. Again, Barchas *et al.* (1967) found that melatonin increased the sleeping time of mice given hexobarbital, whilst chicks, when injected, immediately assumed the roosting position which was maintained for some 45 minutes.

Role of the Pineal in Seasonal Breeding

This discussion of the details of hormonal secretion by the pineal, and of cyclical events in the synthesis of melatonin, has drawn our attention from the main theme of this section, namely the role of the pineal in coordinating the events in the breeding cycle, a role that has been emphasized most strongly by Reiter (1973a) in the light of his studies on the hamster.

Gonadal Atrophy Hibernation

This species exhibits two interesting physiological features, namely the power to hibernate and the loss of sexual competence that take place during the same period. Studies on artificial shortening or lengthening of the diurnal light-dark rhythm leave us in little doubt that it is the shortening of the days that acts as the stimulus or at least the

---

* Sympathetic stimulation certainly increases N-acetyltransferase activity, and this decreases to its original value in spite of a maintained stimulus (Volkman and Heller, 1971). According to Klein and Weller (1973) the sympathetic receptor is β-adrenergic, the induced transferase activity being blocked by propanolol.

conditioner for the onset of both these events, a shortening of the light-period to 12·5 hr daily resulting in a 5–6-fold reduction in testicular weight and atrophy of the accessory organs with reduction in serum LH and FSH concentrations in males, whilst females become anovulatory with uterine atrophy under the same conditions. The regression of the gonads induced by the altered lighting requires an intact pineal (Czyba et al., 1964; Reiter, 1969). Superior cervical ganglionectomy or transection of the nervi conarii also prevents the effects of a shortened light-period.

Role of Melatonin. Because the enzymic activities necessary for synthesis of melatonin vary cyclically (p. 532), it has been assumed that melatonin is concerned in this gonadal atrophy; acutely injected melatonin, however, has not produced consistent anti-gonadal actions, but as Tamarkin et al. (1976) have argued, it may be that prolonged exposure to the hormone is necessary. Thus Turek et al. (1975) found that silastic implants *did* cause gonadal regression; another factor is the time of day at which the hormone is administered; and Tamarkin et al. found that male Syrian hamsters, on a long light-period regimen, would suffer gonadal regression if injected once daily with melatonin during the afternoon, with lowered serum LH and FSH, whereas injections in the morning had no effect. Corresponding changes in the female were observed, those treated in the afternoon becoming acyclic.*

Insensitivity to Melatonin. What may be very important for the understanding of pineal involvement in seasonal rhythms, is the finding that, if regression of the gonads is induced by artificial darkness, this does not last as long as the darkness is maintained, but instead the gonads re-acquire their powers of spermatogenesis spontaneously. It is as though the target-organ for the pineal hormone became ultimately insensitive so that the inhibition of gonadotrophin release was eventually lifted. Thus, naturally, the entry into the burrow, which subjects the animal to prolonged darkness, leads to regression of the gonads, and this serves two important functions, namely the suppression of reproductive activity in the winter and this latter permits the onset of hibernation (Hoffman, 1964). As Spring approaches the inhibitory action of the pineal diminishes, the gonads are activated and hibernation ceases (Fig. 10.15).

---

* Reiter et al. (1975) have shown that the gonadal atrophy of hamsters caused by a short-light cycle may be prevented by repeated injections of melatonin or 5-methoxytryptophol, and this also prevented the loss of pituitary LH that occurs in these short-light period animals. Thus neither of these indoles is responsible for the antigonadotrophic action of the pineal, in fact they have the reverse action in this species.

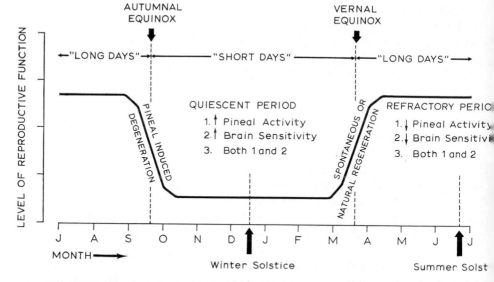

Fig. 10.15. Theoretical role of the pineal gland in seasonal reproductive phenomena in the golden hamster. During the winter months ("short days"), reproductive organ growth is held in check by an increased pineal activity and probably by an increased sensitivity of the brain to pineal substances. During the "long day" period (summer), the neuroendocrine-gonadal axis may be refractory to the pineal influence. (Reiter, *Annu. Rev. Physiol.*)

Diurnal Rhythms

Prolactin Secretion

The release of prolactin into the blood of the rat takes place cyclically, with a diurnal rhythm, and the same is true of other hormones such as LH (Dunn *et al.*, 1972) and testosterone (Kinson and Liu, 1973). These diurnal rhythms may be interfered with by alterations in the light–dark cycle, in which event profound changes in physiological function may be induced, such as the constant oestrus of rats in continuous light, whilst constant darkness produces ovarian atrophy. So far as prolactin is concerned, animals maintained on a 12 hr light–dark cycle exhibit a diurnal rhythm of prolactin secretion, with a nadir at 11.30 a.m. and apogee at 11.30 p.m. This rhythm may be reversed by constant light.

An interesting finding is that pinealectomy in the rat, which blocks the prolactin secretion and gonadal atrophy that follow maintenance in constant darkness, does not affect the diurnal rhythm of prolactin secretion of rats maintained on a 12 hr light–dark cycle, so that, so far as the diurnal rhythm is concerned, the pineal gland is not involved.

Non-Involvement of Pineal. However, the rhythm is controlled by the sympathetic input into the brain by the superior cervical ganglion since ganglionectomy reverses the cycle, and Kizer *et al.* (1975) have concluded that the pineal is basically concerned with long-term changes, manifest in seasonal breeding; the trigger for these seasonal changes will be the light–dark period, but the diurnal variations in secretion, at any rate of prolactin, are not linked to the pineal, presumably being set in motion by sympathetic input to the brain activating a stimulatory noradrenergic pathway for prolactin release.

REFERENCES

Antón-Tay, F., Chou, C., Anton, S. and Wurtman, R. J. (1968). Brain serotonin concentrations: elevation following intraperitoneal administration of melatonin. *Science*, **162**, 277–278.

Antón-Tay, F., Sepulveda, J. and González, S. (1970). Increase of brain pyridoxal phosphokinase activity following melatonin administration. *Life Sci.*, **Pt. II, 9**, 1283–1288.

Antón-Tay, F. and Wurtman, R. J. (1969). Regional uptake of ^{3}H-melatonin from blood or cerebrospinal fluid by rat brain. *Nature*, **221**, 474–475.

Axelrod, J., Sedvall, G. and Moore, R. Y. (1967). Photic and neural control of the 24-hour norepinephrine rhythm in the rat pineal gland. *J. Pharmacol.*, **157**, 487–492.

Axelrod, J., Wurtman, R. J. and Snyder, S. H. (1965). Control of hydroxyindole O-methyltransferase activity in the rat pineal gland by environmental lighting. *J. biol. Chem.*, **240**, 949–954.

Barchas, J., Da Costa, F. and Spector, S. (1967). Acute pharmacology of melatonin. *Nature*, **214**, 919–920.

Benson, B., Matthews, M. J. and Rodin, A. E. (1972). Studies on a non-melatonin pineal anti-gonadotrophin. *Acta Endocrinol.*, **69**, 257–266.

Bindoni, M. and Raffaele, R. (1968). Mitotic activity in the adenohypophysis of rats after pinealectomy. *J. Endocrinol.*, **41**, 451–452.

Blask, D. E. and Reiter, R. J. (1975). Pituitary and plasma LH and prolactin levels in female rats rendered blind and anosmic: influence of the pineal gland. *Biol. Reprod.*, **12**, 329–334.

Blask, D. E. *et al.* (1976). Prolactin-releasing and release-inhibiting factor activities in the bovine, rat, and human pineal gland: *in vitro* and *in vivo* studies. *Endocrinology*, **99**, 152–162.

Carlyle, A. and Carter, S. B. (1961). The influence of light on the occurrence of ovulation in the rat. *J. Physiol.*, **157**, 44–45P.

Cheesman, D. W. (1970). Structural elucidation of a gonadotropin-inhibiting substance from the bovine pituitary gland. *Biochim. biophys. Acta.*, **207**, 247–253.

Cheesman, D. W., Osland, R. B. and Forsham, P. H. (1977). Suppression of the preovulatory surge of luteinizing hormone and subsequent ovulation in the rat by arginine vasotocin. *Endocrinology*, **101**, 1194–1202.

Collu, R., Fraschini, F. and Martini, L. (1971). The effect of pineal methoxyindoles on rat vaginal opening time: *J. Endocrinol.*, **50**, 679–683.

Czyba, J. C., Girod, C. and Durand, N. (1964). Sur l'antagonisme épiphyso-hypophysaire et les variations saisonnières de la spermatogènes e chez le hamster doréc (Mesocricetus auratus). *C.r. Soc. Biol., Paris*, **158**, 742–745.

Daane, T. A. and Parlow, A. F. (1971). Serum FSH and LH in constant light-induced persistent estrus: short-term and long-term studies. *Endocrinology*, **88**, 964–968.

Deanesley, R. (1938). The reproductive cycle of the golden hamster (*Cricetus auratus*). *Proc. Zool. Soc.*, **108A**, 31–36.

Dunaway, J. E. (1966). The effect of pinealectomy on PMS-induced ovulation in immature rats. *Anat. Rec.*, **154**, 340–341.

Dunn, J. D., Arimura, A. and Scheving, L. E. (1972). Effect of stress on circadian periodicity in serum LH and prolactin concentration. *Endocrinology*, **90**, 29–33.

Everett, J. W. (1948). Progesterone and estrogen in the experimental control of ovulation time and other features of the estrous cycle in the rat. *Endocrinology*, **43**, 389–405.

Farrell, G., Powers, D. and Otani, T. (1968). Inhibition of ovulation in the rabbit: seasonal variation and the effects of indoles. *Endocrinology*, **83**, 599–603.

Fiske, V. M., Bryant, G. K. and Putnam, J. (1960). Effect of light on the weight of the pineal in the rat. *Endocrinology*, **66**, 489–491.

Fraschini, F., Collu, R. and Martini, L. (1971). Mechanisms of inhibitory actions of pineal principles on gonadotropin secretion. In *The Pineal Gland*. Ciba Symp., 259–278.

Fraschini, F. and Martini, L. (1970). Rhythmic phenomena and pineal principles. In *The Hypothalamus*. Eds. Martini, Motta and Fraschini. Academic Press: N.Y., 529–549.

Fraschini, F., Mess, B. and Martini, L. (1968). Pineal gland, melatonin, and the control of luteinizing hormone secretion. *Endocrinology*, **82**, 919–924.

Fraschini, F., Mess, B., Piva, F. and Martini, L. (1968). Brain receptors sensitive to indole compounds: function in control of luteinizing hormone secretion. *Science*, **159**, 1104–1105.

Gittes, R. F. and Chu, E. W. (1965). Reversal of the effect of pinealectomy in female rats by multiple isogenic pineal transplants. *Endocrinology*, **77**, 1061–1067.

Hoffman, R. A. (1964). Speculations on the regulation of hibernation. *Ann. Acad. Sci. Fenn. Ser A4, Biol.*, **17**, 202–216. (Quoted by Reiter, 1973.)

Hoffman, R. A., Hester, R. J. and Towns, C. (1965). Effect of light and temperature on the endocrine system of the golden hamster (*Mesocricetus auratus*, Waterhouse). *Comp. Biochem. Physiol.*, **15**, 525–533.

Ift, J. D. (1962). Effects of pinealectomy, a pineal extract and pineal grafts on light-induced prolonged estrus in rats. *Endocrinology*, **71**, 181–182.

Kamberi, I. A., Mical, R. S. and Porter, J. C. (1970). Effect of anterior pituitary perfusion and intraventricular injection of catecholamines and indoleamines on LH release. *Endocrinology*, **86**, 1–12.

Kamberi, I. A., Mical, R. S. and Porter, J. C. (1971). Effects of melatonin and serotonin on the release of FSH and prolactin. *Endocrinology*, **88**, 1288–1293.

Kinson, G. A. and Liu, C.-C. (1973). Effects of blinding and pinealectomy on diurnal variations in plasma testosterone. *Experientia*, **29**, 1415–1416.

Kizer, J. S., Zivin, J. A., Jacobowitz, D. M. and Kopin, I. J. (1975). The nyctohemeral rhythm of plasma prolactin: effects of ganglionectomy pinealectomy, constant light, constant darkness or 6-OH-dopamine administration. *Endocrinology*, **96**, 1230–1240.

Klein, D. C., Berg, G. R. and Weller, J. (1970). Melatonin synthesis: adenosine 3',5'-monophosphate and norepinephrine stimulate N-acetyltransferase. *Science*, **168**, 979–980.

Klein, D. C. and Weller, J. L. (1970). Indole metabolism in the pineal gland: a circadian rhythm in N-acetyltransferase. *Science*, **169**, 1093–1095.

Klein, D. C. and Weller, J. L. (1973). Adrenergic adenosine 3',5'-monophosphate regulation of serotonin N-acetyltransferase activity and the temporal relationship of serotonin N-acetyltransferase activity to synthesis of ³H-N-acetylserotonin and ³H-melatonin in the cultured rat pineal gland. *J. Pharmacol.*, **186**, 516–527.

Klein, D. C., Weller, J. L. and Moore, R. Y. (1971). Melatonin metabolism: neural regulation of pineal serotonin: acetyl coenzyme A N-acetyltransferase activity. *Proc. Nat. Acad. Sci. Wash.*, **68**, 3107–3110.

Lerner, A. B., Case, J. D. and Heinzelman, R. V. (1959). Structure of melatonin. *J. Amer. Chem. Soc.*, **81**, 6084–6085.

Marczynski, T. J. *et al.* (1964). Sleep induced by the administration of melatonin (5-methoxy-N-acetyltryptamine) to the hypothalamus in unrestrained cats. *Expenentia*, **8**, 435–437.

McCord, C. P. and Allen, F. P. (1917). Evidences associating pineal gland function with alterations in pigmentation. *J. exp. Zool.*, **23**, 207–224.

McGuire, J. and Möller, H. (1966). Differential responsiveness of dermal and epidermal melanocytes of Rana pipiens to hormones. *Endocrinology*, **78**, 367–372.

Mess, B., Tima, L. and Trentini, G. P. (1973). The role of the pineal principles in ovulation. *Progr. Brain Res.*, **39**, 251–259.

Moore, R. Y. *et al.* (1968). Central control of the pineal gland visual pathways. *Archs. Neurol.*, **18**, 208–218.

Moszkowska, A. and Elels, I. (1971). The influence of the pineal body on the gonadotropic function of the hypophysis. *J. Neuro-Visc. Res.*, Suppl. X, 160–176. (Quoted by Orts, 1977.)

Moszkowska, A., Kordon, C. and Ebels, I. (1971). Biochemical fractions and mechanisms involved in the pineal modulation of pituitary gonodotrophin release. In *The Pineal Gland*, Ciba Symp., 241–258.

Motta, M., Fraschini, F. and Martini, L. (1967). Endocrine effects of pineal gland and of melatonin. *Proc. Soc. exp. Biol. Med., N.Y.*, **126**, 431–435.

Nagle, C. A., Cardinali, D. P., De Laborde, N. P. and Rosner, J. M. (1974). Sex dependent changes in rat retinal hydroxyindole-O-methyl transferase. *Endocrinology*, **94**, 294–297.

Orts, R. J. (1977). Reduction of serum LH and testosterone in male rats by a partially purified bovine pineal extract. *Biol. Reprod.*, **16**, 249–254.

Orts, R. J. and Benson, B. (1973). Inhibitory effects on serum and pituitary LH by a melatonin-free extract of bovine pineal glands. *Life Sci.*, **12**, 513–519.

Osland, R. B., Cheesman, D. W. and Forsham, P. H. (1977). Studies in the mechanism of suppression of the preovulatory surge of luteinizing hormone in the rat by arginine vasotocin. *Endocrinology*, **101**, 1203–1209.

Pavel, S. (1970). Tentative identification of arginine vasotocin in human cerebrospinal fluid. *J. clin. Endor. Metab.*, **31**, 369–371.

Pavel, S. (1973). Arginine vasotocin release into cerebrospinal fluid induced by melatonin. *Nature, New Biol.*, **246**, 183–184.

Pavel, S., Calb, M. and Georgescu, M. (1975). Reversal of the effects of pinealectomy on the pituitary prolactin content in mice by very low concentrations of vasotocin injected into the third cerebral ventricle. *J. Endocrinol*, **66**, 289–290.

Pavel, S., Dumitru, I., Klepsh, I. and Dorcescu, M. (1973/74). A gonadotropin inhibiting principle in the pineal of human fetuses. *Neuroend.*, **13**, 41–46.

Pavel, S., Gheorghiu, C., Calb, M. and Petrescu, M. (1975). Reversal by vasotocin of pinealectomy and constant light effects on the pituitary melanocyte-stimulation hormone (MSH) content in the mouse. *Endocrinology*, **97**, 674–676.

Quay, W. B. (1963). Circadian rhythm in rat pineal serotonin and its modifications by estrous cycle and photoperiod. *Gen. comp. Endocrinol.*, **3**, 473–475.

Quay, W. B. (1964). Circadian and estrous rhythms in pineal metabolism and 5-hydroxyindole-3-acetic acid. *Proc. Soc. exp. Biol. Med., N.Y.*, **115**, 710–713.

Quay, W. B. (1966). 24-hour rhythms in pineal 5-hydroxy-tryptamine and hydroxy-indole-O-methyl transferase activity in the macaque. *Proc. Soc. exp. Biol. Med., N.Y.*, **121**, 946–948.

Quay, W. B. (1970). Endocrine effects of the mammalian pineal. *Amer. Zool.*, **10**, 237–246.

Reiss, M., Davis, R. H., Sideman, M. B. and Plichta, I. S. (1963). Pineal gland and spontaneous activity of rats. *J. Endocrinol.*, **28**, 127–128.

Reiter, R. J. (1969). Pineal function in long term blinded male and female golden hamsters. *Gen. comp. Endocrinol.*, **12**, 460–468.

Reiter, R. J. (1973a). Comparative physiology: pineal gland. *Annu. Rev. Physiol.*, **35**, 305–328.

Reiter, R. J. (1973b). Pineal control of a seasonal reproductive rhythm in male golden hamsters exposed to natural daylight and temperature. *Endocrinology*, **92**, 423–430.

Reiter, R. J. and Ellison, N. M. (1970). Delayed puberty in blinded anosmic female rats: role of the pineal gland. *Biol. Reprod.*, **2**, 216–222.

Reiter, R. J. and Sorrentino, S. (1970). Reproductive effects of the mammalian pineal. *Amer. Zool.*, **10**, 247–258.

Reiter, R. J. and Sorrentino, S. (1971). Factors influential in determining the gonad-inhibiting activity of the pineal gland. *The Pineal Gland*, Ciba Symp., 329–344.

Reiter, R. J., Vaughan, M. K., Blask, D. E. and Johnson, L. Y. (1975). Pineal methoxyindoles: new evidence concerning their function in the control of pineal-mediated changes in reproductive physiology of male golden hamsters. *Endocrinology*, **96**, 206–213.

Ronnekleiv, O. K., Krulich, L. and McCann, S. M. (1973). An early morning surge of prolactin in the male rat and its abolition by pinealectomy. *Endocrinology*, **92**, 1339–1342.

Rowan, W. C. (1925). Relation of light to bird migration and developmental changes. *Nature*, **115**, 494–495.

Ruzsas, C., Trentini, G. P. and Mess, B. (1977). The role of the pineal gland in the regulation of LH-release in rats with different types of the anovulatory syndrome. *Endocrinology*, **70**, 142–149.

Saxena, B. B., Demura, H., Gandy, H. M. and Peterson, R. E. (1968). Radioimmunoassay of follicle stimulating and luteinizing hormones in plasma. *J. clin. Endocrinol.*, **28**, 519–534.

Schwartz, N. B. (1969). A model for the regulation of ovulation in the rat. *Rec. Progr. Horm. Res.*, **25**, 1–55.

Shiino, M., Arimura, A. and Rennels, E. G. (1974). Effects of blinding, olfactory bulbectomy, and pinealectomy on prolactin and growth hormone cells, with special reference to ultrastructure. *Amer. J. Anat.*, **139**, 191–207.

Tamarkin, L., Westrom, W. K., Hamill, A. I. and Goldman, B. D. (1976). Effect of

medial basal hypothalamus on postcoital reflex ovulation in the rabbit. *Endocrinology*, **99**, 959–962.

Yasaki (1960). *Annot. Zool. Jap.*, **33**, 217.

CHAPTER 6

Bengtsson, L. P. (1973). The role of progesterone in human labour. In *Endocrine Factors & Labour*. Ed. A. Klopper and J. Gardner. C.U.P.

Bosu, W. T. K., Johansson, E. D. B. and Gemzell, C. (1973). Peripheral plasma levels of oestrogen and progesterone in pregnant rhesus monkeys treated with dexamethasone. *Acta. endocr. Copenh.*, **74**, 338–347.

Bryant, G. D. (1972). The detection of relaxin in porcine, ovine and human plasma by radioimmunoassay. *Endocrinology*, **91**, 1113–1117.

Challis, J. R. G., Davies, L. J. and Ryan, K. J. (1973). The relationship between progesterone and prostaglandin F concentrations in the plasma of pregnant rabbits. *Prostaglandins*, **4**, 509–516.

Challis, J. R. G. et al. (1972). A possible role of oestrogens in the stimulation of prostaglandin $F_{2\alpha}$ output at the time of parturition in a sheep. *J. Reprod. Fert.*, **30**, 485–488.

Chamberlain, W. E. (1930). *Amer. J. Roentgenol. Radium Therap.*, **24**, 621–625. (Quoted by Hisaw and Zarrow.)

Chamberlain, W. E. (1937). *Proc. Calif. Acad. Med.*, **41**, 96. (Quoted by Hisaw and Zarrow.)

Crelin, E. S. (1969). The development of the bony pelvis and its changes during pregnancy and parturition. *Trans. N.Y. Acad. Sci.*, **31**, 1049–1058.

Dickson, A. D. and Bulmer, D. (1961). Observations on the origin of metrial gland cells in the rat placenta. *J. Anat.*, **95**, 262–273.

Eglinton, G. et al. (1963). Isolation and identification of two smooth muscle stimulants from menstrual fluid. *Nature, Lond.*, **200**, 960–995.

Ferguson, J. K. W. (1941). A study of the motility of the intact uterus at term. *Surg. Gynaec. Obstetr.*, **73**, 359–366.

Fitzpatrick, R. J. (1977). Dilatation of the uterine cervix. Ciba Symp. No. 47, pp. 31–39. Elsevier, Amsterdam.

Fylling, P. (1970). The effect of pregnancy, ovariectomy and parturition on plasma progesterone level in sheep. *Acta endocr. Copenh.*, **65**, 273–283.

Fylling, P. (1971). Premature parturition following dexamethasone administration to pregnant ewes. *Acta. Endocrinol.*, **66**, 289–295.

Gustavii, B. (1977). Human decidua and uterine contractility. Ciba Symp. No. 47, pp. 343–358. Elsevier, Amsterdam.

Hempel, K. H., Fernandez, L. A. and Persellin, R. H. (1970). Effect of pregnancy sera on isolated lysosomes. *Nature, Lond.*, **225**, 955–956.

Liggins, G. C. (1969). Premature delivery of foetal lambs infused with glucocorticoids. *J. Endocrinol.*, **45**, 515–523.

Liggins, G. C. and Grieves, S. (1971). Possible role for prostaglandin $F_{2\alpha}$ in parturition in sheep. *Nature, Lond.*, **232**, 629–631.

Maurizio, E. and Ottaviani, G. (1934). Comportamento delle reti litatiche sottosierose e muscolari dell' utero della donna durante la gravidanza. *Ann. Ost. Ginecol.*, **56**, 1251–1277.

Miyakawa, I., Ikeda, I. and Maeyama, M. (1974). Transport of ACTH across human placenta. *J. clin. Endocrinol.*, **39**, 440–442.

Miyakawa, I., Ikeda, I., Nakayama, M. and Maeyama, M. (1976). Plasma concentrations of cortisone, progesterone and unconjugated oestradiol in women with live anenecphalic foetuses before, during and after labour. *J. Endocrinol.*, **69**, 291–292.

Pickles, V. R. (1957). A plain muscle stimulant in the menstruum. *Nature, Lond.*, **180**, 1198–1199.

Sharma, S. C. and Fitzpatrick, R. J. (1974). Effect of oestradiol—17β and oxytocin treatment on prostaglandin F alpha release in the anoestrous ewe. *Prostaglandins*, **6**, 97–105.

Sherwood, C. D. and O'Byrne, E. M. (1974). Purification and characterization of porcine relaxin. *Arch. Biochem. Biophys.*, **160**, 185–196.

Zarrow, M. X. and O'Connor, W. B. (1966). Localization of relaxin in the corpus luteum of the rabbit. *Proc. Soc. exp. Biol. Med.*, **121**, 612–614.

CHAPTER 7

Goodman, I. and Hiatt, R. B. (1972). Coherin: a new peptide of the bovine neuro-hypophysis with activity on gastrointestinal motility. *Science, N.Y.*, **178**, 419–421.

Meites, J. *et al.* (1972). Recent studies on functions and control of prolactin secretion in rats. *Rec. Progr. Horm. Res.*, **28**, 471–516.

Selye, H. and McKeown, T. (1934). Further studies on the influence of suckling. *Amer. J. Physiol.*, **60**, 323–332.

Zinder, O., Hamosh, M., Fleck, T. R. C. and Scow, R. O. (1974). Effect of prolactin on lipoprotein lipase in mammary gland and adipose tissue of rats. *Amer. J. Physiol.*, **226**, 744–748.

CHAPTER 8

Baker, H. W. G. *et al.* (1976). Testicular control of follicle-stimulating hormone secretion. *Rec. Progr. Horm. Res.*, **32**, 429–469.

Dierschke, D. J., Weiss, G. and Knobil, E. (1974). Sexual maturation in the female rhesus monkey and the development of estrogen-induced gonadotropic hormone release. *Endocrinology*, **94**, 198–206.

McCormack, C. E. and Meyer, R. K. (1964). Minimal age for induction of ovulation with progesterone in rats: evidence for neural control. *Endocrinology*, **74**, 793–794.

Price, D. and Ortiz, E. (1944). The relation of age to reactivity in the reproductive system of the rat. *Endocrinology*, **34**, 215–239.

Raiti, S. *et al.* (1969). Measurement of immunologically reactive follicle stimulating hormone in serum of normal male children and adults. *Metabolism*, **18**, 234–240.

Schalch, D. S., Parlour, A. F., Boon, R. C. and Reichlin, S. (1968). Measurement of human luteinizing hormone in plasma by radioimmunoassay. *J. clin. Invest.*, **47**, 665–678.

Vandenbergh, J. G., Drickamer, L. C. and Colby, D. R. (1972). Social and dietary factors in the sexual maturation of female mice. *J. Reprod. Fert.*, **28**, 397–405.

Wilkinson, M., de Ziegler, D., Cassard, D. and Ruf, K. B. (1977). In-vitro studies of the effects of oestrogen pretreatment on the sensitivity of the immature female rat pituitary gland to stimulation with gonadotrophin releasing hormone. *J. Endocr.*, **74**, 11–21.

SUBJECT INDEX